System Analysis and Modelling in Air Transport

Demand, Capacity, Quality of Services, Economics, and Sustainability

Milan Janić

Delft University of Technology
Faculty of Civil Engineering and Geosciences, Transport and Planning Department
Delft, The Netherlands

CRC Press
Taylor & Francis Group
Boca Raton London New York

CRC Press is an imprint of the
Taylor & Francis Group, an **informa** business
A SCIENCE PUBLISHERS BOOK

First edition published 2021
by CRC Press
6000 Broken Sound Parkway NW, Suite 300, Boca Raton, FL 33487-2742

and by CRC Press
2 Park Square, Milton Park, Abingdon, Oxon, OX14 4RN

CRC Press is an imprint of Taylor & Francis Group, LLC

Library of Congress Cataloging-in-Publication Data

Names: Janić, Milan, author.
Title: System analysis and modelling in air transport : demand, capacity and quality of services, economics, and sustainability / Milan Janić, Delft University of Technology, Faculty of Civil Engineering and Geosciences, Transport and Planning Department, Delft, The Netherlands.
Description: Boca Raton : CRC Press, Taylor & Francis Group, 2021. | Includes bibliographical references and index. | Summary: "Air Transport System Analysis and Modelling is a unique book dealing with analysis and modelling of the demand, capacity, quality of services, economics, and sustainability of three main components of the air transport system--airports, ATC (Air Traffic Control), and airlines. The existing and prospective modelling approach embraces the illustrative analytical and simulation models supported by corresponding applications and real-life examples. Graduates, researchers, consultants, engineers, experts, and other practitioners dealing with analysis, modelling, planning, design, and operations of the air transport system will find this book of interest and useful"-- Provided by publisher.
Identifiers: LCCN 2020031441 | ISBN 9780367321604 (hardcover)
Subjects: LCSH: Aeronautics, Commercial--Planning. | Airlines--Management. | Aeronautics--Systems engineering. | Operations research. | Industrial efficiency--Mathematical models.
Classification: LCC HE9780 .J27 2021 | DDC 387.7068/4--dc23
LC record available at https://lccn.loc.gov/2020031441

ISBN: 9780367321604 (hbk)

Typeset in Times New Roman
by Radiant Productions

In memoriam to my wife Vesna continuing to inspire me

Preface

The air transport system consists of three main components—airports, airlines, and ATC/ATM (Air Traffic Control/Management). Airports and ATC/ATM operate as the system infrastructure. They provide space, facilities, and equipment for serving the airlines as their users. The airlines operate flights with different aircraft types, transport passengers and freight/cargo shipments on their way between origins and destinations. In such circumstances, the airline flights represent demand for airports and ATC/ATM that serve it by their capacities. The air passengers and freight/cargo shipments represent demand for airlines and airports, which is served by their corresponding capacities.

The air transport demand and capacity serving it have been growing throughout the past few decades in all three components. This growth has been driven by the system's external and internal driving forces. The former have mainly been the overall economic growth, globalization of trade and tourism, and governmental regulation inside and outside the system. The latter has generally been liberalization and deregulation of the air transport markets contributing to consolidation of the airline industry and emerging the innovative business models. Consolidation of the airline industry has resulted in the emergence of rather strong alliances of the conventional/legacy airlines operating the hub-and-spoke networks. In these networks, higher flights frequencies have been offered between hub and particular spoke airports. Emergence of the innovative business models is related to development of LCCs (Low Cost Carrier(s)), which have operated exclusively the point-to-point networks. These have included the number of regional but also in many cases the hub airports. Opposite to their conventional/regional counterparts, LCCs have operated single aircraft types. Such developments have generally resulted in decreasing airfares, which has additionally stimulated growth of demand.

Under conditions of growing demand, the fixed aviation infrastructure—airports and ATC/ATM—have not always been able to handle it safely, effectively, and efficiently. Consequently, many airports and parts of airspace have been frequently overloaded, causing congestion of delays of categories of corresponding users. Both have slowed down particular operations and processes, increased complexity, consequently deteriorating the planned quality and economics of services.

At the same time, the growth of the air transport system has increasingly impacted the environment and society. The former has been mainly in terms energy/fuel consumption and emissions of GHG (Green House Gases) and land use. The latter has related to the congestion and delays, noise, and local air pollution around airports, and air traffic incidents/accidents. As expressed in monetary terms, all these have been considered as costs/externalities. Regarding to the system's obvious effects/benefits in terms to contribution to the social-economic welfare at local, regional, national, and international scale, balancing between its effects/benefits and impacts and their costs as externalities have been increasingly relevant and important. Consequently, the sustainable development implying balancing the system's effects/benefits and impacts/costs in the short-, medium-, and long-term period has become one of its additional but main driving forces.

In order to deal with the above-mentioned developments, the academics, consultants, industry agents and others have developed the various comprehensive "tools" for the analysis, modelling, and planning of demand, capacity, quality of services, economics, and sustainability of the air transport systems and its particular main components.

This book represents a substantive update of the author's previous book: *Janić, M., (2000), Air Transport System Analysis and Modelling: Capacity, Quality of Services and Economics, (Transportation Studies, Volume 16), Gordon and Breach Science Publishers (Amsterdam, The Netherlands).*

The book elaborates demand, capacity, quality of services, economic and sustainability of the particular main components of air transport system—airports, airlines, and ATC/ATM. Compared to its predecessor, new contributions regarding sustainability have been added.

For each of component, the real-life cases and the corresponding primarily analytical models are presented, aiming at being as illustrative as possible. The main criteria for selection of particular models have been the author's judgement about their suitability to reflect the real-life cases, time-independent generosity, and the availability of corresponding data for application. The models are elaborated in terms of the basic structure, main assumptions, inputs, and results.

According to the author's knowledge, this book still represents a unique example due to it simultaneously dealing with all three components of the air transport system in the above-mentioned context. As such, it could be of interest for a relatively wide readership dealing with the analysis and modelling of characteristics of the air transport system and its components: advanced academics, consultants, and the particular air transport industry actors/stakeholders.

The book has been written during the time period in which the air transport system and its components, i.e., global aviation industry, have been facing quite unusual conditions. The presented material generally reflects its past, present, and prosperous development characterised by the short-, medium-, and long-term growth. The manuscript is finalized under conditions when the most of aviation industry has been severely affected by the fast spreading and aggressive global impact of SARS-CoV-2 virus (Coronavirus 19). At present it is quite uncertain how soon after the end of impact the system and its components will recover and what it will look like in both qualitative and quantitative terms. Nevertheless, the basic principles and methods for analysis and modelling of their demand, capacity, quality of services, economics, and sustainability will remain in place and so this book will be of use for the readers.

April 2020 **Milan Janić**

Ljubljana (Slovenia)
Delft (The Netherlands)

Contents

List of Acronyms and Abbreviations

ACAS	Airborne Collision Avoidance
ACI	Airport Council International
ADS-B	Automatic Dependent Surveillance Broadcasting
AEA	Association of European Airlines
AF	Air Fare
A/G	Air/Ground
AIS/AIM	Aeronautical Information Services/Aeronautical Information Management
AM	Airspace Management
AMAN	Arrival Manager
ANS	Air Navigation Services
ANSP	Air Navigation Service Provider
APM	Available Passenger Miles
APW	Area Penetration Warning
ASAS	Airborne Separation Assistance Assurance
ASK	Available Seat Kilometres
ATAG	Air Transport Action Group
ATC	Air Traffic Control
ATC/ATM	Air Traffic Control/Management
ATCC/ATMC	Air Traffic Control/Air Traffic Management Centre
ATFM	Air Traffic Flow Management
atm	air transport movement
ATO	Air Traffic Organization
ATS	Air Traffic Services
AZ	Airport Zone
BA	British Airways
BH	Block Hour
BRT	Bus Rapid Transit
CAA	Civil Aviation Authorities
CARD	Conflict and Risk Display
CDB	Central Business District
CDTI	Cockpit Display of Traffic Information
CFMU	Central Flow Management Unit
CIA	Central Intelligence Agency
CNS	Communication and Navigation Surveillance

CPDLC	Controller Pilot Data-Link Communications
CTAS	Centre/TRACON Automation System
dB(A)	A-weighted decibel
DAP	Downlink of Aircraft Parameter
DMAN	Departure Manager
DME	Distance Measuring Equipment
DNL	Day-Night Level
DNS	Doppler Navigation System
EC	European Commission
ECAC	European Civil Aviation Conference
EEA	European Environmental Agency
EQA	Equivalent Aircraft
EU	European Union
FAA	Federal Aviation Administration
FAG	Final Approach Gate
FCFS	First-Come-First-Served
FDP	Flight Data Processing
FH	Flight Hour
FL	Flight Level
FMS	Flight Management System
FRA	Free Route Airspace
GAM	Global Airport Monitor
GBP	Great Britain Pound
GDP	Gross Domestic Product
G/G	Ground/Ground
GHG	Green House Gases
GIS	Governmental Inspection Service)
GMBM	Global Market-Based Measures
GMT	Greenwich Mean Time
GPS	Global Position System
GS	Glide Slope
HAA	High Altitude Airspace
HF	High Frequency
HL	Hyperloop
HSR	High Speed Rail
IATA	International Air Transport Association
ICAO	International Civil Aviation Organization
IFR	Instrument Flight Rules
ILS	Instrument Landing System
IMC	Instrument Meteorological Conditions
INS	Inertial Navigation System
IPCC	Intergovernmental Panel on Climate Change
ISQ	Indicator of Service Quality
ITS	Information Technology and Systems
LAA	Low Altitude Airspace
LCC	Low Cost Carrier
LRT	Light Rail Transit

LVLASO	Low Visibility Landing and Surface Operating (Program)
MCT	Minimum Connection Time
MJ	Mega Joule
MSAW	Minimum Safe Altitude Warning
MTOW	Maximum Take-off Weight
NAS	National Airspace System
Next Gen	Next Generation
NMAC	Near Mid-Air Collisions
NS	Nederlandse Spoorwegen
OATCC	Oceanic Air Traffic Control Center
O-D	Origin-Destination
PAX, pax	passenger
PCI	Per Capita Income
RFL	Reference Field Length
RPK	Revenue Passenger Kilometres
RLatSM	Reduced Lateral Separation Minima
PRM	Precision Runway Monitor
PRT ULT	Personal Rapid Urban Light Transit
RNP	Required Navigation Performance
RPK	Revenue Passenger Kilometres
RPM	Revenue Passenger Miles
RVSM	Reduced Vertical Separation Minima
R/T	Radio/Telephone
RTL	Rapid Train Link
SAR	Search and Rescue
SESAR	Single European Sky ATM Research
SID	Standard Instrument Departure
SLOP	Strategic Lateral Offset Procedure
SLR	Space Load Ratio
STAR	Standard Arrival Route
STCA	Short-Term Conflict Alert
SYSCO	System Supported Coordination
TCAS	Traffic Alert and Collision Avoidance
TMA	Terminal Manoeuvring Area
TRB	Transportation Research Board
TRM	Trans Rapid Maglev
TSU	Traffic Service Unit
UK	United Kingdom
UTC	Coordinated Universal Time
U.S.	United States
VFI	Vertical Flight Inefficiency
VFR	Visual Flight Rules
VMC	Visual Meteorological Conditions
VOR	VHF Omnidirectional Range
WAAS	Wide Area Augmentation
WB	World Bank
WP	Way Point
WVDS	Wake Vortex Detector System

Chapter 1
Introduction

1.1 Air Transport System

The air transport system is a sub-system of the transport system which includes road, rail, and intermodal inland and the barge and sea water transport. The system generally consists of three main components—airports, airlines, and ATC/ATM (Air Traffic Control/Management) sub-systems. Each of these generally consist of two major components—demand and supply/capacity. The airport demand consists of users, i.e., air passengers, freight/cargo shipments, mail, and airline aircraft. The airport supply/capacity includes the ground infrastructure in terms of space, facilities and equipment arranged in the airport airside and landside area. Airports also provide the physical connection between air and other ground transport modes (Horonjeff and McKelvey, 1994; Janic, 2000; Janic, 2009). The airline demand consists of users, i.e., air passengers, freight/cargo shipments, and mail. The airline supply/capacity is represented by the aircraft of different payload capacities and range performing flights to serve demand. The ATC/ATM sub-system's demand consists of the airline flights carried out in the controlled airspace between airports. The ATC/ATM sub-system's supply/capacity is provided by the ground, space, and on-board facilities and equipment considered (ultimately) as some kind of "infrastructure" together with the controlled (Janić, 2000; ICAO, 2016). Consequently, in the given context, the aircraft have two roles in the air transport system: first, they appear as the airline supply/capacity while performing flights, i.e., transport services to users; second, these flights appear as demand for airports and ATC/ATM (Janić, 2000). The three sub-systems operate according to the internal and external national and international regulations by engaging highly trained personnel/employees of different profiles and levels of expertise aiming at serving demand door-to-door generally safely, efficiently, and effectively. In addition, they consume different types of energy/fuel for their operations.

The main performances of airports, airlines, and the ATC/ATM sub-system as the subjects of further consideration, i.e., analyzing and modelling, are their demand and supply/capacity, their relationships materialized as the quality of services provided to the above-mentioned users, economics, and sustainability, and the latest in terms of the effects and impacts on the environment and society. Figure 1.1 shows the scheme of consideration of the above-mentioned air transport sub-systems.

- Demand is characterized by volume, time, and spatial characteristics, in this case air transport services, to be provided during a given period of time under given conditions.

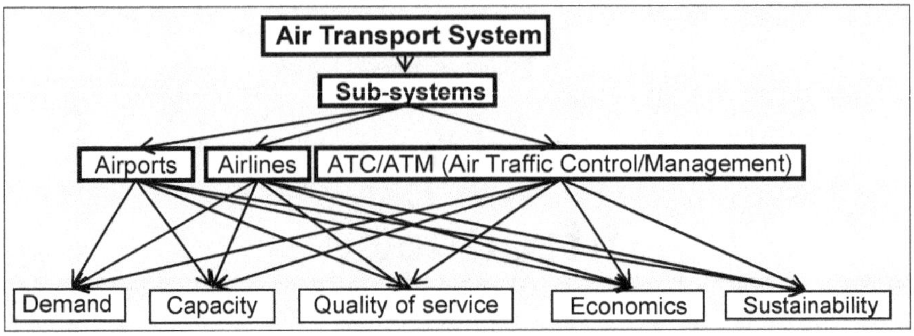

Figure 1.1 Scheme of consideration of the air transport system.

- Supply/Capacity is defined as the capability of the air transport system's sub-systems to produce a certain volume of services in order to satisfy the above-mentioned expected volumes of demand during a given period under given conditions.
- Quality of services is defined as the users' satisfaction with the consumed/used services. Because both production and consumption of transport services in the air transport system occur simultaneously, the rate of consumer satisfaction can be estimated only after the end of these processes. In such context, the relationship between demand and capacity, and consequently the resulting quality of service, can be different. For example, when the demand is constant in space and time, the quality of service can be increased simply by increasing the capacity of particular sub-systems and their components. However, when the demand fluctuates considerably and grows unpredictably, augmentation of the capacity may only be sufficient to sustain the quality of service at the previous/benchmarking level (Manheim, 1979).
- Economics refers to the costs, revenues, and profits of providing air transport services. Generally, production of a larger quantity of services with higher capacity imposes higher total cost on producer(s), and vice versa. The services of a higher quality also cost more. For example, the flights carried out by larger aircraft along the longer routes will in total cost more than otherwise. In addition, if the number of flights increases, their total costs will be higher. However, in many cases, the average cost per service will decrease if the total volume of produced services increases. For example, the average unit cost per passenger- and/or seat-kilometer will be lower if the larger aircraft fly along the longer (non-stop) routes. Also, at particular airports, the average unit cost per service will be lower if the total volume of services increases, and vice versa. Decreasing of the average unit cost with increasing of the volumes of output has been generally defined as "economies of scale" and "economies of density". Both phenomena have proven to be inherent characteristics of the air transport system and its above-mentioned sub-systems. The prices of air transport services are usually set up according to different criteria, though the basic one has been recovering the operators' costs. The others have been maximizing the passenger welfare, keeping the existing and instigating additional air travel demand, etc. (Janić, 2000).
- Sustainability can be considered as increasing positive effects/benefits of own and overall social-economic welfare and mitigating impacts/costs on the environment and society during the medium- to long-term period under conditions of continuous

system growth. The own welfare implies the air transport system's financial, economic, and operational stability in providing services. The overall social-economic welfare implies constant or increasing contribution to the overall employment and GDP (Gross Domestic Product) in the widest sense. Mitigating impacts on the environment implies reducing or maintaining the present level of energy/fuel consumption and related emissions of GHG (Green House Gases), and land use. That on the society implies reducing or not increasing the noise around airports, congestion and delays, and the rate of air traffic incidents/accidents (Akerman, 2005; Janic, 2007).

The above-mentioned demand, supply/capacity, and economic performance are elaborated separately for three sub-systems. The sustainability is elaborated in an aggregate level based on the overall sub-systems' effects/benefits and impacts/costs on the society and environment.

1.2 Airports

1.2.1 Demand, Capacity, and Quality of Services

According to the U.S. CIA (Central Intelligence Agency) World's Fact Book (Airports), the total number of airports is 41788, of which 13513 are in the U.S., 4093 in Brazil, 1714 in Mexico, 1467 in Canada and 1218 in Russia. This total number includes all airports, aerodromes and airfields, both civilian and military. They are paved or unpaved, including also closed or abandoned infrastructure, facilities, and equipment (https://www.cia.gov/library/publications/the-world-factbook/rankorder/2053rank.html). A substantially lower number of 17678 commercial airports considered to handle airlines, freight/cargo, and business aircraft has been reported by ACI (Airport Council International). The trends on the number of passengers, freight/cargo (freight and mail) and aircraft movements, both domestic and international, have been presented for about 2500 airports in 175 countries worldwide (ACI, 2018; https://store.aci.aero/product/annual-world-airport-traffic-report-2018/). In addition, the total number of international airports is reported to be 1282, of which 163 are in Africa, 311 in South and North America, 319 in Asia, 498 in Europe, and 51 in Oceania (https://www.bts.gov/content/number-us-airportsa/; https://en.wikipedia.org/wiki/List_of_international_airports_by_country#Number_of_International_Airports). Figure 1.2 shows an example of the relationship between the annual number of handled passengers and the number of atms (air transport movements) at 25 of the world's largest airports (1 air transport movement is 1 departure or 1 arrival).

As expected, in both domestic and international traffic, the number of handled passengers increases with an increase in the number of aircraft movements. This indicates that larger aircraft with higher load factors have been increasingly operating at these airports during the observed period.

The quality of services at airports at the global scale can generally be expressed by their average annual on-time performance, i.e., punctuality, defined as the proportion of arriving and departing flights operated within 15 minutes of their scheduled arrival and departure times. In some way, this can generally reflect the quality of services provided to both airlines and air passengers. Figure 1.3 shows the example for the airports worldwide (OAG, 2019).

As can be seen, the average annual on-time performance has generally decreased with increasing airport traffic. This implies that the airports with larger traffic volumes could expect lower punctuality of those flights delayed less that specified—

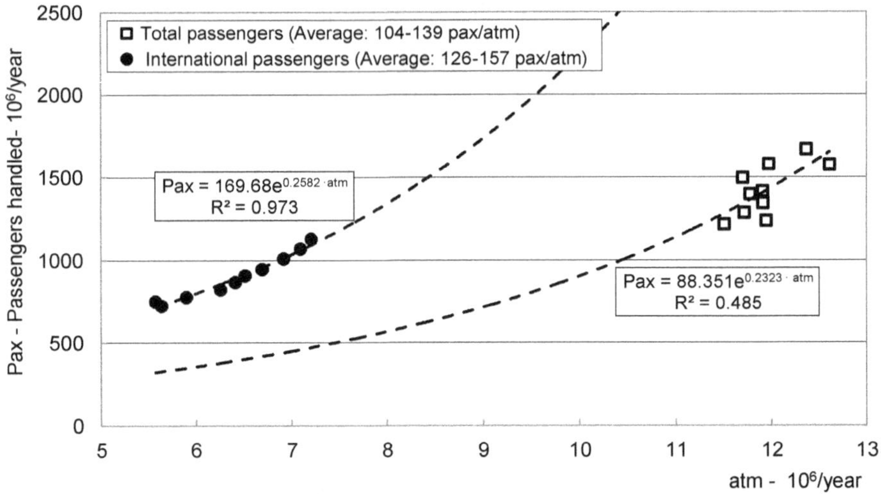

Figure 1.2 Relationship between the annual number of pax (embarked and disembarked passengers) and atm (air transport movements)—Case of 25 largest airports ranked by the total and international passengers (Period: 2008–2017) (ICAO, 2005/2018; ICAO, 2010).

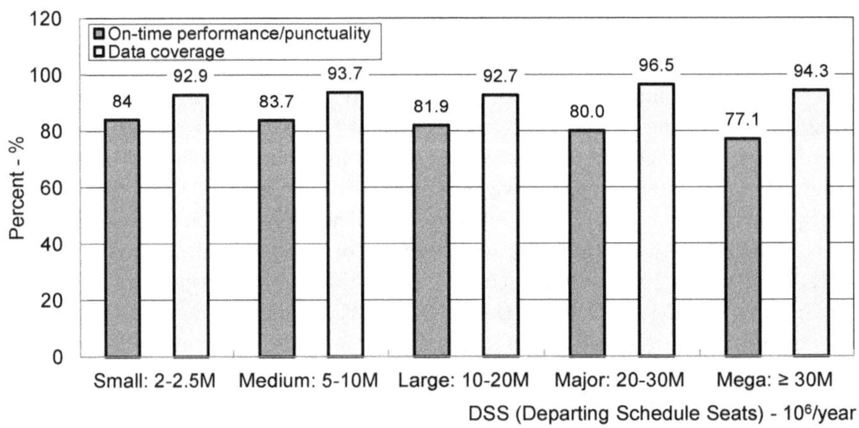

Figure 1.3 Punctuality of the world's airports depending on the volumes of traffic (Period: 2018) (OAG, 2019).

15 minutes at either arrival or departure. These facts have been also strongly supported by the high coverage of airports in the considered samples of 92–97% (OAG, 2019). In addition, Table 1.1 gives the airport-related delays at 34 selected airports in Europe and U.S. (EEC/FAA, 2019).

As can be seen, during the observed period, at the European airports, the average number of IFR flights was 5.1 million/year while imposing an average delay of 0.84 min/flight. The proportion of flights delayed more than 15 minutes was 2.5%/year with an average delay of 33 min/flight. At the U.S. airports, the average number of IFR flights was 8.6 million/year with an average delay of 1.4 min/flight. The proportion of flights delayed more than 15 minutes was 2.2%/year with an average delay

Table 1.1 The airport-related delays—Case of 34 selected airports in Europe and U.S. (Period: 2008–2017) (EEC/FAA, 2019).

Region/year	Indicator			
	Number of flights (IFR) (10^6/year)	Delayed flights > 15 min (%)	Delay per flight (min)	Delay per delayed flight (min)
Europe				
2008	5.5	2.8	0.9	32
2010	5.0	3.3	1.2	36
2013	4.8	1.6	0.5	33
2015	5.0	2.3	0.74	32
2017	5.2	2.7	0.84	31
U.S.				
2008	9.3	2.6	1.9	74
2010	8.6	1.6	1.0	66
2013	8.4	2.6	1.5	57
2015	8.2	1.8	1.1	63
2017	8.4	2.4	1.7	71

of 66 min/flight. These figures indicate that the U.S. airports have handled about 70% more flights but at a cost of about 67% and 100% longer average delay per flight and per delayed flight, respectively. At the same time, the proportion of these (longer than 15 minutes) delayed flights was lower at the U.S. airports for about 14% (EEC/FAA, 2019). For some comparison, in China, 202 airports handled about 8.6 million flights in 2014. The flight arrival punctuality was 68.4%, with an average delay of 21 min per flight (Cook et al., 2019).

1.2.2 Economics

The airport economics relate to their revenues, costs, and profits/losses. To illustrate this, Figure 1.4 shows an example of these economics for the world's airports in the specified year (2013) in the aggregate form (ICAO, 2015).

It can be noted that the profits were higher than the costs in all regions, which indicates that airports worldwide were generally profitable. In particular, the European airports had the highest revenues and costs, followed by those of the Asia-Pacific and North America regions. Those in the Asia-Pacific region were the most profitable followed by those in the Europe and North America regions. The African airports had the lowest three aggregate characteristics. In addition, Table 1.2 gives an example of development of the average economics of the world's airports over time (ACI World, 2016/2018).

As can be seen, during the observed period, both the average total revenues and costs per handled passenger were decreasing. The decrease of the revenues (total, aeronautical, and non-aeronautical) was at a lower rate than that of the costs, which resulted in increasing of the average profits per passenger.

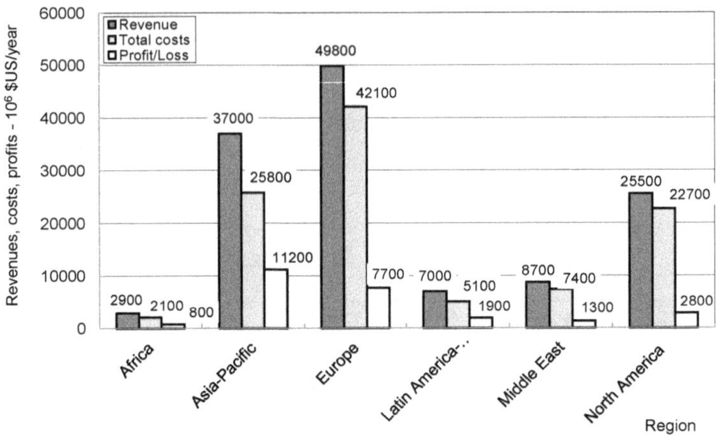

Figure 1.4 Economics of the world's airports—revenues, cost, and profits (Period: 2013) (ICAO, 2015).

Table 1.2 Example of the average economics of the world's airports over time (ACI World, 2016/2018).

Characteristic/Average	**Time/Years**		
	2015	**2016**	**2017**
Revenues (Total) ($US/pax)	20.36	19.20	17.27
• Aeronautical • Non-aeronautical	11.78 8.58	11.23 7.97	10.15 7.12
Costs (Total) ($US/pax)	16.82	15.58	13.55
Profits (Revenues-Costs) ($US/pax)	3.54	3.62	3.72

$US – U.S. dollar; pax – passenger.

1.3 Airlines

1.3.1 Demand, Capacity, and Quality of Services

During the period 1991–2007, the world's air passenger and freight/cargo traffic had been growing by an annual average rate of 4.5% and 5.7%, respectively (ACI, 2006; AIRBUS, 2006; Boeing, 2007). Despite being affected by the various large-scale disruptive events, such as the September 11th terrorist attack on the U.S., regional wars (Iraq, Afghanistan, Syria), the economic and political crises (2008/2009), and the global epidemic diseases, which temporarily undermined the past growth rates, at that time the system's traffic was expected to continue to grow at the average rates of 4–5% in the passenger and 5.5–6.5% in the freight/cargo segment (ICAO, 2005/2018). And what have the system and its traffic looked like during that period of regarding the expectations, i.e., 2005/07–2017/18? At present, about 5000 airlines operate in the world. However, only 153 national/flag and 146 LCCs (Low Cost Carrier(s)) are ICAO coded. Of these, almost all (290) are the members of IATA (International Air Transport Association) (https://www.iata.org/about/members/Pages/airline-list.aspx; https://en.wikipedia.org/wiki/Lists_of_airlines/). These airlines have operated growing fleets in terms of the number of aircraft, as shown in Figure 1.5.

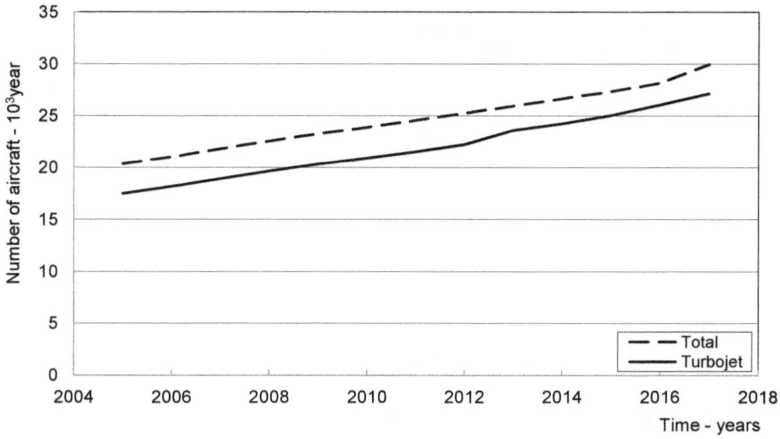

Figure 1.5 Development of the world's commercial aircraft fleet over time (Period: 2005–2017) (ICAO, 2005/2018).

As can be seen, the total number of airline aircraft has increased during the period 2005–2017 for about 47%, i.e., it has grown at an average rate of about 3.2%/year. In addition, in the year 2017/2018, 19803 (66%) of this total have been aircraft with a seating capacity greater than 100 seats. This number is expected to increase to about 45265 aircraft by the year 2037, i.e., by about 2.3 times or at an annual rate of 4.25% in order to serve the annual growth of air passenger traffic (RPK) of 4.4% during the period 2018–2036/37 (AIRBUS, 2018; Boeing, 2017).

During the period 2005–2017, the above commercial aircraft fleet has supported the further growth of the air passenger and freight/cargo traffic, as shown in Figure 1.6(a, b).

As can be seen, the total passenger and freight/cargo traffic has grown by about 98% and 81%, respectively, or at an average annual rate of about 5.8% and 5.1%, respectively. The utilization of the available capacity/supply (i.e., load factor) has increased from about 75% to 81% and decreased from 68% to 62% in the passenger and freight/cargo segments, respectively. The above-mentioned development of air traffic has been stimulated by different driving forces The most important have been the national/regional GDP, air fares and fuel price, PCI (Per Capita Income), population trends, composition of labor force, international trade and investments, and tourism (AIRBUS, 2018; Boeing, 2017; EC, 2017).

Some analytical relationships between the annual volumes of the world's air traffic carried out in both passenger and freight/cargo segment and the selected above-mentioned driving forces during the specified period (2005–2017) are given in Table 1.3 (ICAO, 2006/2018; WB, 2019).

As can be seen, during the observed period, the air passenger traffic has generally increased with increasing of *GDP* and *PCI* and decreased with increasing of *AF*. In case (a), *GDP* had about 3 time stronger positive than *AF* negative influence. In case (b), *PCI* had about twenty five times stronger positive than *AF* negative influence. In both cases, the dependent and the selected independent variables have fitted very well indicated by *F* and *D-W* statistics. The selected independent variables have also been statistically significant, as shown by *t*-statistics (below particular coefficients). In particular, the

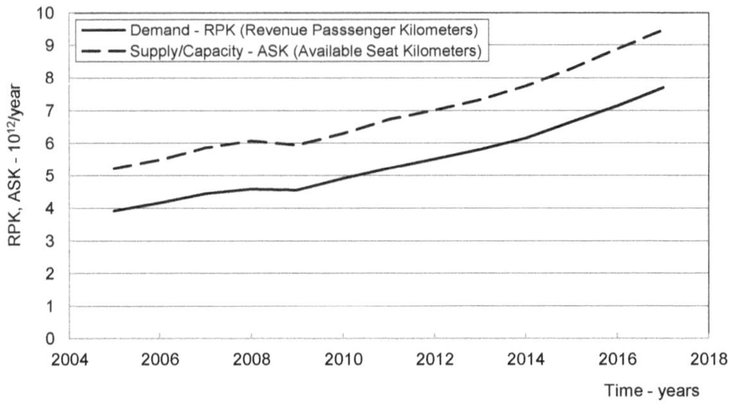

a) Passenger traffic - demand and supply/capacity

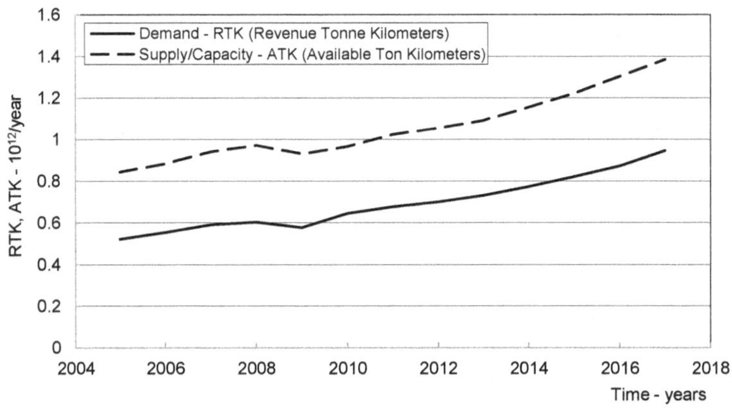

b) Freight/cargo traffic - demand and supply/capacity

Figure 1.6 Development of the world's scheduled domestic and international air traffic over time (Period: 2005–2017) (ICAO, 2006/2018).

R^2 values have indicated the rather strong relationship between the dependent and the selected independent variables, thus indicating these equations as the relative strong bases for eventual forecasting. At the same time, the volumes of freight/cargo traffic have been more than proportionally, i.e., exponentially, driven by GDP. In addition, Figure 1.7 shows the average route length for both air passengers and air freight/cargo shipments during the given period as the ratios between the volumes *RPK/PAX* and *RTK/TC*, respectively (*TC* – Tons of air freight/cargo).

As can be seen, the average route length in both cases has been rather constant during the observed period, about 1900 km (970/980 nm) for passengers and 3900 km (2100 nm) for the air freight/cargo shipments (nm – nautical mile; 1 nm = 1.852 km). This indicates that the volumes of traffic have grown more thanks to an increase in the number of passengers and freight/cargo shipments on board (i.e., load factor) than due to the increases in route length. As can be seen, the average number of passengers per departure has increased more than proportionally during the observed period, from about 85 in the year 2005 to about 111 in the year 2017, i.e., by 30.6%.

Table 1.3 Relationships between the world's air traffic and the main driving forces (Period: 2005–2017) (ICAO, 2006/2018; WB, 2019).

Air traffic type	Regression relationship
• Revenue Passenger Kilometers	
a)	$RPK\,(GDP, AF) = 5.827 + 0.0825 \cdot GDP - 0.0272 \cdot AF$ t (1.28) (10.07) (5.85) $F = 85.862;\ R^2 = 0.945;\ DW = 1.403;\ N = 13$
b)	$PAX(PCI, AF) = 2.612 + 0.377 \cdot PCI - 0.015 \cdot AF$ t (2.12) (6.11) (3.49) $F = 31.024;\ R^2 = 0.861;\ DW = 1.870;\ N = 13$
• Revenue Ton Kilometers	
	$RTK(GDP) = 0.244 \cdot e^{0.015 \cdot GDP}$ $R^2 = 0.829;\ N = 13$

RPK (Revenue Passenger Kilometers – 10^{12}/year); PAX (10^9/year – Total passengers carried); GDP (Gross Domestic Product – 10^{12} \$US/year), PCI (Per Capita Income – 10^3 \$US/year); AF (Airfare – \$US/PAX); RTK (Revenue Ton Kilometers – 10^{12}/year); (PAX – Passenger).

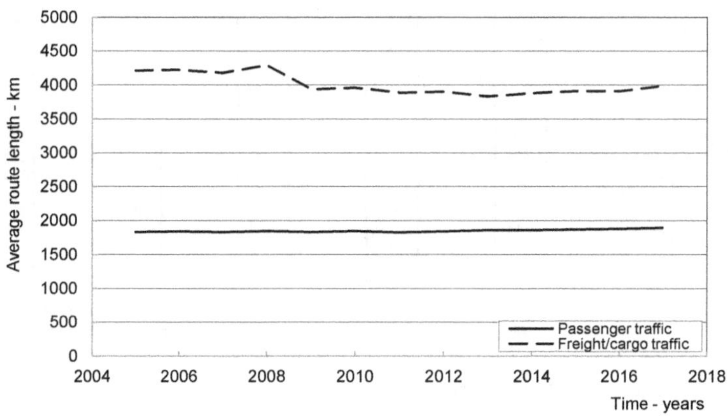

Figure 1.7 Development of the airline route length over time—Case of the world's scheduled air passenger and freight/cargo traffic (Period: 2005–2017) (ICAO, 2005/2018).

Similar to airports, the quality of service at airlines can be generally expressed, in addition to the other measures, by on-time performance, i.e., punctuality, defined again as the proportion of their flights arriving and departing within 15 minutes of their scheduled arrival and departure times. Figure 1.8(a, b) shows an example for the world's selected airlines. As can be seen on Figure 1.8a, the most punctual have been the flights operated by the mainline airlines, followed by those of LCCs (Low Cost Carrier(s)), and mega airlines. In addition, Figure 1.8b shows that the airlines from Africa have been the least and those from Asia Pacific region the most punctual. Those from Latin America and Europe have been better than those from the Middle East and North America (OAG, 2019).

The length of average delays in the particular world regions are given above in Table 1.1 and below in Table 1.5 in the scope of the airports and the ATC/ATM sub-system, respectively.

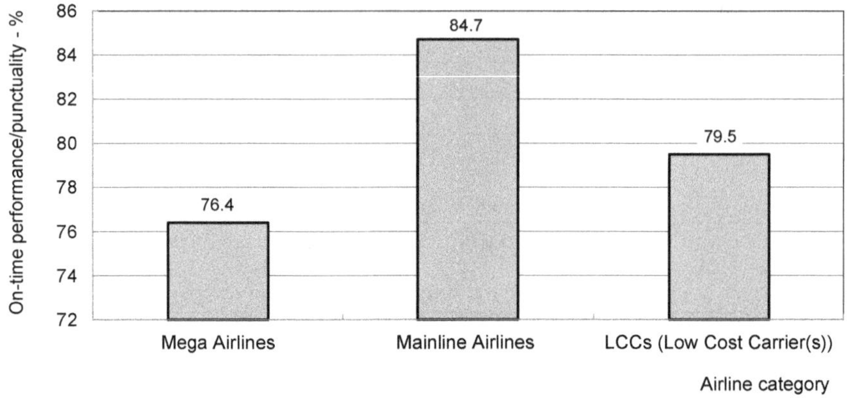

a) On-time performance vs the airline category (20 airlines per category)

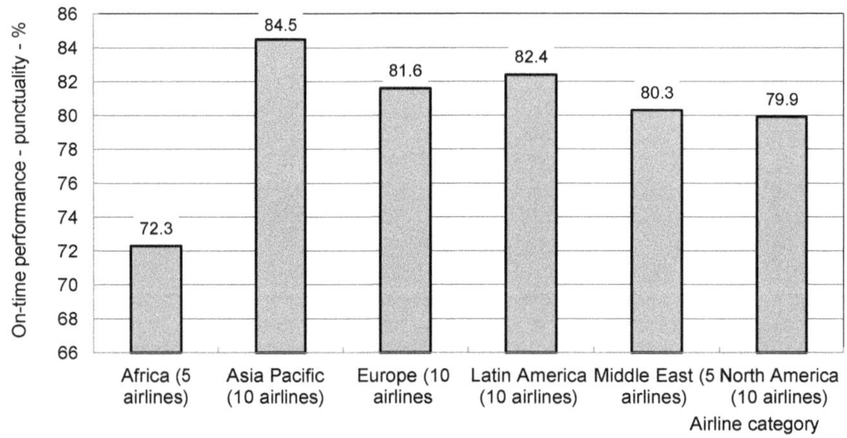

b) On-time performance vs the airline region

Figure 1.8 The airline on-time performance, i.e., punctuality—Case of the selected world's airlines (Period: 2018) (OAG, 2019).

1.3.2 Economics

The airlines economics can be considered at different levels, such as individual airlines, airline categories—mega, mainline, LCC, regions-continents, and the entire world airline industry in both aggregate/total and per unit of output (ASK and PKM) term. In addition, the particular averages can be expressed over time or depending on the volumes of output during given period of time. Figure 1.9(a, b) shows an example of the averages of the world's airline industry (ICAO, 2005/2018).

Figure 1.9a shows that the average revenues were generally higher than the average costs during the observed period of time. The exceptions were the years 2008 and 2009 when both were almost equal. After that, they were increasing in parallel until the year 2012 and then decreasing until the end of the period while the positive gap in favor of the revenues was generally increasing. This indicates the increasing profitability of the world's airlines during that time. Figure 1.9b shows that both the average costs and

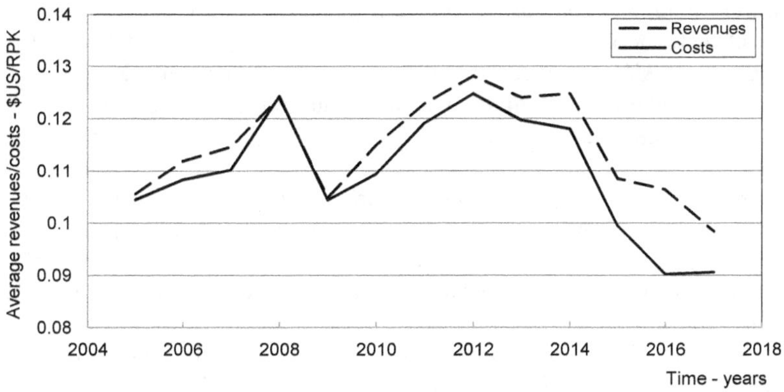

a) Average revenues/costs vs time

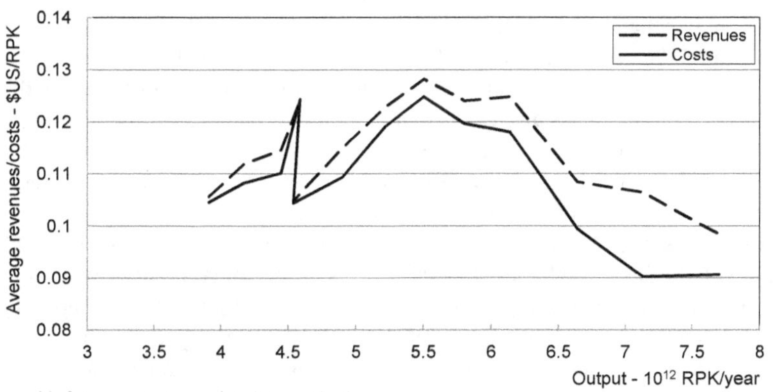

b) Average revenues/costs vs output

Figure 1.9 The average revenues and costs—Case of the world's airline industry (Period: 2005–2017) (ICAO, 2005/2018).

revenues were generally increasing with increasing of the volumes of output from 4.5–5.5 trillion RPKs and then decreasing with increasing of the volumes of output between 5.5 and 7.5 trillion RPKs. This indicated increasing economies of scale at the world's airlines continuously and in parallel reducing airfares and costs. At the same time, the average profits (yields) were (except the years 2008/2009) between 0.0055–0.0162 USD/RPK.

1.4 ATC/ATM (Air Traffic Control/Management)

1.4.1 Demand, Capacity, and Quality of Services

The ATC/ATM (Air Traffic Control/Management) operates as the sub-system of the ANSP (Air Navigation Service Provider) system, which, as a public or private entity, provides Air Navigation Services. In general, an ANSP can provide the following services to its users-airline flights: ATM (Air Traffic Management), CNS (Communication Navigation and Surveillance), MET (Meteorological service for air navigation systems), SAR (Search and Rescue), and AIS/AIM (Aeronautical Information Services/Aeronautical Information Management). These services are provided to the aircraft during all flight phases – during taking-off, en-route, and landing (ICAO, 2016a). At present, 82 ANSP systems in the scope of 166 national CAA (Civil Aviation Authorities) operate in the world (https://en.wikipedia.org/wiki/Air_navigation_service_provider; https://en.wikipedia.org/ wiki/Nationalaviationauthority#List_of_civil_aviation_authorities). Over time, they have provided a safe continuous increase in the number of flights worldwide, as shown in Figure 1.10.

As can be seen, during the observed period the number of handled flights has increased from about 24 million in the year 2005 to about 36 million in the year 2018, i.e., for about 50%, at an average annual rate of 3.3%. At the smaller continental scale, the ATC/ATM systems have possessed different characteristics and performed differently as shown in Table 1.4.

As can be seen, the ANSP is more fragmented in Europe than in U.S. The European area comprises 37 ANSPs with 62 en-route centres and 51 FMPs while the U.S. system controlling about 10% smaller airspace has comprised only one ANSP with 23 en-route centres and 65 TMUs. In addition, the number of controlled airports in Europe has been about 87% greater than that in U.S. Under such conditions, the U.S. system has handled about 47% and 47% more IFR and total daily flights, respectively, with about 32% and 43% less ATC controllers and total staff. Despite the higher control time per flight for about 8%, the U.S. system has also performed more effectively regarding the flight arrival punctuality (3%) and the TMA average delays (10%), but worse regrading ATM/TMI flight delays (19%) (EEC/FAA, 2019). In addition, Table 1.5 gives the characteristics of flight delays between airports, i.e., in the en-route airspace of Europe and U.S.

As can be seen, during the observed period, an average annual number of 5.1 million IFR flights were handled in the European en-route airspace while incurring an average delay of 1.0 min/flight. The proportion of flights delayed more than 15 minutes was 3.5%/year with an average delay of 29 min/flight. At the same time, an average annual number of 8.6 million IFR flights were handled in the U.S. en-route airspace while incurring an average delay of 0.3 min/flight. The proportion of flights delayed more than 15 minutes was 0.7%/year with an average delay of 38 min/flight. Similar to the above-mentioned case of airports, these figures indicate again

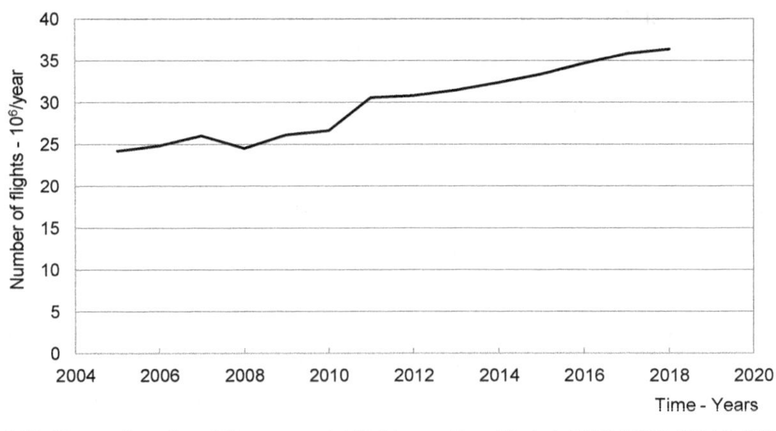

Figure 1.10 The number of world's commercial flights over time (Period: 2005–2018) (ICAO, 2005/2018; https://data.worldbank.org/indicator/is.air.psgr/).

Table 1.4 Characteristics of the ATC/ATM system—Case of Europe and U.S. (Period: 2017) (EEC/FAA, 2019).

Characteristics	**Area/System**	
Capacity—Infrastructure and staff/ employees	**Europe**	**U.S.**
Area (10^6 km^2)	11.5	10.4
Number of civil en-route ANSPs	37	1
Number of en-route facilities	62	23
Number of airports with ATC services	406	217
Number of FMPs (Europe)/TMUs (US)	51	65
Number of ATC controllers	17794	12170
Total ATC/ATM staff	55130	31647
Demand—flights & service quality		
Number of IFR flights (10^6/year)	10.4	15.3
Average daily flights	28457	41874
Average length of flight (nm; km/flight)	591/1095	554/1026
Proportion of GA flights (%)	3.5	19
ATC control time (hr/flight)	1.54	1.66
Arrival punctuality (%)	78.8[1)]	81.1[1)]
ATM/TMI delays (min/flight)	1.73[1)]	2.06[1)]
TMA delay (min/flight)	2.81	2.55

[1)] Flights from 34 main airports; Arrivals delayed by less than or equal to 15 min; ANSP – Air Navigation Service Provider; ATC – Air Traffic Control; FMP – Flow Management Position; IFR – Instrument Flight Rules; TMU – Traffic Management Unit; TMI – Traffic Management Initiative; GA – General Aviation; hr – hour.

Table 1.5 The flight delays in en-route airspace between airports over time—Case of Europe and U.S. (EEC/FAA, 2019).

Region/year	**Indicator**			
	Number of flights (IFR en-route) (10^6/year)	**Delayed flights > 15 min (%)**	**Delay per flight (min)**	**Delay per delayed flight (min)**
Europe				
2008	5.5	5.0	1.4	28
2010	5.0	5.7	1.8	32
2013	4.8	1.3	0.4	31
2015	5.0	2.0	0.6	28
2017	5.2	3.2	0.9	28
U.S.				
2008	9.3	1.1	0.4	38
2010	8.6	0.1	0.1	44
2013	8.4	0.8	0.3	36
2015	8.2	0.7	0.24	35
2017	8.4	1.0	0.35	36

that the U.S. airspace have handled about 70% more flights but on the account of about 3 times shorter and 30% longer average delay per flight and per delayed flight, respectively. At the same time the proportion of these (longer than 15 minutes) delayed flights was for about five times lower at the U.S. compared to the European airspace (EEC/FAA, 2019).

1.4.2 Economics

The economics of the ATC/ATM sub-system can be considered through its total and average revenues, costs, and their differences—profits/losses. In general, the prices of ATC/ATM services bringing its revenues are usually set up to cover costs almost 100%. Therefore, it can be said that the costs reflect revenues. One of the average economics in the given context is the average (unit) ATC/ATM provision cost, usually expressed as the ratio of the system's total provision costs and the volumes of output in terms of the flight-hours controlled. The flight-hours controlled have been estimated as the product between the number of controlled flights and time spend for their controlling. This average (unit) costs cost consists of the corresponding ATC controller employment cost and the supporting cost (non-ATC controllers' employment and the ATC controllers' training and developments, operating, and depreciation/amortization cost) (EEC, 2009/2019; 2019a). Figure 1.11(a, b) shows some cost characteristics of the European and U.S. ATC/ATM system.

Figure 1.11a shows that the average (unit) cost per controlled flight-hour was changing during the observed period in both ATC/ATM systems. It was the highest in the year 2008/2009 when the number of flight-hours was lower due to decreasing of the number of flights caused by the global economic/financial crisis. In addition, this cost was about 44% higher at the European ATC/ATM system during the observed period. Figure 1.11b provides one of the reasons for such difference, i.e., during the observed period, the U.S. ATC/ATM system was permanently carried out much higher volumes of the flight-hours controlled (for about 61–73%) compared to its European counterpart. As well, the average (unit) cost per flight-hour controlled decreased with increasing of the volumes of flight-hours controlled at both systems, thus indicating existence of the economies of scale.

1.5 Sustainability

1.5.1 General

The sustainability of air transport system generally relates to its impacts and effects on society and the environment and their balance in the medium to long period of time. Its operations create the environmental impacts, such as energy/fuel consumption and related emissions of GHG (Green House Gases) and land use. The social impacts are direct congestion and delays, noise, and safety, i.e., traffic incidents/accidents. If internalized, i.e., charged, these impacts represent area considered as externalities. The system's also produces the effects as contributions to the local, national, and international employment and GDP (Gross Domestic Product) and consequently the overall social-economic welfare. Both externalities and GDP-contributions can also be considered in the scope of the system's economic characteristics.

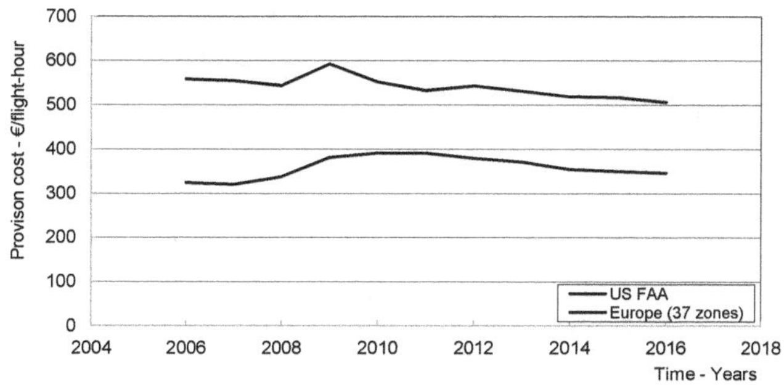

a) Average (unit) cost vs time

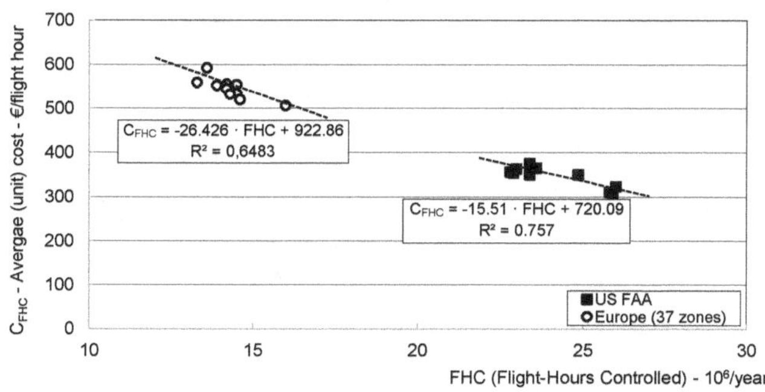

b) Average (unit) cost vs the controlled flight-hours

Figure 1.11 Relationship between the average (unit) cost and the annual number of controlled flight hours over time—Case of European and U.S. ATC/ATM system (Period: 2006–2017) (FHC – Flight Hour) (EEC, 2009/2019; 2019a).

1.5.2 Energy/Fuel Consumption and Emissions of GHG (Green House Gases)

The commercial turbojet aircraft mainly consume Jet A fuel or kerosene as an derivative of crude oil. Their fuel consumption over the past decade and half at the global (the world's) scale is shown in Figure 1.12(a, b, c).

Figure 1.12a shows that after some stagnation at the beginning (2005–2007), the fuel consumption has continuously increased at an average annual rate of 4.7% during the remaining period 2008–2017. Figure 1.12b shows how such increase in the fuel consumption has been driven by increasing of the air traffic volumes in terms of RPK and ATK. While the volumes of RPK and ASK have increased at average annual rates of 7.0% and 6.8%, respectively, the corresponding total fuel consumption has increased at an average annual rates of 2.7% during the period 2005–2017. Figure 1.12c shows the reason for such difference between the growth rates of traffic and corresponding fuel consumption. This is due to the decreasing of the average fuel consumption (g/RPK) by about 30%, i.e., at an average annual rate of 2.5%.

Burning Jet A fuel produces GHGs, of which the most voluminous are CO_2 (Carbon Dioxide) and H_2O (water vapor) emitted at the rates of 3.162 $kgCO_2/kg$ and 1.23 kgH_2O/kg. By multiplying these rates by the above-mentioned fuel quantities, the corresponding total and average emitted quantities of GHG can be estimated (IPCC, 2014). For example, Figure 1.12a shows that the total fuel consumption in the year 2016

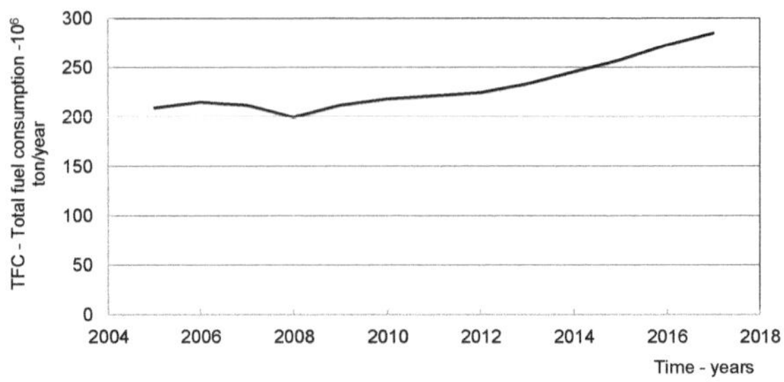

a) Total fuel consumption vs time

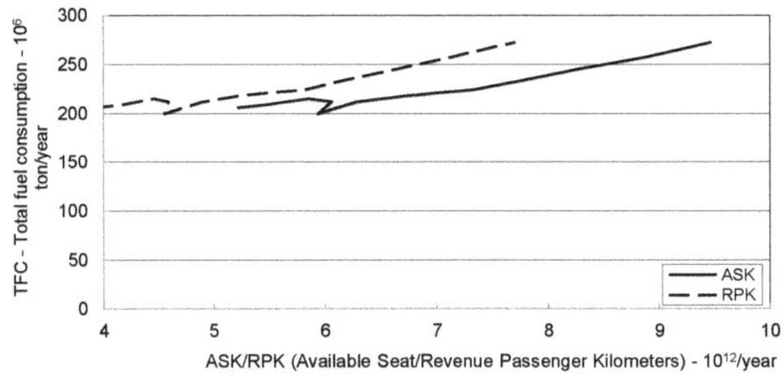

b) Total fuel consumption vs ASK/RPK

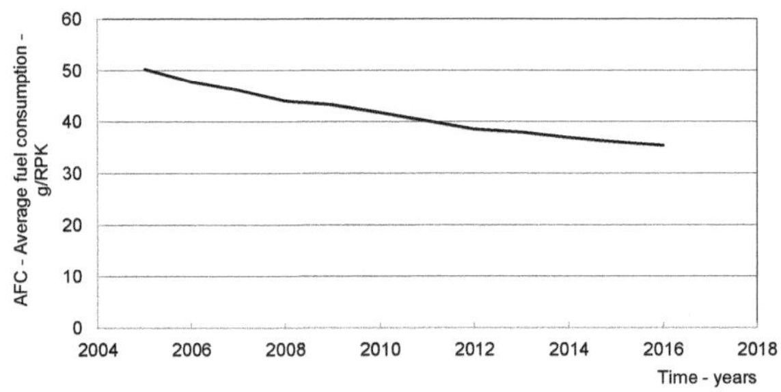

c) Average fuel consumption vs time

Figure 1.12 Fuel consumption over time—Case of the world's commercial air transportation (Period: 2005–2018) (ICAO, 2005/2018; https://www.statista.com/statistics/655057/fuel-consumption-of-airlines-worldwide/).

was 272.552 million tons. This produced 272.552 · (3.162+1.23) ≈ 1197.048 million tons of CO_{2e} (Carbon Dioxide equivalents). The average external cost of emitted CO_{2e} has estimated to be 212 USD/ton (NAE, 2004; IWG, 2016). In the given case, this can give the total costs of emissions of CO_{2e} in the year 2016 of 253774.257 million USD or 0.0061 USD/ASK and 0.0075 USD/RPK.

1.5.3 Land Use

The land use can be considered as an impact of airports on the environment. This land is considered as occupied by the airport airside and landside area, the latter excluding the landside access modes and their systems. Figure 1.13(a, b) shows the examples of land use by 10 U.S. and 15 of the world's most spacious airports.

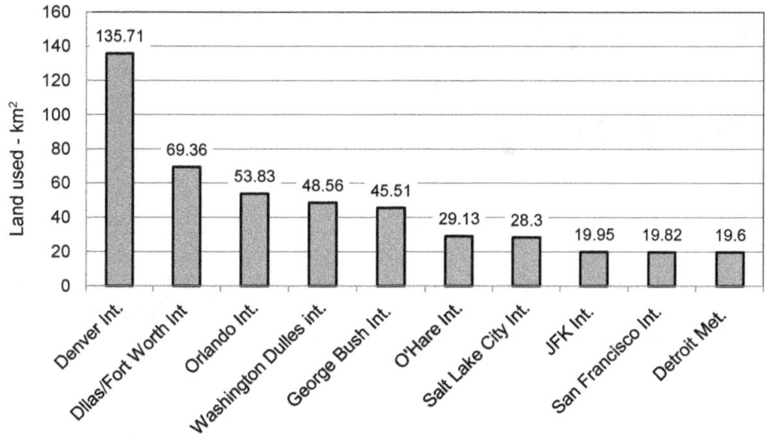

a) The U.S.'s largest airports (Period: 2017)

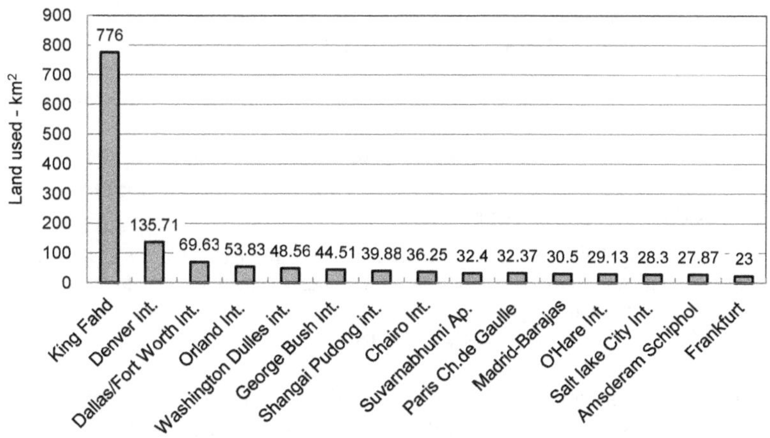

b) The world's largest airports (Period: 2017)

Figure 1.13 Land used by airports—Case of the largest U.S. and world's airports (https://www.worldatlas.com/articles/the-largest-airports-in-the-united-states.html/; https://www.worldatlas.com/articles/the-world-s-10-largest-airports-by-size.html/).

As can be seen, the area of land used is measured in square kilometers. In the U.S., it is the greatest at Denver International and the smallest at Detroit Metropolitan airport. At the world scale, Cairo's King Fahd (Saudi Arabia) is the most and Frankfurt (Germany) is the least spacious airport. In addition, 7 U.S. airports are among the 15 world's most spacious airports. In addition, Figure 1.14(a, b) shows an example of the relationships between the annual number of handled atms and pax and the area of land occupied by 15 world's busiest airports by ATMs.

As can be seen, the number of atms and pax handled at these airports is not correlated with their size, i.e., land used. This indicates that the airports have been planned and designed to handle current but also the future air traffic growth. In general, when the air traffic reaches the airport designed capacity then the intensity of use of available

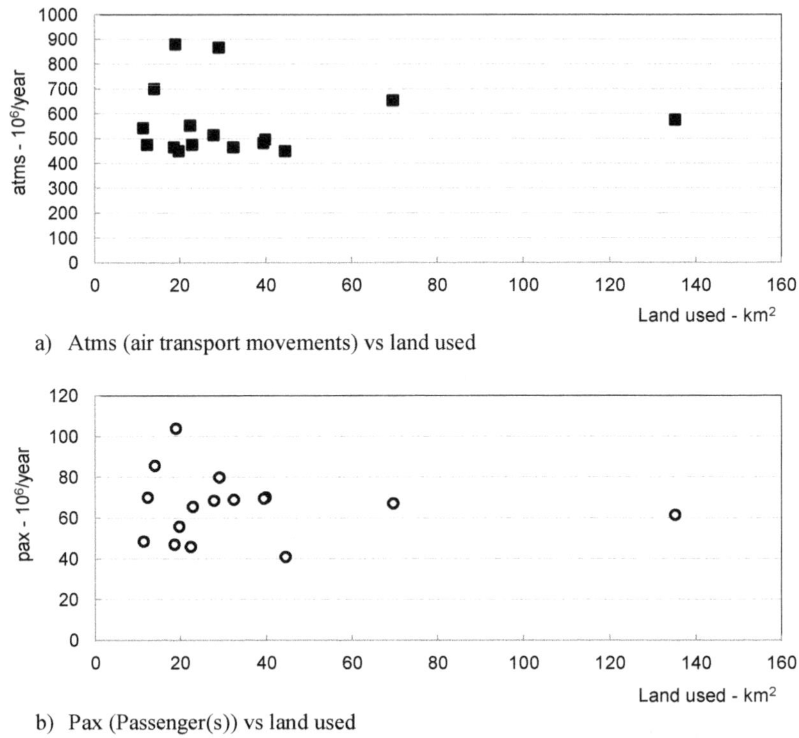

Figure 1.14 Relationship between the number of handled atms, pax, and the land used—Case of the world's most sizeable airports (Period: 2017) (https://aci.aero/data-centre/annual-traffic-data/passengers/2017-passenger-summary-annual-traffic-data/; https://www.worldatlas.com/articles/the-largest-airports-in-the-united-states.html/).

land for providing this capacity will be maximal. In addition to the above-mentioned economics of airports implicitly also reflecting economies of their land use, there may be also diseconomies of their land use, materialized through changing (diminishing) the value of nearby properties exclusively. This happens due the impacts such as the aircraft noise, emissions of GHG, and risks of air traffic incidents/accidents potentially damaging properties and lives of the nearby/local population (Janić, 2016).

1.5.4 Congestion and Delays

If internalized, the average unit cost of flight delays, including the airline operating costs and the cost of passenger time, were estimated to be about 81 €/min in Europe and 75 USD/min in the U.S. (UW, 2011; http://airlines.org/dataset/per-minute-cost-of-delays-to-u-s-airlines/). This implies, for example, that, according to Table 1.1, the cost of airport delays of delayed flights (more than 15 min) would be 33 min · 81 €/min = 2637 € in Europe and 66 min · 75 USD/min = 4950 $US in the U.S. According to Table 1.5, the cost of en-route delays per delayed flight would be 29 min · 81 €/min = 2349 € in Europe and about 38 min · 75 USD/min = 2850 $US in the U.S. These costs can also be considered in the scope of the air transport system economics.

1.5.5 Noise

The aircraft noise is mainly generated in the vicinity of airports by the aircraft take-offs and landings. In addition to the absolute levels of noise by particular aircraft types, the number of population exposed to the certain levels of noise can be an indicator of the air transport system sustainability respecting to noise impact. In general, regarding the aircraft noise, the US FAA has established the level of DNL (Day-Night Level) of 65 dBA as the threshold or "significant" (US DOT/FAA, 2015). The EEA (European Environmental Agency) has established the threshold level of (L_{den}) (Level day-evening-night) of 55 dBA. Above these levels, the aircraft noise is considered to be incompatible with residential areas. Table 1.6 shows the population exposed to the aircraft noise around the U.S. airports over the specified period of time.

As can be seen, despite the increasing population, its absolute number and the corresponding proportions exposed to the above-mentioned threshold noise level have decreased over time, thus indicating the substantial achievements in dealing with the noise burden around the country's airports. This was made possible by stricter aircraft certification, advanced operational procedures (noise abatement procedures), and improved land use compatibility embracing residential, public, commercial, manufacturing, production, and recreational areas (FAA, 1983). In addition, Table 1.7 gives an example of the number of population exposed to the airport noise in the specified noise contours around 47 European airports over time (EEA/EEC, 2019).

Table 1.6 Population exposed to the aircraft noise greater than DNL 65 dBA—Case of U.S. airports (Period: 1975–2014) (US DOT/FAA, 2015).

Year	Population of the country (10^6)	Population exposed to airport noise (10^6)	Proportion (%)
1975	216.0	7.0	3.24
1980	226.5	5.2	2.30
1985	237.9	3.5	1.47
1990	250.1	2.8	1.12
1995	266.6	1.8	0.68
1999	279.3	0.8	0.29
2014	318.6	~ 0.7	0.22

Table 1.7 Area and population exposed to noise around airports over time—Case of 47 European (EEA/EEC, 2019).

Year	Area (km^2)	Population (10^6)	
		$L_{den} \geq 55$ dB	$L_{night} \geq 50$ dB
2005	2250	2.31	0.87
2014	2251	2.27	0.81
2017	2421	2.58	0.98
2040	2172	2.14	0.70

As can be seen, in the year 2017, the total population inside the L_{den} 55 dB and L_{night} 50 dB noise contours of the 47 major European airports were 2.58 and 0.98 million, respectively. This was about 12% and 13% more than in the year 2005 and 14% and 20% more than in the year 2014 for L_{den} and L_{night} contours, respectively. Thanks to the expected technological and operational improvements, it is expected that the number of exposed population within L_{den} and L_{nigh} contours will decrease by about 20% and 30%, respectively, by the year 2040 compared to the year 2018 (EEA/EEC, 2019).

The noise has been charged at many airports worldwide. The noise charges have been frequently included into the airport landing fees based on the aircraft noise categories based on the specified noise level mainly influenced by their size, i.e., MTOW (Maximum Take-Off Weight). In some cases, they have also been transparent, as shown in Figure 1.15.

At three considered airports, the noise charges linearly increase with increasing of the aircraft MTOW (Maximum Take-Off Weight). However, they differed over time at the same and across particular airports. They were the highest and increased at the highest rate at Zurich airport, followed by Frankfurt airport. The charges at Munich airport were higher for lighter and lower for heavier aircraft than at Frankfurt airport, while increased at decreasing rate with increasing of the aircraft MTOW (Zuidberg and Veldhuis, 2008).

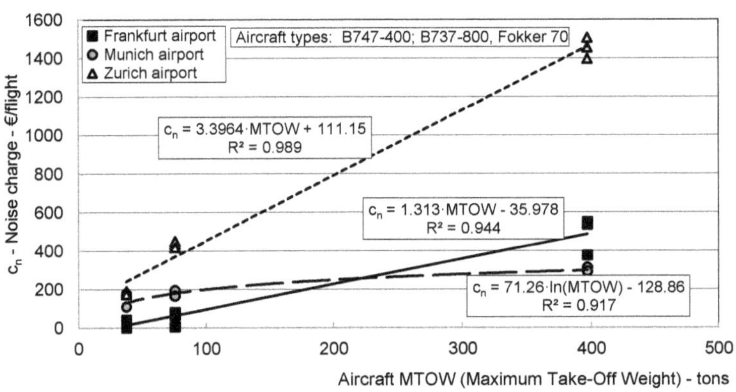

Figure 1.15 Example of the noise charge depending on the aircraft MTOW (Maximum Take-Off Weight)—Case of the selected European airports (Period: 2003, 2006, 2007) (Zuidberg and Veldhuis, 2008).

1.5.6 Air Traffic Incidents/Accidents

The air transport system, like other transport systems, has not been free from air traffic incidents/accidents. In particular, the accidents have resulted in damages of property (aircraft involved), losses of lives and injuries of air passengers and crew on board, and losses of the third parties involved. Therefore, in the narrower sense (not including the eventual environmental damages), these can be considered as the physical impacts on society rather than on the environment, which, expressed in monetary terms, are the corresponding social externalities (Janić, 2007). According to current practice, air accidents have been investigated in order to identify their causes and undertake actions aiming at preventing their occurrence due to the already known (investigated) causes. This has resulted in continuous reduction of the number of these accidents despite increasing of the number flights/departures over time, as shown in Figure 1.16.

During the last 10 years (2008–2018), the rate of air accidents has decreased by about 63%, i.e., from 4.70 to 1.75 events per million annual departures. At least from the statistical point of view, this indicates that, over time, the system has become increasingly safe. Some estimates indicated that the costs of air accidents have been substantial. As an illustration, they were 223 million €/year or 500 €/RPK in the year 2008 in the EU-27 Member States. This amounted to 3.33% under low and 0.875% under high scenario of the total external costs (CE Delft, Infras, Fraunhofer ISI, 2011).

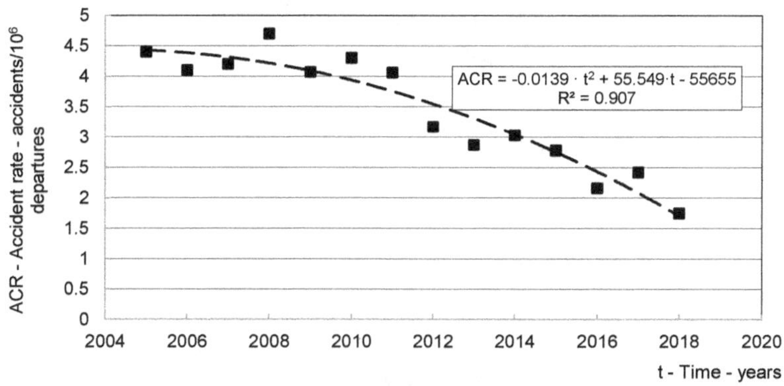

Figure 1.16 Air traffic accident rate over time (Period: 2005–2018) (https://www.icao.int/safety/iStars/Pages/Accident-Statistics.aspx/).

1.5.7 Contribution to Social-Economic Welfare

Contribution of the air transport system to the social-economic welfare consists of its employment and created GDP (Gross Domestic Product). For example, in the year 2014, the number of employees was 62.7 million and the created GDP 2.7 trillion USD worldwide. This gives about 43062 GDP USD/employee-year (ATAG, 2018). Figure 1.17(a, b, c) shows the distribution of two contributions across particular world's regions in the year 2014.

In particular, Figure 1.17a shows that the highest employment was in Asia-Pacific and the lowest in Middle East region. That in Europe was higher than in North America. At the same time, Figure 1.17b shows that the highest GDP was created in Europe, followed by that in North America and Asia Pacific region. The lowest GDP was created in Middle East. However, Figure 1.17c shows that the highest GDP per an employee was achieved in North America followed by Europe and the lowest in Africa region. Some estimates indicate that the total employees worldwide by the air transport system will increase to about 99 million people and generated GDP to 5.9 trillion USD in the year 2034 (ATAG, 2018).

1.6 Concluding Remarks

An overview and limited analysis of performances of the air transport system and its sub-systems—airports, airlines, and ATC/ATM (Air Traffic Control/Management)—during

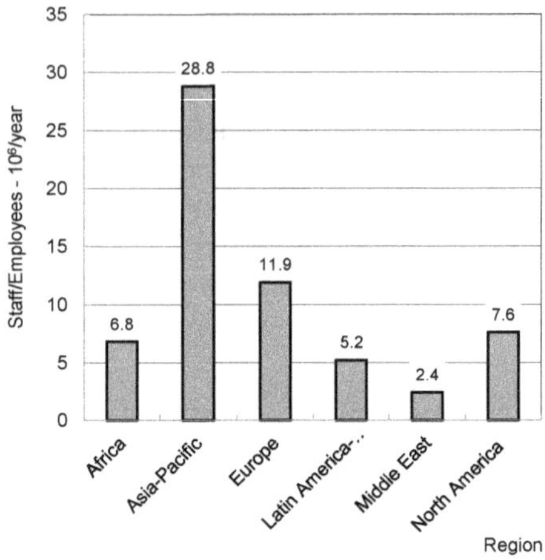

a) Staff/Employment

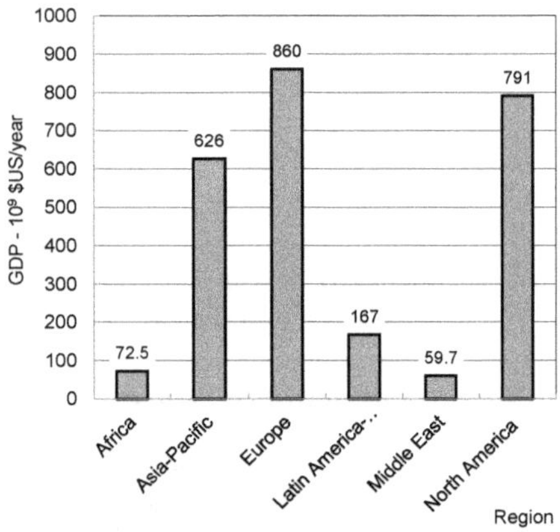

b) GDP (Gross Domestic Product)

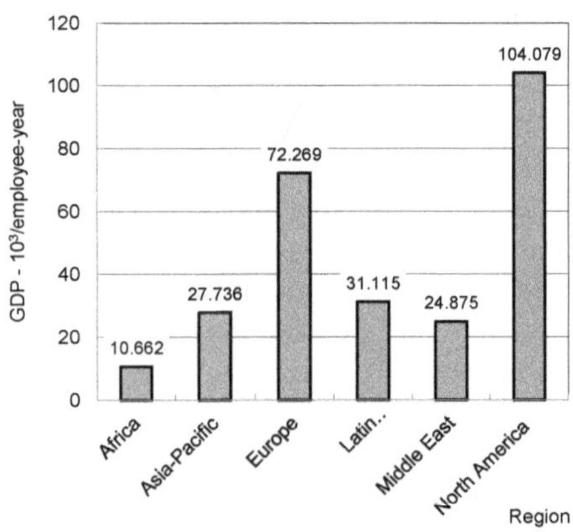

c) GDP/employee

Figure 1.17 Contribution of the air transport system to the social-economics welfare (Period: 2014) (ATAG, 2018).

the specified (past) period of time has been carried out. Their demand, supply/capacity, economics and sustainability have been elaborated mainly at the global (world's) scale, thus providing an overall insight into the performances of these sub-systems as the topic of more detailed analysis and modelling in the forthcoming Chapters. Both demand and supply/capacity of all sub-systems have generally been growing over time despite some fluctuations. They have been driven by different external and internal driving forces, i.e., adapting to each other while providing the services of reasonable quality and economic feasibility to the corresponding users under conditions of growth. Generally, thousands airports handling commercial aircraft and their flights have operated worldwide, but only a certain number, i.e., the larger ones, have always been under focus. At these airports, both the numbers of aircraft movements and passengers have been growing while being provided a reasonable quality of services in terms of the on-time performances expressed by the proportion and duration of the average delays of all flights and those delayed more than 15 minutes. This indicates a relatively good balance between demand and supply/capacity there. As far as economics is concerned, the airports have generally been profitable, implying that their aeronautical and non-aeronautical revenues have covered the operating costs and also generated profits at the total and average scale under conditions of growth. In order to serve growing user demand of both passengers and freight/cargo shipments, the airlines have deployed an increasing number of aircraft, thus offering relatively adequate capacity. This has resulted in carrying out the growing volumes of passenger and ton kilometers while maintaining relatively constant utilization of the aircraft fleet payload capacity, i.e., load factor over the specified period of time. Worldwide, different categories of airlines have had different on-time performance, expressed by the proportion and duration of delays of their flights. Regarding economics, despite fluctuations of both revenues and costs during the specified period of time, the world airlines have generally been profitable. The ATC/ATM sub-system has been permanently upgraded and modernized in order to handle the increasing number of airline flights over the specified period of time safely, effectively, and efficiently. Despite increasing of the number of flights, safety has implied decreasing of the risk of incidents/accidents due to the ATC/ATM reasons. Effectiveness has materialized through reduction of the average delay per handled flight, and efficiency through covering the costs with the revenues almost 100%, relative small fluctuation of both, and reduction of the average cost per unit of the sub-system's output-flight-hour controlled.

Sustainability of air transport system and its sub-systems has been different regarding particular impacts and effects on the environment and society. Regarding the impacts, the total fuel consumption and related emissions of GHG have increased and their average counterparts decreased more than proportionally with increasing of the volumes of air traffic over the specified period of time. Different airports have occupied different areas of land. However, the volumes of handled aircraft movements and passengers on the one hand and the land used on the other have not been in correlation across the considered airports—the largest in terms of land used. This implies that each airport has been rather specific regarding this performance. The population exposed to the specified level of noise has generally decreased around the U.S. airports and increased around their European counterparts during the specified period of time. The higher levels of noise generated by the larger heavier aircraft have been charged at the higher rates than that of their smaller and lighter counterparts at airports practicing charging noise as an impact explicitly. Otherwise, the noise charge has been included in the landing charges.

Costs of flight delays longer than 15 min have shown to be substantial both in Europe and U.S. if the airline and passenger time average unit costs have been included. In addition, the safety performance of air transport system has been continuously improving the specified period of time by reducing the number of accidents per unit of its output despite the growth of the latter. Regarding the effects, the air transport system has generated a substantial number of jobs and amount of GDP. Nevertheless, these have been different in different world's regions, thus reflecting differences in development and efficiency of the air transport systems and overall economics there.

This first chapter has introduced the general characteristics of the air transport system and its sub-systems—airports, airlines, and ATC/ATM. In addition, Chapter 2 describes analysis and modelling of the demand, supply/capacity, quality of services, and economics of airports. Chapter 3 deals with analysis and modelling of the demand, supply/capacity, quality of services, and economics of airlines. Chapter 4 describes analysis and modelling of the above-mentioned four performances of the ATC/ATM sub-system. Finally, Chapter 5 deals with analysis and modelling of the sustainability of air transport systems across its three above-mentioned sub-systems. It should be mentioned that, due to the three sub-systems of the air transport system being considered relatively independently in the forthcoming Chapters, they will be treated as and called the 'systems'.

References

ACI. (2006). ACI Worldwide Air Transport Forecasts 2005–2020: Passengers, Freight, Aircraft Movements, Airport Council International, Geneva, Switzerland.

ACI World. (2016/2018). Airport Economics at a Glance, Airport Council International, Montreal, Canada.

ACI. (2018). Annual World Airport Traffic Report (WATR), Airports Council International, ACI World, Montréal, Québec, Canada.

AIRBUS. (2006). Airbus Global Market Forecast, Airbus Industrie AIRBUS S.A.S., Toulouse, France.

Akerman, J. (2005). Sustainable air transport—on track in 2050. Transportation Research D, 10(2): 111–126.

AIRBUS. (2018). Global Market Forecast: Global Networks, Global Citizens 2018–2037, AIRBUS S.A.S., Blagnac Cedex, France.

ATAG. (2018). Aviation Benefits Beyond Borders, Air Transport Action Group, Geneva Airport, Switzerland.

Boeing. (2007). Current Market Outlook 2007: How Will You Travel Through Life? Boeing Commercial Airplanes: Market Analysis, Seattle, WA, USA.

Boeing. (2017). Current Market Outlook 2017–2036, Boeing Commercial Airplanes Market Analysis, Seattle, WA, USA.

CE Delft, Infras, Fraunhofer ISI. (2011). External Costs of Transport in Europe, Update Study for 2008, CE Delft, Delft, The Netherlands.

Cook, A., Belkoura, S. and Zanin, M. (2017). ATM performance measurement in Europe, the US and China. Chinese Journal of Aeronautics, 30(2): 479–490.

EC. (2017). Annual Analysis of the EU Air Transport Market 2016, Final Report, European Commission, DG MOVE, Brussels, Belgium.

EEA/EEC. (2019). European Aviation Environmental Report 2019, EUROCONTROL, European Environment Agency, Brussels, Belgium.

EEC. (2009/2019). Report on the Operation of the Route Charges System, EUROCONTROL, European Organisation for the Safety of Air Navigation, Central Route Charges Office (CRCO), Brussels, Belgium.

EEC. (2019a). U.S.–Europe Continental Comparison 2006–2016 of ANS Cost-Efficiency Trends, Performance Review Unit on behalf of the European Commission, EUROCONTROL, European Organisation for the Safety of Air Navigation, Brussels, Belgium.

EEC/FAA. (2019). Comparison of Air Traffic Management-Related Operational Performance U.S./ EUROPE—2017, EUROCONTROL, European Organisation for the Safety of Air Navigation, Brussels, Belgium/Federal Aviation Administration, U.S. Department of Transportation, Washington D.C., USA.

FAA. (1983). Noise Control and Compatibility Planning for Airports, Advisory Circular, AC No: 150/5020-1, U.S. Department of Transportation, Federal Aviation Administration, Washington D.C., USA.

Horonjeff, R. and McKelvey, X. F. (1994). Planning and Design of Airports, Third Edition, McGraw Book Hill Company, New York, USA.

https://www.statista.com/statistics/655057/fuel-consumption-of-airlines-worldwide/)

https://www.icao.int/safety/iStars/Pages/Accident-Statistics.aspx/

https://en.wikipedia.org/wiki/Air_navigation_service_provider

https://data.worldbank.org/indicator/is.air.psgr/

https://aci.aero/data-centre/annual-traffic-data/passengers/2017-passenger-summary-annual-traffic-data/

https://www.worldatlas.com/articles/the-largest-airports-in-the-united-states.html/

https://www.worldatlas.com/articles/the-largest-airports-in-the-united-states.html/

https://www.worldatlas.com/articles/the-world-s-10-largest-airports-by-size.html/

https://airlines.org/dataset/per-minute-cost-of-delays-to-u-s-airlines/

https://www.icao.int/sustainability/Airport_Economics/State%20of%20Airport%20Economics.pdf/

https://www.cia.gov/library/publications/the-world-factbook/rankorder/2053rank.html

https://store.aci. aero/product/annual-world-airport-traffic-report-2018/

https://www.bts.gov/content/number-us-airportsa/

https://en.wikipedia/wiki/List_of_international_airports_by_country#Number_of_International_Airports

https://www.iata.org/about/members/Pages/airline-list.aspx

https://en.wikipedia.org/wiki/Lists_of_airlines/

https://en.wikipedia.org/wiki/Nationalaviationauthority#List_of_civil_aviation_authorities/

ICAO. (2001). Aeronautical Telecommunications, Annex 10, Sixth Edition, International Civil Aviation Organization, Montreal, Canada.

ICAO. (2005/2018). Annual Report of the Council 2005–2017, International Civil Aviation Organization, Montreal, Canada.

ICAO. (2015). State of Airport Economics, International Civil Aviation Organization, Montreal, Canada.

ICAO. (2016). Rules of the Air and Air Traffic Services DOC. 4444-RAC/501/12, International Civil Aviation Organization, Sixteenth Edition, Montreal, Canada.

ICAO. (2016a). Manual on Air Navigation Services Economics, Fifth Edition, International Civil Aviation Organization, Montreal, Quebec, Canada.

IPCC. (2014). Climate Change 2014, Synthesis Report—Contribution of Working Groups I, II and III to the Fifth Assessment Report of the Intergovernmental Panel on Climate Change [Core Writing Team, R. K. Pachauri and L. A. Meyer (eds.)], IPCC, Geneva, Switzerland.

IWG. (2016). Technical Support Document: Technical Update of the Social Cost of Carbon for Regulatory Impact Analysis Under Executive Order 12866 (September 2016 Revision), Interagency Working Group on the Social Cost of Greenhouse Gases, Washington, D.C., USA.

Janić. (2000). Air Transport System Analysis and Modelling: Capacity, Quality of Services and Economics, Gordon and Breach Science Publishers, Volume 16, Amsterdam, The Netherlands.

Janić, M. (2007). The Sustainability of Air Transportation: Quantitative Analysis and Assessment, Ashgate Publishing Company, UK.

Janić, M. (2009). The Airport Analysis, Planning, and Design: Demand, Capacity, and Congestion, Nova Science Publishers, Inc. New York, USA.

Janić, M. (2016). Analyzing, modeling, and assessing the performances of land use by airports. International Journal of Sustainable Transportation, 10(8): 683–702.

Manheim, L. M. (1979). Fundamentals of Transport System Analysis: Vol. 1: Basic Concepts, MIT Press, Massachusetts Institute of Technology, Massachusetts, USA.

NAE. (2004). The Hydrogen Economy: Opportunities, Costs, Barriers, and R&D Needs, National Academy of Engineering, The National Academies Press, https:// doi. org/10.17226/10922, Washington D.C., USA.

OAG. (2019). Punctuality League 2019: On-Time Performance for Airlines and Airports Based on Full-Year Data 2018, OAG Connecting the World of Travel, Luton, UK, https://www .oag.com/.

US DOT/FAA. (2015). Statistical Handbook of Aviation (Annual Issues): Departures 1994–95/2014, Bureau of Transportation Statistics/Federal Aviation Administration, Washington, D.C., USA.

UW. (2011). European Airline Delay Cost Reference Values, Final Report (Version 3.2), Department of Transport Studies, University of Westminster, London, UK.

WB. (2019). World Development Indicators (WDI), World Bank, Washington, D.C., USA.

Zuidberg, J. and Veldhuis, J. (2008). Benchmark for Airport Charges and Governmental Taxes for the Years 2003, 2006 and 2007, SEO Economic Research, Amsterdam, The Netherlands.

Chapter 2

Airports

2.1 Introduction

Airports are one of three main components of the air transport system, together with airlines and ATC/ATM (Air Traffic Control/Management) components, which are ultimately considered throughout this book as sub-system or systems. The ACI (Airport Council International) has reported that 3759 airports with scheduled commercial flights operated in the year 2017. Each airport system consists of the airside and landside areas, as shown in Figure 2.1 (ACI, 2019; Janić, 2000).

The airside area includes runways, taxiways, and apron/gate complex. The runways enable the aircraft landings and take-offs, i.e., atms (air transport movement(s)) (1 atm = 1 landing or 1 taking-off). The taxiways connect the runways to the apron/gate complex, enabling the aircraft movement/taxiing between them. The apron/gate complex provides the space-parking stands and/or gates for the aircraft ground handling, which includes embarking, and disembarking air passengers, their baggage, and freight/cargo shipments after landing and embarking the same after the aircraft refuelling and before take-off. The landside area includes the passenger and freight/cargo terminal(s) and the airport landside access modes and their systems. In particular, interfaces in the airside area enable users—air passengers and freight/cargo shipments—to pass from the corresponding terminals to the aircraft, and vice versa. In the case of air passengers, the interfaces include air bridges for the terminal-close and/or buses for the terminal-remote aircraft parking stands. With freight cargo shipments, these are usually trucks that are frequently equipped with loading/

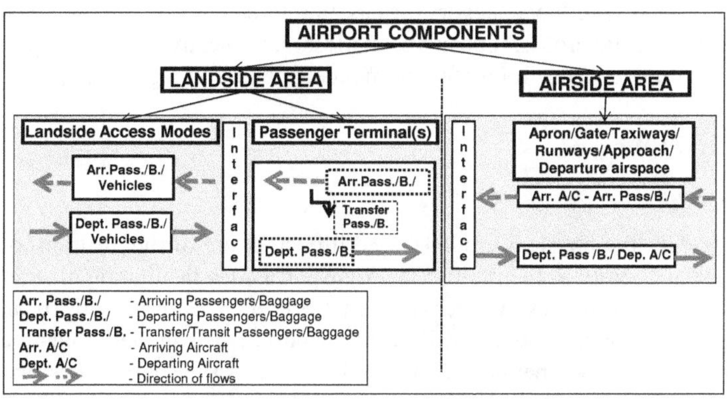

Figure 2.1 Scheme of the main airport components (Janić, 2000; 2013).

unloading devices. In the airport landside area, the landside access modes and their systems and the passenger (and/or freight/cargo) terminals are mutually connected by the appropriate interfaces—car and bus standing/parking areas and the rail stations—enabling the corresponding arriving and departing users to pass between them. In particular, the landside road- and rail-based access modes provide the airport landside accessibility to air passengers, aviation employees, greeters, and others. The former mode includes car and van, and bus system(s). The latter mode includes the streetcar/tramway, LRT (Light Rail Transit), subway/metro (short distance), regional/conventional rail and HSR (High Speed Rail) (medium and long distance), TRM (Trans Rapid Maglev) (currently short distance), and most recently the rather futuristic HL (Hyperloop) (long distance) system. These all connect airports with the specific locations in their catchment areas (in most cases the city centres or CDBs (Central Business District(s))). At some airports, the air freight/cargo airport accessibility is also provided by the corresponding road- and rail-based modes and their systems (Janić, 2013). In general, the airside area can be of different layouts and sizes, mainly depending on the number and horizontal configuration of the runway system(s) influenced by the volumes of air transport movements, weather conditions (prevailing wind directions) and social (noise) and environmental (emissions of GHG (Green House Gases)) constraints. The passenger and freight/cargo terminals and the airport landside access modes and their systems can have different layouts and sizes depending on the volumes and structure of the corresponding demand to be served during the specified period of time (hour, day, and year) under given conditions.

The demand in the airport airside area includes atms (the aircraft or air transport movements) (1 atm is 1 landing or 1 take-off). They are served by the capacity of the above-mentioned components of airside area-runways, taxiways, and apron/gate complex. The demand in the airport landside area mainly consists of the air passengers and freight/cargo shipments. The space and facilities, equipment, and vehicles with specified processing rates/capacities in the corresponding (passenger and freight/cargo) terminals, and at the landside access modes and their systems serve/handle this demand. These service rates/capacities are usually defined by the maximum number of entities/items of demand (passenger and freight/cargo shipments) which can be served/handled during the specified period of time under given conditions.

As soon as the intensity of demand in either area exceeds the available capacity, congestion and delays of the affected entities/items of demand occur. As such, they can be used as indicators of the quality of provided services. In addition, serving/handling the airport demand of either category brings the revenues and imposes costs, whose differences indicate the profits, if positive, and losses, if negative, for the airport operator as the service provider. They represent the main economics of a given airport.

2.2 The System

2.2.1 General

Airports represent the physical ground-based infrastructure of the air transport system. Essentially, they operate as the multimodal transport nodes facilitating air transport and other surface transport modes. As such, they enable users, i.e., air passengers and freight/cargo shipments to change between different transport modes on their door-to-door trips. In addition to airlines and operators of the landside access transport modes (usually road and rail-based), the airport facilities and equipment of sufficient capacity and organization aim to provide safe and secure, effective, and efficient services to aircraft, air passengers, freight/cargo shipments, and vehicles of the landside access modes and their systems. Safety and security imply carrying out services as prescribed without accidents and incidents due to the already known reasons. Effectiveness and efficiency imply providing services at the agreed (prescribed) quality and prices, both acceptable for users on the one hand and covering the airport operator costs on the other (Janić, 2013).

The world's airports are categorized regarding their capabilities to accommodate different aircraft types (categories). For example, both ICAO (International Civil Aviation Organization) and FAA (Federal Aviation Administration) use two-element reference code for categorizing airports. The first element of the ICAO's codes refers to the aircraft RFL (Reference Field Length). This is defined as the minimum field length an aircraft needs in order to take off at the MTOW (Maximum Take-off Weight) under conditions of the standard atmosphere (no wind and sea-level runway). The second code element is represented by the aircraft wingspan and distance between the outside edges of the main gear. Consequently, four codes regarding RFL, varying between less than 800 m (Code I aircraft) to less than 1800 m (Code IV aircraft), have been defined. Regarding the wingspan and the outer main gear wheels' span, the aircraft are grouped into six categories (A, B, C, D, E, F). The wingspan varies from less than 15 m in category A to less than 80 m in category F. The outer main gear wheels' span varies from less than 4 m in category A to less than 16 m in category F (ICAO, 2004; Janić, 2013). In general, the size of particular infrastructure elements in the airport airside area, such as runways, taxiways and apron/gate parking stands, depend on the largest aircraft category expected to operate at the airport. In such case, this aircraft is considered as the "critical aircraft".

2.2.2 Airside Area

The airport airside area consists of the controlled airspace, i.e., airport zone, runways, taxiways, and apron/gate complex.

2.2.2.1 Controlled airspace

The controlled airspace is designated around each airport and close airports to provide safe, efficient, and effective air traffic. This airspace, known as the airport zone, is under jurisdiction of the airport ATC/ATM (Air Traffic Control/Management) unit providing control of the aircraft/flights during final approach and landing, take-off, and manoeuvring on the ground. The airport zone has the shape of a roller, with the circled basis usually of diameter 5 NM (Nautical Miles) and height 1500–2500 ft (1 NM = 1.852 km; 1 ft = 0.305 m). In the U.S. this airspace is categorized as the class D controlled airspace (FAA, 1990; 1999). The aircraft/flights can generally operate there according to the IFR (Instrument Flight Rules) or VFR (Visual Flight Rules) corresponding to IMC (Instrument Meteorological Conditions) and VMC (Visual Meteorological Conditions), respectively. These rules are specified by the vertical ceiling and horizontal visibility whose specified relationships for the U.S. airports are shown in Figure 2.2. As can be seen, the vertical ceiling appears to be most diverse for the horizontal visibility of 3 and 5 (statute) miles and relatively homogenous for the horizontal visibility of 4, 7, and 8 miles (1 mi = 1.609 km). In addition, most U.S. airports operate at the margin between the "high IFR" and the "marginal VFR" conditions (FAA, 1990; 1999; Janic, 2013; NASA, 2001). The quantitative elements of IFR and VFR for the arriving and departing aircraft are described in Section 2.3, which covers airport demand and capacity.

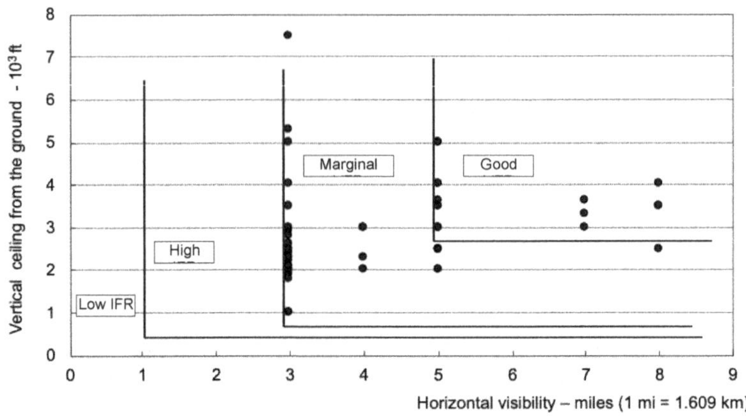

Figure 2.2 Relationship between vertical ceiling and horizontal visibility (boundary conditions)—Case of the selected 75 U.S. airports (FAA, 1999; Janic, 2013; NASA, 2001).

In the airport zone, the IFR aircraft/flights maintain the assigned flight trajectories while the ATC/ATM controls the safe separation between them. The VFR aircraft/flights primarily maintain the assigned flight paths while being responsible for maintaining a safe separation between each other. If both the IFR and the VFR flights are carried out simultaneously, the division of responsibility between the pilots and ATC controllers for the safe separation is the same as with purely IFR aircraft/flights.

2.2.2.2 Runways, taxiways, and apron/gate complex

a) Runways

The runways enable aircraft take-offs and landings. At most commercial airports, aircraft perform precise approach and landing means using ILS (Instrument Landing System) (Cat. I, II, III). The aircraft can operate on a given runway under limited conditions of headwind and tailwind of up to 9–11 km/h (5–6 kts). Operation of the aircraft categories A, B, Con a given runway can be affected by a maximum crosswind of 19 km (10.5 kts) to 37 km/h (20 kts). The aircraft categories D, E, F can operate under across wind of about 46–55 km/h (25–30 knots). Regarding these constraints, the usual objective in designing and locating the airport runways is that they have to be operative for about 95% of the time throughout the year. The crossing runways are designed just to fulfil these usage requirements (ICAO, 2004; Janić, 2013).

The runway length depends on many factors, such as take-off, landing weight, and stage length to be flown by the "critical" aircraft, weather conditions characterized by the above-mentioned prevailing winds and air temperature, the runway elevation and presence of obstacles in its vicinity, and the runway slope and conditions (wet or dry). In general, the higher aircraft weight and stage length lower headwind, higher ambient temperature, greater uphill gradient, higher runway elevation, and presence of obstacles will generally require longer runway(s). Figure 2.3 shows the simplified layout of an airport runway with its basic elements.

Runways are separated from the other components of airport infrastructure. Both ICAO and FAA provide the minimum separation between the runway and

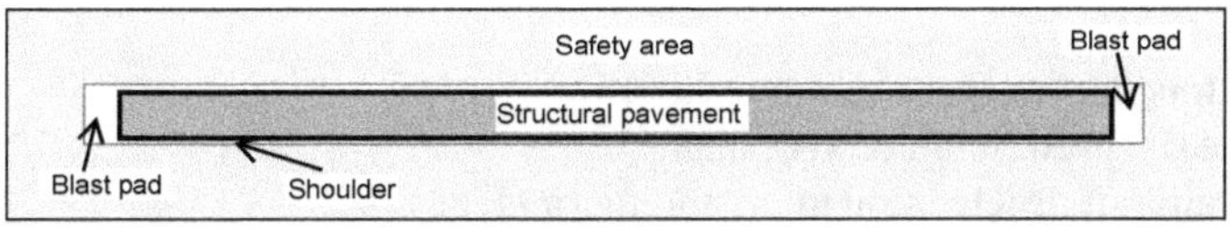

Figure 2.3 Simplified layout of an airport runway (Horonjeff and McKelvey, 1994; Janić, 2013).

taxiway centrelines. The ICAO specifies separation for the instrument runways of 168, 176, 182.5, and 190 m for the airports coded as C, D, E, and F, respectively. The FAA specifies 120, 120, 120, and 180 m for the aircraft categories III, IV, V, and VI, respectively. In both cases, separation of a given runway and the aircraft parking area is 150 m. In addition, the separation distance between the runway and the hold lines where the aircraft wait before take-offs for the aircraft category F and VI is 107.5 m according to ICAO and 85 m according to FAA, respectively. Last but not least, the runway's longitudinal and transverse grades are also standardized regarding the runway/aircraft category (FAA, 1999; Janić, 2013).

b) <u>Taxiways</u>

Taxiways connect the runways and the apron/gate complex near and around passenger and freight/cargo terminals, and the aircraft maintenance areas. The design standards for taxiways include elements such as width, curvature, the minimum separation distances between taxiways and parallel taxiways, the longitudinal and traverse slopes, the sight and distances from the other objects. The U.S. FAA standards imply the width of a taxiway from 7.5 m for Aircraft Group I to 23 m for Aircraft Group VI. The separation distance between the taxiway centrelines is 21 m for Aircraft Group I and 99 m for Aircraft Group VI. The corresponding distances from the fixed or movable objects are 13.5 m to 59 m, respectively. The distances between the parallel taxi-lines vary from 19.5 m to 91 m, respectively. Finally, the distance of taxi-lines from the fixed or movable objects varies from 12 m for Aircraft Category I to 51 m for Aircraft Category IV. These standards are mostly based on the wingspan of a "critical" aircraft and the multiplying factor. The corresponding ICAO's standards are similar to those of FAA (FAA, 1990; 1999; Horonjeff and McKelvey, 1994; Janić, 2013).

In particular, the angle between the centreline of exit taxiway and the runway can influence the aircraft runway exit speed, landing occupancy time, and consequently its capacity. This angle of 90° allows the aircraft exit the runway after landing at the speed of about 25 km/h. The angles of 30°, 45°, or 60°of the high-speed exit taxiways enable the aircraft to exit the runway after landing at speeds up to about 90 km/h (Horonjeff and McKelvey, 1994).

Most landing aircraft touch down the runway at the speed of about 1.3 times greater than their stall speed at 85% of their maximal landing weight. The distance between the landing threshold and the point of touchdown assumed to be fixed for most aircraft varies from 150 to 450 m (i.e., 500–1500 ft). This distance, denoted by (l_{td}), should be added to the distance (l_{dc}) needed by a given aircraft to decelerate from the touchdown speed to the speed of safe exit from the runway. For example, the total distance from the landing threshold to the location of the runway exit taxiways can be estimated as follows (Horonjeff and McKelvey, 1994):

$$L_t = L_{td} + L_{dc} = (v_{td}^2 - v_{ex}^2)/2 \cdot a_{dc} \tag{2.1}$$

where

v_{td} is the touchdown speed (km/h or kts);
v_{ex} is the exit speed (km/h or kts); and
a_{dc} is the aircraft deceleration after landing (m/s^2 or ft/s^2).

The deceleration (a_{dc}) in Eq. 2.1 is usually 1–1.5 m/s^2 (i.e., 3–5 ft/s^2) and sometimes up to 2.4 m/s^2 (8 ft/s^2) (Barker et al., 1999). However, because different categories of aircraft can use the same runways, several dedicated exit taxiways can be available. Some evidence indicates that three high-speed exit taxiways are provided at most commercial airports. The holding bay areas, usually located closer to the runways, enable the aircraft to wait for departures and/or temporarily stop in order to provide smooth traffic circulation.

c) Apron/gate complex

The apron gate/complex consists of the aircraft parking stands and the necessary facilities and equipment to serve aircraft during their turnaround times. In general, the aprons can be dedicated to passenger, freight/cargo, and general aviation aircraft. The particular stands of both can be used for short aircraft stays during their turnaround time as well as for long term parking. The apron/gate complex used by passenger aircraft can contain the stands close and the stands remote from the passenger terminal(s). Configuration of a given apron/gate complex generally depends on the configuration of the passenger terminals complex, which can be a linear, finger or pier, satellite, or transporter concept, and their variations.

Sizing of the apron/gate complex expected to accommodate different aircraft types can be carried out using the EQA (Equivalent Aircraft) factor. It is estimated by dividing the average aircraft seat capacity by 100. As such, it can range between 0.5 (aircraft with 1–60 seats) and 4.8 (aircraft with 421–500 seats) with the benchmarking value of 1.0 calculated for the aircraft with 91–110 seats (FAA, 1988; Janić, 2013).

The average EQA factor for the apron/gate complex at most airports has been calculated as the product between the number of atms and the corresponding EQA factors of particular aircraft types. Then, the obtained EQAs for the specified aircraft types have been multiplied by the aircraft "footprint" in order to determine the total area of the apron/gate complex for particular aircraft categories (FAA, 1988; ICAO, 1999; Janić, 2013). For example, the parking stand size for airbus A380 aircraft is 80 × 80 m, which, with an additional safety "buffer" of 7.5 m on each side, gives the total area of 95 × 95 m. Figure 2.4 shows the relationship between the aircraft size and the required space for a single apron/gate parking stand.

As can be seen, this space linearly increases as the aircraft seat capacity grows. Furthermore, design and sizing of the airport apron/gate complex should also fulfil the requirements as follows: (i) Appropriately fit into current and prospective expansion of the passenger terminals in cases of the terminal-closer apron/parking stands; (ii) Enable efficient and balanced aircraft and passengers movement to/from the aircraft; and (iii) Enable a smooth aircraft manoeuvring in and out of the particular parking gates/stands (deNeufville and Odoni, 2003, Horonjeff and MCKelvey, 1994; Janić, 2013).

2.2.3 Landside Area

The airport landside area contains the landside access modes and their systems, and passenger terminal complex, the latter consisting of one or more terminals.

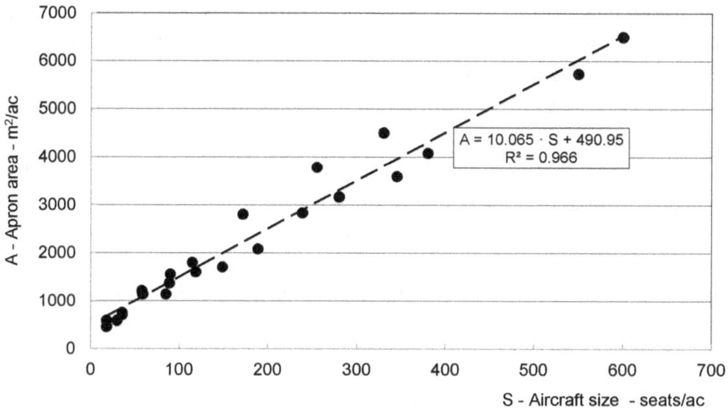

Figure 2.4 Example of the relationship between an apron/gate parking stand and the aircraft size (Janić, 2013).

2.2.3.1 Landside access modes and their systems

Airports are connected with their catchment areas by different landside access transport modes and their systems. In general, these are the road- the rail-based modes. The former includes the systems such as buses and cars/taxis. The latter mode includes the systems such as trams/streetcars, LRT (Light Rail Transit), conventional and HSR (High Speed Rail), and Maglev system. Both modes and their systems use the fixed infrastructure, which is also connected to the wider local, national, and international road and rail networks. Different airports are surrounded by catchment areas of different sizes in terms of area and density of population. The borders of this catchment area are usually measured by the maximum distance and/or accessibility time for a given percentage of passengers and freight shipments with origin and destination within the area. Atmost European airports, these distances and times vary from about 50 km to 100 km and/or one to two hours, respectively.

For example, the landside access systems at the above-mentioned Atlanta Hartsfield-Jackson airport, the world's largest airport in terms of accommodating the annual number of passengers (about 109 million in the year 2019), are particularly illustrative regarding diversity and characteristics of the landside access systems. The first system includes the thirty shuttle services operating the door-to-door and on-demand pick-up services in Atlanta's metropolitan area and the bordering counties (the airport is located 16 km south of the city of Atlanta). The vehicles-vans of seating capacity up to 14 passengers and their baggage depart every 15 minutes from the metropolitan area and every half an hour from the bordering counties. In addition, the RTL (Rapid Train Link) connects the city and the airport by offering the frequent services. The taxi and rent-a-car services are also available. The stations and stops/parking places of all systems are located conveniently around the main terminal. Finally, the airport is fully accessible by the individual cars. For this purpose, about 30 thousand parking spaces are provided around the central terminal complex (http://www.atlanta.com).

The characteristics of transport services provided by the particular airport landside access modes and their systems on the one side and that of users on the other influence their choice and use, i.e., market share. The most important characteristics of access modes and their systems other than private cars influencing their choice are their spatial and time accessibility (i.e., convenience), frequency, reliability, punctuality, and price of services. In addition, some investigations have shown that the higher income passengers prefer the individual car and/or taxi. For example, at most large U.S. airports, the overall share of the individual car has been about 54% (these passengers have also parked their cars at the airport). About 31% of air passengers have been dropped-off by car and about 13% used taxi or shuttle bus services. Finally, overall, only about 2% of passengers used public transport modes (bus and rail-based systems). The main reasons for such low use of these modes at a global scale have been their unavailability at many airports (particularly rail), inconvenient spatial accessibility (i.e., closeness to the final origins/destination in the catchment area), and the convenience for handling baggage. At the airports with more convenient rail systems and services, such as Atlanta Hartsfield-Jackson and Washington National airport, the rail-based mode and its systems have shared about 5–6% and 9–11%, respectively, in the total use of all ground access modes (Hwang et al., 2000; Janić, 2019). At European airports, the market share of individual car use has also predominated over that of the public transport modes and their systems, as shown in Table 2.1.

As can be seen, the above-mentioned choice and consequent market share of particular landside access modes and their systems serving Amsterdam Schiphol airport has been influenced by their characteristics given in Table 2.2 (Janić, 2011; 2019).

Table 2.1 Market share of the landside access modes and their systems—Case of the selected European airports (Period: 2002) (BAA, 2005; SG, 2007).

Airport	Access mode/Share (%)				
	Rail	Bus	Taxi	Private car	Other
London Heathrow	22	12	26	39	1
London Gatwick	21	9	17	50	3
London City	-	24	46	28	5
Amsterdam Schiphol	35	2	15	45	3
Frankfurt Main	17	17	20	45	1

Table 2.2 Some characteristics of the landside access systems—Case of Amsterdam Schiphol airport (Janić, 2011; 2019).

Mode/System	Access time (min)	Frequency (dep/h)	Distance (km)	In-vehicle time (min)	Average fare (€/pax)
Car	5	-	15	25–50	0.46/km[2)]
Taxi	5–10	-	15	25–30	40–45
Bus[1)]	10	6	26	30	6.75
Local rail	20	4–6	15	17.5	3.80 6.40
LRT[3)]	20	1–6	15	15	3.5

[1)] Lines: 370, 198. 300. 310, 199. 197, 97,358, 192,195, 61, 188; Airport Shuttle (SG, 2003/2016);

[2)] Based on the price of fuel of 1.35 €/l (middle class car) (l-liter) (pax – passenger) (http://www.vccr.nl/).

[3)] Proposed (Janić, 2011).

In the given case, convenience of accessibility and the lowest unit average costs have been the main factors of attractiveness of private car, resulting in its 45% market share despite longer in-vehicle time relative to the rather comparable times but cheaper public transport modes and their systems.

2.2.3.2 Passenger terminals

The airport passenger terminals where the dominating airlines operate the "point-to-point" and "hub-and-spoke" networks are of different size, design, and layout. These should provide a smooth, i.e., safe and secure, efficient and effective processing of air passengers and their baggage through the corresponding terminals, i.e., between the airport landside access modes, their systems and the aircraft, and vice versa. In addition to the sufficient capacity of processing facilities and equipment, such services are also achieved by the designs separating these flows horizontally and vertically.

a) Horizontal separation

i) Concepts hosting the airline "point-to-point" networks

At airports hosting dominating airline "point-to-point" networks, i.e., mostly the origin-destination passenger and baggage flows, the apron/gate complex of rectangular shape usually faces the passenger terminal complex designed according to the linear or transporter concept. Combined with the nose-in and/or the nose-out aircraft parking scheme, these concepts enable efficient and effective passage of passengers from the passenger terminal to the aircraft, and vice versa. Figure 2.5 shows the simplified scheme (Janić, 2013).

- Linear concept implies that one or two mutually connected terminals serving domestic and international passengers and their baggage are located face-to-face, i.e., close to the apron/gate complex. At an international airport with a single building, serving domestic and the international passengers and baggage is centralized with the necessary distinctions. In the case of two buildings, these services are decentralized, i.e., one building is used exclusively for serving domestic and another for serving international passengers and their baggage.
- Transporter concept implies that the passenger terminal(s) are located relatively far from the part or entire apron/gate complex. In this case, passengers are transported between the terminal(s) and the aircraft at the remote apron gates/stands by buses and/or mobile lounges. Their baggage is transported by carrying vehicles. In some cases, passengers need to walk between the aircraft and the terminal. Nevertheless,

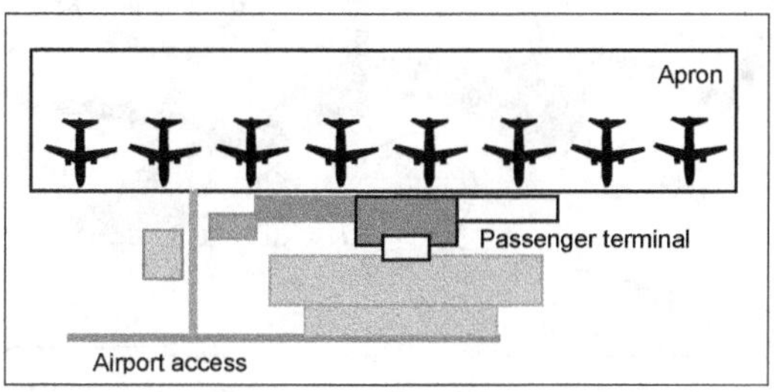

Figure 2.5 Scheme of the passenger terminal linear concept (Janić, 2013).

many regional airports serving predominantly the O-D passengers are designed as a "hybrid" concept, combining the linear and transporter concept. In particular, LCCs (Low Cost Carrier(s)) prefer the transporter-like concept, which enables, in addition to the relatively low handling costs, a shorter turnaround time of their aircraft, which consequently requires a smaller number of the higher-utilized parking gates/stands (Horonjeff and McKelvey, 1994; Janić, 2013).

ii) Concepts hosting the airline "hub-and-spoke" networks

As mentioned above, at the airports hosting predominantly the airline "hub-and-spoke" networks, the passenger terminals are designed generally according to two basic design concepts:

- Finger or pier concept; and
- Satellite concept.

At many airports, these are combined with the linear and sometimes transporter concepts, thus creating "hybrid" configuration(s).

- Finger or pier concept contains a finger or pier directly connected to the central terminal and extending deeper into the existing apron/gate complex towards the airside area. The apron gates/stands are usually arranged along both sides of a finger or pier with the nose-in parking alignment of aircraft of different aircraft types. Figure 2.6 shows the scheme (Janić, 2013).

  This design enables increased flexibility of utilization of fingers or piers. When the flight arrivals and departures are at the same finger of pier, the transit/transfer passenger can proceed between them directly. In cases when arrivals are at one and departures from other finger or pier, the transit/transfer passengers must usually to pass through the central hall located in the middle of entire complex. The O (Origin) passengers can reach the gates of departure flights through the central hall. The D (Destination) passengers can leave the gates of arrived flights directly towards the arrival hall. Building and operating the fingers and/or piers is relatively easy and come with relatively low investment and operational costs, respectively. One ultimate disadvantage can be the constrained space for expanding the apron gate/complex between particular fingers or piers. In some cases, the smooth aircraft entering and leaving parking stands can be compromised. The other can be the relatively long walk of both O-D and transfer passengers between the particular gates/stands. This

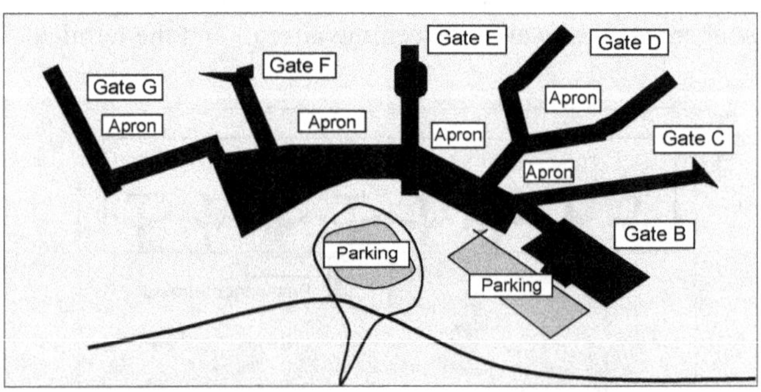

Figure 2.6 Scheme of the finger or pier concept—Case of Amsterdam Schiphol Airport (the Netherlands) (Janić, 2013; http://www.skyteam.com/).

could be particularly inconvenient in cases of delays and limited connecting time to the closest departing flights.

- Satellite concept contains several smaller physically separated (isolated) mutually connected units, i.e., satellites. In addition, they are also connected to the main terminal by the surface and/or the underground fixed and/or mobile connections. These are the fixed constructions, i.e., halls, and/or the surface (or underground) mobile vehicles shuttling between them. Figure 2.7 shows the simplified scheme (Janić, 2013).

The satellites are usually of round or rectangular layout, the latter similar to the fingers or piers, but without the fixed connection with the central terminal. At the above-mentioned Atlanta Hartsfield-Jackson airport, the separate units are called concourses, which are mutually connected to the main terminal by the underground automatic people mover. The system consists of nine, four-car trains running within the closed 5.6 km long track approximately every 2 min. Such service frequency provides efficient collection and distribution of passengers from/to the particular concourses and the main terminal. The aircraft are parked around the satellites and/or concourses according to the nose-in parking scheme. Within this concept, each satellite or concourse can operate independently with the centralized check-in. The security control can be either centralized in the main terminal or decentralized in the given satellite/concourse. In addition, the walking distances within the satellites or concourses as well as between these and the main terminal could sometimes be relatively long and time consuming. Furthermore, passing from one satellite unit to any other often takes place through the main terminal and, as such, can also be time consuming. Nevertheless, the aircraft can manoeuvre easier in and out of particular apron gates/parking stands.

b) Vertical separation

Serving air passenger and baggage flows in the passenger terminal efficiently and effectively has been achieved, in combination with the horizontal, by their vertical separation. This implies that processing of more voluminous and heterogeneous departing, arriving, and transit/transfer, both domestic and international, passengers and

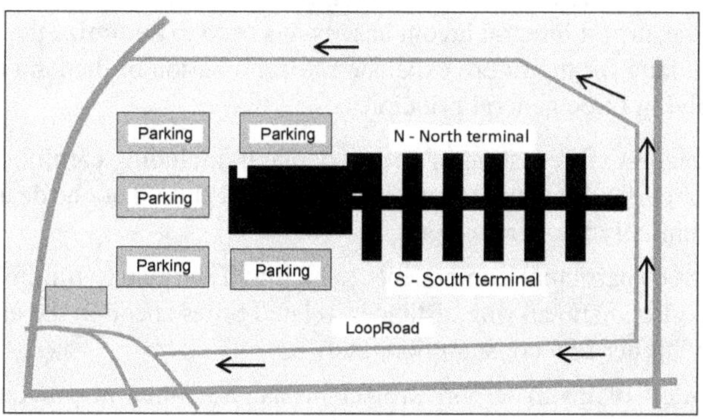

Figure 2.7 Scheme of the satellite concept—Case of Atlanta Hartsfield-Jackson airport (U.S.) (Janić, 2013; http://www.skyteam.com/).

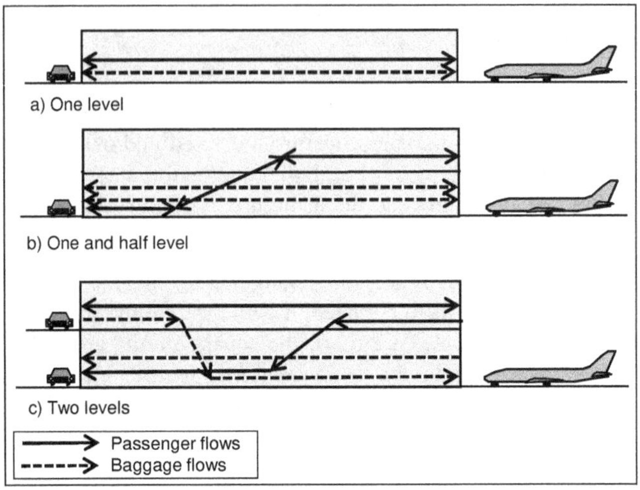

Figure 2.8 Simplified scheme of the vertical distribution (separation) of passenger and baggage flows in an airport passenger terminal (Horonjeff and McKelvey, 1994; Janić, 2000; 2013).

their baggage is arranged at different levels. Figure 2.8 shows the self-explained scheme (Ashford and Wright, 1992, Janić, 2000; 2013).

2.2.4 Integrated Layout

The integrated airport layout comprises its airside and landside area. The size of particular components is governed by standards mainly related to the shape and the number of runways, taxiways, aprons, and related facilities. The selection of particular standards depends on the size of relevant ("critical" or the biggest) aircraft and the predicted volumes of air passenger and freight/cargo demand. In addition, the number and orientation of runways, the most land demanding components given the "critical" aircraft, depends on the required usability factor of a given airport regarding the prevailing weather (wind and ceiling) conditions (not less than 95% during the year) and positioned obstacles around the approach and departure areas (ICAO, 2004; Janić, 2013).

In general, the airport integral layout has always tried to minimize the land taken and provide a reserve land for the future expansion. Minimization of the land taken has been achieved by applying three general principles:

- Relatively accurate forecasting of airport demand, including careful selection of the standards for particular components, i.e., the airport should not be designed and built for aircraft unlikely to operate there;
- Arrangement of particular components, which will ultimately minimize the aircraft taxiing and other manoeuvring distances, related times spend in the airside area, and unnecessary higher fuel consumption; and
- Implementation of given airport project in stages, aimed at preventing the over-capacity which could compromise its utilization.

The above-mentioned principles can be materialized through design of six typical (theoretical) airport layouts (configurations), such as: (i) a single runway;

(ii) two parallel runways, both used for landings and take-offs; (iii) two parallel runways, of which one is used for landings and another for take-offs; (iv) two converging runways, each used for both landings and take-offs depending on the prevailing wind; (v) two parallel plus one crossing runway each used for landings and take-offs; and (vi) two pairs of parallel runways of which two outer runways are used for landings and two inner runways are used for take-offs (Horonjeff and McKelvey, 1994; Janić, 2013). However, most real airports have not fully followed some of these theoretical layouts in terms of size of taken land. In general, the smaller regional airports operate a single runway while the large airports typically operate more than one runway. Typically, with both categories of airports, the runway length varies from about 2000 m at the regionals to 4000 m at the large international hubs. At the latter airports with several runways, a given pair of runways can cross or be parallel to each other. The crossing runways are positioned at a certain crossing angle. Depending on the prevailing wind, one runway can be used for both arrivals and departures, except under conditions of no wind when one can be used exclusively for arrivals and another for departures. Parallel runways can be spaced closely or widely. Under IMC, the closely-spaced parallel runways usually operate as the single runway. Under VMC, they can simultaneously be used for parallel arrivals, paired departures, or mixed operations. The widely-spaced parallel runways can be used in the "segregated" and "mixed" mode. The former implies the use of one runway exclusively for landings and the other exclusively for take-offs. The latter implies that both runways are simultaneously used for both landings and take-offs. At many large airports, one or two crossing runways are added to the existing (predominantly used) parallel runways. These runways are used relatively rarely, given the specific crosswind. Finally, two pairs of closely- or widely spaced parallel runways can intersect with each other. Figure 2.9 shows the simplified schemes of the integrated layouts of the selected airport (FAA, 2004; Janić, 2013; SG, 2007).

a) <u>Amsterdam Schiphol airport</u>

The Amsterdam Schiphol airport (The Netherlands) with the layout shown in Figure 2.9a occupies the area of land of 2787 ha. The airport operates six runways, of which three are parallel: 18C/36C (3800 × 45 m), 18L/36R (3400 × 45 m), and 18R/36 'Polder' (3800 × 60 m). The other three runways are 06/24 (3500 × 45 m), 09/27 (3450 × 45 m); and 04/22 (2014 × 45 m). The parallel runways are sufficiently spaced in order to enable their independent operation exclusively under IMC and IFR. In addition, there are 176 aircraft parking stands, one passenger terminal, and seven air cargo terminals (SG, 2007). The passenger terminal complex of the finger/pier type is located in the area surrounded by four runways, which enables balanced taxiing time after arrivals and before departures from any of these runways. However, taxiing time to/from Runway 18R/36L Polder, because of its relative dislocation from the main runway system, can be substantial, sometimes for about 20–30 minutes (SG, 2007; Janić, 2013).

b) <u>San Francisco International airport</u>

The San Francisco International Airport (US) with the layout shown in Figure 2.9b is one of the largest airports in the US, occupying land of about 1600 ha. The airport operates two pairs of parallel runways, i.e., 1L/R and 28R/L. Dimensions of the runways 1L/28R are 11879 × 200 ft (3618 × 61 m) and that of the runways 1R/28L are 3231 × 61 m. Each pair of runways is spaced for 228.75 m, which is sufficient for simultaneous arrivals

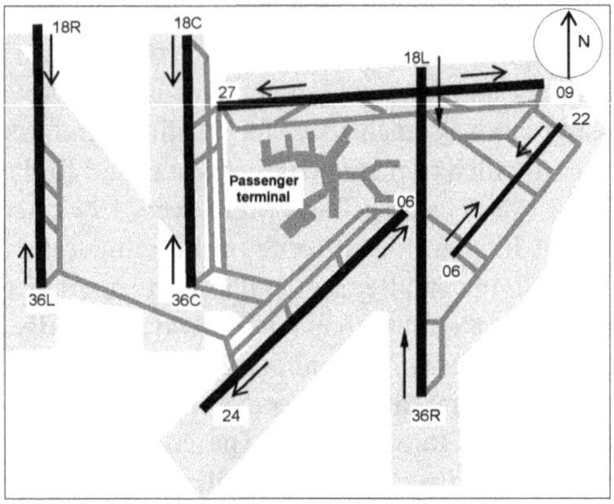

a) Amsterdam Schiphol airport(SG, 2007)

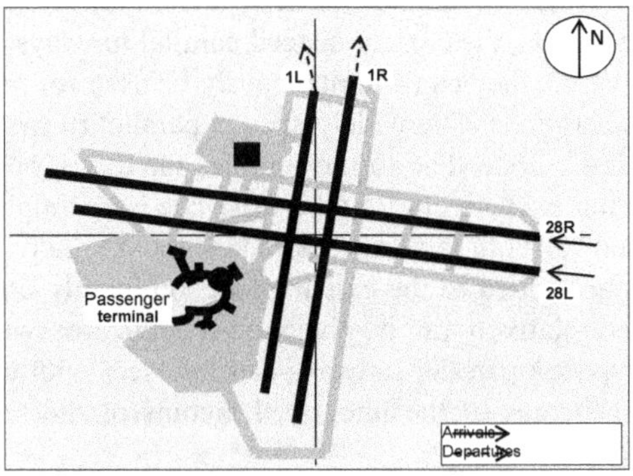

b) San Francisco International Airport (FAA, 2004)

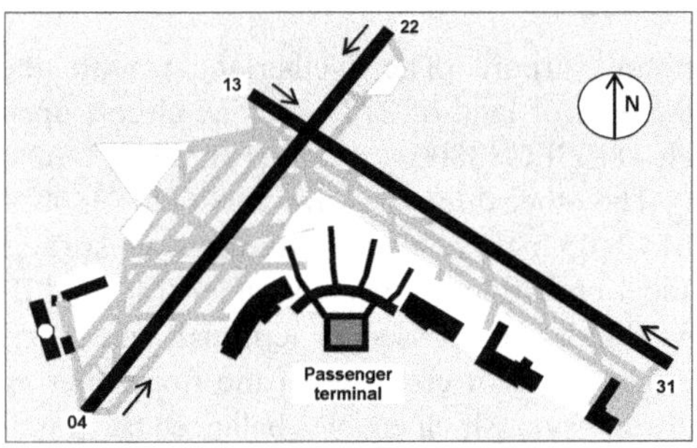

c) NY LaGuardia airport (FAA, 2004)

Figure 2.9 Scheme of the integrated layout of selected airports (Janić, 2013).

and departures under VMC but not under IMC. Under the preferable VMC, the runways 28R/L are used for the paired arrivals and the runways 1L/R for the paired departures. Depending on the mixture of the arrivals and departures, the paired departures can be realized between the successive pairs of arrivals. When the weather deteriorates below the specified minima (20% of the time during the year), both pairs of runways operate as a single runway (the West Plan for Runways 28R/L and the Southeast Plan for Runways

1L/R). The passenger terminal complex is located close to the intersection of both pairs of parallel runways, which, from the operational perspective, enables relatively short taxiing times for the landing aircraft and longer taxiing out times for the take-off aircraft using the runways 28L/R. If the pair of Runways 1R/L is used for take-offs, the taxiing out times will be comparable to those of the landing aircraft using the Runways 28L/R. In addition, this airside configuration offers flexibility for expansion of the landside area and the passenger terminal complex on each side of the runways' intersection. This also implies underground connection between the passenger terminals on different sides of the runways. Adding a new runway in either direction requires the additional land to be acquired and a connection with the existing runway system to be established (FAA, 2004; Janić, 2013).

c) <u>NY La Guardia (US)</u>

The New York (NY) LaGuardia airport,occupying land of about 2600 ha, is one among the three largest airports serving the New York area. Figure 2.9c shows the scheme of its layout. The airside area contains the two right angle-crossing runways, each of length 2135 m, and the associated taxiways. In this case, the passenger terminal complex is located closer to the inner side of the intersection of two runways. This enables the aircraft shortest taxing distances and times overall but a certain imbalance between the taxing in and taxing out distances and corresponding times exists depending on the runway in use. In general, regarding the airport surrounding, there is the lack of available space for the airport expansion either by extending the existing runways or eventually by building a new one(s). However, it seems that there is still a space for expanding the landside area alongside the Runway 4/22 at both the inner and the outer side of the runways' intersection (FAA, 2004; Janić, 2013).

2.2.5 Supporting Facilities and Equipment

The airport supporting facilities and equipment mainly embrace the ITS (Information Technology and Systems). These generally contain deploying and application of the advanced sensor, computer, electronics, and communications technologies aimed at maximizing efficiency, effectives, and safety of operating the airport infrastructure and services in real time.

The main components of an airport ITS are: Airport ITS Communications Systems, Airline and Airside Operations Systems, Airport Landside Operations Systems, Airport Safety and Security Systems, Airport Facilities and Maintenance Systems, Airport Development Systems, and Airport Administration Systems (ACC, 2008; Janić, 2013).

The airport ITS perform three basic interrelated functions: data collection, data processing, and distribution (dissemination) of information to the end users. They use this information to make decisions on executing the particular tasks in the scope of the given processes. Specifically, the function "data collection" implies collecting information from the sources such as the field devices, providers of the various services, and other relevant sources. The function "data processing" implies making the collected data usable to support given tasks. The function "distribution (dissemination) of information" implies making the information available to given users, such as the airport operator, airlines, the ATC/ATM, operators of the ground access systems, the airport service, users such as passengers and freight shippers, the airport employees, etc. (ACC, 2008; Janić, 2013).

The airport ITS generally serves to improve efficiency, effectiveness, and safety of airport operations in both the airside and landside areas. Efficiency implies increasing profitability, effectiveness means providing quality of services to users as planned, and safety implies preventing occurrence of incidents and accidents due to the known reasons. Essentially, the operations intended to serve aircraft, air passengers and their baggage, and freight/cargo shipments can be performed faster and more accurately, which accelerates the processes and consequently enables larger throughput at the corresponding airport components. The shorter time of operations, in combination with allocating some of these operations directly or indirectly to users (for example, providing screens with information on flights and baggage claim devices to air passengers) reduces the required airport staff and consequently raises the overall airport productivity. In parallel, the faster operations carried out for a given volume and intensity of demand diminish the number of units of demand simultaneously occupying a given space, which, in turn, for a given space standard(s) of quality of service, reduces the size of the overall space, thus making its utilization more efficient. Furthermore, the necessary self-screening and scanning of passengers and their baggage does not compromise the planned performance of their overall processing as well as the peoples' overall feelings of comfort. At the same time, the risk of making mistakes and misjudgments decreases, which also decreases the risk of compromising security later on (ACC, 2008; Janić, 2013).

2.2.6 Staff/Employees

The airport staff/employees perform the particular above-mentioned operations in handling the air passengers, aircraft, and freight/cargo shipments. Some investigations have indicated that the number of airport employees has been closely related to the number of handled air passengers. Figure 2.10(a, b) shows the selected examples of the relationships between the number of airports employees and the number of passengers handled (FZ AG, 2006/2018; SDG, 2015; Sewill, 2009; SG, 2008/2019).

Figure 2.10a shows the linear relationship between the number of airport employees and the annual number of handled passengers at 14 UK airports. As can be seen, about 1 million air passengers have been handled by about 1000 airport employees. Figure 2.10b shows the linear relationship between the number of airport employees and the number of handled air passengers at Zurich airport (Switzerland). In this case, about 800 staff members have been engaged to handle about 1 million air passengers during the observed time period. At Amsterdam Schiphol, the relationship between the number of handled air passengers and the number of employees has decreased and then increased more than proportionally with increasing of the number of handled air passengers during the observed period. This has resulted in about 2500 engaged staff/1 million air passengers at the volume of about 65 million air passengers/year and 2800 engaged staff/1 million air passengers at the volume of 85 million air passengers/year during the observed time period.

2.3 Demand and Capacity

2.3.1 Demand

2.3.1.1 Air passengers and air transport movements

The above-mentioned airport infrastructure and other resources are set up to serve demand consisting of the airline aircraft/flights usually called the atms (air transport

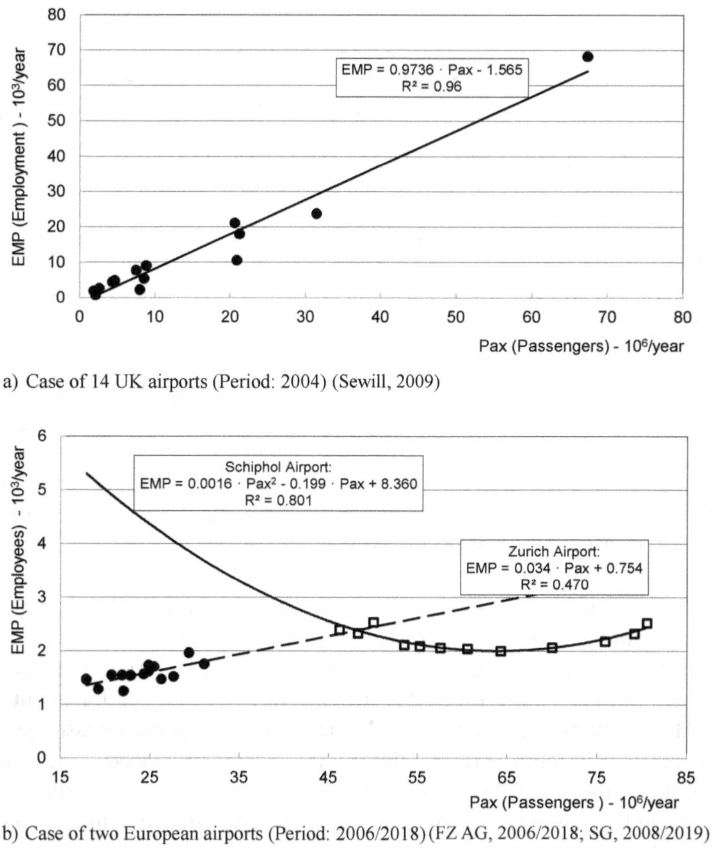

Figure 2.10 Relationships between the number of staffs and the number of air passengers.

movement(s) –1 atm is equal to 1 landing or 1 take-off), air passengers, and freight/cargo shipments. The data on these components of airport demand have been regularly recorded over time—per day, month, and year. In general, the particular components of airport demand have influenced each other. Figure 2.11 shows an example of the development of particular components of airport demand—airport passengers and air transport movements—at 20 world largest airports.

As can be seen, the annual number of airport passengers has generally increased with the growth of the number of atms at decreasing rate. For example, 600 thousands atms handled at an airport has served 78.255 million passengers per year. The average number of passengers has been about 130 pax/atm. In addition, Figure 2.12(a, b, c) shows examples of the relationship between particular components of demand at large U.S. and European airports.

Figure 2.12a shows that the annual number of atms at London Heathrow airport has slightly increased during the first part and remained relatively constant during the second part of the observed period. One of the reasons has been that the atm demand has increased to a certain level and then remained constrained by the airport runway system capacity (AOA, 2014; HAL, 2008/2019; https://www.heathrow.

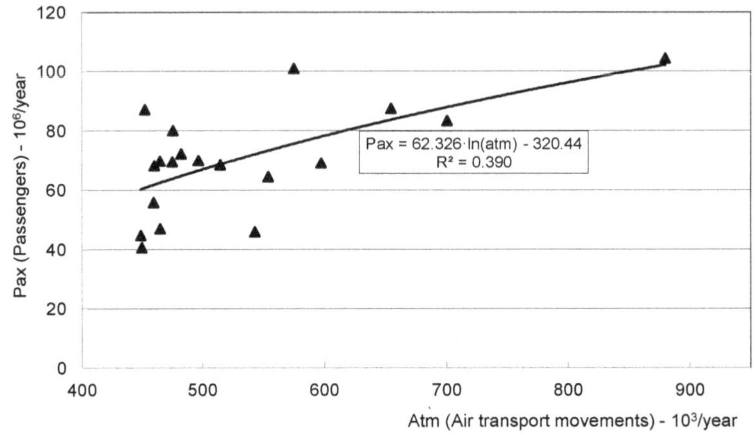

Figure 2.11 Relationship between the number of airport passengers and air transport movements—Case of the 20 world's largest airports (Period: 2017) (https://www.internationalairportreview.com/article/110871/top-20-busiest-airports-world-aircraft-movements/https://www.internationalairportreview.com/article/32311/top-20-largest-airports-world-passenger-number/).

com/company/company-news-and-information/company-information/facts-and-figures/). The annual number of atms at Atlanta Hartsfield-Jackson airport has been relatively constant during the first and slightly increasing during the second part of the observed period. All the time, this demand has been higher for about 55–70% than that at Heathrow airport. Figure 2.12b shows that the annual passenger demand has been increasing at both airports during the observed period. The number of handled passengers has been 25% lower at Heathrow than Atlanta Hartsfield airport. Figure 2.12c shows the relationship between the number of airport passengers and atms at both airports during the observed period. As can be seen, the number of passengers at Atlanta airport has increased alongwith an increase in the number of atms, with an average of 142 pax/atm. At London Heathrow airport, the number of passengers has increased despite stagnation of the number of atms, due to, as mentioned above, constraints of the airport runway system capacity. Consequently, the aircraft size and the number of passengers per atm have increased from about 145 pax/atm in the year 2010 to 168 pax/atm in the year 2019.

2.3.1.2 Connectivity

An important factor influencing the airports' demand is their connectivity. In many research reports, the airport connectivity is defined as the synthesized measure, including the number of destinations, the airline service frequency, type of connection (direct and indirect), minimum connecting time, and the maximum circuity as the ratio of passenger distance flown to the direct distance (ACI Europe, 2019; OAG, 2019).

a) European airports

In general, connectivity of the European airports can be direct, indirect, airport as the sum of two previous, and hub. Figure 2.13 shows the simplified schemes.

In particular, direct connectivity implies the number of destinations, i.e., airports to which a given airport is connected by the number of direct flights during the specified

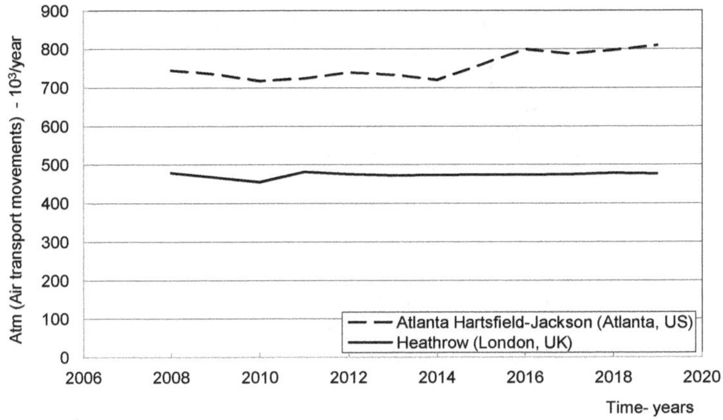

a) Air transport movements

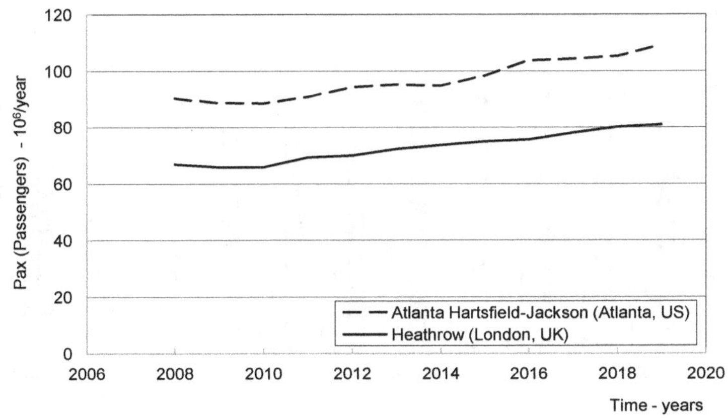

b) Airport passengers

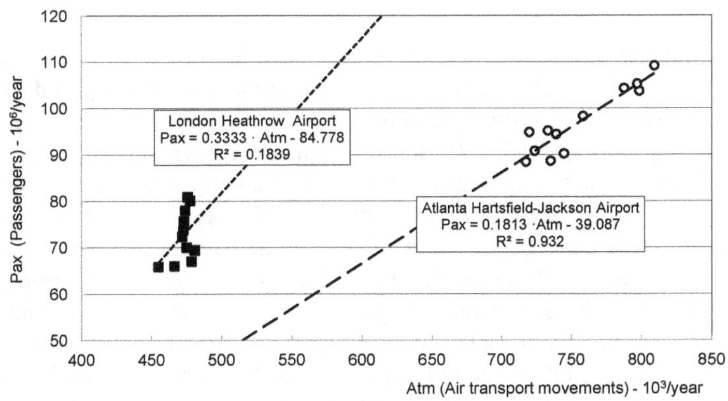

c) Airport passengers vs air transport movements

Figure 2.12 Development of air transport demand at the selected European and U.S. airports—Case of Atlanta Hartsfield-Jackson airport (Atlanta, Georgia, U.S.) and Heathrow airport (London, UK) (Period: 2008–2019) (ATL, 2008/2019) (Excluded military, taxi, general aviation).

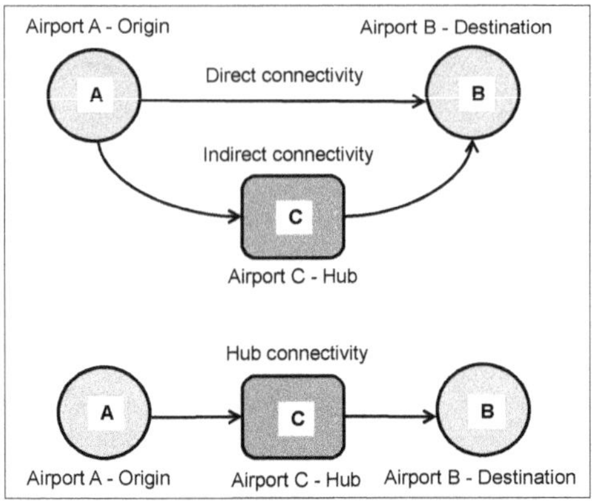

Figure 2.13 Schemes of the different types of airport connectivity (ACI Europe, 2019).

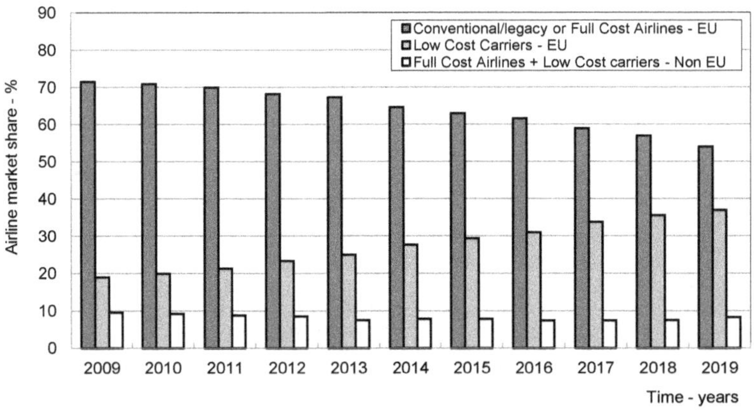

Figure 2.14 Development of market share of different categories of airlines—Case of direct connection of European airports (ACI Europe, 2019).

time period. Indirect connectivity implies the number of destination airports connected by a single flight from the origin airport through the intermediate hub airport during a given time period. Airport connectivity takes into account both direct and indirect connectivity of the airport in question. Specifically, the hub connectivity is measured by the number of connecting flights while taking into account a minimum and maximum connecting time, and weighing the quality of connections by the detour involved and the connecting times. Figure 2.14 shows an example of the relative market shares of particular categories of airlines in direct connectivity of the European airports over time (ACI Europe, 2019).

As can be seen, the conventional/legacy or full cost airlines have maintained the highest but decreasing relative market share in direct connections of the European airports during the observed period. At the same time, that of LCCs has increased, implying that they have added many of their own in addition to taking over some of the connections from their full cost counterparts. The relative market share of the non-EU (European Union) airlines has been relatively constant during the observed period, indicating their rather stricter and more regulated access to the EU airports. In this context, a question of how LCCs have managed to make such takeover of market share from the full cost airlines arises. Figure 2.15 shows one of the reasons for the relationship between the market share of LCCs in direct connectivity of the European airports and their seat capacity supply during the observed time period (ACI Europe, 2019; https://www.aci-europe.org/44-industry-data/40-airport-traffic.html/).

As can be seen in the given case, there has been a strong correlation between the relative gains in the market share in direct airport connections and increase in the seat capacity supply of LCCs during the observed period. Such gain has occurred at a decreasing rate, indicating a decreasing influence of the capacity supply as one of the main driving forces in the given context. Figure 2.16(a, b) shows the airport and hub connectivity of the selected largest European airports (ACI Europe, 2019).

Figure 2.16a shows that the number of indirect connections at all considered airports has been about 1 (Madrid Barajas) to 3 (London Heathrow) times higher than their direct counterparts. In addition, the total number of connections has been highest at Heathrow and lowest at Madrid Barajas airport. Figure 2.16b shows that, regarding the hub connectivity, Frankfurt airport has been connected the strongest (the major airlines Lufthansa and Star alliance partners), followed by Amsterdam Schiphol and Paris Charles de Gaulle airport (the major airlines there has been Air France-KLM). Table 2.3 gives the annual number of passengers, airport, and hub connections for these airports (ACI Europe, 2019; https://www.aci-europe.org/44-industry-data/40-airport-traffic.html/).

As can be seen, the strongest in terms of the number of passengers handled has been London Heathrow (UK) and the weakest Zurich airport (Switzerland). In terms of the annual number of airport (direct and indirect) connections, the strongest has again been London Heathrow and the weakest Istanbul airport (Turkey). The strongest hub airport has been Frankfurt Main (Germany) (the lowest ratio direct/hub connections) and the weakest was London Heathrow airport (the highest ratio direct/hub connections).

b) U.S. airports

The connectivity of U.S. airports has been estimated by the connectivity index, defined as the total number of all possible connections between the inbound and outbound scheduled domestic flights at the considered airports within MCT (Minimum Connecting Time) of

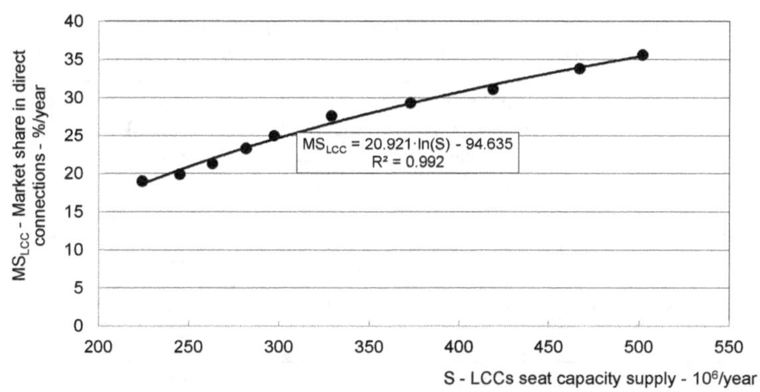

Figure 2.15 Relationship between the market share and supplied seat capacity in direct connections European airports—Case of LCCs (Period: 2009–2019) (ACI Europe, 2019; https://www.aci-europe.org/44-industry-data/40-airport-traffic.html/).

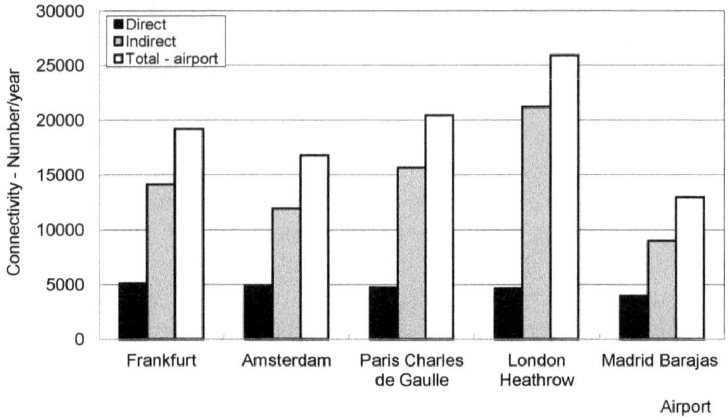

a) Airport connectivity

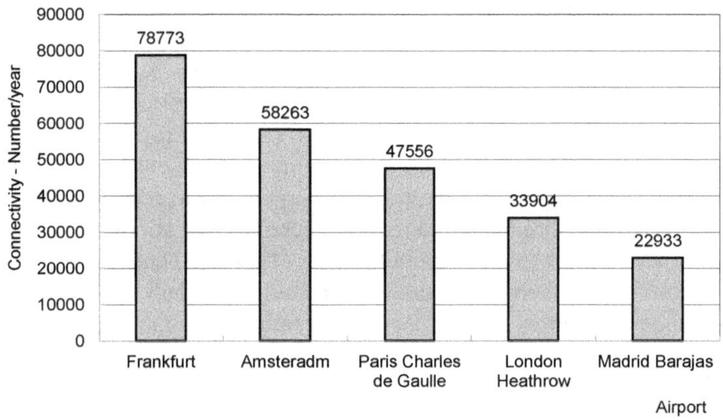

b) Hub connectivity

Figure 2.16 Airport connectivity—Case of the selected European airports (Period: 2019) (ACI Europe, 2019).

Table 2.3 The number of passengers, airport, and hub connections at European airports (Period: 2019) (ACI Europe, 2019; https://www.aci-europe.org/44-industry-data/40-airport-traffic.html/).

Airport	**Number of passengers** (10^6/year)	**AC (Airport connections** (10^3/year)	**HC (Hub connections)** (10^3/year)	**Ratio AC/HC** (–)
London Heathrow	80.866	25.925	33.904	0.765
Amsterdam Schiphol	71.707	16.832	58.263	0.289
Frankfurt Main	70.556	19.243	78.773	0.244
Paris Charles de Gaulle	76.171	20.469	47.556	0.430
Istanbul	52.462	10.229	41.539	0.246
Munich	47.946	14.786	36.058	0.410
Madrid-Barajas	61.705	12.976	22.933	0.566
Zurich	31.464	10.486	18.392	0.570

3 h. In addition, this index takes into account exclusively single connections to/from the selected airports and the maximum circuity of 120 (the ratio of passenger distance flown along the indirect and direct distance) (OAG, 2019). Figure 2.17 shows an example of the relationship between the connectivity index and the market share of flights at 25 top U.S. domestic mega hub airports (A mega hub is considered to be the airport with the highest ratio of the possible scheduled connections and the number of served destinations).

As can be seen, the connectivity index has not been strongly influenced by the market share of flights of the dominant airline. In addition, Figure 2.18 shows the relationship between the annual number of boarded passengers and the connectivity index at these 25 U.S. mega hub airports.

As can be seen, and as intuitively expected, the higher airport connectivity index reflecting the higher supply of flights to the higher number of destinations has supported the increased number of passengers boarding these flights at given airports. In this case,

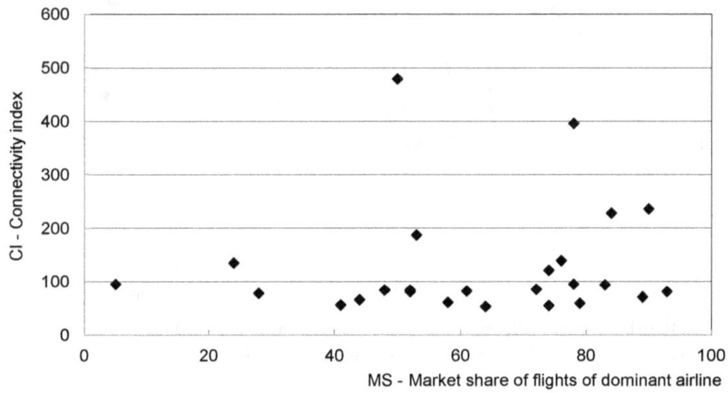

Figure 2.17 Relationship between the connectivity index and the market share of dominant airline—Case of 25 U.S. domestic mega hub airports (Period: 2018) (OAG, 2019).

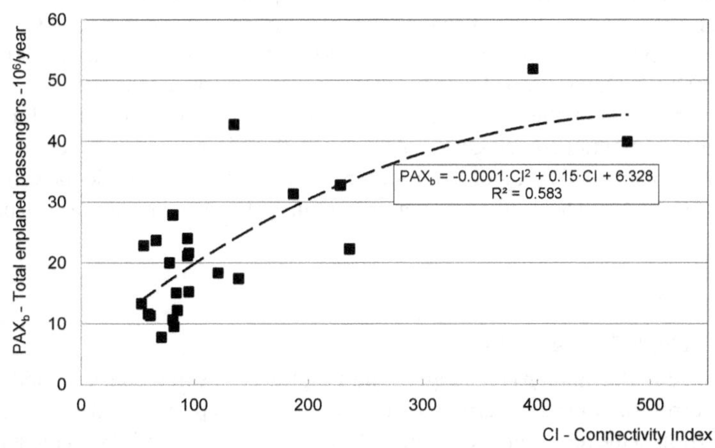

Figure 2.18 Relationship between the number of boarding airport passengers and the connectivity index—Case of 25 U.S. mega hub airports (Period: 2018) (OAG, 2019; USDT, 2019).

this number of passengers has increased at a decreasing rate with the growth of the airport connectivity index.

c) World's airports

Figure 2.19 shows example of the relationship between the connectivity index and the market share of flights at 50 largest world's airports.

As can be seen, similar to the case of U.S. mega hubs, there is no significant correlation between the international connectivity index and the market share of dominant airlines across these 50 airports.

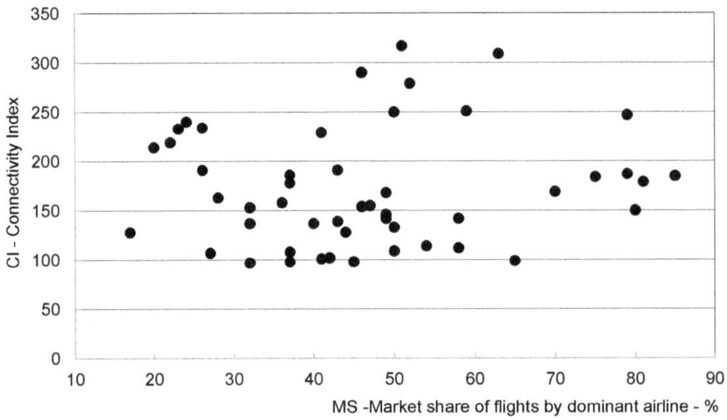

Figure 2.19 Relationship between the international connectivity index and the market share of dominant airline—Case of the world's largest airports (Period: 2018) (OAG, 2019a).

2.3.1.3 Structure of demand

The airport demand, consisting of atms (the airline flights), air passengers, and freight/cargo shipments, can be structured according to different criteria. In terms of the time regularity and seasonality, this can be the scheduled and non-scheduled (charter) demand. Regarding the spatial pattern, this can be the O-D (Origin-and-Destination) and transit/transfer demand. As far as the airlines' business models are concerned, the airport demand can be that carried out by the full cost airlines and LCCs. In general, the large hub airports handle all above-mentioned categories of the aircraft, airline, air passenger, and freight/cargo demand. The smaller regional airports mainly handle the O-D demand frequently carried out by LCCs. The further elaboration mainly focuses on the airport passenger demand.

a) Scheduled passengers

i) O-D (Origin-Destination) passengers

The O-D (Origin-Destination) passengers and freight/cargo shipments start and end their trips at the corresponding pair of airports. The scheduled O-D passenger demand represents the main portion of the total demand (usually it has been greater than 50%) at most large hub airports. This indicates that these airports have been growing thanks to them attracting the substantive volumes of O-D passengers. In general, these airports have been characterized by: (i) the central geographical location; (ii) substantive number of population, employment, GDP (Gross Domestic Product), and PCI (Per Capita Income), and (iii) presence of the economically stable large airlines and their alliances with the substantive market shares of their flights in the total number of atms during the specified time period. Figure 2.20(a, b) shows the examples of influence of some of the above-mentioned economic and operational characteristics on the number of passengers handled at the selected U.S. and European airports.

As can be seen, the total number of airport passengers, of which the O-D segment amounts to about 40–60%, has increased more than proportionally with the growth of the GDP of the airport catchment area and the market share of the dominant airline. Furthermore, the O-D passenger demand at these and particularly smaller regional airports has been substantially driven by LCCs. Table 2.4 gives the example for the selected European and North American airports.

As can be seen, there has been a wide difference between these airports in terms of the proportion of the seat capacity of LCCs. In Europe, the highest proportion of this seat capacity has been at London Gatwick and Stansted airport. In North America, it has been at Fort Lauderdale airport.

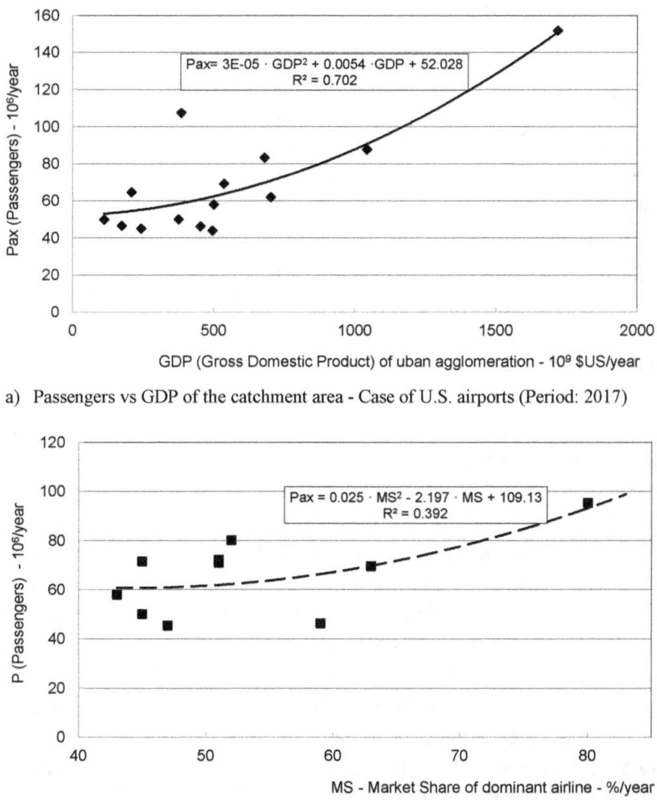

a) Passengers vs GDP of the catchment area - Case of U.S. airports (Period: 2017)

b) Passengers vs the market share of dominant airlines- Case of busiestEuropean airports(Period: 2018)

Figure 2.20 Relationship between the selected economic and operational characteristics of the selected airports on the number of handled passengers (OAG, 2019; 2019a).

Table 2.4 The numbers of passengers and the seat capacity share of LCCs—Case of the selected European and North American airports (Period: 2018) (https://centreforaviation.com/analysis/reports/lccs-increasingly-attracted-to-primary-airports-459531/).

Airport	Passengers (2018) (10^6/year)	Share of LCCs seat capacity (%)
Europe		
• London Heathrow	80.1	2.4
• London Gatwick	46.1	62.2
• London Stansted	28.0	96.7
• Frankfurt	69.5	5.5
• Amsterdam	71.1	22.2
• Paris CDG	72.2	13.0
• Paris Orly	33.1	34.5
• Madrid	57.9	20.6
• Copenhagen	30.3	32.9
North America		
• New York JFK	59.3	26.4
• New York Newark	43.4	13.9
• Chicago O'Hare	79.8	6.4
• Los Angeles LAX	87.5	22.3
• Atlanta	107.3	14.8
• Miami	45.0	3.5
• Fort Lauderdale	36.0	72.9
• Toronto	47.1	23.3

ii) Transit/transfer passengers

Transit passengers arrive at and depart from a given airport by the same aircraft/flight. Usually, these passengers stay on board the aircraft during its turnaround time, but only if this turnaround does not involve aircraft re-fuelling and/or undertaking of some security activities. Otherwise, they have to leave the aircraft temporarily. Transfer passengers arrive at a given airport by one aircraft/flight and depart from the same airport by some other aircraft/flight without breaking their trips between the given O-D airports. The number of transit/transfer passengers increased after many large airlines introduced the hub-and-spoke networks at their main hub airports. In general, the hub-and-spoke network operations imply scheduling the "waves" of incoming and the "waves" of outgoing mutually connected flights at the given hub airport during the relatively short period of time (2–3 hours). Usually, these "waves" have a repetitive character during the day (Janic, 2013). Within a given wave, some passengers from the incoming flights transfer to the outgoing flights carried out by the same airline or its alliance partners during the time of about a half, one, or two hours, called MCT (Minimum Connection Time). This is standardized depending on the type and combination of the incoming and the outgoing flight(s). The combinations of the incoming and outgoing flights can be: short-short haul, long-long haul, short-long haul and vice versa. The national and international flights can

Table 2.5 Minimum Connection Time (MCT)—Case of the selected European airports (AEA, 2004; HC, 2014).

Airport	Minimum Connection Time (MCT) (min)	Combination of incoming/ outgoing flights
Amsterdam (KLM)	40 50	Short-haul–Short-haul Any other combination
Copenhagen (SAS)	30 45	Domestic-Domestic Any other combination
Frankfurt (Lufthansa)	45	Any combination
London Heathrow (British Airways)	45 45 75	Inside Terminal 1 Inside Terminal 5 Terminal 1–Terminal 5
Madrid	60 45	International-Domestic Any other combination
Paris CDG (Air France)	45	Any combination
Rome Fiumicino (Al Italia)	60 45	International-Domestic Any other combination

be in all above-mentioned combinations. Table 2.5 gives some examples of MCT (AEA, 2004).

As can be seen, MCT is generally shorter for the short- and the medium-haul national and international-continental flights and longer for the continental-intercontinental connecting flights. The MCT is influenced by many factors. The most important has been the configuration of passenger terminals and the airline/airport apron aircraft parking strategy. The apron parking stands/gates of the incoming aircraft-flights have been as close as possible to the stands/gates of the outgoing aircraft-flights in order to enable as short as possible passengers' walking distance and transfer time, simplifying and speeding up their passing through the transfer area and resulting in less required space there, and less interactions with the O-D passengers. This has enabled shortening of the aircraft turnaround times. In addition, MCT is shortened when the passenger transfer has been carried out in the terminals with a "common roof" (see Table 2.5).

The number of transfer/transit passengers has primarily been influenced by the scale and scope of the hub-and-spoke networks of dominant airlines. Figure 2.21(a, b) shows the annual number of total and transfer/transit passengers at two large European airports, London Heathrow and Amsterdam Schiphol, during the observed period.

As can be seen, at both London Heathrow and Amsterdam Schiphol airport, the total number of passengers has increased at a higher rate than that of transfer/transit passengers. The transfer/transit passengers at Heathrow airport have made up between 34% and 38% and at Schiphol airport between 30% and 43% of the total number of passengers during the corresponding observed time periods.

b) Non-scheduled (charter) passengers

The non-scheduled (charter) passengers use flights scheduled for specific purposes. Usually, these flights are not listed in the public timetable(s) (Janić, 2013). Tour operators and/or specialized airlines commonly provide air transport services for the groups of passengers, mostly tourists, traveling on an all-in basis. These services are not programmed

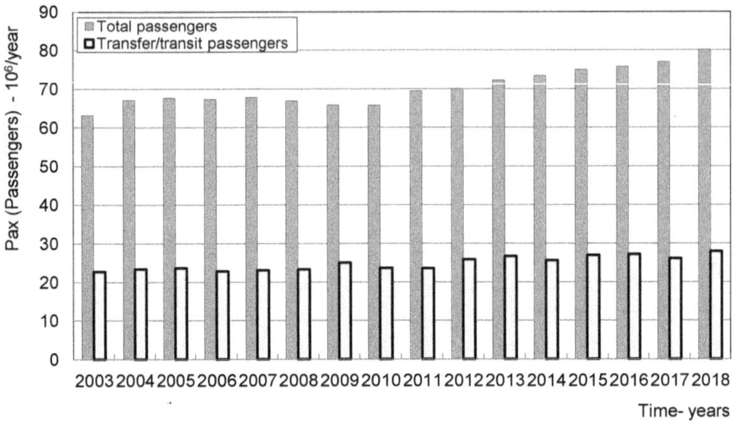

a) London Heathrow airport (Period: 2003-2018) (https://www.heathrow.com/company/ investor-centre/reports/traffic-statistics/)

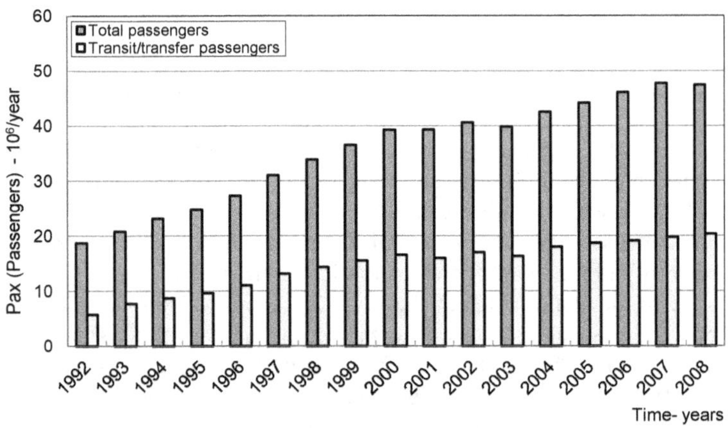

b) Amsterdam Schiphol airport (Period: 1992-2008) (https://www.anna.aero/2009/10/09/ have-cake-and-eaten-it-has-abolition-of-departure-tax-helped-klm-and-amsterdam/)

Figure 2.21 Development of the total and transit/transfer passengers over time—Case of two large European airports.

and are rarely additionally accessible to individuals. In Europe, the traditional air charter markets between the northern European countries and the Mediterranean basin, including North Africa, have sustained and even expanded after deregulation of the EU air transport market. The average share of the non-scheduled passengers in the total numbers of passengers and RPK (Revenue Passenger Kilometres) in Europe had reached about 17% and 36%, respectively (ITA, 2001; Janić, 2013). The non-scheduled passenger airport traffic has had a rather strong seasonal aspect. Nevertheless, in the cases when their numbers have been expected to be a relatively stable and certain, the tour operators and (their) airlines tended to increase the regularity of flights in terms of their departure times and capacity while at the same time maintaining a certain flexibility in the flight frequencies. In the past decade and a half, the LCCs have also taken over a substantial portion of the non-scheduled traffic and converted it into scheduled passenger traffic by adjusting the seat capacities with respect to seasonality throughout the year. Figure 2.22 shows an example of development of the total and non-scheduled passenger traffic at Amsterdam Schiphol airport.

As can be seen, the total annual number of passengers changed more substantially in the absolute terms in comparison to the non-scheduled passenger number counterpart, which remained relatively stable during the observed period. One of the reasons has been the substantial entry of LCC_S in the given airport market, which, in terms of the market share in the total number of flights, increased from 2% in the year 2000 to about 23% in the year 2016 (Janić, 2013; SG, 2005/2016).

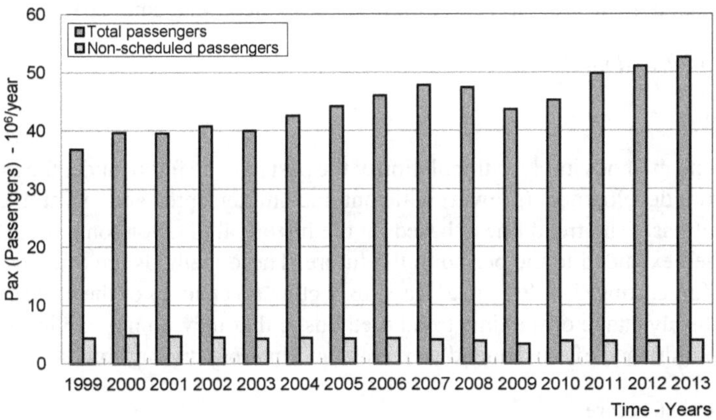

Figure 2.22 Development of the total and non-scheduled (charter) passenger traffic—Case of Amsterdam Schiphol airport (Period: 1999–2013) (SG, 1999/2013).

2.3.2 Modelling Demand

2.3.2.1 Categories, requirements, and recommendations for methods

The airport authorities, aviation agencies, airlines, industry associations, academic researchers, and other professional communities have carried out modelling of the airport passenger and freight/cargo demand, usually including their analysis and forecasting,with the goal of providing the current and prospective medium- to long-term safe, effective, and efficient operations by providing the sufficient capacity in both airside and landside areas. Frequently, the outputs have reflected an inherent inclination to protect and preserve some interests of the agencies producing these analyses and forecasts. Various methods/ models, such as time trends, econometric models, scenarios, ratios, market surveys, and expert judgments, have been used (BCG, 2004; Janić, 2013; TRB, 2002). The experience has shown that the methods for analysis and forecasting of the airport demand should fulfil a range of the requirements and recommendations as follows (Janić, 2013; TRB, 20021; Wadud, 2011):

- The data should be available for updating the methods for analysis and forecasting of the airport demand;
- The methods should be simple rather than complex because the latter does not necessarily translate into the more accurate results;

- Different methods should be applied for analysis and forecasting demand at the given airport;
- The methods that work well under certain conditions could lose their relevance and predictive power due to the inherent dynamics at a given airport; and
- The analysts need always to be motivated and follow the common sense story.

In addition, the methods should be convenient for estimating the volumes of atm, passengers, and freight shipments either directly or indirectly, and simultaneously, as the derivatives of each other. In particular, the time trend and econometric models are explained in more detail because they are most frequently being used by the academic, consultancy, and professional communities worldwide. They are presented through description, analytical structure, and examples of application (Janić, 2013).

2.3.2.2 Time trend methods

a) Description

The time trend methods imply extrapolation of the past into the future under the assumption that the future development (growth) will continue uninterrupted and in a similar fashion to the recent past. The trend line is based on the historical data for some basic period of time and then extended to the period in the future. These methods are often used for the short-term forecasting (i.e., for 1 to 2 years) expecting no changes of the basic conditions. The main disadvantage of the time trend methods is that they do not explicitly take into account the influence of the demand's external and internal driving forces.

b) Analytical structure

In general, two categories of the time-trend methods are commonly used: (i) the methods with a constant growth rate; and (ii) the methods with the variable growth rate. Most frequently, the growth of an airport demand is considered as unconstrained in the medium- to the long-term future. However, over time, many airports have been faced with an increasing problem of generating substantial externalities, such as noise, local air pollution, and land use. This increase in externalities has raised the question of the future growth of these airports, i.e., instigated the ideas of setting up the caps (quotas) on the particular externalities and consequently constraining the volumes of airport traffic. In such cases, the given caps would be set up for a target year. The analytical models handling development in both the constrained and unconstrained cases can be as follows:

i) Unconstrained case, i.e., the constant growth rate

The basic analytical structure of the method with a constant growth rate can be summarized as follows:

$$Q_n = Q_n \cdot (1 + i)^n \cdot \prod_{k=1}^{K}(1 + i_k)^n \tag{2.2}$$

where

Q_n is the airport traffic in the sub-period *(n)* counted from the starting (base) sub-period *(0)* (atm; passengers; tons of freight shipments);

Q_0 is the airport traffic in the base period (0) (atm; passengers; tons of freight shipments);

i is the constant growth rate of the airport traffic in each sub-period of a given time horizon;

i_k is the constant rate of the *(k)*-th factor, which may compromise the growth of a given airport traffic;

K is the number of factors that may affect the traffic at a given airport; and

n is the number of sub-periods of a given time horizon (years).

According to Eq. 2.2, in order to forecast the airport traffic for *(n)* future years, the corresponding data in the base year *(0)* is required and the expected growth rate *i*, which can be taken from the past or set up for the future period and extrapolated ahead, is also needed. In addition, the rate of compromising the airport traffic (r_k) can be specified under conditions of reasonable predictability. One such factor can be the competition from other airports, such as that between large European hubs. In addition, this can be handled by varying the airport traffic growth rate *(i)*.

ii) Constrained case, i.e., the variable growth rate

The basic analytical structure of the method with the variable growth rate can be as follows:

$$Q_n = i \cdot Q_0^{\frac{C-Q_0}{K}} \tag{2.3}$$

where

C is the cap, i.e., the absolute quantity of demand expected to be handled at a given airport at the end of a given time period (atms; passengers; tons of air cargo).

The other symbols are analogous to those in Eq. 2.2. According to Eq. 2.3, this logistic difference equation reflects the rather slow growth at the beginning of the period, almost exponential growth in the middle, and the slowing down in growth of the airport demand when it approaches the prescribed cap, i.e., saturation. In such case, the growth rate *(i)* takes the values greater than 1. Otherwise, if the rate (*i*) takes the value less than 1, Eq. 2.3 reflects the conditions of traffic decline almost to zero at the given airport. This may happen when the major airline collapses or abandons the given airport and the new airlines do not resume operations.

c) Application example

Figure 2.23 shows an example of the application of trend-based unconstrained and constrained methods to the scenarios of development of the annual number of atms at the London Heathrow airport (UK) (Janić, 2020).

The airport currently operates two parallel runways in the airside area. Because the current runway system capacity of $\mu(t)^* = 480000$ atm/year has come to saturation, the long debate and plans for implementing the third parallel runway has been taking place. Under such conditions, the observed period in Scenario I on Figure 2.23a is divided into two sub-periods: t_1 2017–2025/26, and t_2 2025/26–2050. In the year 2025/26 the new third runway would be implemented. Therefore, during the first sub-period, the airport operational capacity and consequently the atm demand will stay constrained. In the second sub-period, the above-mentioned three-runway system annual capacity of $\mu(t) = 769420$ atm/year would enable the airport unconstrained growth of the atm demand, which will be driven by both external and internal demand-driving forces at an average

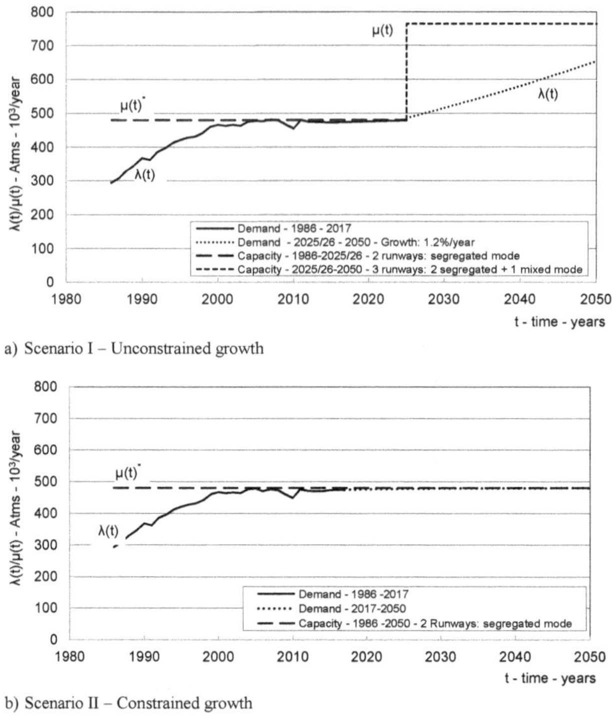

Figure 2.23 Development of the atm demand and the runway system capacity—Case of London Heathrow airport (UK) (Period: 2017–2050) (Janić, 2004; 2020).

annual rate of 1.2% during the rest of the observed period (t_2 2025/26–2050). Scenario I, shown in Figure 2.23b, is based on maintaining the cap on the runway system capacity of $\mu(t)^*$ = 480 thousands atm and consequently the atm demand close to or at that level during the entire observed period (*t*2017–2050).

2.3.2.3 Econometric methods

a) Description

The econometric methods/models are based on considering the airport demand-driving forces. In these methods, the causal relationship between the dependent variable representing the airport demand and the independent variables representing the driving forces of these demands need to be established by the regression analysis in the form of the functional relationship. In case of the rather strong interrelationship, the particular independent variables—driving forces—need to be forecasted and then the dependent variable—airport demand—must be estimated. In general, forecasting these independent variables is usually carried out by the time-trend methods or some other econometric-type causal models.

The most frequently-used independent demand-driving forces relate to the airport catchment area and the service characteristics, such as population, employment, GDP, PCI, the average fares or yield (the airline revenue per passenger kilometre), and some transport service-related variables, such as the travel distance, time, and the capacity (flights, seats, etc.). Based on the historical data during the given time period, the causal relationship between the airport demand as the dependent variable and the above-mentioned demand-driving forces as the independent variables can be estimated using the regression least-square regression technique. In addition to the obvious advantages, the main disadvantage of these methods has been an inherent assumption that the past relationships will likely be sustained in the future (Janić, 2013; Wadud, 2011).

b) Analytical structure

The analytical structure of the econometric methods is presented by the relationship between the overall and/or the segmented airport passenger demand and its main-demand driving forces.

i) O-D (Origin-Destination) passenger demand

Let *(h)* and *(j)* be two mutually connected airports (regions) ($j = 1, 2, 3, .., N$). The given airport is *(h)* and the other airport is *(j)*. In some cases, the airport (region) *(j)* may also refer to the larger geographical (metropolitan) area with a few "clustered" airports. The analytical structure of the econometric method for estimating the airport passenger demand between the airport *(h)* and the airport *(j)* can be as follows:

$$Q_{hj} = a_0 \cdot (GDP_h \cdot GDP_j)^{a_1} \cdot (Y_{hj} \cdot L_{hj})^{a_2} \cdot \tau_{hj}^{a_3} \cdot S_{hj}^{a_4} \cdot P_{hj}^{a_5} + \sum_{k=1}^{K} b_k \cdot D_k \tag{2.4}$$

where

- Q_{hj} is the airport passenger demand between the airports *(h)* and *(j)* (the number of passengers in both directions during the given period of time);
- $GDP_{h/j}$ is the GDP (Gross Domestic Product) of the catchment areas around the airports *(h)* and *(j)*, respectively (usually billions $US);
- Y_{hj} is the average airline yield between the airport *(h)* and *(j)* (¢/RPK or ¢/RPM -Revenue Passenger Kilometre or Mile);
- L_{hj} is the two-way shortest distance between the airports *(h)* and *(j)* (km or mi);
- τ_{hj} is the total travel time between the airports *(h)* and *(j)* (hours);
- S_{hj} is the supply of airline transport capacity between the airports *(h)* and *(j)* (the total number of supplied seats in both directions during the given period of time (day, week, month, year));
- P_{hj} is the competitive power of the airport *(h)* regarding the region-market of the airport *(j)* compared to the competitive power of the other airports in the same and/or the nearby region(s) (market share of the flight frequencies (seats) between the airport *(h)* and the region *(j)* in the total number of frequencies (seats) between all competing airports and the region *(j)*);
- D_k is a dummy variable taking the value "1" if a disruptive event of type *(k)* has considerably affected the overall passenger demand and the value "0", otherwise (a disruptive event could be terrorism, epidemic diseases, regional wars, bad weather, etc.); and
- a_i, b_k is the coefficient to be estimated by estimating the regression equation ($i = 1 - 6$; $k = 1 - K$).

The main forces in Eq. 2.4 drive the passenger demand at the airport *(h)* related to the airport *(j)*. In cases of the short and medium-haul distances between two airports, the variables (L_{hj}), (τ_{hj}), (C_{hj}), and (P_{hj}) need to be modified in order to take into account the eventual competition from the surface transport modes, such as HSR (High Speed Rail) or TRM (Trans Rapid Maglev), if they are available. The variables (GDP_h) and (GDP_j) represent GDP of the corresponding regions surrounding the airports *(h)* and *(j)*, respectively. During the time of overall economic growth, these variables increase over time and seemingly contribute positively to the growth of airport passenger demand, and vice versa. This implies positive elasticity, i.e., the coefficient, a_1. In some cases, the variables, such as PCI, population, employment, trade, investments, tourism, and exchange of the other services between the given airport and the other regions, can be combined and replace those of GDP. The variable (Y_{hj}) (yield) relates to the weighted average yield of all airlines operating between the airports *(h)* and *(j)* during the given period of time. Its product with the variable (L_{hj}), i.e., the two-way travel distance, gives the average return airfare (AF_{hj}). The variable (L_{hj}) represents the shortest (the great circle) distance between the airports. In general, when the airfare as the "travel's resistance factor" increases, the number of passengers will generally decrease, and vice versa, which, in theory, reflects the negative elasticity, i.e., the negative coefficient (a_2).

The variable (τ_{hj}) represents the travel time between the airports *(h)* and *(j)*. It includes the non-stop flying time as well as the schedule delay, i.e., the waiting time for the convenient departure at the airport *(h)*. Since it is based on the shortest distance, which usually does not change over time, this variable can be influential only if the aircraft's technology changes (for example, if faster regional jets replace the slower turbo-props) and/or if the number of flights significantly changes. Otherwise, it is seemingly not particularly relevant. The airport passenger demand tends to increase with decreasing of distance, i.e., the shorter travel time implies the negative elasticity, i.e., the coefficient (a_3). The variable (S_{hj}) represents the airline supply of transport capacity, which is usually expressed by the number of seats or only by the flight frequency offered between the airports *(h)* and *(j)* by all airlines. Since the supply of capacity (at a reasonable price) intends not only to satisfy but also to stimulate passenger demand, the latter is theoretically expected to rise with increasing of this capacity, and vice versa. This generally implies positive elasticity, i.e., the positive coefficient (a_4). The variable (P_{hj}) represents the competitive position (i.e., the competitive "power") of the airport *(h)* in comparison to the other relatively close competing airports in the same or neighbouring regions. For example, the possible competitive cases can be the four largest London airports, two largest Paris airports, and/or three largest New York airports. Under such circumstances, this variable refers to the generalized access cost, the airfare, and the departure frequency at the airport *(h)* compared to that at the other competing airports (Hess et al., 2005; Janić, 2013). In general, the higher competitive power attracts a higher passenger demand, thus reflecting the positive elasticity, i.e., coefficient (a_5). A dummy variable D_k $(k = 1, 2, 3, ..K)$ expresses impact of the various disruptive events on the airport passenger demand, generally implying negative elasticity, i.e., the negative coefficients (b_k). The above-mentioned econometric method can be estimated using the time-series data for particular O-D markets *(hj)* $(j = 1,.., J)$ for the period of at least 10 to 15 past years (TRB, 2002). Consequently, from Eq. 2.4, the total airport passenger demand for the given time period can be estimated as follows:

$$Q_h = \sum_{j=1}^{J} Q_{hj}, \text{ for } j = 1, 2,.., J \tag{2.5}$$

where

J is the number of airports connected with the airport (h) during the specified time.

The other symbols are analogous to those in the previous Eqs.

ii) Transit/transfer passenger demand

The modified Eq. 2.4 can be used for analysis and forecasting of the transit/transfer passenger demand at the airport *(h)*. Table 2.6 shows the necessary conversion of the particular variables in order for them to be applicable to the transit/transfer passenger demand at the airport *(h)*.

Theoretically, the airport transit/transfer demand is expected to have positive elasticity regarding the GDP of the passenger origin and destination regions and the airport competitive power, and negative elasticity with respect to the airline yields, i.e., airfares. It can have either positive or negative elasticity with respect to the supply of transport capacity. In the former case, this reflects the airlines intention to stimulate demand and eventually carry out the capacity supply competition by increasing the number of seats for the transfer/transit passengers. In the latter case, this can reflect the intention of airlines to increase load factor of their flights, similar to the case of O-D passenger demand. According to the notation in Table 2.6 and use of the relevant data, the appropriately modified Eq. 2.4 can be applied in a similar fashion to the O-D model in order to estimate the airport transit/transfer passenger demand during the specified period of time (Janić, 2013).

iii) Non-scheduled (charter) passenger demand

The specifically modified Eq. 2.4 can also be applied to analysis and forecasting of the airport non-scheduled (charter) passenger demand. In such case, the dependent variable Q_{hj} represents the passenger demand between the airports-regions *(h)* and *(j)*, where the latter region is considered to be an attractive tourist (leisure) destination. The independent variable (GDP_h) reflects the leisure trip driving force, which this time can be *PCI* (Per Capita Income) expressing the purchasing power of the individuals from the region *(h)*. The variable GDP_j can be considered as irrelevant. The product of variables $(Y_{ij} \cdot L_{ij})$ can reflect the average airfare despite the fact that it has usually been hidden inside the price of the entire leisure package(s). However, it can also be transparent if the specialized (charter) airlines sell individual seats. The variable (τ_{hj}) can also be considered as less relevant because the leisure travellers have been less sensitive to the travel time. The variable (P_{hj}) can also have similar relevance. Finally, the variable (S_{hj}) can have very

Table 2.6 Conversion of the variables in Eq. 2.4 to handle transit/transfer passenger demand at the airport *(h)*.

Variable	O-D passenger demand	Transit/transfer passenger demand
Passengers	Q_{hj}	$Q_{i/h/j}$
GDP	GDP_h, GDP_j	GDP_i, GDP_j
Yield	Y_{hj}	$Y_{ij/h} = Y_{ih} + Y_{hj}$
Distance	L_{hj}	$L_{ij/h} = L_{ih} + L_{hj}$
Travel time	τ_{hj}	$\tau_{hj} = \tau_{ih} + \tau_{hj}$
Seat capacity	S_{hj}	$S_{ij} = S_{ih} + S_{hj}$
Competitive power	P_{hj}	$P_{ij/h}$
Disruptive event(s)	D_k	$D_{k/ij/h}$

similar relevance as in Eq. 2.4, i.e., the number of seats offered by all (charter) airlines flying between the airport *(h)* and the tourist region *(j)* during the given time period (season). The influence of particular variables on the non-scheduled (charter) passenger demand is similar to the case of its O-D and transfer/transit counterparts. This modified method can be applied in a similar fashion to its original in Eq. 2.4 by using the relevant data during the specified period of time (Janić, 2013).

iv) LCC passenger demand

Analysis (and forecasting) of the airport low-cost passenger demand can be carried out by the modified above-mentioned Eq. 2.4. This implies keeping only the independent variables as the main demand-driving forces,like the case of the non-scheduled (charter) demand such as *(PCI)* and the airlines supply capacity (S_{ih}). The airline "yield" reflecting the airline airfares can be assumed always acceptable for the attracted passenger demand and, therefore, seemingly not particularly relevant in this case. The method also requires the relevant data on the dependent and independent variables for the specified time period in order to make an estimation (Janić, 2013).

v) Total passenger demand

The total airport passenger demand can be estimated by summing up their particular segments based on Eq. 2.4–2.5 as follows:

$$Q_h = \sum_{j=}^{J} (Q_{hj} + Q_{hj/ns} + Q_{hj/lc}) + \sum_{i=1}^{I} \sum_{j=1}^{J} Q_{i/h/j}, \text{ for } i \neq h \neq j \quad (2.6)$$

where

$Q_{i/h/j}$, $Q_{hj/ns}$, $Q_{hj/lc}$ is the airport transfer/transit, non-scheduled (charter), and low-cost passenger demand at airport *(h)* connected to the airports *(i)* and *(j)* (passengers/day, week, month, year); and

I, J is the number of origin and destination airports, respectively, of the air passengers handled at the airport *(h)*.

c) Application example

i) The case airport–background

An application of the above-mentioned econometric methods for analysis and forecasting of the airport passenger demand was carried out for Amsterdam Schiphol airport (The Netherlands). To this end, the time series data on the total passenger demand for the time period 1992–2004 and its market segmentation was used, as shown in Figure 2.24.

As can be seen, the total passenger demand increased during the observed period. This was possible thanks to hosting more than one hundred airlines, which connected the airport to about 200 destinations in 90 countries worldwide. For a long time, the dominant airline was KLM, its subsidiaries and partners in Wings Alliance (including Northwest, MAS, and China Southern airlines). In the year 2004, the airline and most of its subsidiaries and partners were incorporated into Sky Team alliance led by Air France, after merging of Air France and KLM. During the observed period, the host airline KLM and its alliance partners developed the hub-and-spoke network at the airport, which, in addition to contributing to the above-mentioned growth of the total, also contributed to increasing of the transfer passenger demand. Its market share in the total scheduled passenger demand was increased from about 26% in the year 1992 to about 42% in the year 2004. The market share of transit passengers was very low and even decreasing,

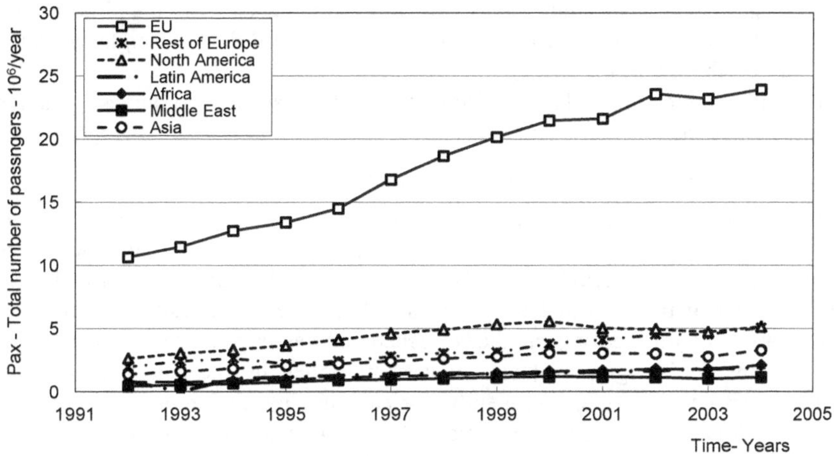

Figure 2.24 Development of the total passenger demand and its segments—Case of Amsterdam Schiphol airport (Period: 1999–2004) (Janić, 2013; SG, 2005).

from about 2.3% in the year 1992 to nearly 0.5% in the year 2004 (Janić, 2013; SG, 2005).

In order to apply the mentioned econometric models to the given case, the aggregation, i.e., clustering of the airport market(s), i.e., the sets of incoming and outgoing routes/markets, was carried out regarding the geographical regions and availability of the relevant data. Consequently, the airport markets with the O-D passenger demand were clustered into seven clusters as follows (Janić, 2008b; 2013):

The Netherlands	• EU (15 until 2004 and 25 Member States after) • Rest of Europe (22 until 2004 and 12 Central, Eastern European, and Mediterranean countries after) • North America (U.S. East Coast and Central) • Latin America (Brazil) • Africa (Kenya, South Africa) • Middle East (United Arab Emirates, Israel) • Asia (China, Japan, Korea, Malaysia, Singapore)

The airport markets containing the transit/transfer passenger demand were clustered similarly as that with the O-D passenger demand as follows (Janić, 2008b; 2013):

EU	• EU
EU	• Rest of Europe
EU + Rest of Europe	• North America
EU + Rest of Europe	• Latin America
EU + Rest of Europe	• Africa
EU + Rest of Europe	• Middle East EU + Rest of Europe – Asia

Clustering the airport markets with the non-scheduled (charter) passenger demand was carried out regarding the long-standing tradition of the Dutch leisure travellers as follows: (i) The Netherlands—Mediterranean basin (South of Europe, North Africa,

Middle East), and (ii) The Netherlands—the other continents. Also, the airport markets with the O-D passenger served by LCCs were clustered into a single cluster: the Netherlands—the rest of the EU.

It should be mentioned that the airport did not have substantial domestic (national) markets (SG, 2005).

ii) Data on the dependent and independent variables

The time series data on the annual number of passengers at the Amsterdam Schiphol airport, including the information on the independent variables, i.e., the demand-driving forces, for the period 1992–2004, were used. The dependent variable(s), the annual number of the O-D and transit/transfer passengers, was estimated as follows: the available data on the total number of scheduled passengers (without those carried by LCCs) in each geographical market was multiplied by the annual proportion of transfer/transit passengers. Consequently, for each market, the annual number of O-D and transfer/transit/passengers was obtained. The annual numbers of non-scheduled (charter) and LCCs passengers were obtained directly from the available data (SG, 2005).

The data on GDP of particular regions and countries were obtained from the UN (United Nations) statistical databases (UN, 2005) (in 10^9\$US, current prices). Where deemed reasonable, PCI (Per Capita Income) instead of GDP data were used (in 10^3\$US/head). In addition, the data on the annual international trade, investments and services where used where appropriate (in 10^9\$US).

The average airfares in the particular airport markets were estimated as the product of the weighted average yield of the incumbent airline KLM, its Wing alliance partners, and the other AEA airlines operating at the airport, and the two-way shortest distance between the airport and the main airport (city) of a given region (market) (in \$US/passenger) (AEA, 2004; Janić, 2008b; 2013; KLM, 2004; http://www.airdistances.com). For the O-D passenger demand, the average one way distances were estimated to be $L_{ih}/2$ = 650 km in the EU, $L_{ih}/2$ = 1560 km in the rest of European, $L_{ih}/2$ = 6700 km in the North American, $L_{ih}/2$ = 9200 km in the Latin American, $L_{ih}/2$ = 6700 in the African, $L_{ih}/2$ = 3300 km in the Middle Eastern, and $L_{ih}/2$ = 8550 in the Asian market(s). The airfare for transfer passengers was assumed to be the same as the airfare for O-D passengers. This was considered as a reasonable assumption since many airlines offered the same (or even lower) airfares in order to compensate the inconvenience of changing the flight at the hub airport. The average airfares in the non-scheduled (charter) markets were estimated similarly, this time by using the average yield of the charter operations of AEA airlines on the distances of about $L_{ih}/2$ = 1700 km in the Mediterranean markets and $L_{ih}/2$ = 7700 km in the other intercontinental markets (AEA, 2004). The airfares for LCCs were assumed to be approximately $(L_{ih} \cdot Y_{ih})$ = \$US70/pax during the period 1997–2005. The number of seats offered to the scheduled and non-scheduled O-D passengers in the given markets in both directions was estimated as the product of the average number of weekly flights and the average aircraft/flight seating capacity. The latter was obtained by dividing the number of passengers by the corresponding load factor. For the scheduled O-D and transit/transfer passenger demand, this number was: s_{hj} = 116 in the EU and the rest of European, s_{hj} = 169 in the Middle Eastern, s_{hj} = 264 in the North American, and s_{hj} = 324 seats/flight in the Latin American, African, and Asian markets. The seat capacity of a flight scheduled by LCCs during the given period of time (1997–2005) was estimated to be s_{hj} = 140 seats. For the non-scheduled (charter) passenger demand, the average number of seats per flight was s_{hj} = 190 in the Mediterranean and s_{hj} = 330 in the other intercontinental markets. The number of seats per flight offered to transfer passengers was estimated using the above-mentioned methodology. The airport competitive power (P_{ih}) was estimated as the proportion of number of destinations and the number of flights at the given airport in the corresponding totals offered at four competing airports: London Heathrow, Frankfurt Main, Paris Charles de Gaulle, and Amsterdam Schiphol (SG, 2005).

iii) Some illustrative results

Some illustrative results from application of the method in Eq. 2.4 by using the input data from the given case are given in Table 2.7.

As was expected, the O-D passenger demand in the airport markets between the Netherlands and EU member states and the Netherlands and North America was mainly driven by increasing of *GDP* at both market sides, supply of the airline seat capacity, the airport market power, and generally decreasing airfares. In both markets, the regression equation was significant (*F*-statistics) with the relatively high explanatory power of the independent variables (R^2 statistics). Furthermore, the particular independent variables were statistically significant at 5% level (*t*-statistics below the corresponding variables). Specifically, in the EU market, development of LCCs affected the O-D passenger demand

Table 2.7 Results from application of the econometric method—Case of Amsterdam Schiphol airport (Janić, 2008b; 2013).

Market cluster	Econometric relationship
O-D passengers	
The Netherlands – EU (15 Member States until 2004 and 25 after)	$Q_{h1} = 4613.18 \cdot (GDP_h \cdot GDP_1)^{0.082} \cdot AF_{h1}^{-0.889} \cdot S_{h1}^{0.770} \cdot S_{lc}^{-2.370}$ t 2.990 3.851 –2.961 3.772 –2.970 $R^2 = 0.893$; $F = 16.749$; $DW = 1.448$; $N = 13$ (2.7a)
The Netherlands – North America (U.S. East Coast and Central Area)	$Q_{h2} = 1640.59 \cdot (GDP_h \cdot GDP_2)^{0.483} \cdot AF_{h2}^{-0.831} - 0.059D1$ t 1.501 3.316 –1.292 –1.336 $R^2 = 0.481$; $F = 4.712$; $DW = 0.488$; $N = 13$ (2.7b)
Transit/transfer passengers	
EU/EU	$Q_{1h1} = 2.541 \cdot (GDP_1 \cdot GDP_1)^{0.383} \cdot AF_{11}^{-0.896} \cdot S_{11}^{0.744} \cdot P_{h1}^{0.484}$ t 0.484 4.680 –5.094 5.838 1.087 $R^2 = 0.998$; $F = 1259.806$; $DW = 2.460$; $N = 13$ (2.7c)
Europe* – North America	$Q_{1h2} = 0.00435 \cdot (GDP_1 \cdot GDP_2)^{0.245} \cdot AF_{12}^{-0.460} \cdot S_{12}^{0.731} \cdot P_{h2}^{2.312}$ t 1.144 1.434 –1.682 3.766 3.567 $R^2 = 0.997$; $F = 360.324$; $DW = 2.22$; $N = 13$ (2.7d)
Non-scheduled (charter) passengers	
The Netherlands – Mediterranean basin (South of Europe, North Africa, Middle East)	$Q_{hc} = 625.173 \cdot PCI^{1.018} \cdot AF_{h1}^{-0.367}$ t 5.131 8.343 –1.642 $R^2 = 0.910$; $F = 51.517$; $DW = 2.26$; $N = 11$ (2.7e)
LCCs passengers	
The Netherlands – EU (15 until 2004 and 25 Member States after)	$Q_{h1/lc} = 0.0147 \cdot PCI^{2.664} \cdot S_{h1/lc}^{0.785}$ t 2.161 4.021 13.943 $R^2 = 0.999$; $F = 4044.791$; $DW = 2.228$; $N = 9$ (2.7f)

AF = $L \cdot Y$; $S_{h1/lc}$ – the seats offered by LCCs (seats/week); * The geographical area consisting of the EU member and other European states; PCI – Per Capita Income (the Netherlands) (10^3 $US).

carried out by the full cost airlines. In the North American markets, the airfares were statistically insignificant, including a dummy variable explaining the impacts of disruptive events on the airport demand. In the former case, the airfares of KLM/Northwest alliance dominated in the market. In the latter case, the dummy variable seemingly could not catch up the impact of some disruptive events (for example, the September 11 2001 terrorist attack on the U.S.). The very similar relationships and influence of particular demand driving forces were in the airport transit/transfer passenger markets between the Netherlands and EU member states and the Netherlands and North America.

In the airport non-schedule (charter) passenger markets, the main demand driving forces were increasing *PCI* and decreasing airfares. The econometric relationship was significant (*F*-statistics) and both independent variables possessed the relatively high explanatory power (R^2-statistics). This demand was elastic with respect to *PCI* and inelastic with respect to the airfares that appeared as statistically insignificant (*t*-statistics). A possible explanation was that an increase in *PCI* provided a higher disposable budget for holidaying overseas while the airfares were hidden within the holiday packages and as such did not have a significant influence on people purchasing them simply due to the fares not being transparent.

The airport LCCs demand in the markets between the Netherlands and the rest of Europe is mainly driven by *PCI* and the airline capacity supply. The econometric relationship significant (*F*-statistics) and the selected independent variables had a very high explanatory power (R^2 statistics). In addition, each independent variable was also statistically significant at 5% level (*t*-statistics given below the corresponding variables). Specifically, this demand was particularly elastic with respect to increase in the local-airport *PCI* as an indicator of the available budget for traveling. Under such conditions, the low airfares were always acceptable, and consequently not relevant.

2.3.2.4 Scenario, ratio, market surveys, and expert judgement methods

a) The scenario-based methods usually indicate the assumed variations of the future conditions influencing the airport passenger demand. For example, particular variables in the econometric models can take a range of the future values based on the scenarios of their development rather than on their exact values. The disadvantage of these methods is that the range between the high and the low values of particular variables can sometimes be so large and as such cause the analysis and forecast to lose their practical value.

b) The ratio methods imply the ratios between the passenger demand at a given airport and the total national air transport demand. The main disadvantage of these methods relates to the conditions of their application, i.e., the lack of resources and expertise for using some other more sophisticated methods (Janić, 2008b; 2013; TRB, 2002).

c) The market surveys methods imply collection of information about passenger behavioural patterns. They include the origins and destinations, the airport's choice in a given metropolitan area, the choice of airport ground-access mode, the trip purpose, and the other factors used for predicting the future behaviour and consequently the passenger demand. The main disadvantage of these methods has been their complexity, time consuming nature, and the costs of collecting the relevant data. In addition, since the data have usually been collected over a relatively short period of time, they could not be sufficiently reflective for the longer-term purposes (Janić, 2008b; 2013; TRB, 2002).

d) <u>The expert judgment methods</u> have been a component of almost all above-mentioned analysing and forecasting methods. In these methods, the assumptions on the future development and the values of particular variables have been mainly the matter of informal judgment. For example, the Delphi method is often used for attaining a consensus on the judgments from different experts about the values of particular variables relevant to the airport demand analysis and forecasting. In addition, these judgments have also been used for selection of the methods, time period as the base, the analytical forms of the models, and the sources of data (Janić, 2008b; 2013; TRB, 2002).

2.3.2.5 Demand as input for planning and design

The airport annual atm and passenger demand estimated by the above-mentioned methods for the given past, current and future time period has been used as the planning and design parameter for the airport airside and landside area infrastructure, facilities, and equipment. As such, this annual demand needs to be converted into the appropriate planning and design parameters. In general, these annual demand values need to be converted into PH (Peak Hour) values. These have usually been estimated for the selected design 5-th, 10-th, or in the case of airport master planning, 20-th year-period ahead. After the design year has been set up with the corresponding demand, what follows is estimation of PM (Peak Month), ADPM (Average Day of Peak Month), and PH (Peak Hour) of ADPM demand. The PH demand can be estimated as follows (FAA, 1988; 2004; Horonjeff and McKelvey, 1994; Janić, 2008b; 2013; TRB, 2002):

$$PH = f_2 \cdot ADPM = f_2 \cdot PM/n = f_2 \cdot (f_1 \cdot Q)/n \tag{2.8}$$

f_2 is the factor of *PH* (Peak Hour) (%);
f_1 is the factor of *PM* (Peak Month) (proportion of the annual demand concentrated in PM (Peak Month) (%);
Q is the annual volume of the airport traffic in the design year (pax/year);
ADMP is the Average Day of Peak Month (pax/month);
PM is the *PM* (Peak Month) (pax/month); and
n is the number of days per month.

In Eq. 2.8, the factor (f_1) is usually determined by the historical data, including prospective local variations. The factor (f_2) has been estimated by different methods. Some past estimates indicate that it has amounted to 12–20% of ADPM. Figure 2.25 shows an example of the relationship between the factor (f_2) and the airport annual passenger demand.

As can be seen, the factor (f_2) decreases more than proportionally with increasing of the airport annual passenger demand. One of the causes has been tendency of the larger demand to spread over the longer time periods during the day. In addition, the number of peaks during the day can be different. At small airports, these are usually the morning and the afternoon-evening peaks. At the larger hub airports, several peaks occur during the day depending on the number of scheduled "waves" of the incoming and outgoing flights of the dominant airline's hub-and-spoke network (de Neufville and Odoni, 2003; FAA, 1988; 2004; ICAO, 2004; Janić, 2013).

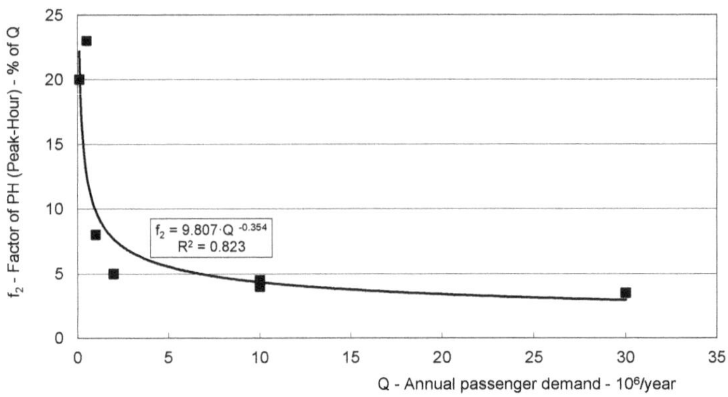

Figure 2.25 Proportion of PH (Peak-Hour) in the airport annual passenger demand (FAA, 1988; 2004; Janić, 2013).

2.3.3 *Capacity*

2.3.3.1 General

The term "capacity" implies a throughput of a given service facility. In such context, the capacity of an airport can be expressed by the maximum number of given entities which can be accommodated/served during a given period of time under given conditions. In the airport airside area, the service facilities with their capacities are the runways, taxiways, and apron/gate complex. The served entities are the aircraft operations or atms (air transport movements), i.e., landings and take-offs. In the airport landside area, the service facilities and their capacities are the passenger and freight/cargo terminals, and the road and rail-based airport landside access modes and their systems. The served entities in the passenger and freight/cargo terminals are the arriving, departing and transfer/transit passengers and their baggage, and freight/cargo shipments, respectively. These entities served by the airport landside access modes and their systems are the O-D passengers, airport and other employees, visitors, and greeters, and freight/cargo shipments (Janić, 2000). London Gatwick airport (UK), the world's busiest single-runway airport, is an illustrative example in the given context. The simplified airport layout is shown in Figure 2.26.

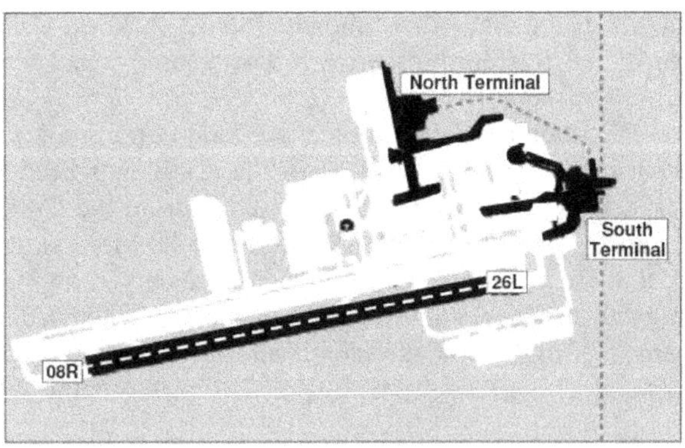

Figure 2.26 Simplified scheme of London Gatwick airport (https://www.gatwickairport.com/business-community/; https://en.wikipedia.org/wiki/Gatwick_Airport).

As can be seen, the airport operates as a single runway (08R/26L) of length L = 3255 m handling all aircraft categories performing the short-, medium-, and long-haul flights. The runway is equipped with Category III ILS (Instrument Landing System). The declared runway capacity is λ = 55 atm/h. The runway operations during the night are restricted from 11 pm to 7 am, which gives an operation time of τ = 16 h/day for 365 days/year. This gives the airport annual capacity of $M = 365 \cdot \tau \cdot \mu = 365 \cdot 16 \cdot 55 = 321000$ *atm/year*, which is between the recommended annual capacity of single runway ($M = 365 \cdot \tau \cdot \mu = 365 \cdot 13.15 \cdot 50 = 240000$ *atm/year*) and the closely-spaced dual parallel runways ($M = 365 \cdot \tau \cdot \mu = 365 \cdot 15.5 \cdot 60 = 340000$ *atm/year*) (Horronjeff and McKelvey, 1994). The apron/gate complex contains: N_g = 115 gates. With the above-mentioned capacity, the airport has handled an increasing number of the atms and passengers, as shown in Figure 2.27(a, b).

Figure 2.27a shows that the number of passengers has increased to about 46 million per year thanks to the number of atms increasing to 280 thousand per year.

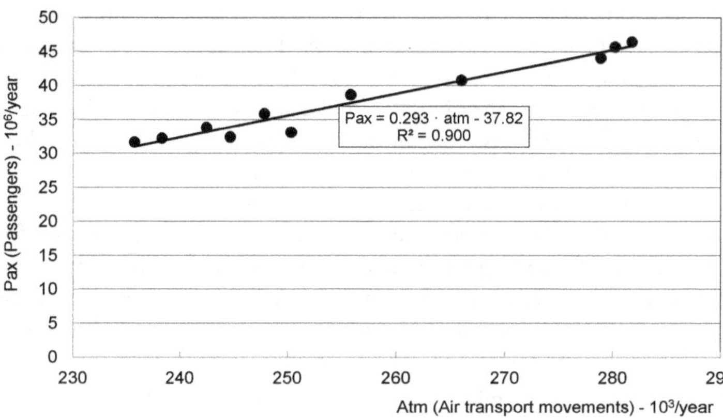

a) Annual number of Pax (Passenger(s)) vs Atms (Air Transport Movement(s))

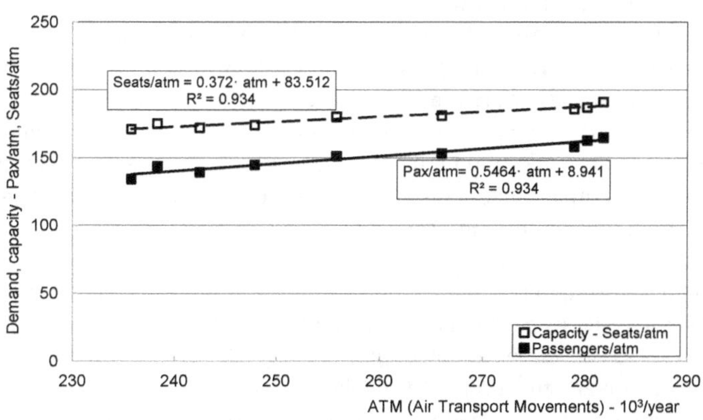

b) Seats (Capacity), Pax (Passengers)/atm vs the annual number of atms (Air transport movement(s))

Figure 2.27 Development of traffic at the large congested single-runway airport—Case of London Gatwick airport (UK) (Period: 2009–2019) (GAL, 2009/2019).

Figure 2.27b shows that the aircraft seat capacity and the number of passengers per atm has in parallel increased with the growth of the number of atms during the observed period. This indicates that the airlines have maintained a relatively constant average load factor (GAL, 2009/2019).

The airport landside access modes and their systems should currently provide the annual capacity for about 42.32 million O-D passengers and 24.43 thousand airport employees. This gives an annual total of 42.344 million users. Regarding the relatively stable market shares of about 46% car, 11% bus, and 43% rail, these available landside access modes need to provide the necessary annual capacities of at least 19.5 million (car), 4.658 million (bus), and 18.208 million (rail) passengers (Jacobs, 2015).

2.3.3.2 Airside capacity

Two of the general concepts of airport airside capacity are known as the "ultimate" and "practical" capacities. The "ultimate" capacity is expressed by the maximum number of entities (atms) that can be served during a given period of time under conditions of constant demand for service. The "practical" capacity is defined as the maximum number of atms that can be served during a given period of time under conditions when the average delay imposed on each atm does not exceed the specified level. The various environmental and social constraints (noise, emissions of GHG, land use) can also be specified as additional conditions (Horonjeff and McKelvey, 1994; Janić, 2000; TRB, 1990). The "ultimate" and "practical" capacities of the airport airside area and its components—runways, taxiways, and apron/gate complex—are usually estimated and used as planning and operational parameters for a specified time period, such as year, day, and/or hour. The former two are usually considered as planning parameters. The latter is considered as an operational parameter. Since the particular airside components are located in serial order and operate as mutually dependent, the airport airside capacity is determined by the capacity of the "bottleneck" element. Also, any temporary or permanent "bottleneck" causes imbalance between the demand and its service rate, i.e., the capacity, and thus causes congestion and delays of users, i.e., aircraft/atms.

The capacity of an airport airside area includes the capacity of its main components. These are as follows: (i) the runway system with the adjacent airspace enabling the incoming aircraft landing after the final approach and the outgoing aircraft take-off before climbing along the departure path; (ii) the taxiway system, which enables aircraft taxiing between the runway system and the apron/gate complex, just after landing and before take-off, and (iii) the taxi-lines with the apron/gate complex, which enable aircraft entering, leaving, and/or parking on the particular parking apron/gates, respectively.

Since the above-mentioned components are positioned and followed by both incoming and outgoing aircraft in serial order, the runway system with the adjacent airspace plays the role of the entry/exit of the airside area. This implies that the runway system "ultimate" capacity critically influences the capacity of the entire airside area. In addition, the serial order of particular components requires balancing of their "ultimate" capacities in order to prevent or alleviate periodical and/or permanent "bottlenecks" that may cause congestion and the aircraft/atm delays.

The past and current investigation and practice indicates that the "ultimate" airport airside capacity depends on many factors that may be constant and/or variable over time. The most important constant factor is the airport layout, characterized by the number and directions of runways, taxiways, and the runway instrumentation. The most time-variable factor is weather, usually characterized by ceiling, visibility, and wind. Specifically, at U.S. airports, depending on the weather conditions, the aircraft landings can be carried out either under IMC or VMC. They both require application of the specific/different ATC/ATM separation rules between landing aircraft/atms. They depend on the aircraft wake vortex categories depending on their MTOW. Table 2.8 gives the aircraft wake vortex categorization by ICAO and FAA.

The aircraft speed increases with the increase in aircraft take-off weight, which implies that the minimum separation rules between the sequences (pairs) of landing or taking-off aircraft are different. Table 2.9 gives the example of the distance-based separation rules between landing aircraft.

As can be seen, the IFR separation rules applied under IMC are about 40% stricter than the VFR separation rules applied under VMC. In both cases, these separation rules enable trailing aircraft to avoid the wake vortices generated by leading aircraft in particular landing sequences.

In order to modify, i.e., investigate, possible refinement of the above-mentioned distance-based separation rules in Table 2.9, extensive research has been undertaken by the European EUROCONTROL, the U.S. FAA, and the aviation industry. Essentially, in addition to MTOW, the aircraft wing span has also been taken into consideration, resulting in re-categorization of the ICAO aircraft wake vortex categories from 4 in Table 2.9 to 6 as follows: CAT A – "Super Heavy", CAT B – "Upper Heavy", CAT C – "Lower Heavy", CAT D – "Upper Medium", CAT E – "Lower Medium", and CAT F –

Table 2.8 The aircraft wake vortex categorization (FAA, 2014; ICAO, 2016).

Category	MTOW (ICAO) (10^3 kg)	MTOW (FAA) (10^3 kg)	Aircraft type	Indicated speed (Knots)
A	Light/MTOW < 7	Small/MTOW < 7	Dornier 228, Cessna Citation	< 91
B	Light/MTOW ≤ 7	Small/MTOW ≤ 7	Dornier 228, Cessna Citation	91–120
C	Medium/MTOW 7–136	Large/MTOW 7–136	ATR 42/72, A319, A320, B737	121–140
D	Heavy/MTOW ≥ 136	Heavy/MTOW ≥ 136	B767, B777, B787, A300, A330, A340, A350, B747,	141–165
E	Super heavy	Super heavy	A380, An-225	≥ 166

Table 2.9 The FAA minimum distance-based separation rules between arriving aircraft (nm) (nm – nautical mile) (FAA, 2014; 2018).

VFR/VMC					IFR/IMC				
i/j	**Small**	**Large**	**B757**	**Heavy**	***i/j***	**Small**	**Large**	**B757**	**Heavy**
Small	1.9	1.9	1.9	1.9	**Small**	2.5(3)	2.5(3)	2,5(3)	2.5(3)
Large	2.7	1.9	1.9	1.9	**Large**	4.0	2.5(3)	2.5(3)	2,5(3)
B757	3.5	3.0	3.0	2.7	**B757**	5.0	4.0	4.0	4.0
Heavy	4.5	3.6	3.6	2.7	**Heavy**	6.0	5.0	5.0	4.0

"Light". Some of aircraft types in particular categories are as follows: *CAT-A* (AIRBUS A-380-800...); *CAT-B* (AIRBUS A-330-200/300, A-340-300/400/500/600, A350-900, BOEING 747-400/8, B 777-200/777-200ER/300/200LR/F, B787-Dreamliner, IL-96....); *CAT-C* (AIRBUS A-300-600/300/310, B757-200/300, B767-200ER/300ER/400, DC-10, MD-11, IL-76, TU-950; *CAT-D* (AIRBUS A-319/320/321, AN-12, BOEING B737-600/700/800/900, MD-81/90, TU-204...); *CAT-E* (ATR-42-300/320/500, ATR-72/201, Boeing B717-200, B737-200/300/400/500, Bombardier CRJ-100/700, Dash 8Q400, EMBRAER ERJ-135/145/170/190..); *CAT-F* (DESSAULT Falcon 10/20, DORNIER 328, EMBRAER Brasilia, BEECH 400 Beechjet, RAYTHEON BAe-125-700/800, BRITISH AEROSPACE Jetstream 32/41, GATES LEARJET Learjet 35/60, SAAB 340, PIAGGIO P-180 Avanti, CESSNA 650 Citation 3/6/7/525,.....). Table 2.10(a,b) gives the RECAT minimum separation rules for the landing and taking-off aircraft (EEC, 2015; https://www.skybrary.aero/index.php/RECAT_-_Wake_Turbulence.Re-categorisation#Re-categorisation/).

In addition, the ICAO minimum distance-based separation rules between successive take-offs from the same runway are as follows: 1 min between the aircraft flying on the track diverting by at least 45 degrees immediately after take-off; 2 min when the leading aircraft is 74 km/h (40 kt) or more faster than the following aircraft and both are on the same track; and 5 min until the vertical separation between the leading and trailing aircraft on the same track is established and while it does not exist. In addition, the minimum time-based separation rules between departing aircraft from the same runway are as follows: 1 min if the leading aircraft in the departing sequence changes the course by at least 45 degrees immediately after take-off; 2 min between the aircraft departure sequence following the same track; and 5 min in the

Table 2.10 The RECAT minimum separation rules (EEC, 2015).

a) Arriving aircraft (nm) (nm – nautical mile)						
i/j	**CAT A**	**CAT B**	**CAT C**	**CAT D**	**CAT E**	**CAT F**
CAT A	3	4	5	5	6	8
CAT B	3	3	4	4	5	7
CAT C	3	3	3	3	4	6
CAT D	3	3	3	3	3	5
CAT E	3	3	3	3	3	4
CAT F	3	3	3	3	3	3
b) Departing aircraft (s) (s – second)						
i/j	**CAT A**	**CAT B**	**CAT C**	**CAT D**	**CAT E**	**CAT F**
CAT A	80	100	120	140	160	180
CAT B	80	80	80	100	120	140
CAT C	80	80	80	80	100	120
CAT D	80	80	80	80	80	120
CAT E	80	80	80	80	80	100
CAT F	80	80	80	80	80	80

case of no vertical separation between the departure sequence on the same track when the trailing aircraft passes through the level of leading aircraft. Furthermore, an arriving aircraft always has priority for landing over departing aircraft from the same runway. The minimum separation rules between these aircraft is 3 min and 5 min, depending on the track/course of departing aircraft and the position of the arriving aircraft at the moment when the departure is to start (ICAO, 2016).

The minimum distance-based separation rules in Tables 2.9 and 2.10a can be divided by the aircraft approach speeds in Table 2.8 in order to obtain the minimum time-based separation rules between particular sequences of landing aircraft. The ICAO has recommended that these speeds have to be calculated as the product of the factor 1.3 and the aircraft stall speed in the aircraft landing configuration and MLW (Maximum Landing Weight) (ICAO, 2016).

Figure 2.28 shows the influence of two above-mentioned categories of separation rules and landing speeds in Tables 2.9 and 2.8, respectively, on the "ultimate" landing capacities under IMC and VMC conditions and their relationship at the selected U.S. airports (FAA, 2014a).

As can be seen, the IFR and VFR capacities are strongly linearly related. The IFR landing capacities generally amount to about 70% of the corresponding VFR landing capacities. In Europe, independently of the weather conditions, the landings are carried out exclusively according to IFR. As well, Table 2.11 shows the total "ultimate" capacity of the runway systems at 10 largest European and U.S. airports (EEC, 2005; 2018; FAA, 2014a; NASA, 2001).

As can be seen, European airports have a smaller number within their runway systems. These operate exclusively IFR operations under IMC conditions. The U.S. airports generally have a greater number of runways, which operate both IFR and VFR operations under IMC and VMC conditions, respectively. Consequently, the "ultimate" capacity of the U.S. airport is higher than that of European airports. The airport airside "practical" capacity is elaborated in the scope of corresponding quality of services.

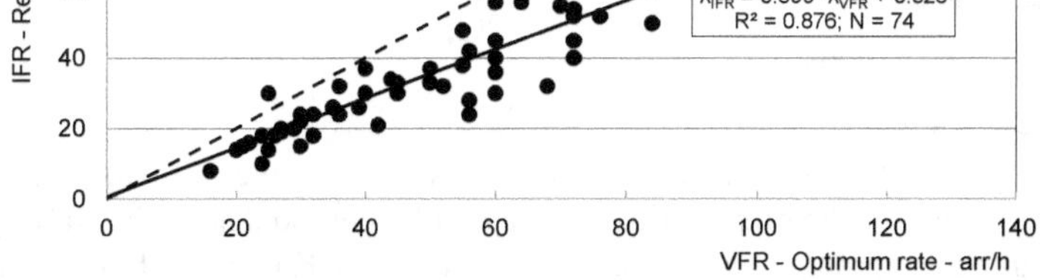

Figure 2.28 Relationship between IFR and VFR landing capacities—Case of the selected U.S. airports (FAA, 2014a; NASA, 2001).

Table 2.11 "Ultimate" capacity of the runway system—Case of the selected European and U.S. airports (EEC, 2005; 2018; FAA, 2014a; NASA, 2001).

	European airports			
Country	**Airport/Code**	**No. of runways**	**Declared IFR capacity** (atm/h)	
France	Paris/CDG	4	112	
Germany	Frankfurt/FRA	3	75–83	
Spain	Madrid/MAD	4	90	
UK	London Heathrow/LHR	2	82	
Netherlands	Amsterdam Schiphol/AMS	6	106	
Germany	Munich/MUC	2	90	
Spain	Barcelona/BCN	3	61	
Italy	Rome/FCO	4	88	
Austria	Vienna/VIE	2	66	
Switzerland	Zurich ZRH	3	68	
	U.S. airports			
State	**Airport/Code**	**No. of runways**	**VFR capacity** (atm/h)	**IFR capacity** (atm/h)
Georgia	Atlanta/ATL	4	180–188	158–162
Illinois	Chicago O'Hare/ORD	7	190–200	136–144
Texas	Dallas Fort Worth/DWF	7	270–279	186–193
California	Los Angeles/LAX	4	137–148	117–124
Colorado	Denver/DEN	6	210–219	159–162
Nevada	Las Vegas/LAS	4	109–113	70–70
Texas	Huston/ IAH	5	120–143	108–112
Arizona	Phoenix/PHX	3	128–150	108–118
N. Carolina	Charlotte/CLT	3	130–131	102–110
Philadelphia	Philadelphia/PHL	4	104–116	96–96

2.3.3.3 Modelling airside capacity

a) Runway system

i) Structure of the models

• Operating scenarios

Many analytical and simulation models/tools have been developed for estimating the "ultimate" capacity of the airport airside area and its components—the runway system, taxiways, and the apron/gate complex (Hockaday and Kanafani, 1974; Janic, 2008; 2008a; Newell, 1979; Odoni et al., 1997; Tošić, 1992; TRB, 1990).

As already mentioned, the "ultimate" capacity of a given airport runway system is expressed by the maximum number of aircraft operations (i.e., atms), that can be accommodated during a given period of time (usually one hour) under conditions of the constant demand for service. This capacity depends on the following factors: the weather conditions, the aircraft mix, restrictions in the use of particular runways, the level of sophistication of the aircraft avionics and the ATC/ATM navigation facilities and equipment, configuration of the runway exits and the associated taxiways, the expected share of IMC and VMC (i.e., IFR and VFR operations), and the management of utilization of the runway system with respect to the operating conditions and type of operations (Horonjeff and McKelvey, 1994). This capacity can be estimated for the conditions when the runway system is exclusively used for landings, take-offs, and mixed operations.

The models for estimating the "ultimate" capacity of a given runway system are based on the defined runway system, the ATC/ATM separation rules, aircraft fleet mix, and scenarios of the runway system's use. Then, the capacity is computed as a reciprocal of the minimum average inter-arrival time between the aircraft passing through a given "reference location", which is assumed to be the runway threshold where all atms pass through (Blumstein, 1960). Figure 2.29 shows the simplified layout of the single runway system.

According to the scheme, after leaving the terminal airspace STARs (Standard Arrival Route(s)), the arriving aircraft follow a common three-dimensional approach path connecting the FAG (Final Approach Gate) and the runway threshold *(T)*. The departing aircraft start take-off at the runway threshold *(T)*, fly over the opposite end of the runway, and climb along a departure path to the assigned altitude. The aircraft maintain constant speeds while on the final approach and departure path(s) according to the values in Table 2.9. However, differences in the aircraft approach speeds produce four pairs of landing sequences *(ij)*, where the aircraft *(i)* is leading and the aircraft *(j)* is the trailing aircraft in the sequence. These are: "slow-slow", "slow-fast", "fast-slow", and "fast-fast". Classification of the particular aircraft types into two groups (fast and slow) is both relative and general at the same time, i.e., the one with the higher speed is always considered as "fast" and the other with lower speed as "slow". Similar categories of speed sequences can be defined for departing aircraft. The ATC/ATM separation rules between the particular landing sequences should be applied along the final approach path. For the, "slow-slow", "fast-slow", and "fast-fast" sequences, the minimum separation rules have to be established at the FAG, and for the "slow-fast" sequence at the landing

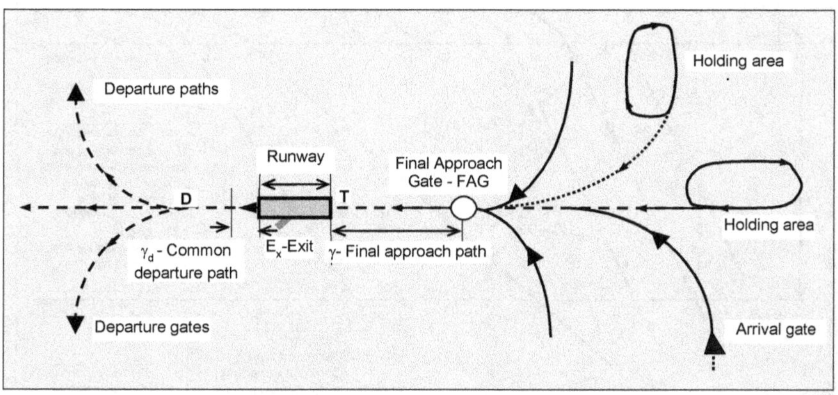

Figure 2.29 Horizontal layouts of the arriving and the departing flight paths to/from an airport single runway (Newell, 1979).

threshold *(T)*. Figure 2.30 shows the time-space diagram for the above-mentioned aircraft sequences.

As can be seen, the horizontal distance between aircraft in the sequence "fast-slow" sequence increases over time, it is constant in the "slow-slow" and "fast-fast" sequences, and decreases in the "slow-fast" sequence. In addition, Figure 2.30 shows that one or more successive take-offs can be carried out safely between the particular consecutive landings if the time gap between them allows that respecting the ATC/ATM minimum separation rules.

- "Ultimate" landing capacity

– Conventional operational procedures

The model for estimating the "ultimate" landing capacity of a single runway implies calculation of the average minimum service time for landing aircraft during a given period of time (τ) (usually 1 h) under conditions of constant demand for service. This time represents the arithmetic mean of the minimum inter-event times, which gives the average number of landings during time (τ) as follows (Janić, 2000):

$$\mu_a = \tau/\overline{\tau_a} = \tau/(1/N) \cdot \sum_{j=2}^{N} \tau_{j-1,j} \tag{2.9a}$$

where

N is the number of inter-event times occurring during time(τ)(s); and

$\tau_{j-1,j}$ is the minimum time interval between *(j-1)*-th and *j*-th event counted at the runway threshold as the "reference location" for calculating the landing capacity.

The denominator of Eq. 2.9a depends of the number, length, and the structure of the inter-event times of the above-mentioned speed-based landing sequences. If the inter-event time between the leading aircraft *(i)* and the trailing aircraft *(j)* in the sequence *(ij)* is $({}_a\tau_{ij})$, and if the proportions of the aircraft types *(i)* and *(j)* in

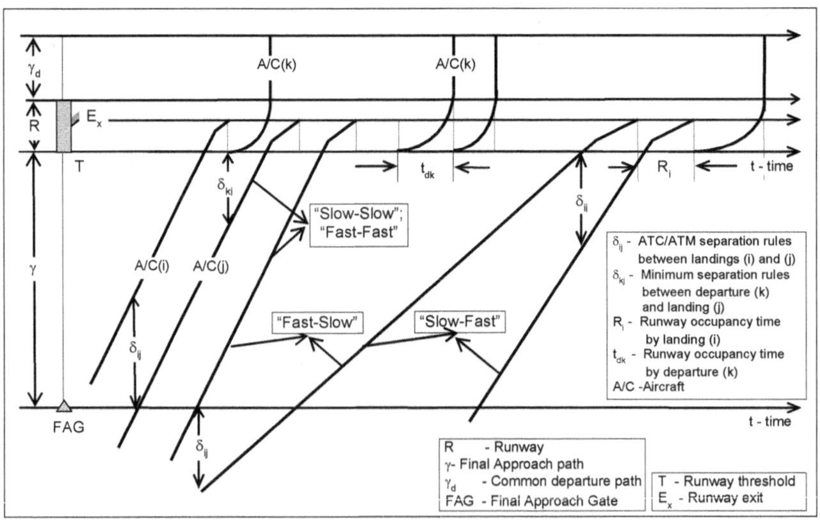

Figure 2.30 The time-space diagram for particular aircraft landing sequences on a single runway (Horonjeff and McKelvey, 1994; Janić, 2000).

the arriving traffic mix are (p_i) and (p_j), respectively, and if *(i)* and *(j)* are the independent events, then the average minimum inter-event time at the runway threshold for all combinations of the landing sequences *(ij)* is determined as follows (Blumstein, 1960):

$$\overline{\tau_a} = \sum_{ij} p_i \cdot \tau_{a/ij} \cdot p_j \tag{2.9b}$$

where

$\tau_{a/ij}$ is the inter arrival time between the aircraft *(i)* and *(j)* at the runway landing threshold (s).

The inter-arrival time $(\tau_{a/ij})$ in Eq. 2.9b depends on the operational procedures applied to the landing aircraft by ATC/ATM. These can generally be conventional horizontal distance-based, time-separation based, innovative horizontal and vertical distance-based, and environmental and social constraints-based.

– Conventional horizontal distance-based separation procedures

These procedures imply exclusive application of the ATC/ATM distance-based separation rules between all landing sequences. In such case, the inter arrival time between the aircraft *(i)* and *(j)* at the runway landing threshold is equal to:

$$\tau_{a/ij} = \begin{bmatrix} \max(R_{ai}; \delta_{ij}/v_j), & for\ v_i \le v_j \\ \delta_{ij}/v_j + \gamma \cdot (1/v_j - v_i), & for\ v_i > v_j \end{bmatrix} \tag{2.10a}$$

where

δ_{ij} is the ATC/ATM minimum longitudinal separation rules between the aircrafts *(i)* and *(j)* measured along the path of the aircraft *(j)*; it is applied either on the runway's threshold (the sequence $v_i \le v_j$) or at the final approach gate (the sequence $v_i > v_j$) (nm);

R_{ai} is the runway occupancy time by the landing aircraft*(i)*(s);

γ is the length of the common final approach path (nm); and

v_i, v_j is the speed of the aircrafts *(i)* and *(j)*, respectively, assumed to be constant along the distance (γ) when $(v_i \le v_j)$ and along the distance $(\delta_{ij} + \gamma)$ when $(v_i > v_j)$ (kts).

Equation 2.10a provides that the ATC/ATM minimum separation rules between landing aircraft are always satisfied either in the air or on the ground.

– Time-based separation procedures

When the ATC/ATM minimum time-based separation rules were applied, the term (δ_{ij}/v_j) in Eq. 2.3a would be replaced by the term $\tau_{aij/min}$, i.e., the ATC/ATM minimum separation time interval applicable to the aircraft sequence *(ij)* either at the runway threshold *(T)* $(v_i <= v_j)$ or at the FAG $(v_i > v_j)$ (Janić, 2008a).

– Innovative horizontal and vertical distance-based separation procedures

○ Mixed horizontal and vertical distance-based separation rules (non-displaced threshold)

The mixture of the ATC/ATM minimum horizontal and vertical distance-based separation rules can be applied between particular sequences of landing aircraft using the specified IFC (Individual Flight Corridors) with two segments, each with different ILS GS

(Instrument landing System Glide Slope) angles. In such case, the minimum inter-arrival time of the aircraft in the sequence *(ij)* can be estimates as follows:

$$\tau_{a/ij} = \begin{bmatrix} \delta_{ij}/v_j, & for\ v_i \le v_j \\ H^0_{ij}/v_j \sin\theta_j, & for\ v_i = v_j \\ H^0_{ij}/v_j \sin\theta_j + \gamma_{ij}\cdot(1/v_j - 1/v_i), & for\ v_i > v_j \end{bmatrix} \tag{2.10b}$$

where

H^0_{ij} is the minimum ATC/ATM vertical distance-based separation rule between the aircraft of the wake vortex categories *(i)* and *(j)* (ft; m); and

θ_j is the GS angle of the aircraft *(j)*, which can be different for the Outer and Inner segment of the final approach trajectory.

The other symbols are analogous to those in Eq. 2.10a.

Equation 2.10b indicates that the ATC/ATM vertical separation rules are applied to the aircraft sequence of the same speed at the landing threshold and to the sequence "fast-slow" at the FAG. The ATC/ATM horizontal separation rules are applied to the aircraft sequence "slow-fast". Under such conditions, at least one separation rule is guaranteed to the safe landing of particular aircraft sequences.

- Vertical distance-based separation rules (non-displaced landing threshold)

The ATC/ATM minimum vertical distance-based separation rules can be applied to the particular landing sequences, in which the aircraft use different GS angles along the entire final approach trajectory. Figure 2.31 shows the simplified scheme.

Depending on the approach speeds and GS angles, there can be twelve different combinations of landing sequences, for which the minimum inter-arrival times at the reference location, i.e., landing threshold, are calculated as follows (Janić, 2007):

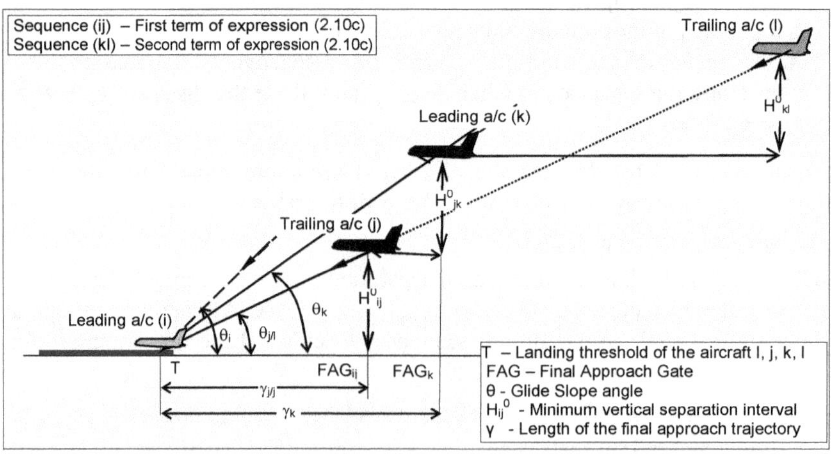

Figure 2.31 Scheme of some cases of applying the ATC/ATM vertical distance-based separation rules between landing aircraft (Eq. 2.10b) (Janić, 2007).

$$\tau_{a/ij} = \begin{bmatrix} H^0_{ij}/v_j \sin\theta_j, & for\ v_i \le v_j\ and\ v_i \sin\theta_i \ge< v_j \sin\theta_j \\ & for\ v_i > v_j\ and\ v_i \sin\theta_i < v_j \sin\theta_j \\ H^0_{ij}/v_j \sin\theta_j + \gamma_i \cdot \tan\theta_i (1/v_j \sin\theta_j - 1/v_i \sin\theta_j), & \\ & for\ v_i > v_j\ and\ v_i \sin\theta_i \ge v_j \sin\theta_j \end{bmatrix} \quad (2.10c)$$

where

γ_i is the length of the final approach path of the aircraft *(i)* (nm); and
θ_i is the GS angle of the final approach path of aircraft *(i)* (^{0}).

Other symbols are analogous to those in the previous Eqs.

In the first term of Eq. 2.10c, the first condition indicates that the horizontal separation remains constant or decreases while the vertical separation remains constant, decreases, or increases. The second condition indicates that the horizontal separation increases but the vertical separation decreases. Under such circumstances, in order to minimize the inter-arrival time $t_{ij/v}$, the ATC/ATM minimum vertical distance-based separation rules should be established at the moment when the leading aircraft *(i)* in the sequence *(ij)* is just at the landing threshold (*T*), as shown in Figure 2.31.

- Vertical distance-based separation rules (displaced landing threshold)

The ATC/ATM minimum vertical separation rules can be applied to the landing aircraft sequences using the same ILS GS angle as shown in Figure 2.32(a, b) (Janić, 2011).

Figure 2.32a shows that the ATC/ATM vertical separation rules are applied to the landing threshold *(T)* of the trailing aircraft *(i)* in the sequences $v_i \le v_j$. Figure 2.32b shows that, in the landing sequence $v_i \le v_j$, the ATC/ATM minimum vertical separation rules are applied at the FAG of the leading aircraft *(i)*. The minimum inter-arrival times between particular landing sequences at the reference location, i.e., the corresponding landing threshold, are calculated as follows (Janić, 2011):

$$\tau_{a/ij} = \begin{Bmatrix} \max[R_{ai}; H^0_{ij}/\sin\theta_j], & for\ v_i \le v_j \\ H^0_{ij}/\sin\theta_j + \gamma_j \cdot (1/v_j - 1/v_i) + \varepsilon/v_j, & for\ v_i > v_j \end{Bmatrix} \quad (2.10d)$$

where

θ_i, θ_j is the ILS GS angle of the leading and trailing aircraft *(i)* and *(j)*, respectively (^{0}); and
ε is the staggered distance of the displaced landing threshold of the "slow" aircraft (nm).

The other symbols are analogous to those in the previous Eqs.

In order to eventually implement the above-mentioned innovative horizontal and vertical distance-based procedures, three sets of conditions need to be fulfilled: supportive technologies and decision support tools should be available and implemented, the aircraft need to be certified for the steeper GS angles, and both ATC/ATM controllers and pilots need to be adequately trained. The technologies supporting the above-mentioned procedures are given in Table 2.12.

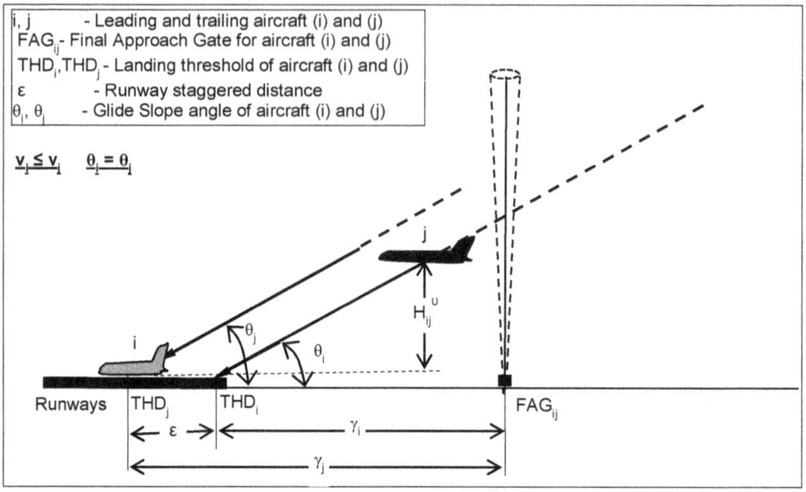

a) Sequence: $v_i \leq v_j$

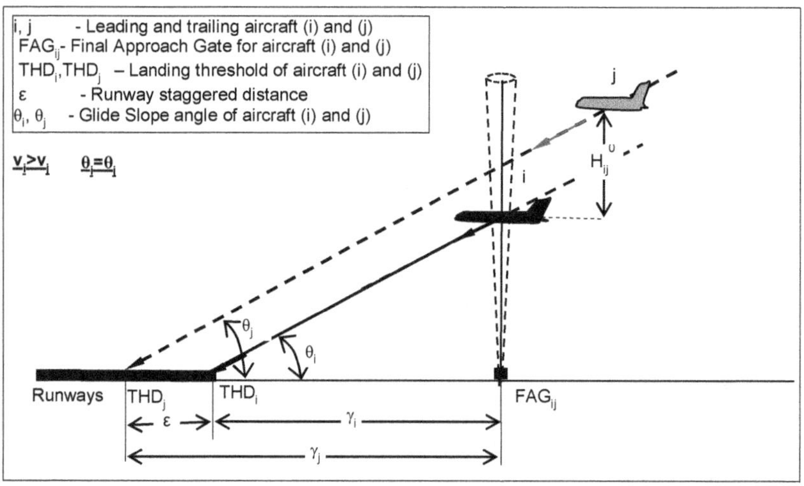

b) Sequence: $v_i > v_j$

Figure 2.32 Scheme of cases of applying the ATC/ATM vertical distance-based separation rules between landing aircraft at displaced threshold (Janić, 2007).

In Eq. 2.10(a, d) the minimum inter-arrival time $(\tau_{a/ij})$ has to be at least equal to or greater than the runway landing occupancy time (R_{ai}) of the leading aircraft *(i)* in the sequence *(ij)*, i.e., $\tau_{a/ij} \geq R_{ai}$. This ensures that only one aircraft occupies the runway at time.

– The environmental and social constraints-based procedures

At many airports, the environmental constraints in terms of noise, air pollution, and land use can affect the above-mentioned runway system capacity. In order to assess these influencing factors, the runway system capacity also needs to be defined regarding these

Table 2.12 Advanced technologies supporting advanced landing procedures (EC, 2005; EC, 2007; http://www.faa.gov/about/initiatives/nextgen/).

Location/ user	Technology	Availability
Air traffic flows management tools by the ATC/ATM	• CTAS (Centre/TRACON Automation System) assists in optimizing the arrival flow and runway assignment; • Integrated Arrival/Departure Manager enables planning in advance and updating the arrival sequences, including replacement of the FCFS (First-Come-First-Served) rule with some other sequencing rule, implying the prioritization of particular aircraft wake-vortex categories as well as the use of the time-based instead of the distance-based separation rules between landing aircraft.	Now Now but still needs improvements
Air traffic surveillance equipment—by ATC/ ATM on the ground	• RADAR of improved precision enables reduction of the minimal separation between aircraft from 3 to 2.5 nm; • PRM – Precision Runway Monitor consisting of a beacon radar and computer predictive displays enables the independent use of dual- and triple-dependent parallel runways spaced less than 4300 ft. • Terminal Wake Vortex Detection System provides information about the wake-vortex behaviour during landings and take-offs.	Now with additional improvements in the medium term Now Medium- to long-term
Improved aircraft avionics – onboard	• FMS 4D – Flight Management System enables more precise following of the time schedule according to the flight plan, which reduces the position error of arrivals at the final approach gate. • Sending 4D trajectory enables the exchange of the permanently updated aircraft current and expected positions obtained from FMS to the external recipients, such as other aircraft-pilots and ATC controller(s) on the ground. • ADS-B – Automatic Dependent Surveillance Broadcasting improves situation awareness both onboard and on the ground and is used independently, but in addition to TCAS and enhanced CDTI (Cockpit Display of Traffic Information). • CDTI – Cockpit Display of Traffic Information provides integrated traffic data onboard the aircraft, which may reduce the separation rules between aircraft. • TCAS – Traffic Alert and Collision Avoidance System shows the spatial relation of two aircraft and provides instructions to avoid potential collisions. • ACAS – Airborne Collision Avoidance System strengthens the quality of information for ASAS, thus enabling the replacement of PRM (above) and reducing position error in all three dimensions. • ASAS – Airborne Separation Assistance (Assurance) System enables airborne surveillance, display of traffic information, and consequently sequencing and merging based on the data from ADS-B. • WVDS (Wake Vortex Detector System) onboard the aircraft enables collection and display of information on the existing wake vortex to both pilots and ATC controllers. • WAAS – Wide Area Augmentation System improves basic GPS accuracy both horizontally and vertically. • LVLASO – Low Visibility Landing and Surface Operating Program reduces, controls and predicts the runway occupancy time.	Now Long term Medium term Medium to long-term Now Medium to long-term Long-term Medium to long-term Medium term Medium term

Table 2.12 Contd. ...

...Table 2.12 Contd.

Location/ user	Technology	Availability
"Mixed" traffic surveillance and conflict alert equipment devices by ATC/ ATM on the ground and onboard	• Distributed Air Ground Solution combines ADS/B, TCAS, and Free Flight devices enabling simultaneous aircraft-ATC traffic surveillance, alerting and resolution of potential conflicts.	Medium-term

constraints. Generally, it can be expressed by the maximum number of atm (landings and/ or take-offs) accommodated at a given runway system under given constraints during a given period of time (1 h). For noise, this is the total generated sound energy, which does not exceed some prescribed limit, i.e., the noise quota or noise cap. For air pollution, this is the prescribed air pollution quota. For the land use, this is the land available for setting up the runway system.

In particular, the noise quota or cap for landings is expressed as the 'energy average sound level', or the 'equivalent continuous noise' index ($L_{a/eq/\tau}$). This index is designed to accumulate all the aircraft sound energy from the multiple noise events, in this case landings, carried out during a given period of time (τ), which can be 1 h, 8 h, or 24 h. At most airports, the concept of ($L_{a/eq/\tau}$) is applied to τ = 16-hours period over the daytime (Ashford and Wright, 1992). The cumulative sound energy contained in ($L_{a/eq/\tau}$) is assumed to be uniformly distributed over the time (τ) and different for the day and the night period. At airports where there is a night flight ban, the noise quota is equal to zero during the ban. However, during the day, it may vary between 57 dB(A) and 85 dB(A) (DETR, 1999). In addition, ($L_{a/*/\tau}$) represents the noise level in *dB(A)* generated by an individual noise event, i.e., landing of the aircraft of type *(*)*, respectively. This noise is usually estimated at the noise reference locations, which may be either the noise certification points or some other selected locations in the vicinity of a given airport (ICAO, 1993).

As the noise quota ($L_{a/eq/\tau}$) is set up according to the maximum level of tolerance of the affected population, the "ultimate" runway landing capacity constrained by the noise quota/cap can be estimated as follows (Janić, 2013):

$$C_{a/\tau} = \tau \cdot 10^{L_{a/eq/\tau}/10} \tag{2.11a}$$

where time (τ) is expressed in seconds. In addition, the average sound energy, i.e., noise of an individual landing, can be expressed as: $N_{a/\tau} = \tau \cdot 10^{L_{a/\tau}/10}$. By dividing the airport noise quota/cap by the average noise per an individual landing experienced at the noise reference locations, the "ultimate" runway landing capacity within the prescribed quota/ cap during time (τ) can be estimated as follows:

$$\mu_{a/\tau} = C_{a/\tau}/N_{a/\tau} \tag{2.11b}$$

where all symbols are as in the previous Eqs.

Similarly, the capacity of the given runway for take-offs subject to the noise and/or other environmental constraints/caps can be estimated.

- "Ultimate" take-off capacity

The model for calculating runway take-off capacity has a similar structure to the model for calculating runway landing capacity. The difference is in determining the times between successive take-offs. Essentially, these times are determined with respect to the fact that the aircraft occupy the runway for some time during take-off. After getting airborne, they follow SID (Standard Instrumental Departure) routes through terminal airspace. The ATC/ATM minimum separation rules applied to the successive take-off aircraft prevent the runway from being simultaneously occupied by two aircraft and provide some spacing along the same SIDs in order to avoid the influence of the wake vortices generated by the leading aircraft in the particular departure sequences. The "reference location" for counting the take-offs is again the runway threshold *T*. The inter-departure time between the two successive take-off aircraft *(i)* and *(j)* is estimated as follows (Janić, 2013):

$$\tau_{d/ij} = max[\tau_{ij/d/0}; \tau_{ij/d/min} - (R_{dj} - R_{di}) - \gamma_d \cdot (1/v_{dj} - 1/v_{di})] \tag{2.12a}$$

where

$\tau_{ij/d/0}$ is the ATC/ATM minimum time-based separation rule applied to the taking-off aircraft *(i)* and *(j)* at the departure threshold (min);

$\tau_{ij/d/min}$ is the minimum required time separation between the aircraft *(i)* and *(j)* at the point where they both leave the airport zone (point *D* in Figure 2.30) (min);

R_{di}, R_{dj} is the runway occupancy time by the departing aircraft *(i)* and *(j)*, respectively;

γ_d is the length of the common departure path of the aircraft *(i)* and *(j)* (min); and

v_{di}, v_{dj} is the average speed of the aircraft *(i)* and *(j)*, respectively, along the distance γ_d (Figure 2.30) (kts).

The probability of occurrence of the aircraft taking-off sequence *(ij),* assumed to be the realization of a pair of the independent events, is equal to: $p_{d/ij} = p_{di} \cdot p_{dj}$, where p_{di} and p_{dj} are the proportions of the aircraft *(i)* and *(j)*, respectively, in the departure traffic mix. The average inter-departure time for all combinations of the departure sequences *(ij)* is equal to:

$$\overline{\tau_d} = \sum_{ij} p_{d/ij} \cdot \tau_{d/ij} \tag{2.12b}$$

Consequently, the take-off capacity is estimated as:

$$\mu_d = \tau / \overline{\tau_d} \tag{2.12c}$$

- "Ultimate" capacity for the mixed operations

At particular airports, some runways can be exclusively used for take-offs. The capacity of these runways can also be achieved under conditions of the constant demand for service during a given period of time (usually 1 h). Such traffic pattern is very common at airports operating as the hubs of particular incumbent airlines, which schedule complexes of flights, each consisting of "waves" of arriving flights followed by "waves" of departing flights. They are scheduled during a relatively short period of time (a half, ¾, 1 or 1.5 h). In some cases, due to some restrictions at the airports operating the independent parallel runways, one runway can be used exclusively for arrivals and the other exclusively for departures. In such a case, the runway system operates in the so-called "segregated" mode. When both runways are used for both landings and the take-offs, the runway system operates in the so-called "mixed" mode (Horonjeff and McKelvey, 1994).

Nevertheless, atmost airports, the runways are simultaneously used for both landings and take-offs, i.e., in the "mixed" mode. The ATC/ATM can apply various tactics for sequencing the different operations at a single runway. Which tactic will be applied depends, on the mix of the landing and the take-off demand and its distribution in time (τ) (Janić, 2000). If the demand for both types of operations is constant, and if each operation is available at the time when it can be realized, the three tactics can generally be applied to maximization of the runway capacity: the alternate tactic, the consecutive tactic, and the random tactic. The alternate tactic enables the realization of at least one take-off between any two successive landings. The consecutive tactic enables continuous serving of the batches of landings before or after serving the batches of take-offs. The random tactic implies a safe realizing of take-offs between the non-interrupted landings while assuming that these landings always have a higher priority than the take-offs. Consequently, according to the time-space diagram in Figure 2.32, at least (n) take-offs can be realized between any two successive landings if the following condition is met (Horonjeff and McKelvey, 1994; Newell, 1979):

$$\tau_{ad/ij} = R_{ai} + \delta_{jk}/v_j + (n-1) \cdot \tau_d \qquad (2.13a)$$

where

$\tau_{ad/ij}$ is the minimum time gap allowing a realization of (n) safe take-offs between the successive landings (i) and (j) (min);

R_{ai} is the runway occupancy time by the landing aircraft (i) (min);

δ_{jk} is the ATC/ATM minimum distance of the landing aircraft (j) from the runway threshold enabling a safe take-off of the aircraft (k) (nm);

v_j is the average speed of the landing aircraft (j) at the distance (δ_{jk}) (nm);

τ_d is the runway occupancy time during a take-off (min).

If the probability of occurrence of a time gap $(\tau_{ad/ij})$ is denoted by (p_d), the total runway capacity (μ_a) can be estimated as follows:

$$\mu = (1 + p_d) \cdot \mu_a \qquad (2.13b)$$

where

μ_a is the "ultimate" landing capacity of a given runway (atm/h).

- <u>"Practical" capacity</u>

When the demand for landings and/or take-offs generally does not exceed the runway system "ultimate" capacity during the longer period of time, the atm delays, when they occur, are stochastic and not particularly long. In such case, in addition to the "ultimate" capacity, the specified average delay(s) per an atm can be used to determine the "practical" runway capacity, which can be declared by the given airport in terms of the number of slots per hour (or 15 min) during the day. To this end, the modified expression for the average delay per an atm derived from the steady-state queuing system theory can be used as follows (Newell, 1979):

$$W^* = \mu_p \cdot (1 + C_s^2)/[2 \cdot \mu_u \cdot (\mu_u - \mu_p)] \qquad (2.14a)$$

where

W^* is the maximum average delay per landing and/or take-off specified for setting up the "practical" capacity (min);

μ_p is the "practical" landing and/or take-off capacity (atm/h);

μ_u is the "ultimate" landing and/or take-off capacity (atm/h); and

C_s is the coefficient of standard deviation of service time of an arrival and/or departure.

After setting the variable $C_s^2 = 0$, the "practical" runway capacity can be estimated as follows:

$$\mu_p = (2 \cdot W^* \cdot \mu_u^2)/(1 + 2 \cdot W^* \cdot \mu_u) \tag{2.14b}$$

where all symbols are analogous to those in Eq. 2.14a.

ii) Application of the models

• Inputs

Application of the above-mentioned runway system capacity models is illustrated on the generic case of single landing runway operating according to the "what-if" scenario. For such a purpose, the following inputs are used:

– The ATC/ATM separation rules

The ATC/ATM distance-based separation rules under IMC/IFR conditions/procedures in Table 2.9 are used to calculate the runway landing capacity as the benchmarking case. In addition, the ATC/ATM minimum vertical distance-based separation rules of $H_{ij}^0 = 1000$ ft are assumed to be applied to all sequences of landing aircraft.

– Characteristics of the aircraft fleet mix

The characteristics of the aircraft types, including the assumed GS angles, are given in Table 2.13.

As can be seen, the average approach speeds are different to those in Table 2.9. According to the assumed scenarios, Small, Large, and B757 aircraft are assumed to approach and land at more than one GS angle, while Heavy aircraft exclusively use a GS angle of $\theta = 3°$. The fleet mix is varied while maintaining the proportion of Small and B757 aircraft constant in all calculations (5%).

– Assignment of GS (Glide Slope) angles

Based on GS in Table 2.13 and respecting the two cases when the mixture of the ATC/ATM horizontal and vertical distance-based separation rules and exclusively the ATC/

Table 2.13 Characteristics of the particular aircraft landing categories (the averages) (EC, 2005; Janić, 2007; 2008; Thompson, 1997).

Aircraft category	Mass M (10^3kg)	Average approach speed v (kts)	GS angle θ (°)	Runway landing occupancy time R_a (s)
Small	20	120	3/4/5.5	30–40
Large	55	130	3/4/–	40–50
B757	117	155	3/4/–	40–45
Heavy	206	155	3/–/–	50–60

ATM vertical distance-based separation rules are applied, the combinations of GS angles for particular landing sequences are given in Table 2.14(a, b).

Specifically, in the case when a combination of the ATC horizontal and vertical distance-based separation rues is applied, the length of common approach path with GS angle of $\theta = 3^o$ for all aircraft is assumed to be $\gamma = 6.16\ nm$ (Janić, 2008). The same length is used when the capacity is calculated for the case of exclusively applying the ATC/ATM vertical distance-based separation rules.

- Results

The results from the model application by using the above-mentioned inputs are given in Figure 2.33. This shows the relationship between the runway landing capacity and the proportion of Heavy aircraft in the fleet mix and type of the ATC/ATM minimum separation rules.

As can be seen, when the current ATC/ATM IFR/IMC separation rules are applied, the landing capacity decreases more than proportionally with an increase in the proportion of Heavy aircraft in the fleet. When the ATC/ATM mixed horizontal and vertical distance-based separation rules are applied, the capacity increases with an increase in the proportion of Heavy aircraft in the mix. The main reason is that the vertical separation applied to the landing sequence "fast-slow" at the FAG reduces the horizontal distances between the aircraft, thus shortening the corresponding inter-arrival times at the landing threshold *(T)*. When only the ATC/ATM minimum vertical distance-based separation rules are applied, the capacity again decreases with an increase in the proportion of Heavy aircraft in the mix. This happens because these aircraft use the smaller GS angles, not allowing shortening of the horizontal distances in the particular aircraft sequences, which, despite their higher approach speeds, increase the corresponding inter-arrival times at the landing threshold *(T)* and consequently decreases the capacity. In addition, the landing capacity is greater from 2–33% when the ATC/ATM vertical instead of the ATC/ATM horizontal and vertical distance-based separation rules are applied. This difference decreases with an increase in Heavy aircraft in the mix. Furthermore, application of the ATC/ATM minimum vertical separation rules enables the highest landing capacity for the wide

Table 2.14 Assignment of GS (Glide Slope) angles in the given scenarios.

a) The ATC/ATM minimum horizontal and vertical distance-based separation rules are applied (^{0})				
i/j	**Small**	**Large**	**B757**	**Heavy**
Small	3/3	3/3	3/3	3/3
Large	3/5.5	3/3	3/3	3/3
B757	3/5.5	3/4	3/3	3/3
Heavy	3/5.5	3/4	3/4	3/3
b) The ATC/ATM minimum vertical distance-based separation rules are applied (^{0})				
i/j	**Small**	**Large**	**B757**	**Heavy**
Small	3/5.5	3/4	3/4	3/3
Large	3/5.5	3/4	3/4	3/3
B757	3/5.5	3/4	3/4	3/3
Heavy	3/5.5	3/4	3/4	3/3

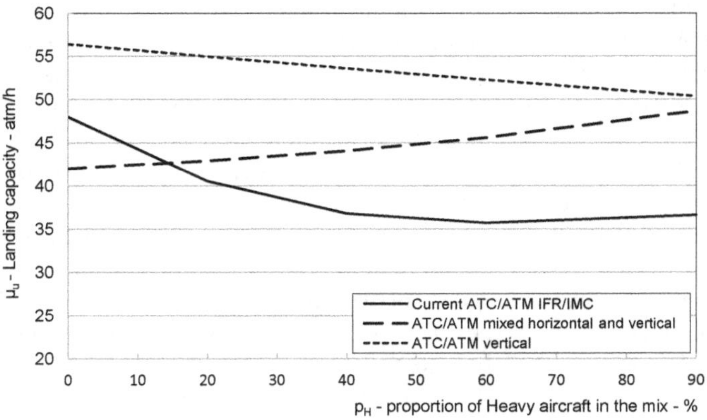

Figure 2.33 Dependence of the runway landing capacity on the proportion of heavy aircraft in the mix and different types of the ATC separation rules (Janić, 2008).

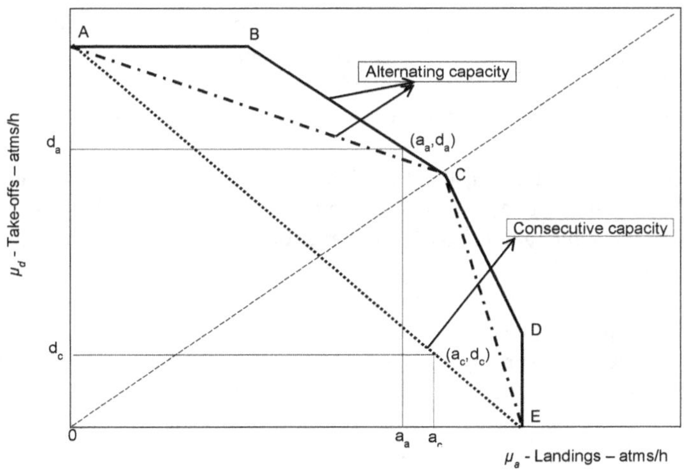

Figure 2.34 Some generic shapes of the runway "capacity envelope" (Janić, 2013; Newell, 1979).

range of proportion of Heavy aircraft in the mix and when the ATC/ATM mixed, and the ATC/ATM horizontal-distance-based separation rules are applied. In addition, the above-mentioned interdependency of the landing and take-off "ultimate" capacity of a given runway is usually represented by the "capacity envelope", as shown in Figure 2.34.

As can be seen, the heavy line *ABCDE* represents the various (practically all) cases of trading-off between the landing and take-off capacity, indicating their possible interdependence. For example, the point *A* of the "envelope" *ABCDE* indicates the maximum take-off capacity and the minimal-zero-landing capacity. The point *E* indicates the opposite case, i.e., the maximum landing capacity and the minimal-zero-take-off capacity. The segment *AB* indicates that, despite the runway operating at the maximum take-off capacity, some landings are still possible without interrupting these take-offs. However, with more landings, the trading-off between landings and the take-offs also increases in terms of the larger number of landings and the smaller number of take-offs (segment *BC* of the "envelope" *ABCDE*). At the point *C*, the number of landings and take-offs is equal. A further increase in the number of landings decreases the number of take-offs, as shown by the segment *CD*. Finally, at the maximum landing capacity, the limited number of take-offs is still possible. In any case, the total runway capacity is equal to the sum of the landings a_a and take-offs d_a. In addition, possible influence of the structure of aircraft wake-vortex categories and the actual degree of possibility for trading-off between landing and the take-offs is indicated by the dotted line *ACE*. Also, the "capacity envelope" *AE* represents the runway capacity when the consecutive tactic is applied, i.e., when all take-offs are carried out before or after all landings. As mentioned above, this usually happens at the hub airports where the incumbent airline(s) schedules the "waves" of successive landings followed by a "wave" of successive take-offs during a relatively short period of time (1 h, ½ h, ¼ h). Like the alternating tactic, the total capacity is the sum of the landings a_c and the take-offs d_c. In general, the shape of the "capacity envelope" for a given runway influenced by the applied tactics of carrying out landings and take-offs depends on their demand during a given period of time. Anyway, in any consideration for the relatively longer period of time (for example, several successive hours during the day), the landing and take-off capacity become equal since, generally, the number of departures cannot be greater than the number of arrivals. The analogous reasoning applies to estimating the capacity of the multi-runway system when the "capacity envelope" can also be used. Figure 2.35 shows such "capacity envelope" for NY JFK airport (U.S.), which operates two pairs of parallel runways, of which one crosses the other two (FAA, 2004; 2014a; Janić, 2013).

As can be seen, in general, the VFR/VMC capacity is greater than the IFR/IMC capacity by about 30–40%. Also, both "capacity envelopes" show the possible trading-off between landings and take-offs, which, as mentioned above, depends on their relative structure of demand during the given period of time (usually 1 h).

Figure 2.36 shows the relationship between the noise of an individual event and the runway landing capacity constrained by the given noise quota/cap.

As can be seen, the runway landing capacity decreases more than proportionally with an increase in the average noise by an individual noise event–landing. For

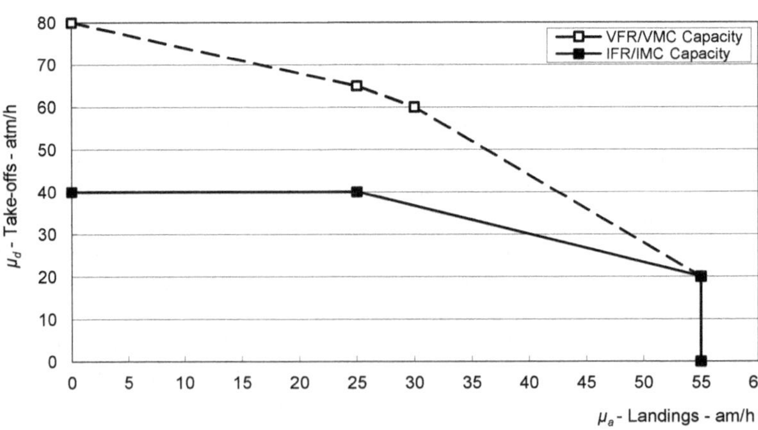

Figure 2.35 "Capacity envelope" of the runway system—Case of NY JFK airport (U.S.) (FAA, 2004; 2014a; Janić, 2013).

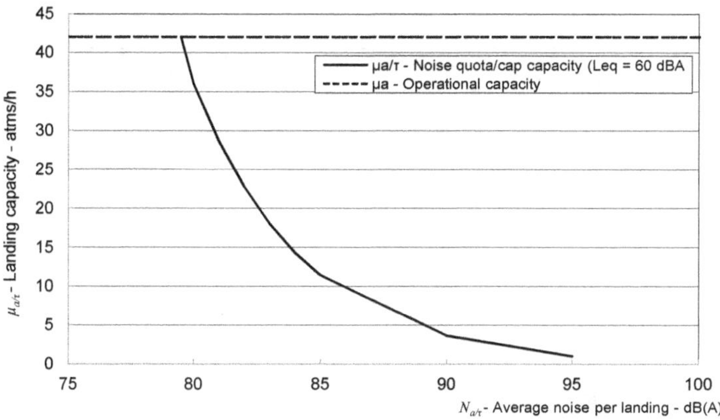

Figure 2.36 Relationships between the airport runway landing capacity, noise quota, and the average noise per an individual noise event–landing (Janić, 2007a).

example, if it is about 90–95 dBA, only a few landings can be carried out within the noise quota/cap of 60 dBA. However, when the average noise per event is 80 dBA, the number of landings within the given noise quota of 60 dBA will be about 36 atm/h, which is close to the runway operational landing capacity of 42 atm/h (The medium-sized aircraft of Category 3 generate certificated noise of about 80 dBA during landing and 74 dBA during take-offs). The noise quota may also act as a real constraint to the airport runway capacity under the specific circumstances, such as, for example, imposition of the severe night limitations or a complete night-flight ban (Ashford and Wright, 1992; Janić, 2007a; Smith, 1993).

Figure 2.37 shows the relationship between the average delay per landing and the runway "practical" landing capacity estimated by Eq. 2.14a.

As expected, under conditions of the completely stochastic aircraft final approach and landing, the "practical" can come closer to the "ultimate" runway landing capacity by increasing the average delay, although at a decreasing rate.

b) Taxiway system

The "ultimate" capacity of the taxiway system at a given airport can be expressed by the maximum number of taxiing aircraft that can pass through the "bottleneck" segment as a link and/or intersection as a node of the taxiway connecting the runway(s) and apron/gate complex. This capacity is influenced by the aircraft distance separation and taxiing speed. In the given context, the models of this capacity and their application are not particularly considered. Some useful material can be found in the literature (Horonjeff and McKelvey, 1994).

c) Apron/gate complex

i) Structure of the models

The "ultimate" capacity of the apron/gate complex can be expressed in two forms. The first is the "static" capacity, referring to the number of available aircraft stands/gates for parking particular aircraft categories. The other is the "dynamic" capacity, which can be expressed by the maximum number of accommodated aircraft at the apron/gate complex

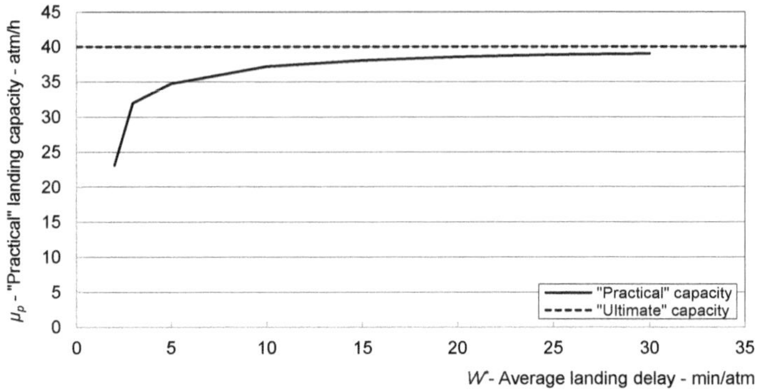

Figure 2.37 Relationship between the average landing delay and the runway "practical" landing capacity.

during a given period of time under conditions of constant demand for service. In general, this "dynamic" capacity depends on the apron/gate complex "static" capacity (i.e., the number of available aircraft parking gates/stands), structure of the aircraft fleet, flight type (domestic, international, originating, terminating, and transit) and the corresponding service time of particular aircraft categories/flights. Consequently, under conditions of the exclusive use of particular gates/stands by the particular aircraft categories and flights, the "ultimate" "dynamic" capacity of the given apron/gate complex can be estimated as follows (Horonjeff and McKelvey, 1994; Janić, 2000).

$$\mu_{apg} = \sum_{i=1}^{N} \sum_{j=1}^{M} \left[\frac{U_{ij} \cdot N_{ij}}{\tau_{ij} + \tau_{g/ij}} \right] \tag{2.15a}$$

where

N, M is the number of different types of the apron/gate parking stands and corresponding aircraft/flights, respectively;

U_{ij} is the utilization of apron/gate parking stand *(i)* by the aircraft/flight of category *(j)*;

N_{ij} is the number of apron/gate parking stand *(i)*, which can be used by the aircraft/flight category *(j)*;

τ_{ij} is the apron/gate stand *(i)* separation time when used by the aircraft /flight category *(j)* (min); and

$\tau_{g/ij}$ is the expected occupancy time of the apron/gate stand *(i)* by the aircraft/flight category *(j)* (min).

When different aircraft types cannot use all available gates, the mix of available gates and aircraft using them are not always properly matched. It such case, it is necessary to compute the capacity for each type of parking stand/gate and then to calculate the overall capacity of the apron/gate complex. In this case, the "ultimate" capacity of an apron/gate complex can be estimated as follows (Ashford and Wright, 1992; Horonjeff and McKelvey, 1994; Janić, 2000):

$$\mu_{apg} \cong min_i \; [N_i/(\tau_i \cdot p_i)] \tag{2.15b}$$

where

N_i is the number of gates that can accommodate the aircraft of class *(i)*;
τ_i is the average gate occupancy time by the aircraft class *(i)* (min/gate);
p_i is the proportion of the aircraft class *(i)* in the traffic mix requesting service at the apron/gate complex.

Utilization of the particular parking gates/stands (U_{ij}) in Eq. 2.15a is usually measured by the proportion of time the specified gates/stands are occupied by the aircraft during the specified period of time (for example during the period of the day). The average gate/stand occupancy time $(\tau_{g/ij})$ is defined as the time between the aircraft's wheel stop at the stand/gate and the time of the aircraft's moving out from the stand/gate. This time, also known as aircraft turnaround time, is primarily influenced by the required aircraft service, type of the aircraft/flights, and the airline schedule. The aircraft service is specified for each aircraft type for the given conditions. Generally, it increases with an increase in aircraft size. In some cases, for a given aircraft type, some service activities are dropped off, thus shortening this time. However, in some other cases, the airline schedule may ultimately extend this time. This is often witnessed at the hub airports where the airlines operate their hub-and-spoke networks with the "waves" of incoming and outgoing flights. These networks may often consist of few sub networks, each characterized by the specific categories of flights, such as short-, medium-haul, and long-haul, complementing each other. Under such circumstances, the turnaround times of the aircraft serving particular sub-networks become dependent on each other (Janić, 2013). In addition, the following conditions generally need to be fulfilled:

$$U_{min} \cdot N_{min} \geq \mu_a \cdot \tau_{s/mmx} \text{ and } N_{min} \geq (\mu_a \cdot \tau_{s/mmx})/U_{min} \qquad (2.15c)$$

where all symbols are analogous to those in Eq. 2.15(a, b).

In addition to the above-mentioned analytical models, the simulation models for estimating the "ultimate" capacity of the apron/gate complex have also been developed. They are based on the deterministic queuing processes occurring at the apron/gate complex. For this purpose, two processes have been recorded during the specified time period: the process of aircraft entering the apron gate after landing and the process of aircraft leaving the apron gate after being served. Both kinds of processes could be described by the time-discrete step-shaped cumulative functions denoted by $(A_g(t))$ and $(D_g(t))$, respectively. Actually, the $(A_g(t))$ and $(D_g(t))$ represent the cumulative number of aircraft entering and leaving the given apron/gate complex by time *(t)*, as shown in Figure 2.38 (Janić, 2000; Newell, 1982).

As can be seen, the value of both curves increases by one whenever an aircraft enters or leaves the system. The vertical difference between two curves at any time *(t)* represents the number of aircraft simultaneously occupying the apron/gate complex. At time *(t)*, it is equal to:

$$n_g(t) = \max[0; A_g(t) - D_g(t)] \qquad (2.16a)$$

The "ultimate" capacity of the apron/gate complex is equal to:

$$\mu_{apg} = max[n_g(t)] \; for \; t \in \tau \qquad (2.16b)$$

where

τ is the length of simulation interval (h).

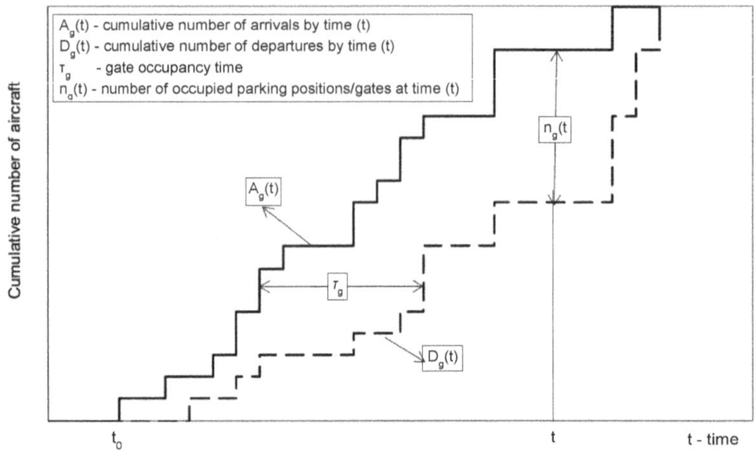

Figure 2.38 Simplified scheme of the aircraft congestion at the apron/gate complex (deterministic case) (Janić, 2000).

Equation 2.16a indicates that opposite to the analytical model (Eq. 2.15(a–c)), the simulation model does not explicitly incorporate the number of available parking stands/gates at the apron/gate complex and their utilisation factor. However, with slight modifications of this model, these characteristics may easily be taken into account (Janić, 2000; Newell, 1982).

ii) Application of the models

- Inputs

The model for estimating the "ultimate" capacity of apron/gate complex in Eq. 2.15(a, b) is illustrated using the case of 7 U.S. airports. The relevant inputs are given in Table 2.13(a, b).

- Results

By using the inputs in Table 2.15(a, b) in the application of the above-mentioned Eq. 2.15a, the "ultimate" capacity of the apron/gate complex at the given U.S. airports is calculated. Then, the relationship between this capacity and the number of gates is established and given in Figure 2.39.

As can be seen, the "ultimate" capacity of the apron/gate complex increases more than proportionally with an increase in the number of gates per complex. It should be mentioned that the capacity is based on the gate use by two atms, including an arrival and a departure of one aircraft.

2.3.3.4 Landside capacity

a) General

The capacity of the airport landside area includes the capacity of its main components, such as the landside access modes and their systems, the passenger and freight/cargo terminal complex, and interfaces between them enabling passing of users, such as air passengers, airport employees, senders and greeters, visitors, and air cargo personnel,

Table 2.15 Characteristics of the apron/gate complex and its use—Case of the selected U.S. airports (Bishop, 2012; FAA, 2014a).

a) Fleet structure and the average gate occupancy time				
Aircraft/ Airport	**Heavy**	**B757**	**Large**	**Small**
	Proportion(%)/Average occupancy time (min)			
BOS	4.8/125	9.4/110	69.7/70	16.1/60
JFK	26.2/230	9.7/170	63.0/85	1.1/65
LAX	16.7/180	11.5/115	61.5/70	10.3/50
ATL	6.1/120	12.3/100	80.0/70	1.6/65
DWF	4.6/200	6.5/120	87.0/65	1.9/60
LGA	-	6.2/90	92.5/60	1.2/45
DCA	-	2.6/65	96.7/60	0.7/50

b) Number of gates, the average gate occupancy time, number of turnarounds per gate/day, and gate utilization				
Airport	**Number of gates**	**Average occupancy time**[1] (min)	**Average number of aircraft turnarounds** (–/day)	**Average gate utilization** (%)/16 h/day
BOS	102	74.8	5.9	0.460
JFK	128	131.2	3.9	0.533
LAX	132	91.5	6.6	0.629
ATL	195	78.6	6.8	0.577
DWF	182	74.7	5.	0.459
LGA	72	61.6	6.9	0.443
DCA	46	6.2	8.4	0.553

[1] This is assumed to contain also the time for entry and exit the gate.

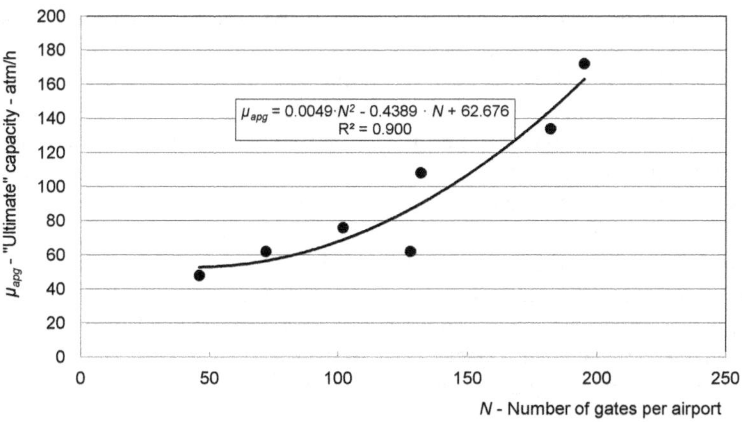

Figure 2.39 Relationship between the "ultimate" capacity and the number of gates of the apron/gate complex: Case of 7 U.S. airports (Bishop, 2012; FAA, 2014a).

between them. The particular above-mentioned components consist of the sub-components operating as the processors, reservoirs, and links. In the airport landside access modes and their systems, processors are the vehicles operating along the road and rail links connecting a given airport and its catchment area. The reservoirs are parking lots, consisting of the holding bays and the other parking areas for particular vehicles. Links are the rail stations, curbs, and freight/cargo docks. At the passenger and freight/cargo terminals, the processors are the servers of customers, reservoirs are the waiting areas for customers, and the links are the areas connecting processors and reservoirs. In particular, the processors in the passenger terminals enable the air passengers to enter and leave the aircraft. Reservoirs provide awaiting area for the next processing phase. Links designated as the passageways, walkways and escalators connect particular processors and reservoirs in the same and/or different terminals.

The capacity of the processors, reservoirs, and links at the airport landside access modes and their systems, passenger terminals, and interfaces can be “static” and “dynamic” “ultimate” and “practical”. The “static” capacity of components with storage capabilities in the airport landside access modes and their systems are the number, length and width of lanes of the roads and tracks along the rail lines, stalls in the car and bus parking lots in front of the given passenger terminal(s), the vehicle slots on the curb areas, the curb areas themselves, etc. In addition, the vehicles operated by particular modes and systems have a storage capability usually expressed by the maximum number of simultaneously (and safely) accommodated passengers. The dynamic “ultimate” capacity of particular components of the airport landside modes and their systems is defined by the maximum number of air passengers that can be served during a given time period under conditions of constant demand for service. This includes the capacity of lines, stations, and vehicles. In such context, the capacity of lines is defined by the maximum number of vehicles (transport units) served at any point of a given line during a given period of time under conditions of constant demand for service. This capacity depends on the vehicle departure frequency, the average travel speed along the line, and the time needed to serve the vehicles at the specified stations (Janić, 2019; Vuchic, 2007). The “static ultimate” capacity of particular components of the passenger terminal(s) can be expressed by the maximum number of passengers that can be simultaneously accommodated in a given area while providing the minimum space standards assigned to each of them. These areas are generally in front of the processors and in the reservoirs. These standards are specified as the planning and design of the airport terminals (Ashford et al., 1997; TRB, 2010; 2011; USDT, 2018; http://go.updates.iata.org/l/123902/2017-03-07/7lrghg/). The “dynamic ultimate” capacity is estimated for the processors and links. For processors, it is expressed by the maximum number of air passengers that can be served during a given time period under conditions of constant demand for service. For links, this capacity is expressed by the maximum number of passengers passing through during a given time period under constant demand for service.

The “static and dynamic practical” capacity of the particular airport landside components reflects their capabilities to serve a given number of customers while imposing the average delay, which does not exceed a given level. In most cases, this capacity reflects the quality of service provided to users under given conditions, which is elaborated in the scope of the airport quality of services.

b) Landside access modes and their systems

The road-based systems connecting airports and their catchment areas are car or van, and bus systems. The main components of these systems in the given context are infrastructure, vehicles, and transport service networks for bus systems.

i) Road-based mode and its systems

- Car or van

The cars and vans used either individually/privately or carrying out the taxi services can be of different size and seating capacity, usually of 4–5 and 6–8 seats/veh, respectively. The infrastructure that these vehicles use includes the road and/or highway networks and the parking spaces. The roads and highway networks are characterized by their "ultimate" and "practical" traffic and transport capacity. The "ultimate" capacity is determined under conditions of constant demand for service, characterized by the vehicles continuously passing through the "reference location" for their counting during a given period of time. The "practical" capacity is determined under conditions when an average delay is imposed on each vehicle passing through the "reference location" during a given period of time. The traffic capacity relates to the vehicles. The transport capacity relates to the users–passengers and/or airport and other aviation employees transported under given conditions. In any case, the "given period of time" is usually 1/4, 1/2 and/or 1 hour. The basic "ultimate" capacity of the single road lane is 2000–2200 veh/h (veh – vehicles; h – hour) (Teodorović and Janić, 2016).

The car or van parking at the airports is possible in the short-term spaces in front of the airport passenger terminals, enabling dropping-off or picking up air passengers, or the long-term ones located further from the terminals, enabling parking for the longer period of time—several hours to several days. On the catchment area side, these are the parking space at or near the individual's home, public garages, etc. These vehicles can be parked in these areas according to parallel, perpendicular, and/or angled parking schemes. *The parallel parking scheme* implies arranging the vehicles in a line parallel to the curb of the street or the longer side of the rectangular parking area. For example, if the "footprint" of each vehicle is a rectangle of size of about 5.50 · 2.50 m (13.75 m^2) and the width of access road is 3.00 m, the total size of parking area can be estimated as follows: $PA = 2.50 \cdot (n \cdot 5.50)\ m^2$, where *(n)* is the number of vehicles successively parked behind each other.

The perpendicular parking scheme, also known as bay parking,implies parking vehicles side-by-side at aright angle to an aisle and/or curb. The vehicle "footprint" varies from 10.125 m^2 to 12.15 m^2. The total size of parking area can be estimated as follows: $PA = 9.90 \cdot (n \cdot 2.25)\ m^2$ or $PA = 9.90 \cdot (n \cdot 2.70)\ m^2$, where *(n)* is the number of vehicles parked side-by-side.

The angle parking scheme implies that vehicles are parked at an angle (usually 30°, 45°, or 60°) compared to the curb and direction of approach. The vehicle "footprint" is about 11.25 m^2. The total size of the parking area can be estimated as follows: $PA \approx 5.00 \cdot (n \cdot 3.50)\ m^2$ where *(n)* is the number of vehicles parked side-by-side at the specified angle. In the U.S., for example, the vehicle "footprint" in the parallel parking scheme is typically 2.8 m width, and 6.1 m length (area: 17.08 m^2). The vehicle "footprint" in the angled and perpendicular parking schemes is usually 2.3–2.7 m width, and 4.9–6.1 m length (area: 11.27–16.47 m^2) (the recommended length for the vehicle "footprint" at the standard perpendicular parking scheme is 5.8 m). In the U.K. (United Kingdom), the typical vehicle "footprint" is 2.4 m width, and 4.8 m length (area: 11.52 m^2). In Hong Kong, the minimum vehicle "footprint" independently of the parking scheme is 2.5 m width, and 5.0 m length (area: 12.50 m^2). Similar methods will be used for the future AVs (Automated Vehicle(s)), which will look like their conventional counterparts with space/seating capacities of 4–5 (cars) and 8–10 (vans) seats/vehicle. They will likely be powered by conventional and alternative fuels, and/or electric energy. In this case, the above-mentioned size of parking spaces in combination of the vehicles' time of staying there enables the estimation of their "ultimate" traffic capacities (Janić, 2019).

- Bus system

Bus systems operating between airports and their catchment areas use buses of different sizes, i.e., seating capacity, which can be typically be up to 50 seats in conventional and 60–80 seats in double decker models. The buses operate along the same roads and highways as their car and van counterparts by sharing the same lanes or using the dedicated ones. In addition, they use the bus stops along these roads and the end stations/terminals in front of the airport passenger terminal(s) and in the centres of cities/urban agglomerations of their catchment areas.

The bus systems usually operate the transport service networks, consisting of lines and routes that cover the most spatially distributed air passenger demand and airport employees as their main users. In some cases, the bus networks can be established between a few airports serving the same catchment area. The examples are three London airports (Heathrow, Gatwick, and Stansted) and three New York airports (La Guardia, Newark, and JFK (John F. Kennedy). Figure 2.40 shows the simplified scheme of such networks serving the access to and between London airports.

As can be seen, this bus network connects two London airports (Heathrow and Gatwick) through particular cities in their catchment area (Janić, 2019).

In this case, the capacity of bus lines/routes and of the entire network is often expressed by the maximum number of offered seats during a given unit of time (usually 1 h). This is expressed as the product of the number of departures and the space capacity per departure along a given line/route, set of lines/routes, and the entire network. Figure 2.41 shows an example of the relationship between the bus system capacity and the space capacity of individual buses (Wright, 2005; Wright and Hook, 2007).

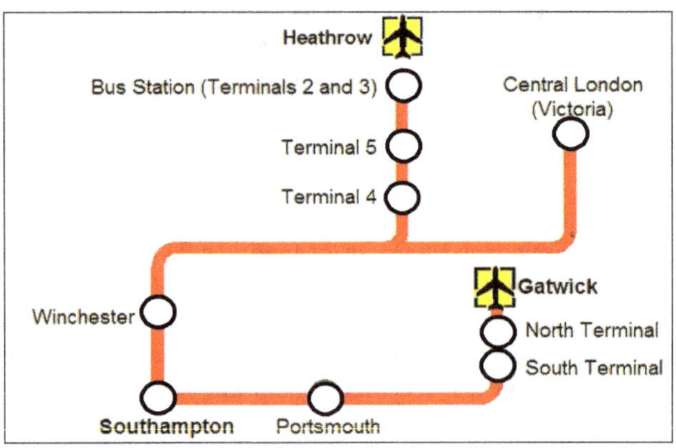

Figure 2.40 Scheme of the bus network serving cities in the multi-airport catchment area—case of London airports (Janić, 2019; https://www.google.nl/search?q=heathrow+central+bus+station+map/).

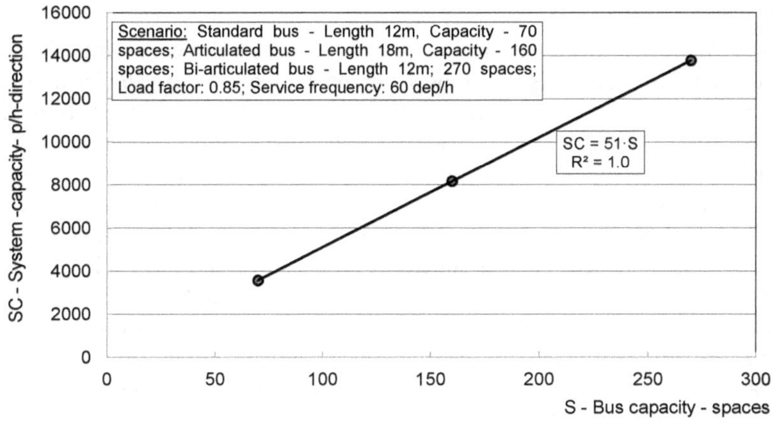

Figure 2.41 Relationship between the bus system capacity and the individual bus capacity under the given operating scenario (p – passenger; h – hour) (Janić, 2019; Wright, 2005; Hook, 2007).

As can be seen, with an increase in the capacity of individual vehicles/buses, the transport capacity along the corridor will also proportionally increase.

ii) Rail-based mode and its systems

Similar to their road-based mode counterparts, the rail-based mode and its systems connecting airports with their catchment areas consist of infrastructure (rail lines, stops, and stations/terminals), vehicles-train sets, and the transport service networks. These components are characterized by the corresponding "ultimate" and "practical" capacities, which can be defined and understood similarly to those in the road-based mode and its systems. In general, the systems within the rail-based mode include the streetcar/tramway & LRT (Light Rail Transit), subway/metro, regional/intercity conventional rail, HSR (High Speed Rail), and TRM (TransRapid Maglev) system.

- Streetcar/Tramway & LRT (Light Rail Transit) System

The streetcar/tramway and LRT systems use the rolling stock-vehicles/trains, consisting of a single or few coupled transit units-cars. The most common composition at the streetcar/tramway vehicles/trains is 1–3 units-cars/vehicle, each with the width 2.30–2.65 m, length of 14–23 m and capacity of 100–180 spaces (22–40 seats). In LRT systems, the vehicle/train compositions usually consist of 1–4 units-cars/vehicle, each with the length of 14–30 m and capacity of 110–250 spaces (25–80 seats) (Vuchic, 2007). As in the case of bus systems, the fleet size of the streetcar/tramway system operating along the line connecting an airport and its catchment area generally depends on the transport service frequency and the average turnaround time of the vehicles operating there. The transport service frequency depends on the volume and time pattern of user/passenger demand during particular hours during the day, the average vehicle size and the expected utilization of the vehicle seating/space capacity, i.e., load factor. The average turnaround time is defined as the time between starting from and returning to the first station/terminal of the line. This time depends on the length of line, average vehicle's travel speed, and the number of and time taken at stops/stations along the line, including the time spent at the first/last stations/terminals.

- Subway/Metro System

The subway/metro systems use the rolling stock—vehicles/trains consisting of a single or a few coupled transit units-cars. The most common composition of vehicles/trains is 3–4 and also 5 units-cars/train-vehicle, each with a width of 2.5–3.2 m, length 14.6–22.9 m, and capacity of 125–280 spaces (47–50 and 62 seats). For example, the Amsterdam 4-axle subway/metro vehicle-train has a length of 18.7 m, a capacity of 179 spaces (50 seats), and a maximum operating speed of 80 km/h. By coupling more units-cars, the capacity of a single vehicle-train can also be increased. For example, the 6-unit subway/metro train set operating in Athens (as mentioned above) has a width 2.65 m, a length 106 m, a capacity of 844 spaces (188 seats), and a maximum speed of 80 km/h. The infrastructure capacity of subway/metro systems can also be the "ultimate" and "practical" traffic and transport and determined and expressed similarly to its streetcar/tramway and LRT system counterparts—for the segments and/or the entire lines and stations along them including, the first/last ones/terminals. The fleet size of the subway/metro system serving an airport access can be determined similarly to that of the streetcar/tramway and LRT system. For example, at the above-mentioned Athens' Metro Line M3 with the transport service frequency between the city centre (station Syntagma) of 2 veh-dep/h, each taking about 40 minutes to the airport, the required fleet will be: 2 veh-dep/h $\cdot$ (2 $\cdot$ 40 min/dep)/ 60 $\approx$ 3 veh (Janic, 2019; Vuchic, 2007).

- Regional/Intercity Rail

The conventional/intercity rail systems provide transport services to their users/ passengers, generally at the maximum speeds of up to 160 km/h. The regional rail systems usually operate between towns and cities in addition to providing accessibility to the airports included in the regional rail networks. They have less frequent stops than their streetcar/tramway, LRT, and subway/metro counterparts, including those along the lines connecting the airports. In the intercity rail systems referring to the express passenger train services covering longer distances than the regional trains, the airports included in the long-haul intercity rail network can be one of the rare stations between any pair of urban/city agglomerations. For example, these systems enable the landside accessibility to 22 European airports complementing in some way the above-mentioned streetcar/ tramway, LRT, and subway/metro systems (Janić, 2019; https://en.wikipedia.org/wiki/ Airport rail_link#Europe/). Similarly to the streetcar/tramway, LRT, and subway/metro systems, the main components of the regional/intercity conventional rail systems, whose capacity is relevant in the given context, are the vehicles, rail lines constituting the networks spreading between particular urban/city and sub-urban areas, and stations along the lines and stations/terminals at their ends.

The rolling stock-vehicles/trains used for the regional/intercity conventional rail transport services, including those meant for airport access, have different seating capacities. For example, the trains performing the intercity services connecting Amsterdam Schiphol airport and the rest of the country operated by NS (Nederlandse Spoorwegen) (Dutch Railways) consist of four or six units-cars, with capacities of 391 and 571 spaces, respectively. These trains operate at the speed of 140 km/h (https://www.ns.nl/). The infrastructure capacity of the regional/intercity conventional rail systems can be traffic and transport and can be expressed similarly to the streetcar/ tramway, LRT and the subway/metro systems. The fleet size of these systems depends on the same factors as with their streetcar/tramway and subway/metro counterparts.

For example, the transport service frequency between Amsterdam and Schiphol airport by the intercity Dutch VIRM trains has been 4 dep/h. The average travel time is 14 min implying that the train's turnaround time is 2·14 min = 28 min. Consequently, the required fleet size is: 4 veh-dep/h · [(28 min/dep)/(60 min/h)] ≈ 2 veh (Janić, 2019).

- HSR (High Speed Rail)

The HSR (High Speed Rail) systems have been developing worldwide (Europe, Far East-Asia, and USA (United States of America)) as rather innovative systems within the railway-based transport mode, particularly as compared to their above-mentioned conventional counterparts. The main components in the given context are again the rolling stock and infrastructure—rail lines, stops, stations/terminals, and networks. In the given context, the HSR rolling stock/trains are generally distinguished from the conventional ones by their optimized aerodynamic shape, fixed composition and bi-directional set, and compatibility with the existing and dedicated infrastructure (track and loading gauge, platforms, catenary, etc.) (UIC, 2010). In general, the length of HS train sets is about 200–380/390 and the capacity is 320–935 seats/set. Their maximum design speed is 250–350 km/h. For example, the maximum operating/cruising speed of TGV (Train à Grande Vitesse) is 320 km/h. The Japanese Shinkansen and German ICE (Inter-City-Express) trains currently operate at a maximum speed of about 300 km/h. The forthcoming more advanced European AGV (Automotice Grande Vitesse) and Japanese Fastech 360Z trains are expected to operate at an average speed of 350 km/h and 360 km/h, respectively (Janić, 2016; 2019).

The size of rolling stock of the HSR systems connecting airports with their respective catchment areas can be determined by assuming that the HS rail line connects large urban agglomerations and the large airport(s) in between them. An example is the Thalys HS train system operating transport services between Paris (France), Brussels (Belgium), and Amsterdam (The Netherlands) with a transport service frequency of 1 dep/h. Figure 2.42 shows the simplified network scheme.

The trains also stop and pass through Amsterdam Schiphol airport (The Netherlands) at the same transport service frequency –1 arr/dep/h. As such, the system generally serves several categories of passengers: rail with origin-destination Paris-Brussels-Amsterdam,

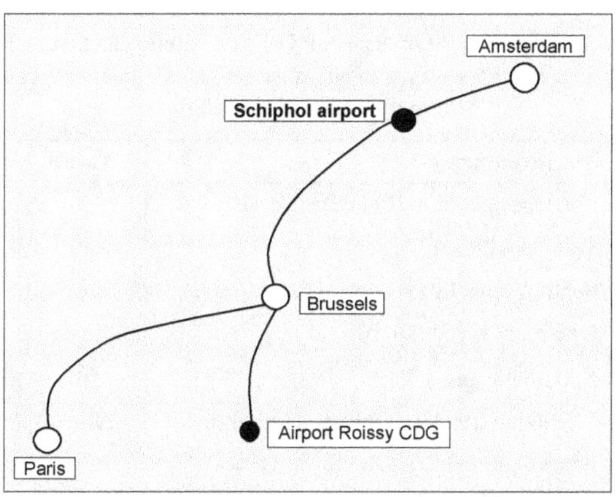

Figure 2.42 Simplified scheme of HSR connecting airports—Case of Thalys Paris-Amsterdam line (https://www.ertmssolutions.com/ertmscamcorder-chosen-by-thalys-for-etcs-compliance-field-testing/).

air originating at Paris, Brussels, and Amsterdam using Schiphol airport, and those taken over from airlines operating between Paris and Amsterdam. The one-way travel time between Paris and Amsterdam takes 3 h and 23 min (i.e., 3.38 h) and the turnaround time 6 h and 46 min (i.e., 6.77 h), including the stops at both end stations/terminals. Under such conditions, the required HS train fleet size to operate along the entire Paris-Amsterdam line is equal to 1 dep/h · 6.77 h/dep ≈ 7 vehicles (train sets).

The HSR system infrastructure components, such as stations/terminals, lines/routes, and networks, are also characterized by their "ultimate" and "practical" capacities. These are influenced by the operational rules and procedures providing safe separation of trains while operating along the lines and at the stations in the same and/or different directions. These rules specify the minimum time separation between occupying the same section of the line(s) and/or of the station(s), thus mainly influencing their corresponding capacities.

- TRM (TransRapid Maglev) system

The TRM (TransRapid MAGLEV (MAGnetic LEVitation)) as an additional HS (High Speed) system is based on Herman Kemper's idea of magnetic levitation, dating from the 1930s. The magnetic levitation enables the suspension, guidance, and propelling of MAGLEV vehicles by magnets rather than by the mechanical wheels, axles, and bearings used by the HS (High Speed) wheel/rail vehicles. Two forces—lift and thrust or propulsion—both created by magnets enable the operation of TRM vehicles/trains. Table 2.16 gives the main relevant characteristics of TRM08 vehicles/trains operating between Shanghai Pudong International airport and the city of Shanghai (China).

In addition, Figure 2.43 shows the simplified scheme of the line and some operational characteristics of the above-mentioned TRM system.

The size of the TRM system fleet or rolling stock is influenced by similar factors as that of its HSR counterpart. Consequently, based on the transport service frequency of 4 veh-dep/h and the turnaround time of 2·7.33 min = 14.66 min/dep, the required fleet is: 4 veh-dep/h·14.66 (min/dep)/60 min ≈ 1 veh (train set). The traffic capacity of a given TRM system's line mainly depends on the separation rules and procedures between successive TRM vehicles moving in the same direction and their average operating speed. The transport capacity of a given line is influenced by its traffic capacity and the TRM vehicle/train seating capacity and their utilization (load factor) (Janić, 2019).

Table 2.16 Technical/technological performances of TRM 08 rolling stock/vehicles (Janić, 2014; 2019; Powell and Danby, 2013; http://www.maglevboard.net/en/facts/systems-overview/transrapid-maglev/transrapid-maglev-shanghai).

Performances	Value
Carriages per train (linked)	2
Total length (m)[1)]	50
Seating capacity (average) (spaces/seats)	200
Operating speed (km/h)	400–450
Maximum speed (km/h)	500
Maximum acceleration/deceleration (m/s^2)	0.8–1.5
Lateral tilting angle (^{0})	12–15
Levitation air gap (mm)	8

[1)] Meter.

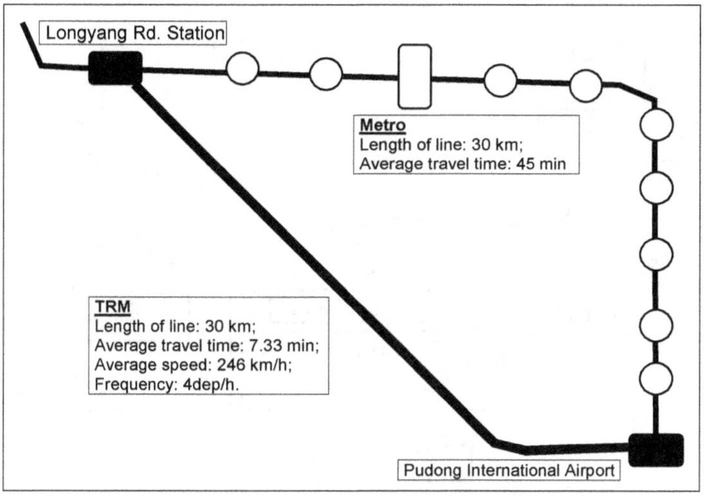

Figure 2.43 Simplified scheme of the MAGLEV system connecting Pudong International airport and Longyang Rd. Station (Shanghai, China) (https://www.travelchinaguide.com/cityguides/shanghai/getting-around.htm/).

c) Passenger terminal complex

The "static ultimate" capacity of the airport passenger terminal(s) is estimated for reservoirs, i.e., the areas where passengers generally wait for processing by the particular processors. As such, this capacity can be expressed by the maximum number of passengers simultaneously being in a given area while having a certain minimum amount of space. This minimum space will be considered in the scope of quality of service in the passenger terminals. Table 2.17 gives the total area of largest passenger terminals, the corresponding number of passengers, and the average "static ultimate" capacity at five airports (https://www.businesstraveller.com/airlines/2013/08/07/five-of-the-worlds-largest-airport-terminals/).

The average "static ultimate" capacity in Table 2.17 is estimated by assuming that the passengers in each terminal spend about 1 h and that the airports operate 18 h/day.

Table 2.17 Size of the largest passenger terminals—Case of five airports (Period: 2018) (https://www.businesstraveller.com/airlines/2013/08/07/five-of-the-worlds-largest-airport-terminals/).

Airport	**Area of passenger terminal** ($10^3 m^2$)	**Number of passengers** (10^6/year)	**Average "static ultimate" capacity** (m^2/pax-h)
London Heathrow	632.064	80.126	51.8
Hong Kong International	710.000	74.517	62.6
Tokyo Narita	821.000	87.132	61.9
Beijing Capital International	1383.580	100.983	90.1
Dubai International	1972.474	89.188	145.3

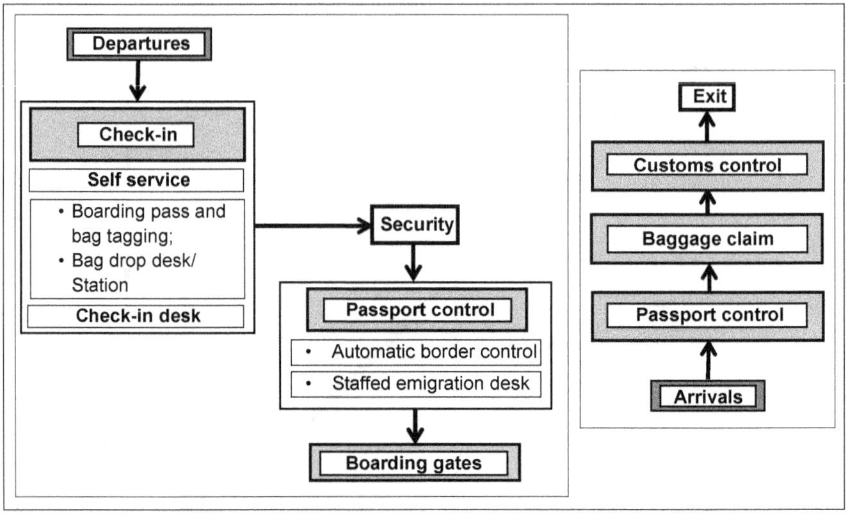

Figure 2.44 The main processes of handling passengers in the airport passenger terminal.

The "dynamic ultimate" capacity of processors in the passenger terminal(s) is expressed by the maximum number of passengers that can be served during a given period of time under conditions of constant demand of service. For both departing and arriving passengers, these processors are positioned in serial order, as shown in Figure 2.44.

With departure passengers, the check-in process includes obtaining a boarding pass and delivering checked baggage for transport to the baggage compartment of the aircraft. With new developing technologies, many passengers have started to use some form of electronic check-in, including online and self-serve kiosks at an airport or at remote locations. The security process aims to reduce potential safety threats with maximum efficiency. In general, the security includes screening of both passengers and their belongings. The passport control process implies checking the identity and entitlement of passengers boarding the international flights. The boarding gates as processors enable passengers to directly board the aircraft connected to the gates by air bridges or the buses transferring them to the remote apron/gate parking stands. With arrival passengers, the passport control is carried out to check if they are allowed to enter the country. The "ultimate" capacity of processing baggage is influenced by the time of delivering it from the arrived aircraft to the baggage claim area. Then, it is the capacity of feeding bags onto the carousel, the number of bags presented on a carousel, and number of passengers in the active baggage claim area. The customs desks enable checking of the bag contents at a certain rate, i.e., "dynamic ultimate" capacity. Table 2.18 gives some examples of the "dynamic ultimate" capacity based on the processing rates of particular processors in the terminal(s) (Horonjeff and Mc Kelvey, 1994; Janić, 2013; TRB, 2009).

The above-mentioned processing times and capacities relate to the single processor. At most airports, the number of these facilities is greater depending on the number and type of flights (domestic, international, charter), total passengers, and specified waiting time, which have to be simultaneously processed during the specified period of time, i.e., 1.5–2.5 h before their schedule departure times (Janić, 2013).

Table 2.18 Example of the processing times and processing rates in the passenger terminals (Horonjeff and McKelvey, 1994; Janić, 2013).

Departing passengers	**Processing time** (min/pax)	**"Dynamic ultimate" capacity** (pax/h)
Check-in	2.0–4.0	15–30
Security	0.5–1.0	60–120
Passport control	0.5	120
Boarding gate	0.5	120
Arriving passengers		
Passport control	0.5–1.0	60–120
Baggage claim	2.7–5.5	11–22
Customs control	0.5–1.0	60–120

2.3.3.5 Modelling landside capacity

Modelling of the "static" and "dynamic ultimate" capacity of the airport landside systems are presented for the road-based and the rail-based mode by using analytical models (Janić, 2019; Vuchic, 2007).

a) Road-based mode and its systems

Capacity of the road-based airport landside access systems consists of two components: the capacity of road links (i.e., network) connecting the airport with its catchment area and the capacity of vehicles transporting air passengers and other users to/from a given airport (Janić, 2019; Vuchic, 2007).

i) Car/taxi

The "dynamic ultimate" capacity of a given road can also be traffic and transport. They are expressed by the maximum number of vehicles (and persons) that can pass and be transported, respectively, through a given "reference location" during a given period of time (for example 1 h) under the given ("ideal") traffic conditions and constant demand for service. This traffic capacity can be estimated as follows (Janić, 2000; 2019):

$$\mu_{road/traffic} = n \cdot 3600 \cdot v/(L_v + d) \qquad (2.17a)$$

where

n is the number of lanes of the road in the same direction;

v is the average speed of the saturated vehicle flow (m/s);

l_p is the average spacing between the two closest neighboring vehicles measured by the distance from the front of the leading vehicle to the front of the following vehicle (m);

L_v is the average length of a vehicle in a given traffic flow (m); and

d is the vehicle's stopping distance at the speed v including the driver's reaction time, breaking distance, and a factor of safety (m) ($s_d = v^2/2a^-$, where a^- is deceleration rate (m/s^2)) (m – meter; s – second).

The traffic capacity in Eq. 2.17a has been estimated to be about ($n \cdot 2000$) veh/h at the optimal speed of traffic flow of about 50 km/h and the uniform headways between vehicles, including their length and closest distances in the traffic flow, of 25 meters or 1.80s. The capacity of ($n \cdot 2000$) vehicles/h is often used as the ideal "ultimate dynamic" capacity. If all vehicles have $S = 4$ spaces/seats for air passengers the transport capacity of the given road based on Eq. 2.17a is equal as follows: $\mu_{road/transport} = n \cdot 2000 \cdot S = n \cdot 2000 \cdot 4 = n \cdot 8000$ *seats/h*. The above-mentioned capacities are usually modified regarding prevailing conditions along a given road, such as: the shoulders' width, incidence of the commercial vehicles (buses and heavy trucks), road's gradient, sight distance, and other factors specific to the type of roadway. Such modified transport capacity based on Eq. 2.17a can be estimated as follows (Janić, 2019):

$$\mu_{road/transport} = (2000 \cdot w \cdot f_{hv} \cdot f_p \cdot f_w) \cdot S \tag{2.17b}$$

where

w is the capacity adjustment factor for the width of lanes and shoulder (i.e., the lateral clearance);
f_{hv} is the heavy vehicle factor influencing the ideal capacity;
f_p is the adjustment factor for unfamiliar driver population;
f_w is the factor of weather (in case of rain, it can be 0.85);
S is the available seat capacity of a vehicle (seat/veh).

The other symbols are analogous to those in Eq. 2.17a.

The "static ultimate" capacity of a given road can be traffic and transport. These are expressed by the number of vehicles (and persons onboard) simultaneously occupying the given segment or the entire length of road, i.e., traffic and transport density, while maintaining operations at the reasonable speed. The traffic density is estimated as the ratio between the number of counted vehicles during the period of time *(τ)* and the length of observed road section *(L)* as follows:

$$\mu_{stat/tarffic} = \mu_{road/traffic}/L \tag{2.18a}$$

where

L is the length of a given road segment (km).

From Eq. 2.18a, the "static ultimate" transport capacity is equal as follows:

$$\mu_{stat/transport} = \mu_{road/traffic} \cdot S \tag{2.18b}$$

S is the number of spaces/seats per vehicle for air passengers (spaces/veh).

For example, the "static ultimate" traffic capacity of the given lane expressed by reflecting the traffic flow density is estimated by considering: the free-flow speed along the single lane of: $v_f = 35$ km/h, the total travelling distance: $d = 4.022$ km, and the lane capacity: $\mu = 1600$ (veh/h). It is equal to: $\mu_{stat/traffic} = \mu_{road/traffic} \cdot (d/vf) = 1680 \cdot (4.022/35) \approx \mu_{road/traffic}/L = 184\ (veh)$ or $\approx 48\ (veh/km-h)$. In this case, the "dynamic ultimate" traffic capacity is about $\mu_{road/traffic} = 1680$ (veh/h). When all vehicles have the capacity of $S = 4$ spaces/seats, the corresponding "static ultimate" transport capacity is equal as follows: $\mu_{stat/transport} = \mu_{stat/traffic} \cdot S = 48 \cdot 4 == 192\ spaces/km-h$.

The "static ultimate" capacity of car and van parking areas along the road/highway and at the airport terminals can be generally estimated as follows:

$$\mu_{cv} = N_{cv}/\tau_{cv} \tag{2.18c}$$

where

N_{cv} is the number of car or van parking spaces of the given parking area (-); and
τ_{cv} is the average occupation time of a parking place of the given parking area (min/place).

For example, the given airport has a parking area with N_{cv} = 2000 places for cars and vans. About 50% of them are occupied for about τ_{cv} = 4 h and the rest for about τ_{cv} = 8 h. The capacity of given parking area is equal: $\mu_{cv} = N_{cv}/\tau_{cv} = 2000/(0.5 \cdot 4 + 0.5 \cdot 8)$ = = 333 *veh/h.*

ii) <u>Bus system</u>

The bus systems operate between airports and their catchment areas to/from a given airport either according to their fixed timetables or often adjusted to the airport and particular airline timetables. The vehicle capacity can vary from 6 (minibus) to about 50 seats (classical bus) (Ashford et al., 1997; Janić, 2000; 2019). Such a wide range of capacities and adaptability to the airport schedule make the bus system flexible in adapting to the short-term fluctuations in demand on the hourly, weekly, and/or seasonal basis. The "dynamic ultimate" capacity of a given bus line connecting an airport and its catchment area can be traffic and transport. The former relates to the service frequency scheduled along the line, which can be estimated as follows (Janić, 2019):

$$f_{bus}(\tau) = \tau/\overline{h} \tag{2.19a}$$

where

τ is the time period (τ) in which the service frequencies are scheduled (h); and
$\overline{h}$ is the average time interval between successive bus departures along the line (min; h).

From Eq. 2.19a, the "dynamic ultimate" transport capacity of the given bus line is equal to

$$S_{bus}(\tau) = f_{bus}(\tau) \cdot s_b = [(\tau/\overline{h})] \cdot s_b \tag{2.19b}$$

where

s_b is the average space capacity of a bus deployed along the line per service frequency ((seats + standings)/dep).

For example, the local bus system provides the landside connection/accessibility between both terminals at London Gatwick airport and many local destinations. One of the lines is Metrobus Fastway 20: Horley–Broadfield, where the traffic capacity is: $f_{bl}(\tau) = 60/15 = 4$ dep/h. If each bus offers approximately $s_b = 50$ spaces, the corresponding transport capacity is equal to: $S_{bl}(\tau) = f_{bl}(\tau) \cdot s_b = 4 \cdot 50 = 200$ spaces/h. The service is provided during 24 h per day (https://www.gatwickairport.com/to-and-from/by-coach-or-bus/local-buses/).

The "static ultimate" capacity of the bus system relates to that of the particular bus stops along the lines and end stations/terminals. The "static ultimate" capacity (μ_b) in terms of the number of buses handled per unit of time (usually 1 h) on the given number of gates/platforms (N_b) at the station/terminal can be estimated as follows (Janić, 2019):

$$\mu_b = N_b/\tau_b \tag{2.20a}$$

where

τ_b is the average bus turnaround time at the bus station gate/platform (h/bus-platform).

Consequently, the number of required platforms/gates at the given station/terminal can be estimated as follows:

$$N_b = \lambda_b \cdot \tau_b \tag{2.20b}$$

where

λ_b is the constant intensity of arriving buses at the bus station during the time (τ) (buses/h).

For example, the given bus station/terminal at an airport has $N_b = 15$ gates/platforms where the buses spend about $\tau_b = 0.5$ h during their turnaround time. The buses from different lines arrive at an intensity of $\lambda_b = 30$ buses/h. The "static ultimate" capacity of this bus station/terminal is equal to: $\mu_b = N_b/\tau_b = 15/0.5 = 20$ (*buses/h*)). The required number of gates/platforms at the bus station: $N_b = \lambda_b \cdot \tau_b = 30 \cdot 0.5 = 10$ (*gates/platforms*).

The "static" and "dynamic" "practical" capacity is elaborated in the scope of the quality of services.

b) Rail-based mode and its systems

The rail-based airport landside access modes and their systems consist of the rail lines connecting a given airport with one to the several rail stations in its catchment area, and the rolling stock operating along these lines. The number of rail lines, the capacity of rolling stock, and the train frequency depend on the size of a given airport, reflecting the potential volumes of demand (Ashford et al., 1997). The trains can be composed of varying numbers of cars, depending on the volume and the structure of the expected demand. The carrying capacity of a typical rail car varies from 40 to 70 seats (Vuchic, 2007). There is also a space for standing passengers in each car.

Similar to the road-based bus system, the capacity of rail-based mode and its systems relates to the stations/terminals and lines. This can be "static ultimate" and "dynamic ultimate" traffic and transport capacities. The "dynamic ultimate" traffic capacity of a given line, defined as the maximum number of trains that can be accommodated along a given line during a given period of time (usually 1 h), can be estimated as follows (Janić, 2019; Vuchic, 2007):

$$\mu_{dr/traffic/l}(\tau) = \tau/\overline{\tau}_{min} \tag{2.21a}$$

where

τ is the time interval for which capacity is estimated (h, day); and

$\overline{\tau}_{min}$ is the minimum average time separation between successive trains operating in the same direction along a given line (min/train).

For example, the typical "dynamic ultimate" capacities of rail lines are 14–18 and 22–26 trains/h. Nevertheless, the typical capacity of an open HSR line is considered to be: $\mu_{dr/traffic/l}(\tau) = 13–15$ trains/h. The number of HS (High Speed) Shinkansen "Nozomi" services (Japan) during the peak hours has been scheduled to be 10 dep/h.

From Eq. 2.21a, the "dynamic ultimate" transport capacity of the given lie is equal to:

$$\mu_{dr/transport/l}(\tau) = \mu_{sr/l}(\tau) \cdot S_t \tag{2.21b}$$

where

S_t is the average number of seats per train operating along the given line (seats/dep).

Based on Eq. 2.21b, the required number of rolling stocks/trains to carry out at the service frequency at the level of traffic capacity of the given line $(\mu_{dr/l}(\tau))$ can be estimated as follows (Janić, 2019; Vuchic, 2007):

$$F_r(\tau) = \mu_{dr/traffic/l}(\tau) \cdot \tau_{r/l} \tag{2.21c}$$

where

$\tau_{r/l}$ is the average turnaround time of the trains along the given line (h/train).

For example, if the service frequency at the level of capacity along a given line is $\mu_{dr/l}(\tau) = 15$ trains/h, and if the average turnaround time per train is $\tau_{r/l} = 4$ h, the required number of trains will be: $F_r(\tau) = 15 \cdot 4 = 60$ trains. The "static ultimate" capacity of the rail stations/terminal can be modelled analogously to that of the bus system.

The "static" and "dynamic" "practical" capacity is elaborated in the scope of the quality of services.

c) Passenger terminal complex

In general, the consistent methodology for estimating the "static" and "dynamic" "ultimate" capacity of the passenger terminal complex has not been developed. Instead, different models have been developed in order to estimate the capacity of the above-mentioned particular components, such as processors, reservoirs, and links (Janić, 2000; Tošić, 1992). The main inputs for these models, in addition to demand, are the processing rates of particular components. In such context, the "static ultimate" capacity is estimated for reservoirs where the air passengers and their companions wait for the next service phase. For the departing passengers, these reservoirs are the central terminal hall(s), in front of ticketing and check-in counters, the security checking area(s), and the hold-room(s). For the arriving passengers, these are the baggage claim areas around the baggage claim device(s) in the baggage hall(s), the secondary examination queue areas, and the GIS (Governmental Inspection Service) area(s), including the custom, immigration, and the health desks areas (see Table 2.18 and Figure 2.42) (Ashford et al., 1997). The number of occupants waiting in any of the above-mentioned areas at a given moment of time is estimated as follows (Janić, 2000; TRB, 2010; 2011; http://go.updates.iata.org/l/123902/2017-03-07/7lrghg/):

$$N_{t/st} = A/A_0 \tag{2.22a}$$

where

A is the size of a given reservoir, i.e., waiting area (m^2); and
A_0 is the standard unit of space intended for standing and/or seating occupants in a given reservoir, i.e., waiting area within the terminal (m^2).

In addition, when the number of occupants $(N_{t/st})$ and the space standard (S_0) are known, the size of the entire waiting area (A) reflecting its static "ultimate" capacity can be computed using Eq. 2.22a. The "dynamic ultimate" capacity of the processors, i.e., service facilities, consisting of several identical units, can be estimated as follows (Janić, 2000; TRB, 2010; 2011; http://go.updates.iata.org/l/123902/2017-03-07/7lrghg/):

$$M_{t/d} = n \cdot \mu_{pr} \tag{2.22b}$$

where

n is the number of processing/service units operating within a common processor, i.e., service facility; and

μ_{pr} is the average service rate of a single service unit; this can be estimated as a reciprocal value of the average service time per customer (the customer can be a passenger and/or his/her bag) (pax or bags/min) (see also Table 2.16).

Equation 2.22b indicates that the processing rate of particular units of a given processing/service facility can be standardized, which depends of the technical/ technological characteristics of the facility. However, in most cases, these rates fluctuate as influenced by the performance of the operating staff and the characteristics of customers requesting service. These can be, for example, as mentioned above, domestic, international, and charter passengers, usually having varying numbers of pieces of baggage of different weights (Janić, 2000; 2013).

Specifically, the "ultimate static" capacity of a baggage claim device can be expressed by the number of simultaneously handled passenger bags. This number mainly depends on the type and size of the mechanized device and the average size of a bag and is typically from about 78 to 318 pieces. The "dynamic ultimate" capacity of the baggage claim device depends on two factors: the average speed of the belt, which is usually 0.45 m/s; the rate of loading, i.e., the intensity of bags arriving onto the device, and the rate of emptying the device dependent on the speed of passengers' recognizing and retrieving their bags. The "ultimate" capacity of links in a given passenger terminal(s) is specified mainly during their design and construction. This capacity can also have "static" and the "dynamic" dimensions. The maximum number of occupants (passengers) simultaneously being in a given link represents the link's "static" capacity. The maximum number of passengers passing through the specified "reference location" of a given link during a given period of time under a constant demand for service represents the link's "dynamic" capacity (Ashford et al., 1997; Horonjeff and McKelvey, 1994; Janić, 2000; 2013).

The capacity estimates of the above-mentioned particular components of the passenger terminals can be used as inputs for more complex models intended to the planning and design of the terminals. These models are based on the network analysis, queuing theory, and the fast-time simulation. These models can examine the detailed behaviour of the passenger and baggage flows while being processed in a given terminal building and consequently investigate their rather complex interactions, which are relevant for design of the required space for both passengers and the related service facilities (Horonjeff and McKelvey, 1994; Janić, 2013; Tosic, 1992; TRB, 2010; 2011).

2.3.3.6 Balancing airside and landside capacity

Balancing the airport airside and the airport landside capacity can be carried out in different ways. One of them can be balancing their "dynamic ultimate" transport capacities, as follows:

$$\textstyle\sum_{k=1}^{K} \mu_k \cdot S_k = \mu \cdot S \tag{2.23}$$

where

K is the number of landside access modes serving a given airport (-);

μ_k is the "dynamic ultimate" traffic capacity of the airport landside access system *(k)* (veh/h);

S_k is the average seat capacity of a vehicle operated by the access system *(k)* (seats/veh);

μ is the "ultimate" capacity of the airport airside area (atm/h); and

S is the average seating capacity of an atm operating at the airport during a given time period (seat/aircraft).

2.4 Quality of Services

2.4.1 General

The quality of service provided by an airport can be analysed and modelled separately for its airside area and landside area. In the airside area, it primarily concerns the aircraft operations/atms in terms of their congestion and delays. These are mostly related to the runway system and, to a somewhat lesser degree, to the taxiways and apron/gate complex. In the landside area, it relates to the service quality provided to air passengers by the landside access modes and their systems and in the passenger terminal complex. In both areas, this quality of services is provided under given conditions.

In particular, during the last few decades that airports have been growing, privatizing parts of their services, and consequently increasingly competition between each other, providing the highest possible competitive quality of services to both airlines and air passengers has been of growing importance. This has also been used by many airports as a powerful competitive tool, together with the airlines' destinations and corresponding flights frequencies. In general, the airports aim to maintain a standardised quality of services at the desired (planned) levels. The airlines intend to know how the services are rated compared to those at their own and hubs of their rivals. The air passengers, as users of services of both airports and airlines, for whom the total time and price of the trips are not exclusive and decisive factors in choosing the airport, airline and air route are also interested in the quality of airport services regarding both the terminals and the airport access modes and their systems.

The airport quality of services has generally been evaluated using two approaches. The first is empirical estimations/evaluations, the other is analytical investigations.

- Empirical estimations/evaluations usually express the quality of service by the "level of the users' satisfaction". The users/air passengers, airport operators, different marketing agencies, consultancy firms, research institutes and individuals dealing with the analysis, research, planning and designing of the airport operations have carried out these empirical estimations/evaluations. In most cases, the airport operators (suppliers of services) have been in a position to control the processes of producing services in real time (on-line). Since these services have been produced and consumed simultaneously, their quality can be evaluated only at the end of production process. This is carried out by interviewing the air travel-experienced air passengers. In some cases, these passengers have been asked to perceive the quality of services in advance (i.e., before starting a journey) in order to give some input to the production process. A representative case has been GAM (Global Airport Monitor), which regularly (annually) measures air passenger satisfaction through 80000 interviews at 52 major airports worldwide, regarding a wide range of the service quality attributes (https://www.iata.org/en/pressroom/pr/2002-06-12-24/). The attributes commonly evaluated have been: the overall passenger convenience, signposting, ground transportation, speed and efficiency of check-in staff, lounges and waiting areas, special services for overseas visitors, custom and immigration services, passport and visa inspection, baggage delivery, baggage carts, shopping, restaurants, availability of connecting flights, availability of low fares, ease of making connections, and on-time departures. The results have been useful not only for comparisons and ratings between the airports themselves but also for estimating the local performances of the quality of services in order to adapt them to the users' requests as far as possible (Janić, 2000).

- Analytical investigations of the airport quality of services have been based on modelling operations and processes generating the services for the users—airlines and air passengers. The most frequent have been analysis and modelling of the aircraft/flights, i.e., atms congestion and delays in the airside area and the quality of services provided to the air passengers in the passenger terminal complex. In the airside area, most investigations have resulted in setting up the runway "practical" or "declared" capacity based on the maximum average delay per an atm estimated for the given relationship between demand and "ultimate" capacity. In the landside area, the results of different investigations have been summarised as the recommendations containing the space and time standards to be assigned and guaranteed to each air passengers while passing through the passenger terminal complex (Janić, 2013; https://www.iata.org/en/services/consulting/airport-pax-security/level-of-service/).

2.4.2 Airside Area

2.4.2.1 Quality of services of the runway system

At the global scale, the quality of services at airports have generally been measured by the average annual on-time performance, defined as the proportion of arriving and departing flights/atms operated within 15 minutes of their scheduled arrival and departure times. Figure 2.45 shows this for airports worldwide (OAG, 2019).

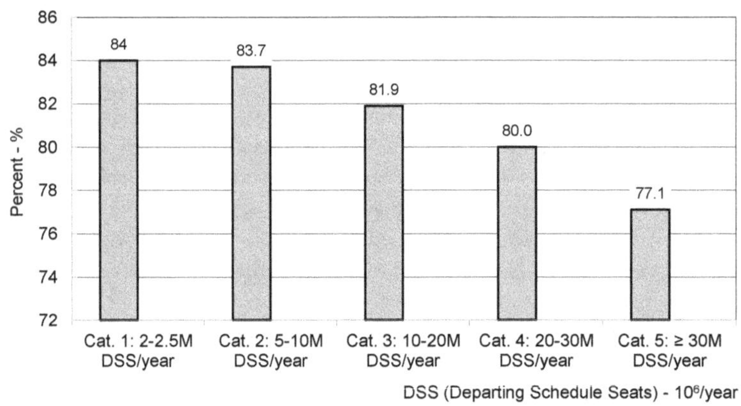

Figure 2.45 Punctuality of the world's airports depending on the volume of supplied capacity (Period: 2018) (OAG, 2019).

As can be seen, the average punctuality has generally been around 80%, and slightly decreasing with increasing airport traffic. This is expressed by the DSS (Departing Schedule Seats).

In addition, the quality of services of the airport runway systems has been measured by the delays longer than 15 min. These have generally been caused by the longer discrepancy between the airport demand and supply. On the demand side, particularly at large airports, the demand, including its time concentration, has generally grown due to development of the airline "hub-and-spokes" networks. On the supply side, many airports have been faced with a runway capacity shortage, making them unable to properly adapt to these conditions. Figure 2.46 shows an example of one such airline flight schedule pattern at a large hub airport.

In this case, discrepancy between demand and capacity is expressed by their ratio as: $\rho = D/C$, where D is demand and C is capacity. As can be seen, 8 waves of incoming and outgoing flights are scheduled during the day. Often $\rho > 1.0$, indicating the presence of arrival and departure delays due to longer lasting discrepancies between demand and capacity. When $\rho < 1.0$ delays also occur but only due to the short-lasting discrepancy between demand and capacity. In the former, the delays last longer than 15 minutes and, as such, are counted. In the latter case, they will be shorter than 15 minutes and, as such, not counted. Figure 2.47(a, b, c) confirms the above-mentioned statements.

Figure 2.47a shows that, during the observed period, the average delay per delayed departure and arrival was between 50 and 80 min. The delay per delayed departure was higher than that per delayed arrival, with both generally increasing with an increase in the annual number of operations/atms. This indicates the tendency of delays to propagate between interconnecting flights. Figure 2.47b shows the relatively strong relationship between arrival and departure delays. Figure 2.47c shows that the proportions of delayed departures have been higher than those of arrivals, with both increasing with an increase in the annual number of operations/atms.

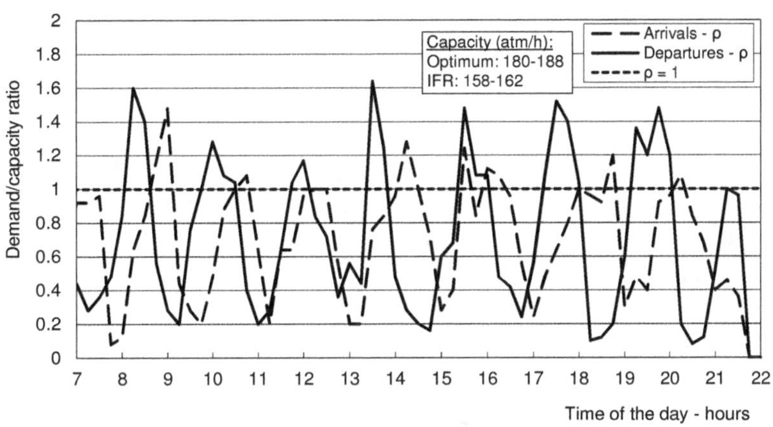

Figure 2.46 Demand/capacity ratio over time—Case of Atlanta Hartsfield-Jackson international airport (U.S.) (Period 2007) (FAA, 2007).

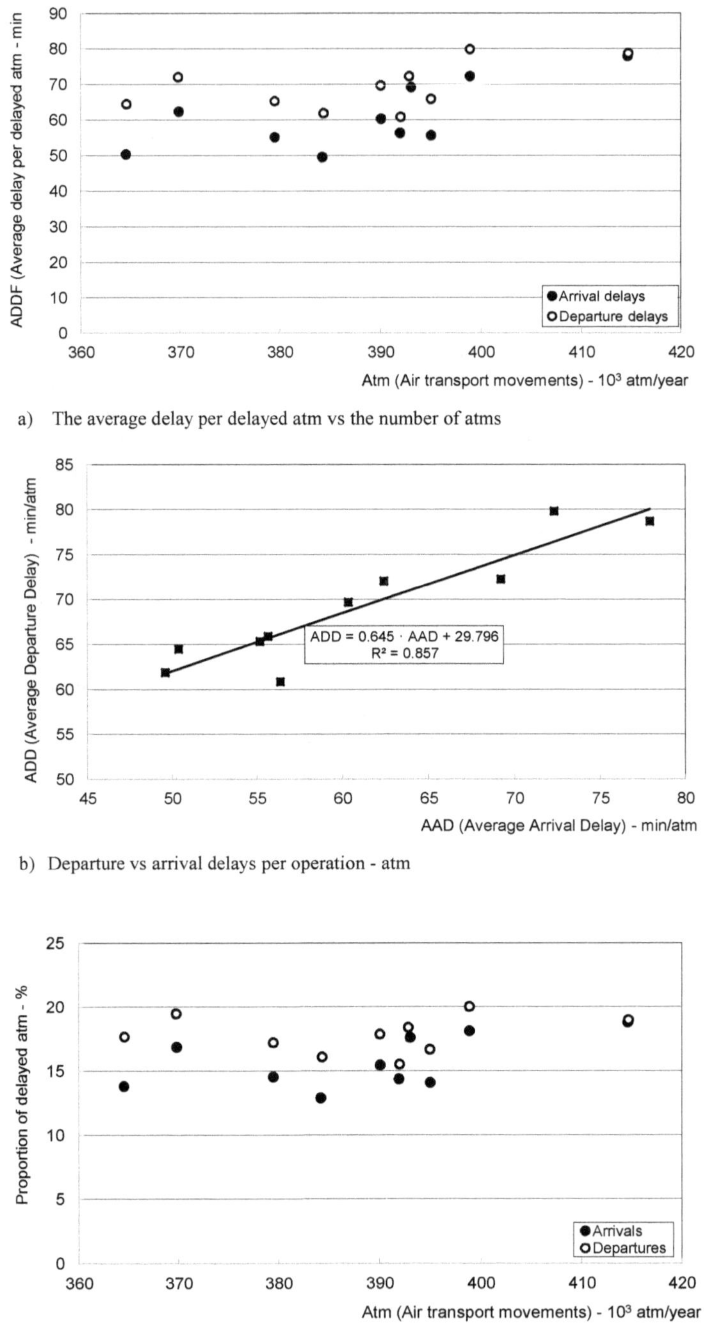

a) The average delay per delayed atm vs the number of atms

b) Departure vs arrival delays per operation - atm

c) Proportion of delayed vs the number of operations - atms

Figure 2.47 Characteristics of the airport delays as measures of the quality of service of runway system—Case of Atlanta Hartsfield-Jackson international airport (Period: 2010–2019) (https://www.transtats.bts.gov/HomeDrillChart.asp).

2.4.2.2 Modelling quality of services of the runway system

a) Generic case

Modelling the quality of services of an airport runway system mainly relates to congestion and delays of landings and take-offs, i.e., atms under given conditions. This is generally based on the queuing system theory (Janić, 2013; de Neufville and Odoni, 2007). In such case, the airport runway with the adjacent airspace is always assumed to operate as the single server system. In cases when two or more runways are used for carrying out independent operations, the system is considered to operate as the multi (two)-server queuing system. In cases where landings and take-offs simultaneously request the service of a single runway, the former are always given the "ultimate" priority. The runway service rate is adopted to be its "ultimate" capacity for either type of atms/customers. The number of waiting places where the landing and/or taking-off queues can be formed is assumed to be unconstrained. The taxi-way network has always been considered as the component connecting the runway and apron/gate complex and, as such, is usually modelled in the scope of modelling the entire airport operating as the service system. The apron/gate complex has always been modelled as the multi-server queuing system where the gates/parking stands are assumed to be the servers. The processing rate of each gate/stand is computed as the reciprocal of the aircraft ground service (turnaround) time (Horonjeff and McKelvey, 1994). The number of waiting places where the landing and/or taking-off queues can be formed is assumed to be unconstrained. The delays imposed on landings are performed in the holding stacks located in the vicinity of landing runway(s). Theoretically, this space is unlimited. The take-off delays are realised on the ground, either at the gate/parking stands with the aircraft engines turned-off (ground-holding delays) or along the taxi-ways just before departure with the engines turned-on. The space where both types of departure delays are carried out is also assumed to be unlimited.

Delays are the consequence of congestion, which happens whenever the demand for service exceeds the capacity of the service system. This becomes over-saturated for shorter or longer time periods depending on some of the above-mentioned modelling conditions. Dependent on the rate and intensity of the over-saturation, two typical situations may occur. Two cases can happen: the first is when the overall server capacity is higher than the demand. In this case, despite the fact that there is plenty of capacity overall, the congestion and delays may happen due to the instant (short-time) exceeding of the server's processing rate by the arrival (demand) rate. When the arrival rate increases close to the server's processing rate, the frequency of instantaneous (short-term) exceeding of the server's rate by demand rate significantly increases, causing the delay to customers to grow disproportionately faster than the arrival rate (de Neufville and Odoni, 2007; Janić, 2013). The other type of queuing situation occurs when the arrival rate is exceeding the server's processing rate for a longer time period. In this case, three phases of the developing congestion and delays can be noticed. First, if the demand rate is less than or equal to the server processing rate, any congestion cannot be registered. The facility throughput is equal to demand rate. Second, when the demand rate exceeds the server-processing rate, the queue starts to develop and increases over time. The system's throughput is equal to the server's processing rate, e.g., to its "ultimate" capacity. Finally, when the arrival rate falls again below the server's processing rate, the throughput will stay at the level of this capacity until the queue vanishes. After that, the throughput is again equal to the demand rate (Newell, 1982).

The generic model for estimating delay of an atm interfering with the other approaching to the same airport runway and simultaneously requesting service can be estimated as follows (Newell, 1982):

$$W_{i+1} = \max[0; W_i + S_i - (T_{i+1} - T_i)] \text{ } for \text{ } i \in N \tag{2.24}$$

where

W_i is the delay of the aircraft/flight *(i)* (min/atm);
S_i is the service time of the aircraft/flight *(i)* (min/atm);
T_i, T_{i+1} is the arriving time of the aircraft/flight *(i)* and *(i + 1)*, respectively; and
N the number of aircraft considered.

Equation 2.24 is the well-known recurrence relation from theory of the time-dependent or the transient queues (Newell, 1982). In this case, FCFS service rule is applied to the given number of the arriving (and/or departing) aircraft/flights, i.e., atms. Adding up the W_i's and dividing the sum by N gives the average delay per an aircraft, i.e., the main outcome of interest. The average delay can also be computed by carrying out a series of experiments regarding the different number of aircraft/flights involved.

b) Stochastic case—the average delays when demand is lower than capacity

The analytical models based on the theory of steady-state queuing systems have been used for a long time to estimate the aircraft/flight delays served on the single runway. In these models, the generic expression for estimating the average delay per an arriving aircraft/flight requesting service is as follows (Horonjeff and McKelvey, 1994; Janić, 2013):

$$W_a = [\lambda_a(\sigma_a^2 + 1/\mu_a^2)]/[2(1 - \lambda_a/\mu_a)] \tag{2.25a}$$

where

W_a is the average delay of an arriving aircraft/flight (time units);
λ_a is the average intensity of demand, i.e., the aircraft/flight arrival rate (atms/unit of time);
μ_a is the average aircraft/flight service rate as the reciprocal of the mean service time for arrivals ($\mu_a = 1/\bar{t}_a$; where $\bar{t}_a$ is the minimum average service time per an arrival (time units)); and
σ_a is the standard deviation of the arrival service time (time units).

In Eq. 2.25a the average service time for arrivals ($\bar{t}_a$) is the larger value among the runway occupancy time and the average separation time between the successive arrivals/flights while approaching to a given runway. As the reciprocal of this average service time, the average services rate (μ_a) represents the runway "ultimate" landing capacity. The analogous expression with the modified subscripts calculates the average delay of a departing aircraft/flight. In most cases, the arrival aircraft/flight flow is considered as a Poisson process and the service time as the stochastic variable with the general probability distribution (Janic, 2000; Newell, 1982). The above-mentioned model has shown to be of very high practical value mainly due to its simplicity and transparency, and particularly after obtaining the empirical evidence on the correctness of its results in the most general sense. When a single runway is simultaneously used for both landings and take-offs, a higher priority is always assigned to landings. Operations of the same type have always been served according to FCFS service rule. Under such conditions, the average delay per departure can be estimated as follows (Horonjeff and McKelvey, 1994):

$$W_a = [\lambda_d(\sigma_j^2 + j^2)/[2(1 - \lambda_d \cdot j)] + g \cdot [(\sigma_f^2 + f^2)/[2(1 - \lambda_a \cdot f)] \quad (2.25b)$$

where

- W_d is the average delay of departure aircraft (time units/dep),
- λ_d is the average intensity of departures (atm/unit time),
- λ_a is the average intensity of arrivals (atm/unit time),
- j is the average time between successive departures (time units);
- σ_j is the standard deviation of (j) (min);
- g is the average intensity of occurring gaps between successive arrivals (number/unit of time);
- f is the average time in which departure cannot be realised (min); and
- σ_f is the standard deviation of (f) (min).

Equations 2.25(a, b) are valid only when the intensity of both arrival and departure demand is constant and lower than the runway capacity, i.e., when it operates below the saturation. Figure 2.48 shows the example of applying Eq. 2.25a for estimating the landing delays at London Heathrow airport (Janić, 2013).

The average arrival delay in Figure 2.48 increases more than proportionally with an increase in the intensity of demand. If the average delay as the criterion for estimating quality of service is considered to be 5 min, the runway "practical" capacity will be 34 atm/h, which is a quite similar to the throughput at the Frankfurt Main airport (Germany) when operating under IFR conditions. If the average delay is 15 min, the runway "practical" capacity will be 37 atm/h, which is close to the "ultimate" landing capacity at the London Heathrow airport (UK). If this average delay is guaranteed for a period of several hours during the day, the corresponding throughput can represent the "sustainable practical" capacity, which is usually used in the slot allocation procedure(s). A similar reasoning can be used to estimate the departure delays (Janić, 2000; 2013).

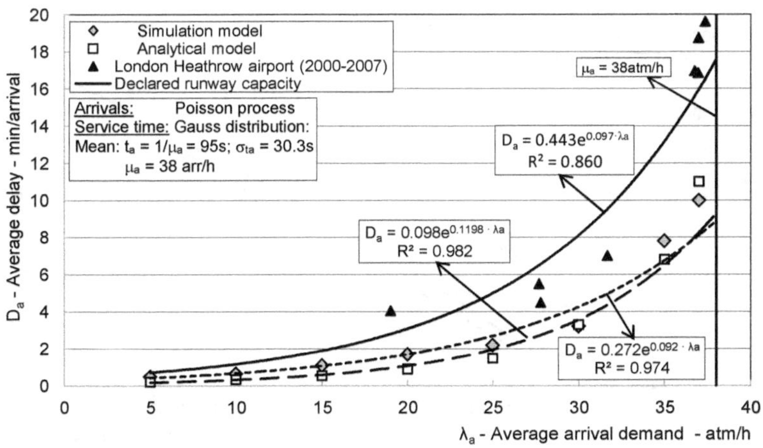

Figure 2.48 Relationship between the average delay and intensity of arrival demand—Case of London Heathrow airport (UK) (Janić, 2013).

c) Deterministic case - The average delays when demand is higher than capacity

The above-mentioned steady-state queuing models have proven to be inappropriate for modelling the delays of arrivals and departures in cases when the when the demand has significantly exceeded the runway "ultimate" capacity (see Figure 2.46). For such cases, the deterministic queuing model based on fluid approximation has been proposed (Newell, 1982). This model does not take into account the short-lasting discrepancies between demand and capacity under conditions when the demand is actually lower than the capacity. It takes into account only the queues and delays when a given runway as a server is over-saturated. The process taking place at the congested runway can be presented in typical graphical form, as shown in Figure 2.49.

The horizontal axis represents the time of day and the vertical axis represents the cumulative counts of total demand and capacity in terms of atms. The demand expresses the cumulative number of atms—landings and take-offs over time. The cumulative number of served atms represents the capacity. The vertical difference between two cumulative curves at a given time approximates the congestion, i.e., the number of atms in the queue. The horizontal distance between the two curves indicates the delay of last atm in the queue. As can be seen, at around 13:00 h, the queue consisted of about 40 atms, either landings or take-offs, with the waiting time of the last one in the queue of about 48 minutes. The area between the curves represents the total atms waiting (queuing) time during the observed period.

Modelling congestion and delays of atms using the above-mentioned curves is based on the assumption that they are mutually independent. In the given context, the curves $(A(t))$ and $(D(t))$ may relate only to a single realization or be the averages of many realizations of the two processes at a given airport runway during a given time period. The queue of atms at a time (t) can be approximated as follows (Newell, 1982):

$$Q(t) = \max[0; A(t) - D(t)] \quad for\ t \in \tau \tag{2.26a}$$

The queues $(Q(t))$ during the short time increment (Δt) (5, 10, and/or 15 minutes) are estimated as follows:

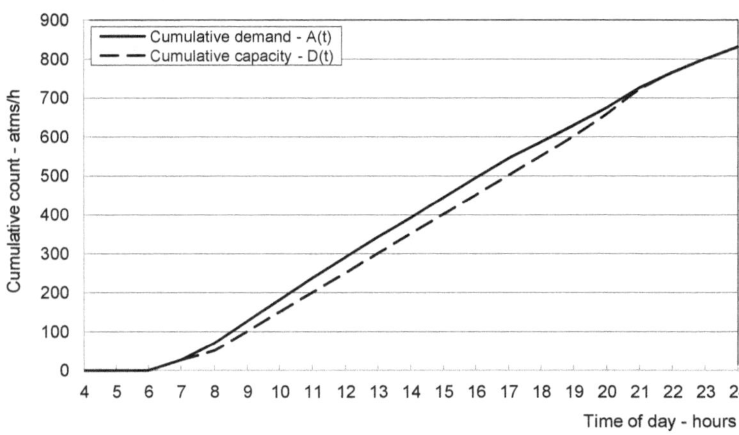

Figure 2.49 Cumulative count of demand, capacity, and resulting congestion and delays when demand is higher than capacity—Case of NY La Guardia airport (U.S.) (Period: US Independence Day – 2006) (Janić, 2013).

$$Q(t) = \max\{0; Q(t - \Delta t) + [\lambda(t) - \mu(t)]\Delta(t)\} \ for \ t \in \tau \tag{2.26b}$$

where

$Q(t - \Delta t)$ is the queue of aircraft/flights at the time $(t - \Delta t)$ (atms);
$\lambda(t)$ is the intensity of the aircraft/flight demand at time (t) (atms/unit of time); and
$\mu(t)$ is the aircraft/flight service rate (i.e., the "ultimate" capacity) at time (t) (atms/unit of time).

From Eq. 2.26b, the delay of the last atm-aircraft/flight waiting in the queue at the time (t) is estimated as follows:

$$w(t) = Q(t)/\mu(t) \tag{2.26c}$$

where all symbols are analogous to those in the previous Eqs.

The average delay per an aircraft/flight waiting in the queue during the congested period (τ) is approximated as follows (Newell, 1982):

$$\overline{w(\tau)} = [1/A(\tau)] \cdot \int_0^{\tau} [A(t) - D(t)] \, dt \tag{2.26d}$$

where

$A(\tau)$ is the number of atms-aircraft/flights requesting service during the congestion period (τ).

Using the inputs in Figure 2.49 in Eq. 2.26(a, b, c), the queues and delays are estimated and shown in Figure 2.50.

As can be seen, during the given day, the aircraft/flight queue started to develop early in the morning at 6:00 a.m., grew over time, and reached the maximum around 8:00 p.m. Since after that time the demand significantly decreased, the long queue also disappeared relatively quickly, one hour before midnight. In this case, the average of $\bar{Q}(\tau) = 35$ atm-flights and the maximum of $Q_{max}(\tau) = 59$ atms-flights were in the queue. The aircraft/flight delays developed during the day were affected by the queues. For example, delay of the last aircraft/flight in the queue changed in proportion with the changes in the queue. The average delay and the maximum delay per an aircraft/flight were about $w(t) = 35$–40 min and $w_{max}(\tau) = 65$ min, respectively.

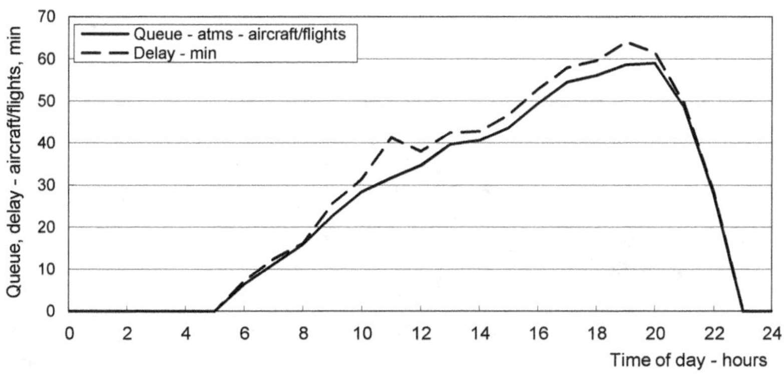

Figure 2.50 Queues and delays—Case of NY La Guardia airport (U.S.) (Period: US Independence Day – 2006) (Janić, 2013).

2.4.3 Landside Area

2.4.3.1 Quality of services of the airport landside access modes and their systems

The airport landside access modes are road- and rail-based. The former involves systems such as individual cars/taxis and mass/public vans/buses. The latter generally includes the streetcars/tramway and LRT, subway/metro, regional/conventional rail, HSR, and TRM. The quality of services provided to air passengers by the landside access modes and their systems is characterised by the attributes such as duration, reliability, and punctuality of the access time, comfort on-board the vehicles, complexity of handling baggage, and the out-of-pocket travel costs.

Considered in terms of the generalized travel costs, which depend also on the trip purpose (business, leisure), the above attributes simultaneously influence the choice of airport ground access mode and its systems. The additional factors affecting choice can be the type of flight—short, medium- and long-haul domestic and international. In addition to the air travellers, the quality of services of the particular modes and their systems also influence the choices of the airport employees, passenger companions, visitors, and greeters.

In particular, the air passengers most frequently use one system of the chosen mode and rarely combinations of systems of the single or few different modes. Under such circumstances, duration of the access time has been of particular importance. Figure 2.51 shows example for European and Asian airports (TRB, 2008).

As can be seen, and as intuitively expected, the airport access time generally increases proportionally (road) and more than proportionally (rail) with an increase in the access distance at both modes. In most cases, the access time by road is greater than that by rail. One of the reasons for this is the prevailing congestion along the roads connecting CBD and airports. In addition to the access time, in-vehicle comfort and convenience of handling baggage have also influenced market shares of the available access modes and their systems. The in-vehicle comfort depends on the vehicle size and the number of air passengers on board: car/taxi with the seat capacity up to 5 persons, van and bus from 10 to 50 seats, train set 140 to 2000 passenger spaces (Janić, 2019). Table 2.19 shows

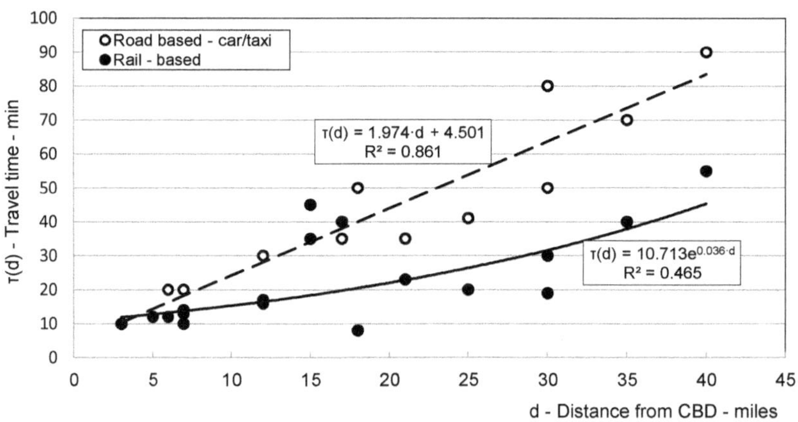

Figure 2.51 Relationship between the airport access time and distance to CBD (Central Business District) (1 mile = 1.609 km) (TRB, 2008).

Table 2.19 The market share of the airport access modes and their systems—Case of the selected U.S. and European airports (TRB, 2008).

U.S. airports		**Market share of access mode/system (%)**		**European airports**	**Market share of access mode/system (%)**
	Rail	**Car/taxi**	**Bus/van**		**Rail**
San Francisco	7	77	16	Zurich	42
NY JFK	8	81	11	Oslo	39
Boston	6	82	12	Amsterdam	35
Reagan National	13	83	4	Copenhagen	33
Oakland	9	85	6	Munich	31
New Orleans	0	85	15	Vienna	30
NY Newark	5	86	9	Paris CDG	28
Atlanta	10	86	4	Frankfurt	27
Denver	0	86	14	London LHR	25
Los Angeles	0	87	13		

examples of the market share of particular airport access modes and their systems at the U.S. and European airports (TRB, 2008).

As can be seen, at the U.S. airports, the market share of car/taxi system is the highest, approximately at and above 80%. The market share of rail mode has been quite symbolic, since at some airports it does not exist. The market share of bus/van system has been diverse at particular airports but the highest up to 15–16%. In addition, the market share of public access modes and their systems at 27 U.S. airports has not been dependent on the volumes of passenger demand there. At the selected European airports, the rail market share has been from 25% to 42% (TRB, 2008).

2.4.3.2 Modelling quality of services of the airport landside access modes and their systems

Modelling the quality of services of the airport landside access modes and their systems mainly relates to the access time models for car, taxi, and selected combinations of the individual (car/taxi) and the public/mass (rail and road-based) systems. These models are presented only in their analytical form.

a) Access time by car

When a potential air traveller uses a car to access an airport, it is assumed that it is available to him (her) at any time because the user is located either at home or in the office when starting their trip. Figure 2.52 shows the simplified scheme.

Under such conditions, the average airport access time (τ_c) can be estimated as follows (Janić, 2000):

$$\tau_c = [d_c/v_0] \cdot C[\rho(d_c)] \tag{2.27a}$$

where

d_c is the average travel distance between the user's temporary location and the airport,

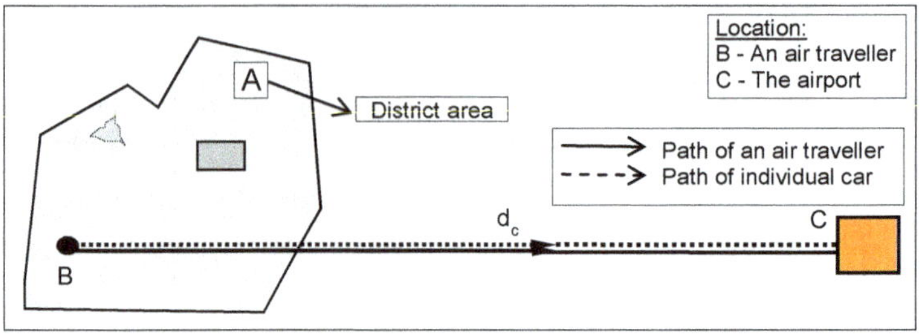

Figure 2.52 Scheme of airport access by car (Janić, 2000).

$v_0(d_c)$ is the free traffic flow operating speed of a car over distance (d_c); this speed usually varies between 20–50 km/h in urban areas, and from 60–90 km/h on freeways;

$C[\rho(d_c)]$ is the factor representing the influence of traffic density along route (d_c) on the free operating speed of a car $[v_0(d_c)]$.

In Eq. 2.27a the factor $(C[\rho(d_c)])$ can be determined as follows:

$$C[\rho(d_C)] = 1 + J \cdot \left[\frac{\rho(d_C)}{1 - \rho(d_C)} \right] \tag{2.27b}$$

where

$\rho(d_c)$ is the traffic load along the lane (d_c), i.e., $\rho(d_c) = q(d_c)/[k \cdot \mu(d_c)]$, where $q(d_c)$ is the prevailing intensity of traffic along the line (d_c) (veh/h); k is the number of lanes along the airport access road ($\rho(d_c)$ is assumed to be constant in time of travelling along the lane d); $\mu(d_c)$ is the saturation capacity of the lane (veh/h); and

J is the factor modifying the expected vehicle delay subject to the road (lane) type, frequency of intersections and the other factors which may cause the delay of a car $(J = 0.01 - 1.0)$.

Equation 2.27b indicates that the airport access time by car is highly dependent on the conditions prevailing along the access road/highway.

b) Access time by taxi

Modelling the airport access time by taxi is based on the following scenario: the potential air traveller is either at home or in the office located in an urban district when he starts going to the airport. He/she calls the taxi, which needs some time $(\tau_{0/T})$ for arrival and pick-up. Then, they proceed directly to the airport along more or less dense streets/roads/highways. Figure 2.53 shows the simplified scheme.

The average time from the moment of calling the taxi to the moment of arriving at the airport can be estimated as follows:

$$\tau_T = \tau_{0/T} + \tau_{a/T} = \frac{E(d_T)}{v_0(d_T)} \cdot C[\rho(d_T)] + \left[\frac{D_T}{v_0(D_T)} \right] \cdot C[\rho(D_T)] \tag{2.27c}$$

where

$E(d_T)$ is the average distance travelled by the taxi from its location and the user's location at the moment of call (km); and

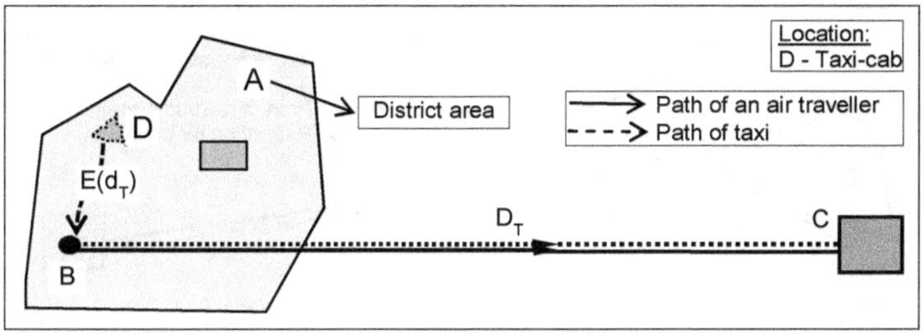

Figure 2.53 Scheme of airport access by taxi (Janić, 2000).

$v_0(d_T)$ is the free traffic flow operating speed of the taxi along the distance $E(d_T)$ (km/h).

The first term in Eq. 2.27c represents the average response time of a taxi to the call for service. The other term represents the average taxi travel time from the user's location to the airport. Both terms take into consideration the traffic conditions prevailing along the corresponding segments of the door-to-door route . The average travel distance of the taxicab $(E(d_T))$ in Eq. 2.27c can be approximated as follows (Daganzo, 1978; Larson and Odoni, 1981):

$$E(d_T) \approx cA \cdot \sqrt{A/(N - N_0)} \tag{2.27d}$$

where

- A is the size of a given urban area (district) where the taxis operate; the shape of this area is assumed to be of fairly compact dimensions, i.e., the "length" is not much greater than the "width"; the major barriers are assumed not to exist in this area (km^2);
- N is the number of taxis licensed to operate between the district (A) and the airport (-),
- N_0 is the number of busy taxis at the moment when the call for service arrives (-);
- c is the constant depending on the metric in use, the assumptions relating to the characteristics of the space distribution of calls for services, and locations of taxis over the district (A) (the calls for service are assumed to be independently and spatially uniformly distributed over district (A). When the taxis are randomly positioned in district (A), the constant (c) is estimated to be 0.52 for the Euclidean and 0.67 for the right-angle travel; when the taxis are located at the centre of the district, the constant $c = 0.38$ in the case of Euclidean and 0.50 in a case of the right-angle travel).

It is also assumed that the on-scene service time of a taxi (i.e., the time it needs to pick-up the user) is negligible in comparison to the total airport access time. The typical average operating speed of taxis is similar to that of cars (Vuchic, 2007).

c) Access time by combination of individual and mass transit system(s)

The access time of an airport is also modelled for the cases when two combinations of the available access systems are used. The first consists of the road-based taxi and mass (public and/or airline bus) system. The other consists of the road-based and mass rail-based system. The use of car instead of the taxi can easily be included in the modelling procedure. Figure 2.54 shows the simplified scheme.

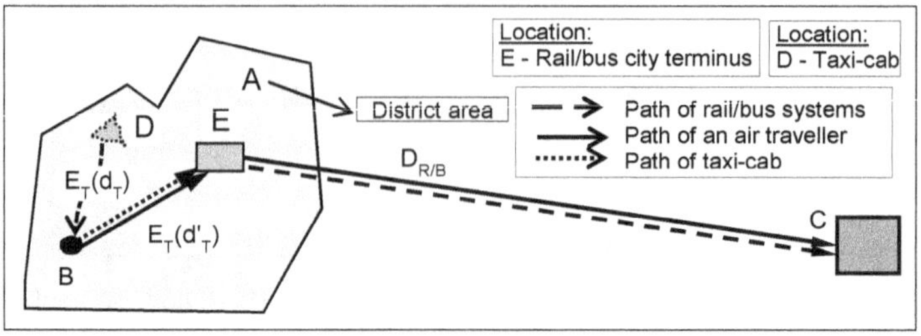

Figure 2.54 Scheme of airport access by individual road-based (taxi) and road or rail-based mass system (Janić, 2000).

i) Combination of the road-based taxi and bus system

The following scenario is applied to modelling the airport access time by combination of the road-based taxi and bus system: the potential air passenger user starts their trip either from his/her home or office by taking a taxi to the urban bus terminal. Then, he/she takes the bus to the airport. As in the previous case, the taxi needs some time to come on-scene and pick-up the user. Then, it takes time to arrive at the urban bus terminal. Since the arrival of the taxi at the terminal and the bus schedule are not co-ordinated, the user will certainly wait some time for the bus departure. Consequently, the average access time to the airport ($\tau_{T/B}$) can be estimated as follows:

$$\tau_{T/B} = \tau_{0/T} + \tau_{1/T} + \tau_{w/B} + \tau_{a/B} \tag{2.27e}$$

where

$\tau_{w/B}$ is the average waiting time of the user for departure from the urban bus terminal to the airport (min); and

$T_{a/B}$ is the average bus travel time from the urban bus terminal to the bus terminal at the airport (min).

The other symbols are analogous to those in Eq. 2.27 (c, d). The times ($\tau_{0/T}$), ($\tau_{1/T}$), and ($\tau_{a/B}$) can be estimated similarly as in Eq. 2.27 (c, d). The time ($\tau_{w/B}$) is estimated after being considered as follows (Larson and Odoni, 1981; Janić, 2013):

$$\tau_{w/B} = 1/2 \cdot [\tau/f_B(\tau)] \text{ or } \tau_{w/B} = 1/4 \cdot [\tau/f_B(\tau)] \tag{2.27f}$$

where

τ is time period (h, day); and

$f_B(\tau)$ is the bus service frequency between the urban terminal and the airport (dep/time period).

ii) Combination of the road-based taxi and the rail-based system

When a potential air traveller uses the combination of the road-based taxi and the rail-based regional/intercity conventional rail system in order to access the airport, the structure of their time is similar to the combination of taxi and bus system. The average travel time of the rail-based system from the urban rail terminal to the rail station (terminal) at the airport can be estimated as follows:

$$\tau_{a/R} = D_R/v_R\,(D_R) \tag{2.27g}$$

where

D_R is the length of rail line connecting the urban and airport railway station (terminal) (km); and

$v_R(D_R)$ is the average rail operating speed along the line (D_R) (km/h).

The scheduled delay at the urban rail terminal can be estimated similarly as in Eq. 2.27f. Since train free operating speed is not influenced by other traffic, the irregularities of the rail schedule are much more rare and smaller than those of the road-based systems (Vuchic, 2007; Janić, 2013).

2.4.3.3 Quality of services in the passenger terminals

After arriving at the airport, the passengers disembark at one of the micro locations, such as car-parking, taxi space, and the rail or bus airport stations (terminals). They pass the distance between these locations either on foot or by one of the different transport systems operating inside the narrow airport area. These local transit systems, extensively operated at the large metropolitan airports, include taxi, bus-shuttle, and the PRT ULTra (Urban Light Transit) system, such as that at London Heathrow airport (UK) (Ashford, 1988; Janić, 2014; 2019). Then, the passengers enter the terminal(s) where they experience some quality of service.

This has been the subject of a lot of efforts aiming at developing an appropriate concept for measuring the quality of service in the passenger terminal(s). These have resulted in developing several concepts, each containing the various space and time standards to be guaranteed to each air passenger under given conditions (Wirasinghe and Shehata, 1988; https://www.iata.org/en/services/consulting/airport-pax-security/level-of-service/).

Despite the fact that there is no clear evidence (except some ergonomic researches performed for other purposes) as to how these space standards are actually determined, they have been widely applied by various organisations, institutions and offices, which have primarily dealt with modernisation of the existing and planning of the new passenger terminals. Some typical values of the time and space standards reflecting the quality of service are given in Table 2.20 (de Neufville and Odoni, 2007; Janić, 2013; www.iata.org/los).

The time standards of quality of services in Table 2.20 aim at guaranteeing that the air passengers spend no longer than the prescribed amount of time in the particular phases of services. These times include the waiting for and the service time, which are both carried out in a space of a certain size. In addition, Table 2.21 gives more detailed standards of the space standards of quality of services in the passenger terminals.

In Table 2.21, 'A' denotes "excellent", 'B' "high", 'C' "good", 'D' "adequate", and 'E' "unacceptable", 'F' denotes "zero" level of service, which applies when the service is broken-down. From the standpoint of air passengers, the actual quality of service in an airport terminal can be expressed by the "level of their satisfaction". This quality can be estimated either by the passengers themselves, the airport and/or airline operators, or researchers, planners and designers dealing with the airport problems (Janić, 2000). The air passengers can perceive the expected quality of services *a priori*, i.e., before starting the trip, or evaluate it *a posteriori,* i.e., after finishing the trip. The airport and/or airline operators (i.e., suppliers of services) usually monitor and control the real-time processes of producing the services. However, since these services are produced and consumed simultaneously, their quality cannot be evaluated immediately, only after completing

Table 2.20 Example of the time and space standards of the airport passenger terminals for economy class passengers by IATA (International Air Transport Association) (www.iata.org/los).

Server/area	Time standard (min/pax)	Space standard (m^2/pax)
Departures		
• **Check-in**		
– Boarding Pass & Tagging – Bag Drop; Desk/Station – Check-in Desk	1–2 1–5 10–20	1.3–1.8
• **Security**	5–10	1.0–1.2
• **Passport control** – Automatic Border Control – Staffed Emigration Desk	1–5 5–10	1.0–1.2
• **Boarding Gates** 50–70% passengers to be seated – Seated Space – Standing Space	- -	1.8–2.2 1.2–1.5
Arrivals		
• **Custom control**	1–5	1.3–1.8
• **Baggage claim** – First passenger to first bag – Last bag (from the first delivery)	0–25 0–15	1.5–1.7
• **Passport control** – Staffed Immigration Desk – Automatic Border Control	5–10 1–5	1.0–1.2

Table 2.21 The space standards of quality of services in the airport passenger terminals (AACC/IATA, 1981; Ashford, 1988; Janic, 2000; 2013).

Area	Level of Service (m^2/pax)					
	A	B	C	D	E	F
• Check-in, Baggage Claim Area	1. 6	1.4	1.2	1.0	0.8	--
• Holding w/o bags, Hold room, Pre-inspection	1.4	1.2	1.0	0.8	0.6	--
• Wait/Circulate	2.7	2.3	1.9	1.5	1.0	--

these processes. The last group of those involved in evaluation of the quality of service apply the similar approach to both air passengers and service suppliers. They usually use the data obtained by the interviews of air passengers and recordings of real-life service processes.

In order to determine the size of particular areas in the passenger terminal guaranteeing the planned quality of service, the relevant space standard in Tables 2.20 and 2.21 should be multiplied by the corresponding number of occupants simultaneously there. The obtained space should then be enlarged to the space required for installing the appropriate service facilities and equipment (de Neufville and Odoni, 2007; Janić, 2013).

In addition to an inherent complexity in determining the total area of an airport passenger terminal and its particular parts, the composition of them into the rather compact physical and technological entity and standardisation of the particular service phases according to specific (local) conditions may additionally complicate the planning procedure. Once an airport terminal is built, the quality of service is expected to be uniform for all passengers over time. However, the inherent variation of the actual demand in comparison to that forecasted may cause deviations of the actual from the planned quality of service. Additionally, this quality may be degraded by an unexpected growth of air traffic and the inability of the airport to properly match. Consequently, a greater number of passengers will spend a longer time in a limited space than planned.

2.4.3.4 Modelling quality of services in the passenger terminal(s)

a) General

In the existing airport terminals the quality of services is expected to be constant and guaranteed according to the space and/time standards determined in advance for given traffic volumes and scenario(s). However, in the real-life situations, the passengers may experience a lower quality of service than planned due to the temporal and/or permanent fluctuations of demand, unpredictable aircraft delays, failures of the service facilities and equipment and increased interactions between passengers themselves. In such cases, both the space and time standards of service quality can be simultaneously deteriorated and interrelated. For example, the space standard can be deteriorated by an unexpected increase in the concentration of occupants in some area, caused by delay of flights. Unpredictable holding of a relatively small number of passengers in an area either due to the internal and/or external reasons may also deteriorate the planned time standard of the quality of services. Of course, both standards of quality can deteriorate simultaneously for the same reasons. For instance, a relatively large number of air passengers can be held for a relatively long time in an area due to the long delays or cancellations of particular flights, significant failures of the service facilities and equipment, as well as other incidental and accidental situations. Under such circumstances, the question arises as to whether the particular standards of service quality can be mutually compensated and for how much. For instance, when the space standard of quality in Table 2.21 is at level 'C', each air passenger in front of any of the processing facilities, such as ticketing, check-in counter and/or baggage claim device, is assigned a space of 1.2 m^2. But the question is, how long can the passenger stay there? In addition, the passenger may stay there surrounded by others. Can any shortening of waiting and service time compensate the temporal deterioration of the guaranteed (planned) space standards? In other words, is it possible to employ the concept of indifference curves to explain the dependability of both attributes of service quality under varying traffic scenarios? One answer to the above questions consists of designing a suitable model to empirically estimate both the time and space attributes of the service quality and then recommend them as given in Table 2.20. The alternative model, based on the deterministic queuing theory, consists of designing the combined time/space indicator of the quality of service as ISQ (Indicator of Service Quality) (Janić, 2000; Newell, 1982). The other indicator has been SLR (Space Load Ratio) as the ratio between the actual and planned ISQ. It is employed to measure the short-term deviations of the actual (experienced) quality of service from the planned one.

In this context, referring to the queuing system theory, the rate *(r)* has been analogous to the queuing system's load factor (ρ) (Newell, 1982).

b) ISQ (Indicator of Service Quality)

The ISQ (Indicator of Service Quality) integrates the "static" (space) and "dynamic" (time) attributes of the service quality. It is defined as the ratio between the cumulative passenger delays in a given area and the total size of that area. The cumulative passenger delays are expressed in the units of passenger time; the size of area is expressed in the square metric units. This is expressed as follows:

$$ISQ = W(A)/A \tag{2.28a}$$

where

$W(A)$ is the cumulative passenger delay carried out in the area (A) (pax-min); and
A is the size of the area in question (m^2).

Equation 2.28a indicates that ISQ expresses the average unit "load" of a given area by the passenger-time, which is uniformly distributed over this area, i.e., each unit of area is approximately equally "weighted". This assumption is real, since frequent observations at many airports have indicated that the air passengers tend to be equally distributed over available space, particularly while waiting in large queues (or crowds) for longer periods of time. In addition, ISQ can be estimated for the entire terminal and the particular service phases performed over the passengers in the terminal as follows: ISQ_d for the passenger terminal; ISQ_s for the areas where the processing of passengers is carried out; ISQ_w for the areas where the passengers wait for the next service phase; ISQ_p for the passageways and walkways; etc.

c) Estimating size of the particular areas

In order to estimate ISQ in Eq. 2.28a the size of corresponding areas need to be estimated. This is the constant value, determined in advance during the planning and building of the passenger terminal(s). For example, the size of the particular areas is estimated as follows:

$$A = \begin{cases} A_0 \cdot N_0, & \textit{for the entire terminal} \\ A_{1l} \cdot N_1, & \textit{for areas around the service facilities} \\ A'_{2l} \cdot N_{2S} + A_{2S} \cdot (N_2 - N_{2S}), & \textit{for waiting areas with seats} \\ A_{3l} \cdot N_3 \equiv L \cdot W, & \textit{for passageways/walkways} \end{cases} \tag{2.28b}$$

where

A_{kl} is the space providing the level of service (l) in the area of type (k) for standing passengers; (A'_{kl}) is analogous to (A_{kl}) for seating passengers ($k = 0$ for the entire terminal; $k = 1$ for the service facilities; $k = 2$ for the waiting areas; and $k = 3$ for the passageways/walkways; $l \equiv A, B, C, D, E, F$, see Table 2.21) ($m^2$/pax);
N_k is the projected number of passengers simultaneously occupying the area of type (k) (pax);
N_{2s} is the number of chairs (seats) in the waiting and/or processing area (seats); and
L, W is the length and width of the passageway (or walkway), respectively (m, m).

a) <u>Estimating load of particular areas</u>

i) <u>Terminal—the concept of dwell time</u>

Estimation of ISQ for the entire passenger terminal is based on the dwell time, defined as the total time of a passenger passing through the terminal (Janić, 2000; Odoni and de Neufville, 1992). Generally, with respect to the activities which have to be performed for the air passengers passing through the terminal, the dwell time consists of three portions: the time, that they spend in queuing and servicing at the service facilities (ticketing and check-in counters, body-check desks, baggage claim devices, etc.), the "slack time" spent in the waiting areas (central hall, restaurants, duty-free shops, etc.), and the time needed for moving through passageways/walkways. Under such conditions, the passenger terminal is assumed to operate as the single server where the queuing processes take place as schematically presented by two cumulative passenger counts shown in Figure 2.55(a, b) (Newell, 1982).

Figure 2.55a shows the service process of departing passengers and Figure 2.55b shows the service process of arriving passengers. In both cases, the horizontal axis

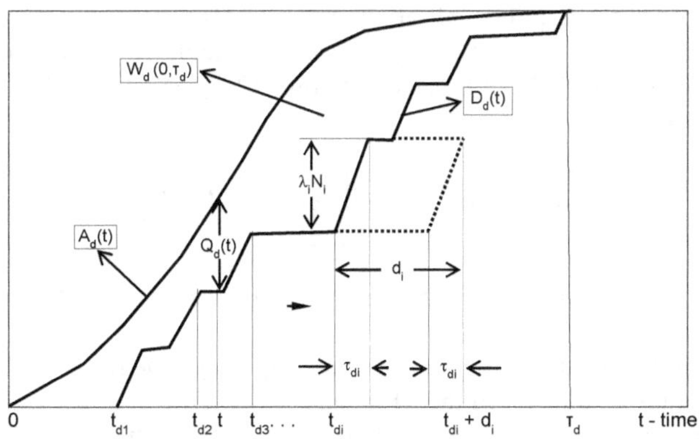

a) Departing passengers

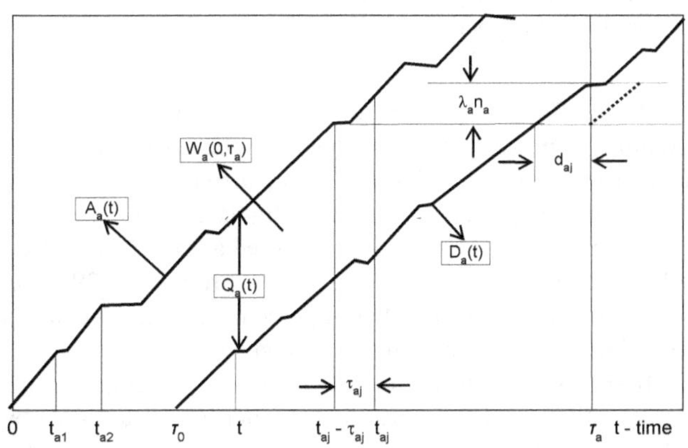

b) Arriving passengers

Figure 2.55 Scheme of congestion of air passengers in an airport terminal (Janić, 2000).

represents the time and the vertical axis the cumulative passenger counts, i.e., the cumulative number of passengers departing and/or arriving at the terminal by time (t) $(t \geq 0)$.

The continuous curve $(A_d(t))$ on Figure 2.53a represents the cumulative number of departing passengers entering the terminal by time (t). This curve is considered continuous and smooth because: a relatively large number of passengers always request to depart the terminal during the observed period $(0, T_d)$ (for example, there may be several thousand passengers departing the airport during a day); they may enter the terminal either individually or in groups of various sizes; each time a new passenger (or a group of passengers) enters the terminal, the curve $(A_d(t))$ increases but its change is assumed to be small compared to its current value (Newell, 1982). Curve $(D_d(t))$ represents the cumulative number of passengers boarding departing flights by time (t) and, thus, leaving the terminal. This is the step-shaped and discontinued curve at the times when the all passengers get on board the flight(s). The number of passengers leaving the terminal at once can vary from around thirty (when boarding a small plane) to several hundred (when boarding several large long-haul aircraft at the same time). Both $(A_d(t))$ and $(D_d(t))$ curves may be considered to be either the averages obtained from different samples of the day-to-day realisations or the specific realisations recorded for the specific day during a week, weekend, month, season, etc. From Figure 2.55a, the total dwell time of all departing passengers passing through the terminal during time $(0, T_d)$ is estimated as follows (Janić, 2000; Newell, 1982):

$$W_d(0,\tau_d) = \int_0^{\tau_d} A_d(t)dt - \sum_{i=1}^{N_d(\tau_d)} [\tau_d - t_{di} + \delta_i \cdot \left(\frac{\tau_{di}}{2}\right) - (1-\delta_i)\cdot(d_i - \frac{\tau_{di}}{2})]\cdot(\lambda_{di}\cdot n_{di}) \quad (2.29a)$$

where

$A_d(t)$ is the cumulative number of departing passengers entering the terminal by time (t) (pax);

$N_d(t)$ is the cumulative number of the departing flights by time (t), $(t \in (0, T_d))$;

λ_{di} is the average load factor of the (i)-th departing flight (-);

n_{di} is the seating capacity of the (i)-th departing flight (pax/flight);

τ_d is the time period in which the departure process is analysed (h, days);

d_i is delay of the departing flight (i) (min/flight); and

δ_i is binary variable taking value "1" if the flight (i) is not and the value "0" if it is delayed.

Equation 2.29a takes into account the eventual impact of delays of departure flights on the cumulative passenger dwell time. It can be easily shown that the departure delays increase the passenger dwell time since they arrive at the airport according to the timetable, not expecting delays of their departure flights. In addition, Eq. 2.29a is valid if: $\tau_d - (t_{di} + d_i) \geq 0$; otherwise, the flight (i) is considered as either cancelled or rescheduled for the first forthcoming period (τ_{d+1}) $(\tau_{d+1} = \tau_d + 1)$. From Eq. 2.29a, the average dwell time per passenger and the average number of passengers simultaneously in the terminal during the time period $(0, \tau_d)$ can be easily computed.

Figure 2.55b shows the dwell time for arriving passengers. The curve $(A_a(t))$ represents the cumulative number of passengers entering the terminal by time (t) after disembarking the arrived flights. The curve $(D_a(t))$ represents the cumulative number of passengers leaving the terminal by time (t). Similarly, as in the case of departing passengers, the main characteristics of dwell time can be estimated by deterministic queuing theory. Thus, the total dwell time of the arrived passengers passing through the terminal during time period $(0, \tau_a)$ can be estimated as follows:

$$W_a(0,\tau_a) = \sum_{j=1}^{N_a(\tau_a)} (\lambda_{aj} \cdot n_{ai})\,[\tau_a - t_{aj} - \left(\frac{\tau_{aj}}{2}\right)] - \int_{\tau_0}^{\tau_a} D_a(t)dt \tag{2.29b}$$

where

t_{aj} is the arrival time of flight (j) $(0 \leq t_{a1} \leq t_{a2} \leq ... \leq t_{aj} \leq ... \leq t_{aN} \leq \tau_a)$; (min);

τ_{aj} is the time which passengers need to disembark the flight (j) and enter the terminal (min); and

τ_a is the time period in which the arrival process is analysed (h, days);

τ_0 is the time needed for the arrived passengers to pass through the terminal $(\tau_0 > t_{a1})$, i.e., the dwell time of an arriving passenger (for the transit/transfer passengers, this time is for passing from the incoming to outgoing flights/gates).

The other symbols are analogous to those in Eq. 2.29a.

Similar to the case of departing passengers, the corresponding averages can be estimated. In cases of delays of some arriving flights, the slope of the curve $(A_a(t))$ will decrease, indicating that the smaller number of passengers will pass through the terminal during time period (τ_a) $(\tau_a > 0)$.

The concept of dwell time can be applied to two levels in order to manage the quality of services provided to air passengers in the airport terminals. First, the quality of services can be improved by shortening the total dwell time and consequently the total trip door-to-door time. This has been particularly important for business passengers, since the other categories of passengers seem to be less sensitive to the length of the dwell time. Both classes of passengers have been sensitive to delays of flights since they generally prolong duration and inconvenience of the entire door-to-door trips. Second, the structure (content) of dwell time can be an important factor influencing the quality of services offered to the passengers while in the terminal.

ii) Waiting/processing area—check-in counters

The time that the passengers spend in the specific service phases while in the terminal is primarily dependent on the temporary relationships between demand for service and available capacity of the service facilities. The service facilities are positioned "in line" (serial order) along the passengers' paths through the terminal building. Demand for service is expressed by the number and type of passengers requesting service at the particular facility per unit of time (Horonjeff and McKelvey, 1994; Janić, 2000).

The capacity of a service facility is dependent on the number of processors (servers) operating in a scope of the service facility, the average processing time per unit of demand (i.e., per passenger) and service discipline ("FCFS-First Come First Served" or other priority service disciplines can be applied). Whenever demand for service exceeds the capacity of the service facility it becomes congested and queues form in front of it. The instant length of queue and delay are dependent on the past and current relationships between demand and capacity. Departing passengers may form queues in front of the following service facilities: ticketing and check-in counters, security control counters, and the entrance of departure lounges. Queues of arriving passengers usually form in front of the immigration (security) desks and around baggage claim devices. Queues can be also formed in front of the custom desks intended for the arriving passengers. Transfer passengers may pass through similar procedures as both arriving and departing passengers. Passengers waiting in front of ticketing and check-in counters and those leaving the baggage claim area carry their baggage with themselves which may make their waiting for service and walking time through the terminal difficult and inconvenient, particularly if there are insufficient baggage trolleys.

The ticketing and check-in counter areas are the typical examples of the waiting/processing facilities of the departing passengers while in the terminal. Figure 2.56 shows a scheme of a typical queuing situation in the check-in counter area.

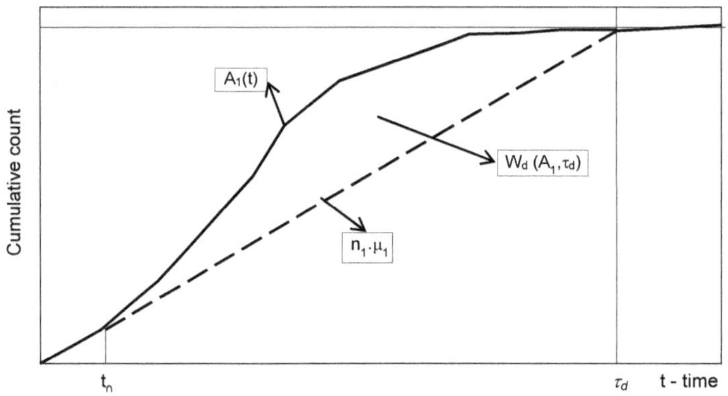

Figure 2.56 Scheme of congestion of air passengers in check-in counter area.

The total time that all passengers spend in the check-in counter area can be estimated similarly as in Eq. 2.29(a, b), as follows:

$$W_d(A_1, \tau_d) = \int_{t_0}^{\tau_d} A_1(t) - D_1(t)]dt = \int_{t_0}^{\tau_d} A_1(t)dt - (n_1 \cdot \mu_1) \cdot \frac{(\tau_d - t_0)^2}{2} \tag{2.29c}$$

where

$(n_1 \cdot \mu_1)$ is the "ultimate" capacity of the service facility; n_1 is the number of service units per service facility of type *"1"* (i.e., the number of either ticketing or check-in clustered counters operating as the single check-in service facility); μ_1 is the average service rate of the service unit of type *"1"* ($\mu_1 = 1/t_1$, where t_1 is mean service time per passenger per service unit) (min/pax); and

$\tau_d - t_0$ is the duration of the "busy" period at the service facility (min, h).

Using Eq. 2.28a and 2.29c, the actual *ISQ* for the check-in counter area can be estimated as follows:

$$ISQ_{1C}(A_{1C}, \tau_d) = \frac{W_d(A_1, \tau_d)}{N_1 \cdot A_{1C}}, \left[\frac{pax - min}{m^2}\right] \tag{2.29d}$$

where all symbols are analogous to those in the previous Eqs.

iii) Combination of waiting/processing area—departure lounge

The departure lounge is another typical example of an area in the airport terminal where a combination of both waiting and processing of departing passengers simultaneously takes place. Figure 2.57 shows the simplified scheme of passenger congestion in the

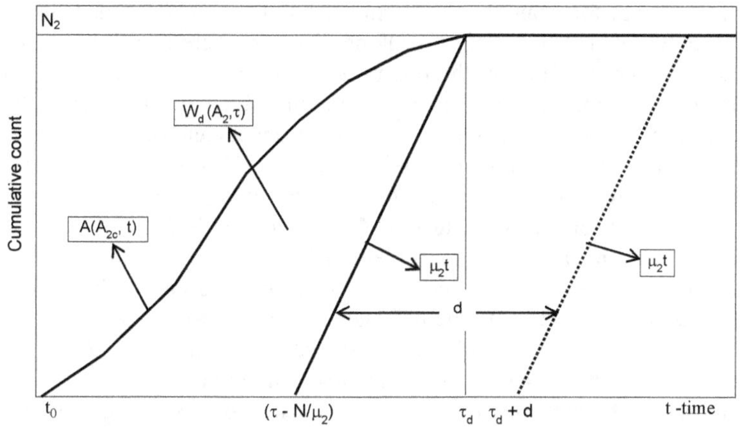

Figure 2.57 Queues and delays of passengers in the departure longue/gate (Janić, 2000).

departure lounge/gate when the departing flight is assumed to be delayed for some time (Janić, 2000; Wirasinghe and Shehata, 1988).

The total time that all passengers spend in the check-in counter area can be estimated similarly as in Eq. 2.29c, as follows:

$$W_d(A_2,\tau_d) = \int_{t_0}^{\tau_d} A(A_{2C},t)dt - \int_{t_0-N/\mu_2}^{\tau_d} (\mu_2 \cdot t)dt + \int_{\tau_d}^{\tau_d+d} Ndt = \int_{t_0}^{\tau_d} A(A_{2C},t)dt - \\ -\frac{N^2}{2\cdot\mu_2} + N_2\cdot d \tag{2.29e}$$

where

N_2 is the number of passengers boarding the flight $[A_2(A_{2c},\tau_d) = N]$;

μ_2 is the rate of boarding the flight, i.e., the rate of empting departure lounge/gate (pax/min); and

d is the anticipated delay of the departure flight (min/flight).

Using Eq. 2.28a and 2.29e, the actual *ISQ* can be estimated as follows:

$$ISQ_{wp}(A_{2C},\tau_d) = \frac{W_d(A_2,\tau_d)}{A_{2S}\cdot N_{2S} + A_{2C}\cdot(N_2 - N_{2S})}, \quad \left[\frac{pax-min}{m^2}\right] \tag{2.29f}$$

where

N_{2s} is the number of seats in the passenger lounge $[A_2(A_{2c},\tau_d) = N]$;

A_{2s} is the space standard for single seat (m^2/seat); and

d is the anticipated delay of the departure flight (min/flight).

iv) Passing through the passageways/walkways

The time taken for passing along the particular passageways/walkways is particularly important for transit/transfer passengers who always have a limited amount of time for passing from the incoming to the outgoing connecting flights. Since the volume of these passengers has been increasing considerably at particular airports, this time becomes of growing importance. This time may also be significant at large airports where the walking distances are very long. At these airports, the passengers pass through long corridors connecting, for example, two terminal buildings or different halls (areas) in the same large terminal. The length of walking distances there is primarily conditioned by terminal configuration and by the organisation and management of the use of the apron/gate complex. Since the terminal configuration is fixed, the use of available apron/gate stands crucially influences passenger walking distances and walking time. The other factor influencing the walking time is walking speed (Tošić, 1992).

In order to estimate the actual *ISQ* for a passageway (or walkway) in an airport terminal, its area is specified by the length (L) and width (W) $(L >> W)$. As mentioned above, this area may either link two airport terminals, the airport satellites with the central terminal complex or the gates equipped with the air bridges located in the long finger (Horonjeff and McKelvey, 1994). The passageway (or walkway) is assumed to provide the space for (N_3) air passengers with the guaranteed space standard 'C' (see Table 2.21). It is also assumed that the area has never come into saturation, e.g., a queue of customers has never formed at its entrances. The average time taken to pass through the passageway in question is denoted (τ_{pw}). The intensity of passenger flow entering the passageway/walkway during this time is assumed to be constant $(\lambda(\tau_{pw}))$. Consequently, the number of passengers simultaneously occupying the passageway/walkway during time (τ_{pw}) can be estimated as follows (Newell, 1982):

$$N_3(\tau_{pw}) = \lambda(\tau_{pw}) \cdot \frac{L}{v[\lambda(\tau_{pw})]} \tag{2.29g}$$

where

$v[\lambda(\tau_{pw})]$ is the average speed of passenger flow through the passageway (m/s).

From Eq. 2.29g, the total time that all passengers spend in the passageway can be estimated as follows:

$$W(A_{3C}, \tau_{pw}) = \int_0^{L/v[\lambda(\tau_{pw})]} N_3(t)dt = \lambda(\tau_{pw}) \cdot \left[\frac{L}{v[\lambda(\tau_{pw})]}\right]^2 \tag{2.29h}$$

In Eq. 2.29h, the average speed of passengers flow through the passageway/walkway, $w[\lambda(\tau_{pw})]$ can be estimated analogously as in the traffic flow theory (Gazis, 1974; Janić, 2000). From Eq. 2.29h, the actual *ISQ* for a passageway/walkway is estimated as follows:

$$ISQ(A_{3C}, \tau_{pw}) = \frac{W(A_{3C}, \tau_{pw})}{L \cdot W}, \quad \left[\frac{pax - min}{m^2}\right] \tag{2.29i}$$

where all symbols are analogous to those in the previous Eqs.

v) "Slack" time

As a component of dwell time, the "slack" time is represented by the gaps of "free time" between the successive services performed for air passengers while in the terminal. This time may be spent either in the bars, restaurants, duty-free shops, walking around, or by sitting/standing in the particular halls. In the given context, *ISQ* of these areas is not particularly elaborated.

e) Estimated projected load of particular areas

From Table 2.18, the unit projected *ISQ* for the terminal or any of its above-mentioned passenger processing areas can be estimated as follows:

$$ISQ_k = \frac{1 pax \cdot \tau_{kl}}{A_{kl}}, \left[\frac{pax - min}{m^2}\right] \tag{2.30}$$

where

τ_{kl} is the time standard of quality of service in the area of type *(k)* (min);

A_{kl} is the space standard of the quality of service *(l)* in the area *(k)* for the single occupant (m^2/pax) (see Table 2.19).

f) SLF (Space Load Factor)

If the designed *ISQ* is known, then it can be compared with the actual *ISQ*. One way is to divide the actual ISQ by the projected *ISQ*. If the obtained ratio is less than one ("1"), the passengers experience higher than the projected quality of service; otherwise they experience a lower quality of services. The ratio between the actual and projected *ISQ* is called *SLR* (Space Load Ratio) *(SLR ≡ r)*. Referring to the queuing system theory, the ratio (r) is analogous to the load factor (ρ). Similar to the factor *(ρ)*, *SLR* or *(r)* can vary between zero and up to and above one. For example, if $r = 0$ there is no load in the given area. The values of *(r)* between 0 and 1 indicate that the occupants experience higher than projected quality of services. If $r = 1$, the occupants get exactly the same as the projected quality of services. If $r > 1$, the occupants experience lower than projected quality of services.

g) Examples of SLR (Space Load Factor)

i) Check-in counter area

The concept of *SLR* in the check-in area is presented by the hypothetical scenario of serving $N_1 = 935$ passengers intending to board 7 departing flights during the peak time period. They are assumed to arrive at the airport by the car/taxi (30%) and bus (70%) during the time interval $\tau = 1.5 - 0.5$ h before the scheduled departure time of each flight. The modal split and the arrival time at the airport terminal influences the shape of cumulative curve of total demand, as shown in Figure 2.58.

After entering the terminal, the air passengers approach the block of *22* check-in counters assumed to operate as the integrated service facility where each counter serves

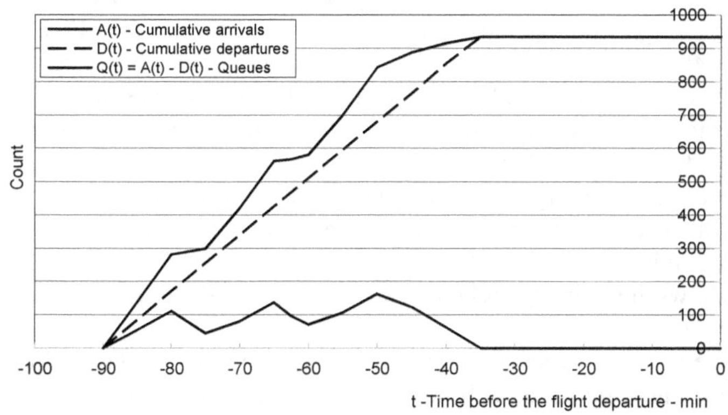

Figure 2.58 Demand, capacity, and congestion at the check-in counters area over time in the given example (Janić, 2000; 2013).

passengers from each flight. The average service time per passenger per check-in counter is assumed to be: t_{1c} = 1.5 min/pax, which is equivalent to the average unit processing rate of: $\mu_0 = 1/t_{1c} = 1/1.5 = 0.67$ pax/min. The "ultimate" capacity of these 22 check-in counters is: $\mu = n \cdot \mu_0 = 22 \cdot 0.67 = 14.7$ pax/min. The space standard is assumed to be at the level 'C', i.e., A_{1c} = 1.5 m^2/pax. The average reference waiting time for the check-in is adopted as 15 min (see Table 2.20). Because the intensity of demand is higher than the capacity of check-in counters, the air passenger congestion in front of the check-in counters immediately starts to develop, then sustains, although varying over time, and finally disappears about 35 min before the flight departures. After being served at the check-in counters, the passengers proceed to the security checking desks and then towards the departure lounges/gates. Figure 2.59 shows *SLR* based on this scenario and recommendations in Table 2.20.

As can be seen, *SLR* changes over time in proportion of changing the check-in queue. At the same time, it is substantively lower than "1" mainly due to the significant difference between the queues and delays in the given example and those suggested/recommended in Table 2.21.

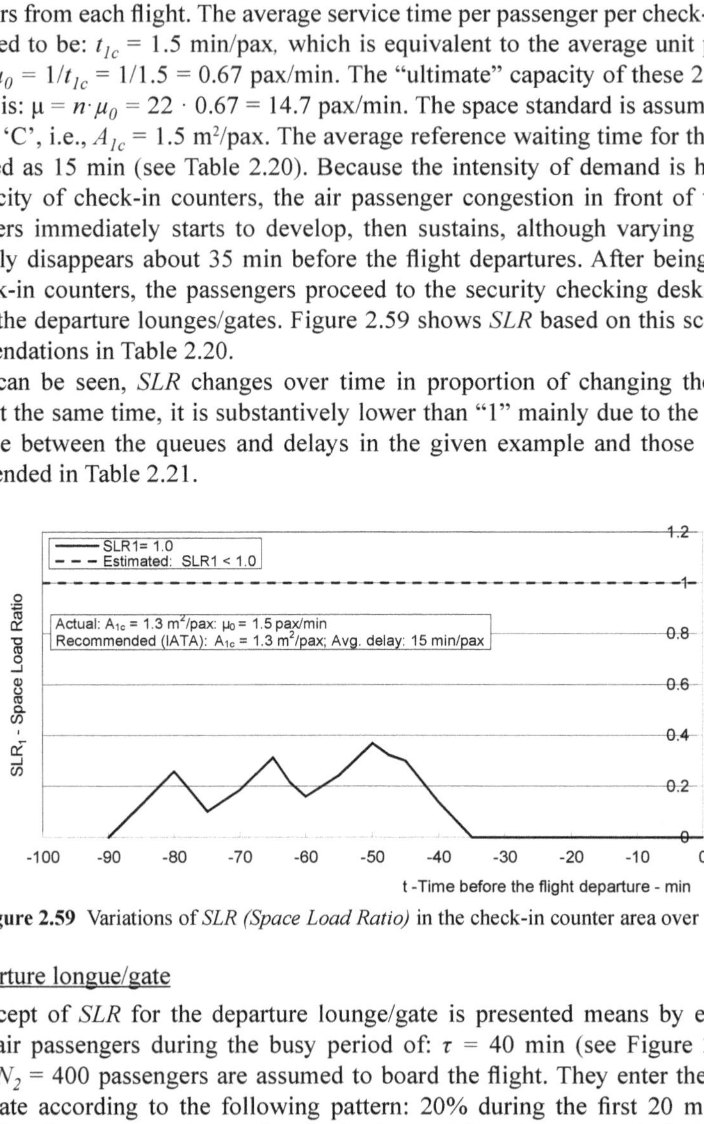

Figure 2.59 Variations of *SLR (Space Load Ratio)* in the check-in counter area over time.

ii) Departure longue/gate

The concept of *SLR* for the departure lounge/gate is presented means by example of serving air passengers during the busy period of: τ = 40 min (see Figure 2.55). The number N_2 = 400 passengers are assumed to board the flight. They enter the departure lounge/gate according to the following pattern: 20% during the first 20 minutes and 70% during the next 10 minutes, i.e., in the interval from 20 to 10 min before closing the gate/flight for boarding; the last 10% of passengers arrive during the last 10 min before the end of the gate/flight boarding. About 15 min is needed for 400 passengers to board the flight. This implies that the rate of emptying the departure lounge is equal to: μ_2 = 400/15 pax/min. The time standard of quality of service is adopted to be: $t_{2/wp}$ = 15 min. Figure 2.60 shows the estimated relationship between the SLR, delay of given flight, and size of lounge/gate area.

As can be seen, *SLR* increases by increasing of the flight delay and decreasing of the area of departure lounge/gate. It deviates from the value of one if the same number of passengers is expected to stay in a smaller space for a longer time, which is intuitively expected.

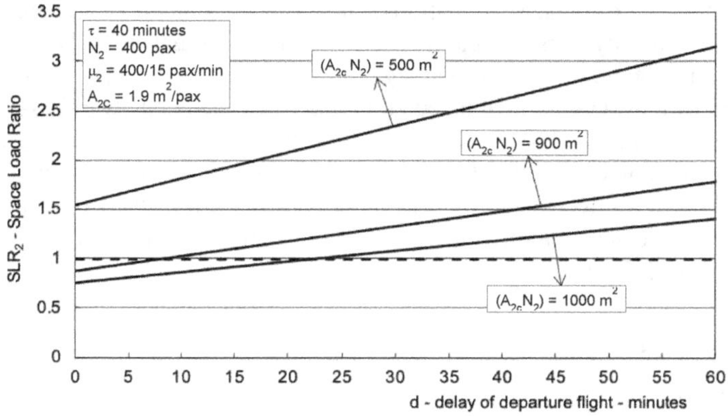

Figure 2.60 Relationship between *SLR (Space Load Ratio)* of the departure lounge/gate and delay of departing flight.

i) Passageways/walkways

The concept of *SLR* for the departure longue/gate is based on the assumption that the given passageway/walkway operates at the low to medium traffic loads, reflecting the typical situations at most airports. The free speed of passengers is assumed to be $v_0 \cong 1.37$ m/s indicating that they are walking through passageway/walkway (Lovas, 1994). Each passenger is assigned the space standard of $A_{3c} = 1.9$ m²/pax, corresponding to the quality level 'C' (see Table 2.21). The average time of occupying the unit space $(A_{3/c})$ is estimated to be $t_{3pw} = 2.77$s ($t_{3pw} \cong (1.9)/(1.37/2) = 2.77$s) (the actual passenger speed is assumed to decrease to a half of the free speed (v_0) due to the increased interaction between air passengers while in the passageway/walkway). Figure 2.61 shows the relationship between the *SLR*, intensity of passenger flows, and width of the given passageway/walkway.

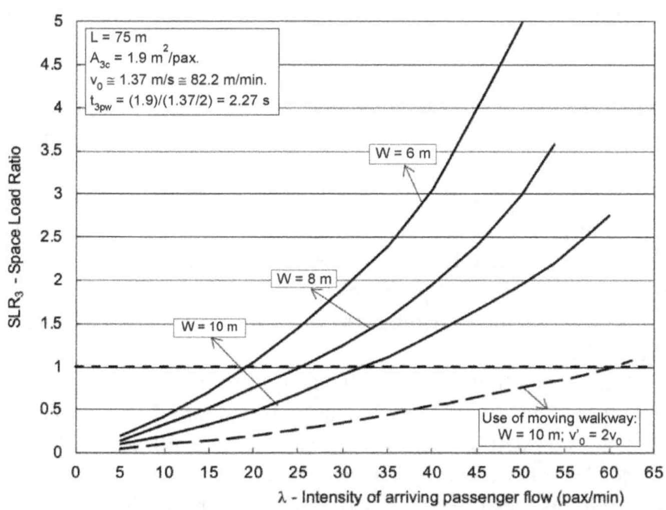

Figure 2.61 Relationship between the space load ratio and the intensity of passenger flows entering the passageway (Janić, 2000; 2013).

As can be seen, the space load ratio *(r)* increases more than proportionally with an increase in the intensity of passenger flow entering the passageway/walkway. This ratio decreases with enlargement of the passageway/walkway by increasing its width. The *SLR*, denoted by a heavy line, will be equal to one when the intensity of passenger flows varies between about λ = 20–30 pax/min. When the average moving speed is doubled by installing moving walkways, the time spent in the passageway/walkway would be shortened. This could decrease *SLR*, i.e., it would reach the value "1" at the intensity of passenger flow entering the passageway/walkway of λ = 60 pax/min (dotted line on Figure 2.59) (Janić, 2000; 2013).

2.5 Economics

2.5.1 General

The airport economics in the given context relate to the airport revenues, costs, and their profits and losses. The airport revenues, costs, and profits generally relate to both airport airside and landside areas, the latter excluding the landside access modes and their systems. Therefore, the economics of the airport landside access modes and their systems are elaborated separately.

The total airport revenues can generally be categorized as aeronautical and non-aeronautical. The aeronautical revenues are obtained from landing fees, passenger fees, aircraft parking fees, handling fees (if it is provided by the airport operator), and terminal rental fees (at the U.S. airports). The non-aeronautical revenues include retail, food and beverage, car hire, advertising, car parking, and recharges (gas, water, electricity, etc.).

The total airport costs can generally be distinguished as the operating and capital costs. The operating costs include the personnel costs and the costs of services. The personnel costs relate to the salaries, pensions and other costs relating to an airport's staff. The costs of services depend on their proportion carried out by the third parties and that carried out by airport staff and other resources. The capital costs make up a large proportion of the total costs. These costs include the interests on debt and depreciation on the airport infrastructure. For example, in the year 2018, the global airport industry realized a total of 161.3 billion $US. In this amount, the aeronautical revenues made up 56%, the non-aeronautical revenues 39.4%, and the non-operational revenues 4.6% of the total. The average aeronautical revenue was 10.15 $US/pax and the non-aeronautical revenue 7.12 $US/pax. The average cost was 13.55 $US/pax. Consequently, the average profits as the difference between the revenues and costs was (10.15 + 7.12) – 13.55 = 17.27 – 13.55 = 3.72 $US/pax (https://aci-economics.com/aef19/conference-report/).

In general, the airport revenues are obtained by charging fees for the services provided to its users. These charges enable the full operating and capital costs to be covered by the revenues. In general, three approaches have been commonly applied as follows: (a) the single-till or the residual method; (b) dual-till as the compensatory method; and (c) hybrid-till (ICAO, 2012).

The single-till approach implies that the charges to the air transport users are based on the total operational and capital costs. These charges are then modified by including the influence of the non-aeronautical revenues. The dual-till approach implies setting up the charges to the users of air transport services exclusively based on the costs of resources for providing these services. Consequently, these charges are not modified regarding the non-aeronautical revenues. The hybrid-till approach combines the single-

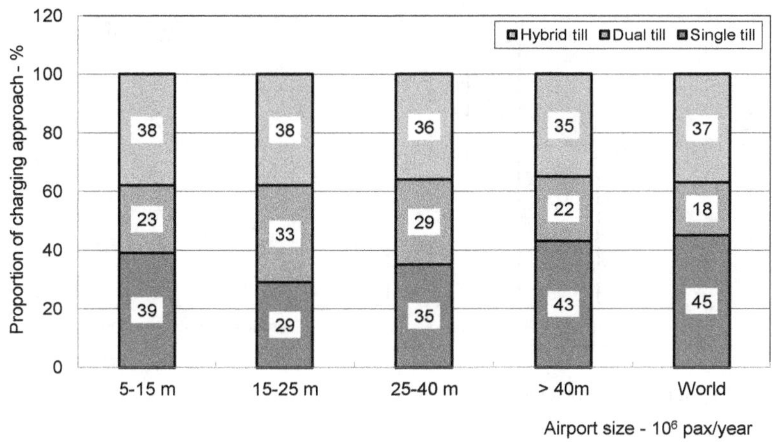

Figure 2.62 Example of the relationship between the proportion of different charging approaches and the airport size (ACI Europe, 2015).

till approach and the dual-till approach. In such case, the airport operator may choose to recover landing costs by using the single-till and the passenger terminal costs by using the dual till approach. Figure 2.62 shows the proportion of using particular charging approaches at the airports of different sizes (ACI, Europe 2015).

As can be seen, the single till makes up from 29% to 43%, dual till from 21% to 33%, and hybrid till approach from 36% to 38%. The averages at the world scale have been 45, 18, and 37%, respectively. These figures indicate that the type of applied till approach generally does not substantially depend on the airport size, in this case expressed by the annual number of passengers handled.

2.5.2 Airside and Landside Area

2.5.2.1 Structure of economics

a) Revenues

As mentioned above, the total airport revenues consist of the aeronautical and non-aeronautical components. For example, for the year 2013 in the global airport industry, the corresponding shares were 60% aeronautical and 40% non-aeronautical revenues. The average structure of the aeronautical and non-aeronautical revenues in this case is shown in Figure 2.63(a, b) (http://www.aci.aero/Publications/ACI-Airport-Statistics/ACI-Airport-Economics-Report-with-Excel-indicator-tables).

Figure 2.63a shows that the passenger and landing charges have the highest shares. Figure 2.63b shows a dispersion of shares of the non-aeronautical revenues across particular categories, with the highest being retail concessions, car parking, and real estate income or rent.

b) Costs

The total airport costs include the operating and capital costs. For example, in the world's airport industry, the total costs were 106500 million $US in the year 2013. The share

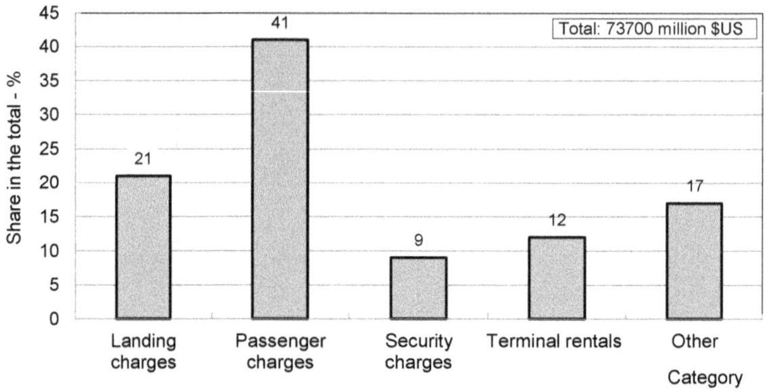

a) Aeronautical revenues

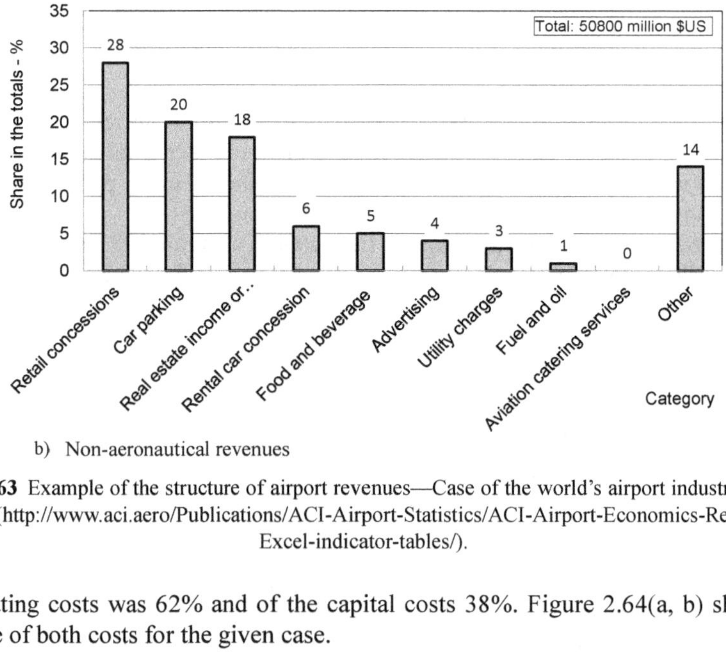

b) Non-aeronautical revenues

Figure 2.63 Example of the structure of airport revenues—Case of the world's airport industry (Period: 2013) (http://www.aci.aero/Publications/ACI-Airport-Statistics/ACI-Airport-Economics-Report-with-Excel-indicator-tables/).

of operating costs was 62% and of the capital costs 38%. Figure 2.64(a, b) shows the structure of both costs for the given case.

Figure 2.64a shows that the highest share in the total operating costs was that of the personnel costs (35%) followed by the share of contracted services due to payments to the third parties (23%), which represents the second-largest component of operating expenses. Figure 2.64b shows that the share of depreciation/amortization costs was higher than that of the interest on debt costs.

c) Profits

The airport profits represent the difference between the total revenues and costs. In addition to being expressed in the absolute terms annually, the profits are also expressed per unit of the airport output—handled passenger or WLU (Workload Unit) (1 WLU = 1 passenger or 100 kg of freight/cargo) (Doganis, 1992). Figure 2.65 shows an example for the large European airport – London Heathrow (UK).

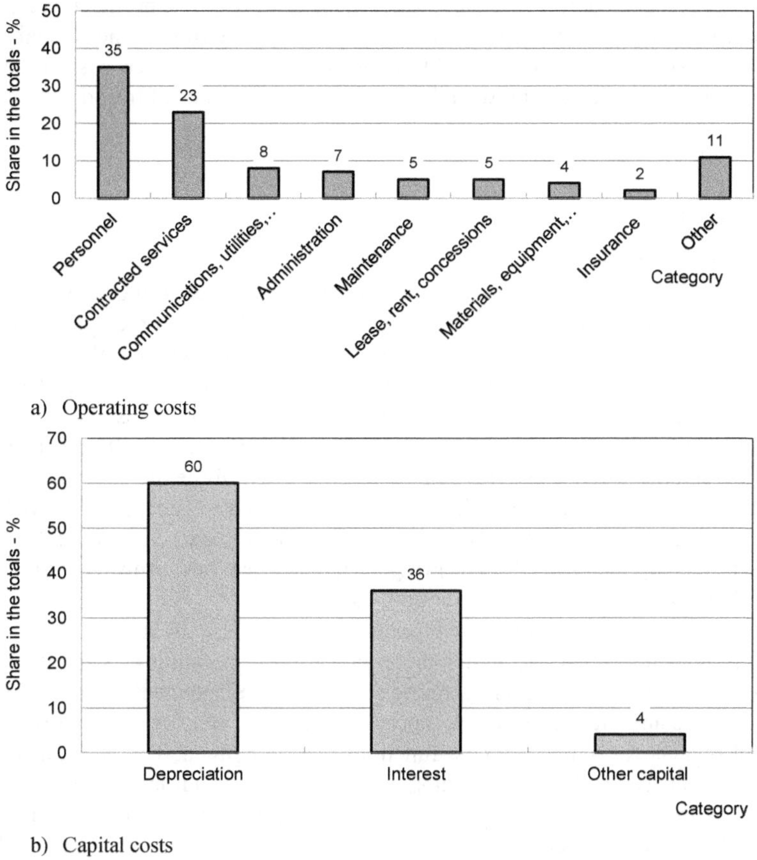

Figure 2.64 Example of the structure of airport costs—Case of the world's airport industry (Period: 2013) (http://www.aci.aero/Publications/ACI-Airport-Statistics/ACI-Airport-Economics-Report-with-Excel-indicator-tables/).

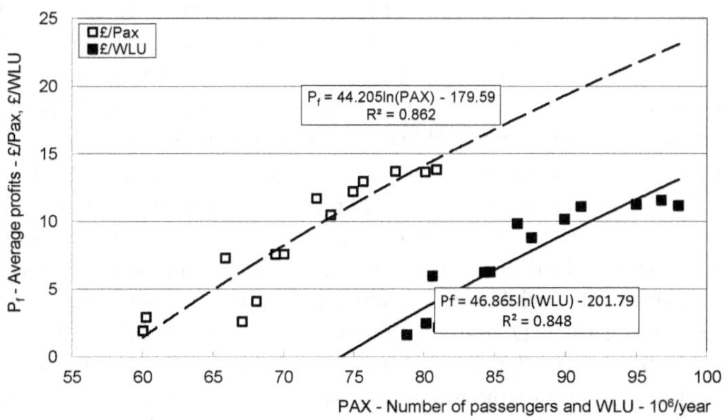

Figure 2.65 Relationship between the average profits and the airport annual output—Case of London Heathrow airport (Period: 2006–2019) (H(SP)L, 2006/2019).

As can be seen, during the observed period, the average profits increased at a decreasing rate with an increase in the airport output in terms of the annual number of handled passengers and WLUs. This indicates that in the given case the airport's profitability has been decreasing with the growth in the volumes of traffic.

2.5.2.2 Modelling economics

a) Airside area

i) Landing charges based on the airport operational costs

The above-mentioned till methods applied by airports to charging services to their users—airlines and air passengers—have generally been based on the operational costs, i.e., the expenses of resources for providing them. This implies that the charge-based revenues aim to recover a part of the operational costs and gain some profits. Eventually, this approach can be used to indirectly indicate the potential needs for expansion of capacity of the airport airside and landside components (Doganis, 1992). This principle has two advantages. First, as strictly applied, this rule will force the "expected demand" to become a "real demand" consisting of the users willing to accept the offered prices. Second, the "real demand" indicates whether the new (additional) capacity of the airport's infrastructure components needs to be provided at that price (new runway(s), taxi-way(s), apron parking stand(s), terminal building(s), etc.).

The market attractive cost-based pricing policy is usually based on the airport cost function, which relates to either the entire airport operating as a single entity or the airport's specific components operating/functioning independently. The main airport infrastructure components, runways, aprons, terminals, etc., are the most common individual objects for which the cost functions can be defined. The cost function may have different analytical forms depending on the operating conditions, which can change over time. With respect to the variability depending on the volumes of demand, the fixed and variable total cost functions for both short-run and long-run periods can be defined. In both cases, the average and marginal short-run and long-run cost can also be estimated (Button, 1993; Doganis, 1992; Janić, 2000).

The total cost function of an airport usually consists of the fixed and variable cost components. The total fixed costs are represented by the fixed payments for the new facility. These include the investments and their costs as debts, interests, and taxes paid during a given period of time. These costs do not change in the short-run or with changes in demand. The total variable costs are the expenses paid for operation of the facilities and equipment during a given time period. These costs change in the short-run and with changes in demand. The average variable cost represents the operator's expenses related to serving a single user during a given period under given conditions. The marginal cost represents an additional cost, which can be considered as an expense that would be saved if an additional (new) user would not be served. Once the cost function has been estimated, two cost-based pricing sub-policies for charging the airport services can be designated: the short-run variable cost pricing policy and the short-run marginal cost pricing policy (Button, 1993; Doganis, 1992; Janić, 2000; Whol and Hendrickson, 1984). In order to model these costs, let $(TC_i(n))$ be the total operator's cost to serve (n) users at the facility (i). The average and the marginal cost are $(AC_i(n))$ and $(MC_i(n))$, respectively. The service facility can operate under two regimes: low to medium (i.e., off-peak) load and heavy (i.e., peak) load. The generic form of the function $(TC_i(n))$ is expressed as:

$$TC_i(n) = \begin{bmatrix} FC_i + a_i \cdot n, & for < n < n_{0i} \\ FC_i + a_i \cdot n + b_i \cdot n^{Ci}, & for\ n_{0i} < n < \mu_i \end{bmatrix} \quad (2.31a)$$

where

FC_i is the short-run fixed cost of facility *(i)* ($US);

a_i, b_i, c_i are the coefficients of the cost function (-);

n is the number of users requesting the service at facility *(i)* (user per unit of time);

n_{0i} is the peak demand at the moment when it starts (user per unit of time);

μ_i is the capacity (throughput) of service facility (i) (users/unit of time).

Equation 2.31 indicates that the total costs $(TC_i(n))$ increase approximately linearly with demand increasing at the low and medium levels and more than proportionately at the heavy level (i.e., power fitted). From Eq. 2.31a, the average cost of facility *(i)* is equal to:

$$AC_i(n) = TC_i(n)/n = \begin{bmatrix} FC_i/n + a_i, & for < n < n_{0i} \\ FC_i/n + a_i + b_i \cdot n^{C_i - 1}, & for\ n_{0i} < n < \mu_i \end{bmatrix} \quad (2.31b)$$

The fixed costs (F_i) in Eq. 2.31b are uniformly distributed over all users; (a_i) is the unit cost per user served under the low to medium traffic conditions; (b_i) is the unit cost per user served under the heavy traffic conditions; and (c_i) is the rate of increase in the service cost per user with an increase in traffic load. The marginal cost per user is equal as follows (Janić, 2000; Whol and Hendrickson, 1984):

$$MC_i(n) = \partial TC_i(n)/\partial n = \begin{bmatrix} a_i, & for < n < n_{0i} \\ a_i + b_i \cdot c_i \cdot n^{C_i - 1}, & for\ n_{0i} < n < \mu_i \end{bmatrix} \quad (2.31c)$$

where all symbols are analogous as in the previous Eqs. Actually, the term (a_i) represents the cost of additional resources consumed to serve a user under low to medium traffic conditions (this is the cost of additional staff, engagement of additional equipment, energy, lighting, etc.). With an increase in demand from the medium to high traffic load, this unit cost becomes higher due to consumption of more resources to serve each user at peak demand. The product $(b_i c_i)$ in Eq. 2.31c quantifies such an increase.

Figure 2.66 shows a scheme of the above-mentioned cost functions for an airport runway as the service facility *(i)*. The landing aircraft represent demand for service (the horizontal axis). The charge for service is shown on the vertical axis. The runway capacity is assumed to be constant. Two down-sloped curves represent the various levels of demand for service. The grouping of horizontal and up-ward sloped curves represents the cost of running the runway operations (Doganis, 1992; Janić, 2000).

When the demand for landings is low (curve D_1D_1), the corresponding charge aims to cover only the marginal cost (a_i) (The fixed cost of the runway FC_i is not included). For example, when demand for landings is at the level (n_1) (the horizontal axis), the charge to each user, i.e., landing aircraft is $(p(Q_1))$ (the vertical axis). Because demand is sensitive to price, setting the charge to cover the average cost (i.e.., this is at higher level) will cause a decrease in demand. If the charge is set up to $(p(n_o))$ $(p(n_o) > p(n_1))$, the demand for landings will decrease from (n_1) to (n_o) $(n_o < n_1)$, and vice versa; if the charge is set at the near-zero level, the demand or landings will not exceed the level (n_i). This shows that the marginal cost-based pricing policy may be efficient for the

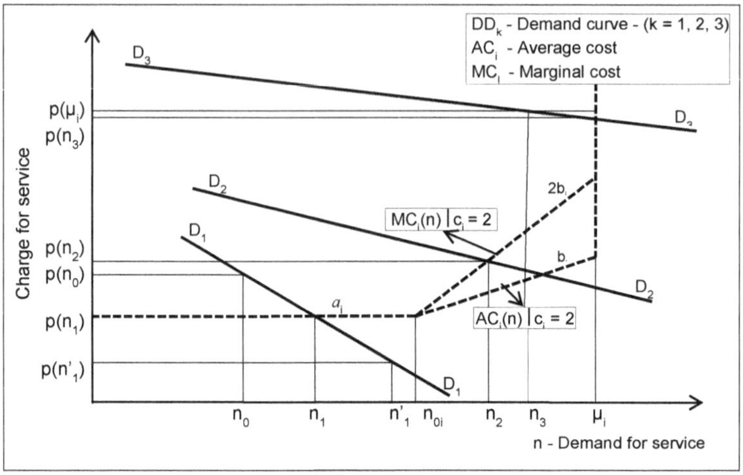

Figure 2.66 Scheme of the cost function and its application to landings at an airport runway (Janić, 2000).

airports with the low-to-medium traffic demand. By setting the low cost-based charge, these airports may attract higher volumes of demand, which can provide better utilisation of the available capacities. The peak traffic conditions start at the level of demand for landings (n_{0i}) when both the average and marginal costs start to increase in line with an increase in demand. The down-sloped curve (D_2D_2) represents the peak demand, which is sensitive to the price. The peak demand imposes higher cost for the airport operator. Under such conditions, the marginal cost pricing policy will impose higher charging rates on particular landings during the peak period (i.e., $p_2(n_2) > p_1(n_1)$, $n_2 > n_1$).

When the demand for landings approaches or crosses the runway capacity (the demand curve D_3D_3 and the runway capacity μ_i), the landings will be charged by the fees equal to the long-run marginal costs. Such charging policy has the following effect: first, the unit charge will be sufficient to cover the construction and operational cost of the new facility (in this case the new runway). Second, the demand for landings will decrease for an amount $(\mu_i - n_3)$ because $p(n_3) > p(\mu_i)$. However, despite such loss of demand, the airport will continue to keep the charge $p(n_3)$ while building the new runway (Doganis, 1992; Janić, 2000).

The above examples indicate that the pricing policy based on the average cost can have some advantages as follows: first, it may instigate demand by setting the relatively low (attractive) charges under the low traffic volumes. Second, the charges are based exclusively on the airport operational/aeronautical costs. Third, it permits the application of different pricing policies during the peaks and off-peaks. Fourth, the charges apparently reflect the need for the new facility because they take into account the "actual" and not the "expected" demand. Finally, this policy may provide better utilisation of the available airport capacities. In addition to these advantages, this pricing policy possesses some disadvantages, which can compromise its efficient application. The main disadvantage seems to be the high sensitivity of this policy to the changes of demand. As mentioned above, the significant changes of demand cannot be controlled by the airports and their charging policies. This is because the development of demand at most airports is mostly affected by the external local and global driving forces (See sub-section above).

ii) Landing charges based on the external costs

In addition to charging users based on their own operational costs, airports can also implicitly or explicitly apply an external cost-based pricing policy by charging for externalities, such as congestion, noise, and air pollution. In case of congestion, this has been proposed as a marketing tool for controlling access to the congested airports (Janić, 2005).

The external cost of a user competing for service at the congested facility depends on its characteristics and the level of congestion at the moment of joining it. In general, the users of services provided by an airport runway as the service facility are the aircraft and the passengers on board. The aircraft can be charged by the cost of airborne delays. The passengers can be charged by the value of time lost due to delay. The cost of aircraft delay depends on the aircraft size and flight operating cost. The cost of passenger time depends on the passenger's socio-economic characteristics, trip purpose (business, leisure), and the time of carrying out the trip (peak, non-peak, morning, night, working day, weekend, etc.). The charging policy based on the external cost due to congestion implies that the charge for landing should be equal to the marginal cost of landing delay. This policy aims to sanction those users who cause any extra delay to both other users and themselves. The marginal delay cost consists of the user's private cost (e.g., passenger and aircraft cost of delay) and an incremental cost, which the new user imposes on the others (Janić, 2005). In order to apply the concept of marginal congestion charging, it is necessary to estimate the total, average, and marginal delay cost-demand function for different conditions. Theoretically, these functions can have the following generic forms:

$$TC_d(\lambda) = a \cdot \lambda^b \tag{2.32a}$$

$$AC_d(\lambda) = a \cdot \lambda^{b-1} \tag{2.32b}$$

$$MC_d(\lambda) = \partial TC_d(\lambda)/\partial\lambda = (a \cdot \lambda^b)/\partial\lambda = a \cdot b \cdot \lambda^{b-1} \tag{2.32c}$$

where

a, b are the coefficients estimated by the calibration of the delay cost-demand function (-); and

λ is the average intensity of demand during a given period of time (atms/h).

The other symbols are analogous to those in Eq. 2.31. The cost functions 2.32 (a, b, c) are developed under the assumption that the average internal cost of delay is approximately equal for all landing aircraft and independent on the level of congestion, as shown in Figure 2.67.

The down-sloped curve *(DD)* represents the price-sensitive demand, which decreases when the landing charges are raised. The other factors affecting demand are assumed to be constant. Equations 2.31(a, b, c) and 2.32(a, b, c) indicate that if $b \geq 1$, $MC_d(\lambda)$ will be greater than $AC_d(\lambda)$ for the factor *(b)*. The difference $[MC_d(\lambda) - AC_d(\lambda)]$ represents the additional cost of delay caused by an additional (marginal) user-landing. The intersection of the curves *(DD)* and $(MC_D(N))$ (point *A* in Figure 2.65) determines the intensity of demand (λ_0) when the additional user joining it should be charged by $(c(\lambda_0))$. This charge consists of the user/landing aircraft cost $c(\lambda_0)$ and the social cost $[c(\lambda_0) - c'(\lambda_0)]$. The latter actually represents the congestion toll *(AB)*. The additional user/landing aircraft should be imposed this toll due to imposing the additional costs on all other users/landing aircraft in the flow $(\lambda > \lambda_0)$ arriving during the congestion period.

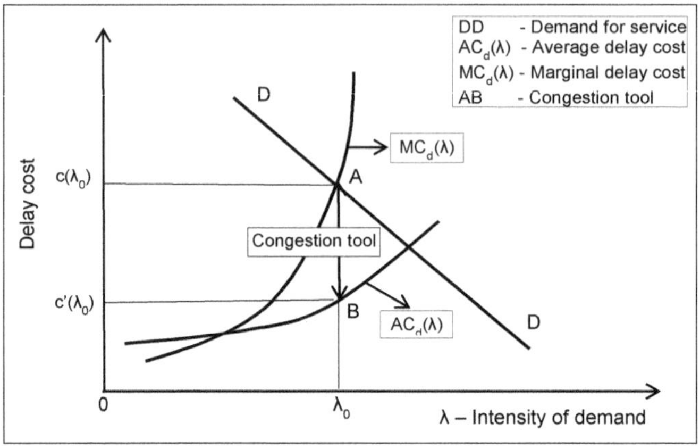

Figure 2.67 Pricing policy based on the external (congestion) cost during landings.

The model in Eq. 2.32(a, b, c) is applied using the data from NY La Guardia airport, one of the three large airports serving New York area (USA) (See also Figure 2.9c). The airport mainly handles domestic U.S. short- and medium-haul flights with about 92% O-D passengers, of which 45–55% are business. At present, 20 airlines operate at the airport by the fleets mostly consisting of the aircraft categories B737/717, A320 (100-150 seat) as well as some smaller regional jets and turboprops (70–110 seats). The greatest market shares in terms of the number of flights and the number of passengers, respectively, belong to US Airways (38%; 14.2%), Delta (18%; 17.2%), and American (17%; 18.5%) Airlines (PANYNJ, 2003; 2003a).

The runway VMC and IMC "ultimate" capacity is about 80 (40/40) and 64 (32/32) atms/h, respectively. The apron/gate complex has 72 aircraft parking stands. Since the lack of slot control, the hourly and daily atm demand has frequently exceeded both capacities and consequently caused frequent and significant congestion and delays. Under such conditions, the auction of slots (i.e., "slottery") was introduced as the demand management tool in the year 2000 in order to mitigate the congestion problem. Due to unavailability of land for the physical expansion of capacity, the options for mitigating congestion and delays under conditions of expected growth can actually be increasing the average aircraft size (it has been 58–62 pax/atm) on one side and the runway capacity by introducing innovative operational procedures and technologies on the other (See also Table 2.13). The former has already happened by introducing B767-400ER (about 280 seats) in the year 2001 under given operational limits on the runways (AIRWISE NEWS, 2001). The latter, yet to take place, is expected to contribute to increasing the VMC and IMC runway capacity by about 10% and 3%, respectively.

Since it is expected that both options will not be able to efficiently cope with potential increase in congestion and delays, introducing congestion charges may again be considered. At present, the airport charges landings based on the aircraft weight with the unit charge of $US6.55/500 kg of the aircraft MTOW (Maximum Take-Off Weight). In addition, each atm carried out between 8:00 a.m. and 9:00 p.m. is charged by the amount of $100 (PANYNJ, 2003; 2003a).

The inputs used for application of the model include data on the atm demand and capacity needed for estimation of congestion and delays under given circumstances, the aircraft operating costs and the average fare per passenger, the latest needed for assessment of the profitability of particular flights. An example of the relationship between the atm demand and corresponding capacity at NY La Guardia airport during an average day is shown in Figure 2.68.

As can be seen, despite being at the level of 80 atms/h, the capacity during the considered average day was 50 atms/h. The aircraft/flight operating cost *(c(S))* has been estimated in dependence on the aircraft seat capacity *(S)* as the regression equation based on the data from the U.S. airlines as follows ($US/Block Hour): (*N* is the number of elements) (The data used related the U.S. airlines). According to the mentioned regression equations, the average total cost per unit of time of an aircraft of 100–150 seats (B737/717) operated at La Guardia airport would be between: $c(S)$ = $US 2209 and $c(S)$ = $US 3307/h or $US 37 and $US 55/min). For B767-400ER with $S = 280$ seats, this cost would be $c(S)$ = $US 6162/h ($US 103/min). The average airfare per passenger at NY La Guardia airport has been obtained by using the data from the U.S. airlines in the year 2002. The regression equation was used to estimate the relationship between the average fare *(F)* and route length *(L)* for different U.S. markets as follows: $F = 9.561 \cdot L^{0.390}$; $R^2 = 0.941$; $N = 15$). In addition, for the average length of flight to/from the airport of about $L = 1200$ km, the average fare has estimated to be about: F = $US 152/pax (Janić, 2005; PANYNJ, 2003; 2003a). The average and marginal costs are estimated by Eq. 4(a, b, c) and shown in Figure 2.69.

Since the demand has mostly been lower than the capacity, queues, delays and their total, average, and marginal costs increased more than proportionally with an increase in the intensity of demand. When the intensity of demand increased to about 40 atms/h, the marginal cost of delays started to increase faster than average cost. At the intensity of demand of about 45 atms/h, the new arriving atm would be charged by the congestion toll of about 225 $US/arm.

i) Landing charges as the airport demand management tool

The airport services pricing policy based on the user's external (congestion) cost can also be applied for demand management at the congested airport. The corresponding models are based on the assumption that each customer freely decides whether to join or

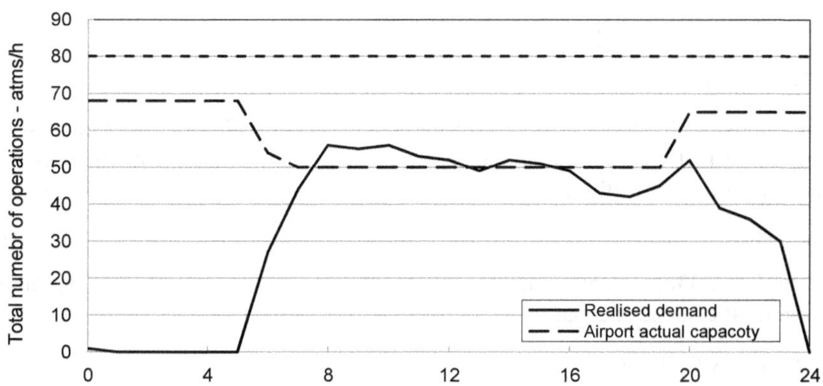

Figure 2.68 Demand and capacity during time of day—Case of NY La Guardia airport (U.S.) (Period: An average day 2001) (FAA, 2002; Janić, 2005; PANYNJ, 2003; 2003a).

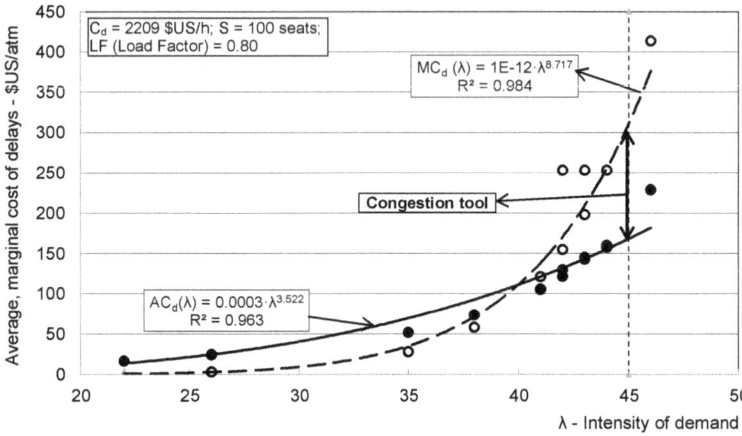

Figure 2.69 Average and marginal costs of delays—Case of NY La Guardia airport (Period: An average day 2001) (Janić, 2005).

not join the queue at the moment of arrival at the congested airport. The main decision criterion is the net benefits that would be obtained by getting service there. These benefits are equal to the difference between the aircraft revenues gained by its landing and the cost imposed on it due to waiting for service. This cost consists of two components: the aircraft operational cost and the external (congestion) cost. According to this policy, an aircraft will decide to join the landing queue if it can gain at least non-negative benefits by landing. Otherwise, the aircraft will give up on joining the queue. In order to model the basic principles of this policy, let *(n)* be the number of aircraft in the landing queue when the additional aircraft arrives. The net benefits of joining this queue can be expressed as follows (Janić, 2000; 2005; Naor, 1969; Rue and Rosenshine, 1985):

$$[c_d \cdot (n+1)]/\mu_a \leq r \tag{2.33a}$$

where

c_d is the cost delay of a newly arrived aircraft ($US/h-atm);
μ_a is the runway landing capacity (atm/h);
q is the queue of aircraft in front of newly arriving aircraft (atm); and
r is the revenue gained by landing ($US/atm).

By imposing the congestion toll on the newly arriving aircraft, its landing should become less attractive. For example, this aircraft will join the landing queue only if the following condition is fulfilled:

$$[c_d \cdot (n+1)]/\mu_a + Toll \leq r \tag{2.33b}$$

From Eq. 2.33b, the congestion tool that will discourage the newly arriving aircraft from joining the queue is equal to:

$$Tool \geq r - [c_d \cdot (n+1)]/\mu_a \tag{2.33c}$$

Once this congestion tool has been set up, the landing queue still acceptable for the newly arriving aircraft should be within the constraints as follows (Janić, 2000; 2005; Naor, 1969):

$$0 \leq n \leq (\mu_d/c_d) \cdot (r - Tool) - 1 \quad (2.33d)$$

where all symbols are as in the previous Eqs.

The model in Eq. 2.33(a, b, c, d) is again applied to the case of NY La Guardia airport. This time, the inputs on the demand and capacity are used for the peak day—30 June 2001 beginning of the Independence Day holiday) (FAA, 2002; Janić, 2005). The cumulative count of demand, capacity and resulting atm queues during this day are shown in Figure 2.70.

The horizontal axis represents time of day and the vertical axis cumulative counts of total demand and capacity—arrivals and departures. The vertical distance between the two curves approximates the number of aircraft in the queue (both arrivals and departures) at given time. The horizontal distance indicates delay of the last customer waiting in the queue. The area between the curves represents the total aircraft waiting (queuing) time, which seems, in given case, to be enormous (Newell, 1982).

The estimated congestion tool deterring the access of new atm to join the queue during the period of congestion is shown in Figure 2.71.

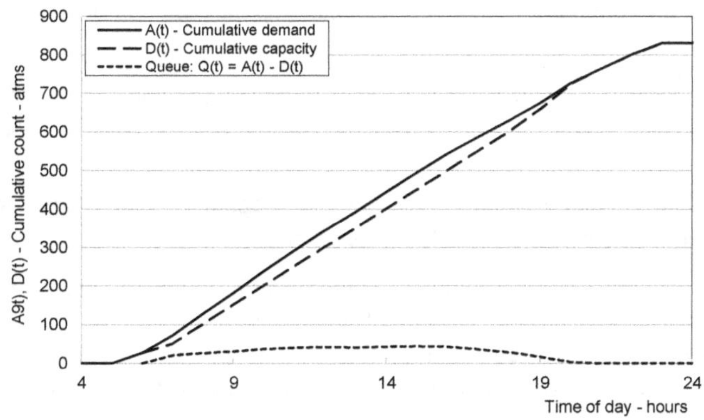

Figure 2.70 Cumulative count of demand, capacity, and congestion queue during time of day—Case of NY La Guardia airport (U.S.) (Period: 30 June 2001) (Janić, 2005).

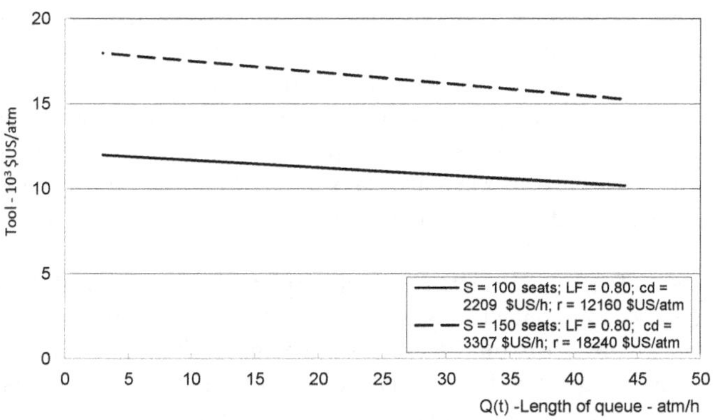

Figure 2.71 Relationship between the airport access tool deterring the flights access and the length of queue—Case of NY LaGuardia airport (U.S.) (Period: 30 June 2001) (Janić, 2005).

As can be seen, when the queue that the new atm intends to join is longer, the congestion deterring tool will be lower and vice versa. This indicates that, if faced with the longer queue, the weaker "barrier" would be required in order to discourage the new access. In addition, this tool would be higher for the atms carried out by larger aircraft with the higher costs of delays.

ii) Implementation of congestion charging

Implementation of congestion charging at many airports has been complex mainly due to the lack of clear criteria for setting up the level of delays, which would be charged. At the same time, as mentioned above, the aeronautical and non-aeronautical revenues have sufficiently covered the airport costs.

In general, congestion and delays usually appear as a consequence of many simultaneous causes, "interrelationship between demand and capacity" being one of the most important. However, this cause may vary between different airports. Such diversity creates diversity of the peaks, causing congestion and delays. This is noticeable in terms of the time pattern, size and duration of peaks, combination of causes, type of operations, and the affected aircraft/flights.

Regarding the time pattern, the peaks causing congestion and delays at airports can be as follows:

- Short and frequent peaks during the day, in which demand exceeds capacity; this commonly happens at the hub airports due to the airline hub-and-spoke operation practice;
- Long and infrequent peaks during the day, when demand exceeds capacity for longer periods; this is characteristic of the non-hub congested airports.

Consequently, determining the relevant level and duration of congestion and delays to be charged may be complex. In general, only the long daily congestion and delays caused by local regular (scheduled and thus expected) positive differences between demand and declared airport capacity should be considered to be a matter of demand management, and consequently the subject of charging (Janić, 2005). In addition, the demand management of the short sharp congestion and delays as the consequence of airline hub-and-spoke operations and/or those due to disruptions of the airport capacity is questionable for eventual charging. In the former case, the integrity of airline schedules could be compromised, which is opposite to the overall objectives to protect the vital interests of actors/stakeholders involved—airports and airlines. These interests embrace maintaining the airport attractiveness despite increased access charges on the one side and non-discrimination of the particular types of users due to the unacceptable access costs (i.e., FCFS – "First Come-First Served" principle should still be guaranteed) on the other. In the latter case, congestion and delays of any kind caused by factors out of the control of actors/stakeholders involved—airports, airlines, and ATC/ATM—should not be charged.

The concept of congestion charging at airports itself seems to be ambiguous. Actually, it imposes additional charges on those customers-flights who presumably could impose additional delay costs on the other succeeding flights during a given period of time. The objective is to deter (i.e., prevent) access of these new customers who would create congestion and delays or significantly contribute to increasing the already existing delays or congestion. First, the 'prevention of access' seems to be in contradiction with the guaranteed right of 'free access' according to the commonly accepted rule "First-Come First-Served" (Corbett, 2002). Second, the charge may be difficult to determine since the real market conditions seem to often be very different to those theoretically assumed, in which the charge based on the cost of marginal delays is supposed to work. If it is too weak, a charge will not be able to reduce congestion simply because demand will not react to it. Otherwise, it may significantly deter sensitive demand and, thus, reduce it below a desirable level, which may compromise the airport revenues. Third, it is not quite clear if and what could be a combination of congestion charges and the other already existing airport aeronautical charges as well as externalities. Fourth, it is not quite clear how the congestion pricing money would be distributed. If it were intended for expansion of the airport infrastructure, then after such expansion has taken place, the sources of funding—congestion and delays—would disappear for a while. Last but not least, it would be difficult to impose the additional charges to the inherently economically and financially vulnerable airline industry.

The above-mentioned reasons have contributed to building up the opposition, i.e., barriers to introducing congestion charging, which could be clustered as follows (Janić, 2005):

- Institutional organisational, political and legal barriers have been maintained by the monopolistic powerful hub airports (Europe) and air carriers (both Europe and U.S.) including the lack of harmonisation across the countries (Europe) and airports—both small and big hubs (Europe and U.S.).
- Reluctance of large airlines and their alliances faced with the financial troubles has always been inherently present in considering acceptance of the concept. In particular, these airlines seem not to be willing to accept charging if other modes would not (both in the U.S. and Europe).

Technological barriers relate to lack of the system for collection of the relevant data on congestion and delays at airports and their causes as well as the data on the actual airport capacity (Europe). Relatively good databases about developments at airports have been set up both in the U.S. and Europe (EEC, 2019; FAA, 2002; https://www.transtats.bts.gov/HomeDrillChart.asp). In addition, there has also been tendency to mix congestion charging with other externalities, such as noise and emissions of GHG. Also, the analogy with other transport modes while developing the 'marginal cost pricing systems' has been frequently considered.

b) Landside area

The economics of the airport landside access modes and their systems is presented by providing the quantitative figures of their costs and revenues. Their detailed analytical models can be found in the corresponding referent literature (Janić, 2019).

i) Road-based mode and its systems

• Car and van

The economics of cars and vans operating as the airport landside access systems generally relate to the capital/investments and operational costs of infrastructure and rolling stock/vehicles and the revenues if they carry out taxi services. Since the infrastructure is also shared with the other non-airport access users, its costs are not considered.

The capital/investment costs for vehicles—cars and vans—are the expenses for their acquirement, which as the fixed annual amounts spread over the period of 10–15 years. As such, together with insurance and different fixed taxes, they represent the fixed component of the total costs. The operational costs of cars and vans mainly include the expenses for fuel/energy and maintenance, and in the case of taxi services the expenses for drivers and other operating/supporting staff. As such, these costs mainly depend on the intensity of car use during the specified period of time (usually 1 year), thus representing the variable component of the total costs. For example, the average cost of privately using cars, including for airport access, amounts to 0.249–0.273 $US/p-km and in cases of performing taxi services 1.554 $US/p-km. This rather great difference could be caused by the inclusion of costs of salaries for taxi drivers (Litman, 2017). The revenues of taxi services depend on their fares. Table 2.22 gives the fares by Medallion taxi services between Manhattan (NYC – New York City) and three airports NY LaGuardia, NY JFK and NY Newark.

Table 2.22 Some estimation of fares of the Medallion taxi services between Manhattan (Empire State Building) (NYC – New York City) and its three airports (TLC, 2016; https://www.taxifarefinder.com/main.php?city=NY/).

Airport	Distance L (mi)	Official fare[1)] F_a($US)
NY LaGuardia	9.4	26.48
NY JFK	19.2	52.12
NY Newark	19.0	62.41

[1)] Fares without influence of traffic congestion; mi – mile (1 mi = 1.609 km).

- Bus system

The economic performances of the bus systems relate to the capital/investment and operational costs, generally for both road infrastructure and vehicles, i.e., buses, and revenues from charging users—air passengers, the airport and aviation employees, and all others. The capital/investment costs include the expenses for building infrastructure-roads/lanes, bus stops, stations/terminals, and supporting facilities and equipment, such as the above-mentioned ITS. For the vehicles (buses) the capital/investment costs are the expenses for their acquirement. The capital/investment costs for both infrastructure and vehicles are usually spread as the fixed amounts over their life cycle (for the former it is usually 20–25 years, for the latter it is 10–15 years). As such, they represent the fixed component of the total costs. The expenses for administration also represent the fixed cost component for the bus operators. Since the infrastructure is shared with the non-airport access bus systems, it is not considered. For example, when the average costs per bus mile are $\overline{c}_{b/i}$ = 250 £ pence, the costs of bus transport between London City and three airports would be Heathrow – 250/100 (£/mi) · 19.3 (mi) = 48.25 £; Gatwick – 250/100 (£/mi) · 27.9 (mi) = 69.75 £; and Stanstead – 250/100 (£/mi) · 38.8 (mi) = 97.0 £ (£ – British pound; mi – mile; 1 mi = 1.609 km). In addition, Table 2.23 gives the structure of average total costs for the selected BRT (Bus Rapid Transit) systems in the U.S. implying that they are not substantially different than in cases when these systems operate as the airport landside access systems.

The costs of BRT systems are covered by revenues. For example, the average fare of 40 BRT systems operating round the world has been 1.25 $US/p (p – passenger). About

Table 2.23 The costs of selected BRT systems (averages) (GAO, 2012; Janić, 2014; 2019; TCRP, 2018).

Cost element	**System**
	BRT
Infrastructure and vehicle costs	
Infrastructure (millions $US/km)	8.98
Vehicle (millions/unit)	0.4–1.0
Amortization period (years) • Infrastructure • Vehicles	 25 12–15
Operating costs	
• $US/pax-km[1)] • $US/veh-km[2)] • $US/pax[3)]	0.12 3.05 3.20

[1)] 5 BRT systems in the U.S.; [2), 3)] Six BRT systems in the U.S.

68% of the systems (27 of 40) have needed subsidies at an average level of 25–30% (Again, for the purpose of comparison, the LRT systems have also needed subsidies at a level of 20–25%) (Janić, 2014; 2019).

i) Rail-based mode and its systems

The economics of rail-based modes and their systems operating as the airport landside access systems relate to their capital/investment and operating costs, and revenues.

Capital/investment costs include the expenses for building, installing, and capital maintaining infrastructure (rail lines), supporting facilities and equipment (power supply and traffic control/management system), and acquiring and capital maintaining the rolling stock (vehicles). The operating costs include the expenses for operations including those for energy, operating staff, and daily maintenance of the systems' components, administration, etc. Table 2.24 gives examples of both average infrastructure/capital and operational costs for the rail-based system operating, assuming that they can operate as the airport landside access systems.

Table 2.24 Examples of the costs of the rail-based mode and its systems (Janić, 2019).

System	**Costs**	
	Infrastructure (10^6 $US/km)	**Operations** ($US/veh-km)
Streetcar/Tramway	12[1)]	2.16[1)]
LRT (Light Rail Transit)	25[2)]	0.372[3)]
Subway/Metro	149[4)]	7.2[5)]
Conventional rail	15.4[6)]	9.2[7)]
HSR	20.5[8)]	11.7[9)]
TRM08	48.3[10)]	6[10)]

[1)] 6 U.S. cities; [2)] 6 U.S. cities; [3)] 7 U.S. cities; [4)] 5 countries; [5)] 16 countries; [6)] UIC (International Union of Railways); [7)] S = 279 seats; L = 50 km; v = 80 km/h; 8) 6 countries; [9)] S = 328 seats; L = 400 km; v = 250 km/h; [10)] Shanghai International airport (China).

As can be seen, there has been high diversity of both categories of costs, mainly due to being taken from different cases operating under different conditions and circumstances. As such, they can be considered only for the illustrative purposes in the given context. In addition, in most cases, the fares providing revenues for Tramway & LRT transport operators have not covered the costs, and consequently required subsidies by the local authorities. For example, the average cost coverage of the LRT systems in the U.S. was about 30% (Garett, 2004; Janić, 2019). However, some LRT systems in Europe have managed to fully-cover their costs. In the subway/metro systems in the U.S., the revenues gained from the fares have covered the costs up to 30%. The rest have been covered by the local and regional subsidies. In the LU (London Underground) system, the fares have fully covered the total costs starting from the year 2012/2013 (TfL, 2015). The conventional rail and HSR have obtained revenues based on the fares covering the costs (JR, 2012). In particular, the fares for users/passengers have been set up to cover the total operating costs if subsidies have not been provided due to any reasons. In the only commercialized case, the costs were covered by charged fares.

References

ACC. (2008). Airport Information Technology & Systems (IT&S): Best Practice Guidelines for the Airport Industry, Airport Consultants Council, Alexandria, Virginia, USA.

AACC/IATA. (1981). Guidelines for Airport Capacity/Demand Management, Geneva, Switzerland.

ACI Europe. (2015). Airport Economics Report—2014, Airport Council International, Brussels, Belgium.

ACI. (2019). ACI Annual Report 2018, Airport Council International, ACI World, Montreal, Quebec, Canada.

ACI Europe. (2019). Airport Industry Connectivity Report 2019, Airport Council International Europe, Brussels, Belgium.

AEA. (2004). Yearbook 2005, Association of European Airlines, Brussels, Belgium.

AIRWISE NEWS. (2001). "Delta Launches B767-400ER at La Guardia", http://news. airwise.com/.

AOA. (2014). Sustainable Airports: Improving the Environmental Impact of the UK's Global Gateways, Airport Operators Association, Edinburgh, UK, www.aoa.org.uk.

Ashford, N. (1988). Level of Service Design Concept for Airport Passenger Terminals: A European View. Transportation Research Record 1199, Transportation Research Board (TRB), Washington DC, USA, pp. 19–32.

Ashford, N. and Wright, P. (1992). Airport Engineering, A Wiley Interscience Publications, John Wiley and Sons, New York, USA.

Ashford, N., Stanton, H. P. M. and Clifton, M. A. (1997). Airport Operations. Second Edition, McGrow Hill Book Company, Boston, Massachusetts, USA.

ATL. (2008/2019). Comprehensive Annual Financial Report, City of Atlanta, Department of Aviation, Georgia, USA.

BAA. (2005). Heathrow Interim Master Plan, The British Airport Authority, London, UK, http://www.baa.co.uk.

Barker, K. L., Hankins III, W. W. and Hueschen, R. M. (1999). Speed Profiles for Deceleration Guidance During Rollout and Turnoff (ROTO), National Aeronautics and Space Administration (NASA), NASA/TM–1999–209829, Langley Research Center, Hampton, Virginia, USA.

BCG. (2004). Airports-Dawn of New Era: Preparing for One of the Industry's Biggest Shake-Ups, Boston Consultancy Group, Munich, Germany, p. 36.

Bishop, C. K. (2012). Assessment of the Ability of Existing Airport Gate infrastructure to Accommodate Transport Category Aircraft with Increased Wingspan for Improved Fuel Efficiency, Report No. ICAT-2012-4, MIT International Center for Air Transportation (ICAT), Department of Aeronautics & Astronautics, Massachusetts Institute of Technology, Cambridge, Massachusetts, USA.

Blumstein, A. (1960). An Analytical Investigation of Airport Capacity, Report No. TA-1358-8-1, Cornell Aeronautical Laboratory Inc., New York, USA.

Button, J. K. (1993). Transport Economics. 2nd Ed., Edward Elgar, P. L., Grower House, England, UK.

Corbett, J. J. (2002). Small communities are concerned about congestion pricing. The Air and Space Lawyer, Summer, 17(1): 17–21.

Daganzo, F. C. (1978). An approximate analytic model of many to many demand responsive systems. Transportation Research, 12: 325–333.

de Neufville, R. and Odoni, A. (2003). Airport Systems: Planning, Design, and Management. McGraw Hill, New York, USA.

DETR. (1999). Oxford Economic Forecasting: The Contribution of the Aviation Industry to the UK Economy, Department of the Environment, Transport and Regions, London, UK.

Doganis, R. (1992). Airport Business. Routledge, Abingdon, UK.

EC. (2005). Optimal Procedures and Techniques for the Improvement of Approach and Landing—OPTIMA, Sixth Framework Program, FP6-2002-Aero 1502880, Deliverable 1.2, European commission, Brussels, Belgium.

EC. (2007). Gaps Identification Airport, TMA, En Route and TBS—Reduced Separation Minima-RESET, Deliverable D4.2, EC 6th Framework Program, European Commission, Brussels, Belgium.

EEC. (2005). European Wake Vortex Mitigation Benefits Study; Identification of WV Constrained Airports, WP1-Deliverable D1, TRSPO3, EUROCONTROL, European Organisation for the Safety of Air Navigation, Brussels, Belgium.

EEC. (2015). "RECAT-EU" European Wake Turbulence Categorisation and Separation Minima on Approach and Departure, Edition 1.1., EUROCONTROL, The European Organisation for the Safety of Air Navigation, Brussels, Belgium.

EEC. (2018). European Aviation in 2040: Challenges of Growth, EUROCONTROL, Brussels, Belgium.

EEC. (2019). CODA Digest-All-Causes Delay and Cancellations to Air Transport in Europe: Q2 2019, EUROCONTROL, The European Organisation for the Safety of Air Navigation, Brussels, Belgium.

FAA. (1988). Planning and Design Guidelines for Airport Terminal Facilities, Advisory Circular, AC No. 150/5360-13, Federal Aviation Administration, US Department of Transportation, Washington DC, USA.

FAA. (1990). Runway Length Requirements for Airport Design, Advisory Circular 150/5325/4A, Federal Aviation Administration, US Government Printing Office, Washington DC, USA.

FAA. (1999). Standards for Airport Markings, Advisory Circular, 150/5340 1H, Federal Aviation Administration, US Government Printing Office, Washington DC, USA.

FAA. (2002). Aviation Policy and Plans (APO)—FAA OPSNET and ASPM, Federal Aviation Administration, Washington D.C., USA.

FAA. (2004). Airport Capacity Benchmarking Report 2004, US Department of Transportation, Office of Airports Airport Planning and Programming, Federal Aviation Administration, Washington D.C., USA.

FAA. (2007). FAA Operations & Performance Data, Department of Transportation, Federal Aviation Administration (http://www.apo.data.faa.gov/), Washington DC, USA.

FAA. (2014). Aircraft Wake Turbulence, Advisory Circular, AC_90-23G, Federal Aviation Administration, U.S. Department of Transportation, Washington D.C., USA.

FAA. (2014a). Airport Capacity Profiles, Federal Aviation Administration, Office of Airports Airport Planning and Programming, Washington D.C., USA.

FAA. (2018). United States Standard for Terminal Instrument Procedures (TERPS), FAA_Order_8260.3D1, Federal Aviation Administration, U.S. Department of Transportation, Washington D.C., USA.

FZ AG. (2006/ 2018). Annual Report, Flughafen Zürich AG, Zurich, Switzerland.

GAL. (2009/2019). Annual Report and Financial Statements for the year ended 31 March 2009/2019, Gatwick Airport Limited, Crawley, West Sussex, England, UK.

GAO. (2012). Bus Rapid Transit: Projects Improve Transit Service and Can Contribute to Economic Development, GAO-12-811, Report to the Committee on Banking, Housing, and Urban Affairs, United States Government Accountancy Office, Washington D.C., USA.

Garrett, A. T. (2004). Light-Rail Transit in America: Policy Issues and Prospects for Economic Development, Federal Reserve Bank of St. Louis, St. Louis, USA.

Gazis, C. D. (1974). Traffic Science. John Wiley and Sons, USA.

HAL. (2008/2019). Annual Report and Financial Statements for the Year Ended 31 December (2008/2019), Heathrow Airport Limited, Hounslow, Middlesex, UK.

HC. (2014). Aviation Strategy, First Report of Session 2013–14, Volume II, Transport Committee, House of Commons, London, UK.

Hockaday, M. L. S. and Kanafani, A. (1974). Development in airport capacity analysis. Transportation Research, 8: 171–180.

Horonjeff, R. and McKelvey, F. X. (1994). Planning & Design of Airports. McGrow Hill Book Company, New York, USA.

H(SP)L. (2006/2019). Annual Accounts, Heathrow (SP) Limited, The Compass Centre, Nelson Road, Hounslow, Middlesex, UK, https://www.heathrow.com/company/investor-centre/reports/annual-accounts.

http://www.airdistances.com

http://www.faa.gov/about/initiatives/nextgen/

http://go.updates.iata.org/l/123902/2017-03-07/7lrghg/

http://www.aci.aero/Publications/ACI-Airport-Statistics/ACI-Airport-Economics-Report-with-Excel-indicator-tables

http://www.maglevboard.net/en/facts/systems-overview/transrapid-maglev/transrapid-maglev-shanghai/

https://www.aci-europe.org/44-industry-data/40-airport-traffic.html/

https://www.anna.aero/2009/10/09/have-cake-and-eaten-it-has-abolition-of-departure-tax-helped-klm-and-amsterdam/

https://www.internationalairportreview.com/article/110871/top-20-busiest-airports-world-aircraft-movements/

https://www.heathrow.com/company/investor-centre/reports/traffic-statistics/

https://www.heathrow.com/company/company-news-and-information/company-information/facts-and-figures/

https://www.iata.org/en/services/consulting/airport-pax-security/level-of-service/

https://www.iata.org/en/pressroom/pr/2002-06-12-24/

https://www.transtats.bts.gov/HomeDrillChart.asp

https://www.google.nl/search? q=heathrow+central+bus+station+map/

https://www.businesstraveller.com/airlines/2013/08/07/five-of-the-worlds-largest-airport-terminals/

https://www.ertmssolutions.com/ertmscamcorder-chosen-by-thalys-for-etcs-compliance-field-testing/

https://aci-economics.com/aef19/conference-report/

https://www.taxifarefinder.com/main.php?city=NY/

https://www.internationalairportreview.com/article/32311/top-20-largest-airports-world-passenger-number/

https://centreforaviation.com/analysis/reports/lccs-increasingly-attracted-to-primary-airports-459531/

https://www.gatwickairport.com/business-community/

https://www.gatwickairport.com/to-and-from/by-coach-or-bus/local-buses/

https://en.wikipedia. org/wiki/Gatwick_Airport

https://www.skybrary.aero/index.php/RECAT_-_Wake_Turbulence_Re-categorisation#Re-categorisation/

https://en.wikipedia.org/wiki/Airport rail_link#Europe/

https://www.ns.nl/

https://atlanta.com/

https://www.vccr.nl/

https://www.skynews.com/

https://www.travelchinaguide.com/cityguides/shanghai/getting-around.htm/

https://www.transtats.bts.gov/HomeDrillChart.asp

Hwang, H. L., Chin, S. M., Hopson, J. and Tagof, F. A. (2000). Evaluating the accessibility of the U.S. airports: Results from the American travel survey. TRB Transportation Research Circular E-C026, Transportation Research Board, US Department of Transport, Washington DC, USA, pp. 288–305.

ICAO. (1993). Aircraft Noise, Environmental Protection, Annex 16, Vol. 1, International Civil aviation Organization, Montreal, Canada.

ICAO. (2004). Aerodromes, Annex 14, International Standards and Recommendation Practices, 4th Edition, International Civil Aviation Organization, Montreal, Canada.

ICAO. (2012). ICAO's Policies on Charges for Airports and Air Navigation Services, Doc 9082, Ninth Edition, International Civil Aviation Organization, Montreal, Canada.

ICAO. (2016). Doc 4444 Air Traffic Management, Procedures for Air Navigation Services, Seventh editions 2016, International civil aviation Organization, Montreal, Canada.

https://www.skybrary.aero/index.php/RECAT_-_Wake_Turbulence_Re-categorisation#Re-categorisation/

ITA. (2001). Air Transport Markets in Europe and the United States: A Comparison, Institute of Air Transport, Paris, France, p. 52.

Jacobs. (2015). Appraisal Framework Module 4: Surface Access: Dynamic Modelling Report Gatwick Airport Second Runway, Airports Commission, Jacobs U.K. Limited, Wokingham, UK.

Janić. (2000). Air Transport System Analysis and Modelling: Capacity, Quality of Services and Economics, Gordon and Breach Science Publishers, Volume 16, Amsterdam, The Netherlands.

Janic, M. (2004). Expansion of airport capacity: Case of London heathrow airport. Transportation Research Record 1888, 7–14.

Janić, M. (2005). Modelling airport congestion charges. Transportation Planning and Technology, 28(1): 1–26.

Janic, M. (2007). A Steeper approach procedure for increasing the ultimate capacity of closely spaced parallel runways. Transportation Research Record-Aviation, 81–90.

Janić, M. (2007a). The Sustainability of Air Transportation: Quantitative Analysis and Assessment, Ashgate Publishing Company, UK.

Janić, M. (2008). Modelling the capacity of closely-spaced parallel runways using innovative approach procedures. Transportation Research C, 16(6): 704–730.

Janić, M. (2008a). A model of runway landing capacity based on the ATC time-based separation rules. Transportation Research Record No. 2052 Aviation, 79–89.

Janic, M. (2008b). Analysis and Forecasting Passenger Demand at the Large Hub Airport. New Transportation Research Progress, Editor Filip N. Gustavsson, Nova Science Publisher, Inc, Hauppage, New York, USA, pp. 93–120.

Janić, M. (2011). Greening Airports: Advanced Technologies, & Operations, Springer, UK.

Janić, M. (2013). The Airport Analysis, Planning, and Design: Demand, Capacity, and Congestion, Nova Science Publishers, Inc. New York, USA.

Janić, M. (2014). Advanced Transport Systems: Analysis, Modelling, and Evaluation of Performances, Springer, UK.

Janić, M. (2019). Landside Accessibility of Airports Analysis, Modelling, Planning, and Design, Springer International Publishing, Springer Nature Switzerland AG, Cham, Switzerland

Janić, M. (2020). An assessment of some pros and cons to airport growth. Proceedings of 99th TRB (Transportation Research Board) Annual Conference, Washington DC, USA, January 11–16.

JR. (2012). Data Book 2011, Central Japan Railway Company, Nagoya, Japan.

KLM. (2004). Annual Report 1998–2003, KLM Dutch Airline, Amsterdam, the Netherlands.

Larson, C. R. and Odoni, R. A. (1981). Urban Operations Research, Prentice Hall, Englewood Cliffs, New Jersey, USA.

Litman, T. (2017). Autonomous Vehicle Implementation Predictions: Implications for Transport Planning, Victoria Transport Policy Institute, Victoria, Canada, www.vtpi.org/ca.

Lovas, G. G. (1994). Modelling and simulation of pedestrian traffic flow. Transportation Research B, 28B(6): 429–443.

Lufthansa. (2003). Annual Report 2002, Deutsche Lufthansa AG CGN IR, Cologne, Germany.

Naor, P. (1969). The regulation of queue size by levying tolls. Econometrica, 37(1): 15–23.

NASA. (2001). Enhanced Airport Capacity through Safe, Dynamics Reduction in Aircraft Separation: NASA's Aircraft Vortex Spacing System (AVOSS), NASA/TM-2001-215052, National Aeronautics and Space Administration, Langley Research Centre, Virginia, USA.

Newell, F. G. (1979). Airport capacity and delay. Transportation Science, 13(3): 201–241.

Newell, F. G. (1982). Applications of Queueing Theory. Chapman and Hall, UK.

OAG. (2019). Punctuality League 2019: On-Time Performance for Airlines and Airports Based on Full-Year Data 2018, OAG Aviation Group Limited, Luton, England, UK.

OAG. (2019a). Megahubs International Index 2018: The World's Most Internationally Connected Airports, OAG Aviation Worldwide Limited, Chicago, Illinois, USA.

Odoni, R. A., Bowman, J., Delahaye, D., Deyst, J. J., Feron, E., Hansman, R. J., Khan, K., Kuchar, J. K., Pujet, N. and Simpson, W. R. (1997). Existing and Required Modelling Capabilities for

Evaluating ATM Systems and Concepts, Final Report No. NAG2-997, International Centre for Air Transportation, Massachusetts Institute of Technology, Massachusetts, Boston, USA.

PANYNJ. (2003). La Guardia Airport: Traffic Statistics, Report, The Port Authority of NY&PJ, New York, USA, p. 2.

PANYNJ. (2003a). Schedule of Charges for Air Terminals, Report, The Port Authority of NY&PJ, New York, USA, p. 40.

Powell, J. R. and Danby, G. (2013). MAGLEV technology development. *In*: Ehsani, M., Wang, F. Y. and Brosch, G. L. (eds.). Transportation Technologies for Sustainability. Springer, New York, NY, USA.

Rue, C. R. and Rosenshine, M. (1985). The application of semi-markov decision processes to queuing of aircraft for landing at an airport. Transportation Science, 19(2): 154–172.

SDG. (2015). Study on Employment and Working Conditions in Air Transport and Airports, Steer Davies Gleave, London, UK.

Sewill, B. (2009). Airport Jobs: A False Hopes, Cruel Hoax, Aviation Environment Federation, London, UK.

SG. (1999/2013). Annual Report, Schiphol Group, Schiphol, The Netherlands.

SG. (2005). Annual Statistical Review (1999–2005), ZG Schiphol, The Netherlands.

SG. (2007). Annual Statistical Review 2006, ZG Schiphol, The Netherlands.

SG. (2005/2016). Annual Results, Schiphol Group, Amsterdam, The Netherlands.

SG. (2008/2019). Annual Results, Royal Schiphol Group, Amsterdam, The Netherlands.

Smith, M. J. T. (1989). Aircraft Noise, Cambridge University Press, Cambridge, USA.

TCRP. (2018). Battery Electric Buses State of the Practice: A Synthesis of Transit Practice, Transit Cooperative Research Program, The National Academies Press, Transportation Research Board, Washington, D.C., USA.

Teodorović, D. and Janić, M. (2016). Transportation Engineering: Theory, Practice and Modelling, Elesevier, Amsterdam, The Netherlands.

TfL. (2015). International Metro Benchmarking, Final Report, Transport for London, London, UK.

TLC. (2016). 2016 TLC Fact book, New York City Taxi & Limousine Commission, New York, USA.

Thompson, S. D. (1997). Terminal Area Separation Standards: Historical Development and Process for Change, Lincoln Laboratory, Massachusetts Institute of Technology, Lexington, Massachusetts, USA.

Tošić, V. (1992). A review of airport passenger terminal operations analysis and modeling. Transportation Research-A, 26A(1): 3–26.

TRB. (1990). Airport System Capacity: Strategic Choices, Transportation Research Board, Special Report 226, TRB, National Research Council, Washington DC, USA.

TRB. (2002). Aviation Demand Forecast: Survey of Methodologies, Transportation Research Board, National Research Council, Circular No E-C040, Washington DC. USA, p. 44.

TRB. (2008). Ground Access to Major Airports by Public Transportation (2008), Airport Cooperative Research Program-Report 4, Transportation Research Board; Washington, D.C., USA.

TRB. (2009). Airport Passenger-Related Processing Rates Guidebook, ACRP (Airport Cooperative Research Program), Transportation Research Board; National Academies of Sciences, Engineering, and Medicine, Washington, D.C. USA.

TRB. (2010). Airport Passenger Terminal Planning and Design, Volume 1: Guidebook, Airport Cooperation Research Report, ACRP Report 25, Vol. 1, Transportation Research Board, Washington, D.C., USA.

TRB. (2011). Passenger Level of Service and Spatial Planning for Airport Terminals, Airport Cooperation Research Report, ACRP Report 55, Transportation Research Board, Washington, D.C., USA.

UIC. (2010a). Necessities for Future High Speed Rolling Stock, Report January 2010, International Union of Railways, Paris, France.

UN. (2005). GDP-all countries for all years, United Nations, Department Of Economic and Social Affairs-Statistical Division, New York, USA, http://unstats.un.org/unsd/snaama/dnllist.asp.

USDT. (2018). Airport Terminal Planning, Advisory Circular, AC No: 150/5360-13A, Federal Aviation Administration, U.S. Department of Transportation, Washington, D.C., USA.

USDT. (2019). Passengers Boarded at the Top 50 U.S. Airports (2018), U.S. Department of Transportation, Bureau of Transportation Statistics, Office of Airline Information (Air Carriers Statistics—Form 41 Traffic), T-100 Market (All Carriers), http://transtats.bts.gov/as/, Washington D.C., USA.

USGAO. (2001). Mass Transit: Bus Rapid Transit Shows Promise, United States General Accounting Office, (Report to Congressional Requesters), Report No. GAO-01-984, Washington D.C., USA.

Vuchic, R. V. (2007). Urban Public Transportation-Systems and Technology, Prentice Hall, INC, Englewood Cliffs, New Jersey, USA.

Wadud, Z. (2011). Modelling and forecasting passenger demand for a new domestic airport with limited data. Transportation Research Record, DOI: 10.3141/2214-08.

Whol, M. and Hendrickson, C. (1984). Transport Investment and Pricing Principles: An Introduction for Engineers, Planners and Economists, John Wiley and Sons, A Wiley Series in Construction Management Engineering, New York, USA.

Wirasinghe, C. S. and Shehata, M. (1988). Departure lounge sizing and optimal seating capacity for a given aircraft/flight mix: (i) single gate, (ii) several gates. Transportation Planning and Technology, 13: 57–71.

Wright, L. (2005). Bus Rapid Transit, Version 2.0, Deutsche Geselischaft fur Technische Zusammenarbeit (GTZ) GmbH, Eschborn, Germany.

Wright, L. and Hook, W. (2007). Bus Rapid Transit Planning Guide, 3rd Edition, Institute for Transportation & Development Policy, New York, USA.

www.iata.org/los

Chapter 3
Airlines

3.1 Introduction

At present, about 5000–5500 airlines certified by the ICAO (International Air Transport Organization) operate in the world, of which about 770–800 are commercial airlines that carry out scheduled flights (that are officially recognised). These latter airlines are characterized by the types of resources and the way that they are used, in terms of the aircraft fleet, staff/employees on board the aircraft and on the ground, energy/fuel used, the air route network, and their operating rules and procedures. The volume of deployed resources, characteristics of air route network, and operating rules and procedures being in place for a given period of time constitute the airline system.

In many cases, the airlines are compared regarding different attributes/criteria valid for the specified period of time (usually one year). In general, they are distinguished regarding the spatial and operational/economic/business configurations of air route networks as the network airlines and LCCs (Low Cost carrier(s)). The network airlines have emerged from the former conventional/legacy airlines which, after liberalization/deregulation of the air transport markets, have consolidated their networks mainly into the hub-and-spoke spatial and operational configurations in addition to frequently entering into alliances with each other. These airlines usually operate heterogeneous fleets consisting of wide-body, narrow-body, and regional aircraft. The LCCs (Low Cost carrier(s)) that emerged after liberalization/deregulation of the air transport markets have operated the point-to-point spatial and operational networks and offered much lower airfares then their network counterparts. These carriers have typically used a single aircraft type in their fleets.

The additional attributes/criteria for the airline comparison have been their fleet size, number of routes, destinations, and countries served, revenues, the number of passenger carried, the RPK/RPM (Revenue Passenger Kilometres/Miles) and freight/cargo RTK/RTM (Revenue Ton Kilometres/Miles) carried out, etc. Table 3.1 gives an example of ranking the 12 world's largest airlines regarding their fleet size in terms of the number of aircraft operated during the specified period of time (usually one year).

As can be seen, the total number of aircraft in the fleets has varied from about 330 to 960. The largest fleets have been operated by four US airlines (3 networks and one LCC) and the smallest one by two European and two Chinese network airlines. All network airlines have operated at least two categories of aircraft—substantially present wide-body, dominating narrow-body, and marginally present regional. An exception is U.S. SkyWest, which has exclusively operated regional aircraft. Other LCCs have exclusively

Table 3.1 The airline fleet size—Case of the world's airlines (Period: 2019) (FG, 2018; https://atwonline.com/datasheet/2019-world-airline-report-world-airline-fleets).

Airline	Country	Total fleet	Aircraft category		
			Wide-bodies	Narrow-bodies	Regionals
American Airlines	USA	957	155	782	20
Delta Air Lines	USA	880	151	729	-
United Airlines	USA	777	191	586	-
Southwest Airlines	USA	753	-	753	-
China Eastern Airlines	China	607	91	513	3
China Southern Airlines	China	603	97	461	19
SkyWest	USA	486	-	-	486
Ryanair	Ireland	456	-	456	-
Air China	China	439	130	306	-
FedEx Express	USA	378	267	111	
Lufthansa	Germany	360	-	198	47
Turkish Airlines	Turkey	332	115	225	-

Table 3.2 The number of served routes—Case of the world's airlines (Period: 2019) (FG, 2018; http://www.airportspotting.com/worlds-largest-airlines/).

Airline	Country	No. of served routes
Ryanair	Ireland	1831
American Airlines	USA	1106
United Airlines	USA	950
EasyJet	UK	945
Delta Air Lines	USA	939
Southwest Airlines	USA	754
China Southern Airlines	China	667
China Eastern Airlines	China	648
Wizz Air	Hungary	615
Air China	China	470
TUI Airways	UK	470
Lufthansa	Germany	360
Turkish Airlines	Turkey	332

operated narrow-body aircraft fleet. In addition, Table 3.2 gives a ranking of the selected world's airlines regarding the number of routes served.

As can be seen, the European LCC Ryanair has served the largest number of routes, followed by two U.S. network airlines. In this case, four LCCs have been among the other nine network airlines. Also, Figure 3.1 shows the relationship between the number of routes served and the number of aircraft in the fleet for the above-mentioned airlines.

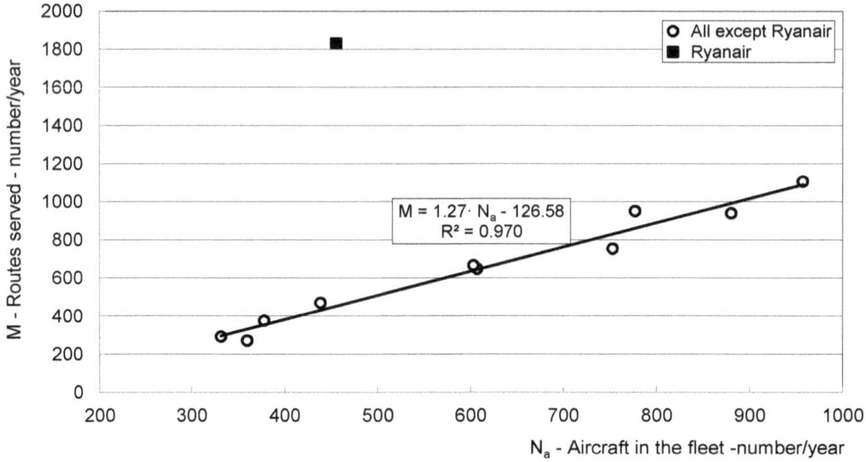

Figure 3.1 Relationship between the number of air routes served and the number of aircraft in the fleet—Case of the world's largest airlines (Compiled from Tables 3.1 and 3.2) (Period: 2019).

As can be seen, there is a relatively strong correlation between the number of routes served and the number of aircraft in an airline fleet, with exemption of LCC Ryanair. This indicates that the airlines with larger fleets generally operate the networks with a larger number of routes, and vice versa. The airline fleets and routes have also influenced the volumes of their output in terms of RPK (Revenue passenger Kilometres) as shown in Figure 3.2.

As can be seen, the volumes of RPK carried out during the specified period of time have increased more than proportionally with an increase in the number of routes in the airline networks. These volumes have also been influenced by the passenger density and length of these routes.

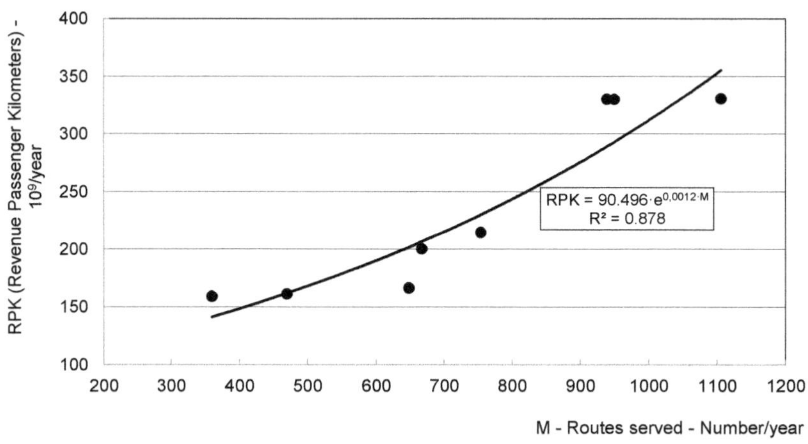

Figure 3.2 Relationship between the annual volume of RPK (Revenue Passenger Kilometres) carried out and the number of air routes served—Case of the world's largest airlines (Period: 2018) (Table 3.2; FG, 2018; https://en.wikipedia. org/wiki/World%27s_largest_airlines/).

3.2 The System

3.2.1 Aircraft

The world's fleet of commercial aircraft has been quite heterogeneous regarding the number of aircraft operating in the particular regions of the world, their size, and the types of propulsion. Figure 3.3(a, b) shows an example of such diversity.

Figure 3.3a shows that, in the year 2018, the largest number of aircraft have operated in the Asia Pacific region and the smallest number in Africa. Some forecasts indicate that 20 years from now (the year 2038), the total number of aircraft will substantially increase, i.e., almost double, while the difference in numbers in the particular regions will remain similar, i.e., again the largest in Asia-Pacific and the lowest in Africa. Figure 3.3b shows that, in the year 2018, the relatively most present have been the narrow-body jets, followed by their wide-body counterparts. The turboprops have been the least present.

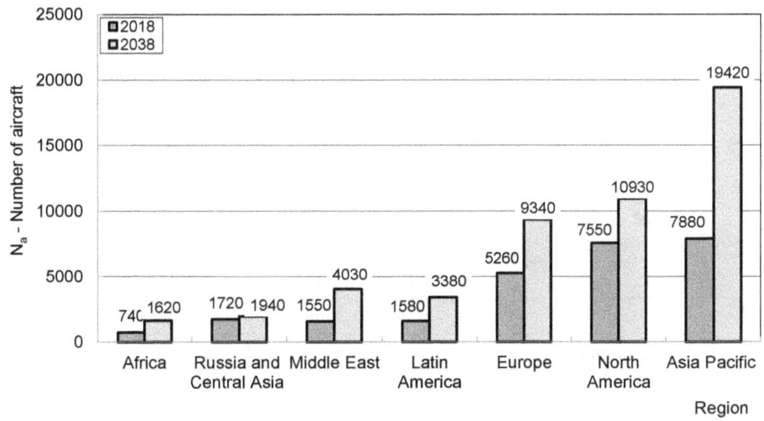

a) Number of aircraft in various regions of the world - 26280 (Period: 2018) and 50660 (Period: 2038)

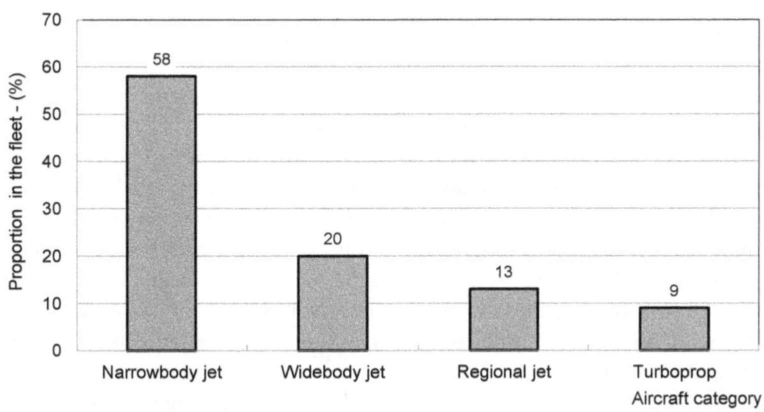

b) Structure (Period: 2018)

Figure 3.3 Characteristics of the current and future world's commercial aircraft fleet (https://www.statista.com/statistics/262971/aircraft-fleets-by-region-worldwide/).

In addition, these aircraft have also been sub-categorized as small, medium, and large. At the main aircraft manufacturers, there have been 23 Airbus, 32 Boeing, 6 Bombardier 6, and 10 Embraer different types (DVB, 2019). Table 3.3 gives some self-explained characteristics of the representative aircraft of three categories: wide-body (A 350-900; B 787-8), narrow-body (A320-200; B737-800), and regional (Embraer 170/190).

Table 3.3 Characteristics of the selected categories of commercial passenger aircraft (DVB, 2019; https://contentzone.eurocontrol.int/aircraftperformance/details.aspx?ICAO=E190&/).

Aircraft type	**A350-900**	**B787-8**
Category	Wide-body	Wide-body
Sub-category	Large	Medium
Seating capacity (standard)	325	242
Range (nm)	7590-8100	7355
Engine options (no. of types)	1	3
Aircraft type	**A320-200**	**B737-800**
Category	Narrow-body	Narrow-body
Sub-category	Medium	Medium
Seating capacity (standard)	150	178
Range (nm)	2630-3300	1305-2680
Engine options (no. of types)	4	1
Aircraft type	**Embraer 175-E2**	**Embraer 190**
Category	Regional	Regional
Sub-category	Medium	Medium
Seating capacity (standard)	80	96
Range (nm)	2060	2400-2450
Engine options (no. of types)	1	1

Some of these characteristics can also be presented in the form of "payload-range" and "productivity-range" diagrams, as shown in Figure 3.4(a, b).

Figure 3.4a shows that the range generally increases with a shrinking aircraft payload and becomes longest when the flights are carried out without a payload, i.e., with the maximum trip fuel. Figure 3.4b shows that the aircraft technical productivity is equal to the product of its payload and block speed and is expressed by the realised ton-miles per block hour. Block speed increases and payload decreases with an increase in the aircraft range. Consequently, the productivity changes due to variations in both block speed and payload. First, it increases with an increase in the block speed at the maximum payload up to a specified range (i.e., the maximum range with the maximum payload). Beyond this, the payload decreases in order to increase range and consequently contributes to decreasing technical productivity at the rather constant block speed. In general, payload is reduced in order to enable more fuel to be loaded for flying longer distances.

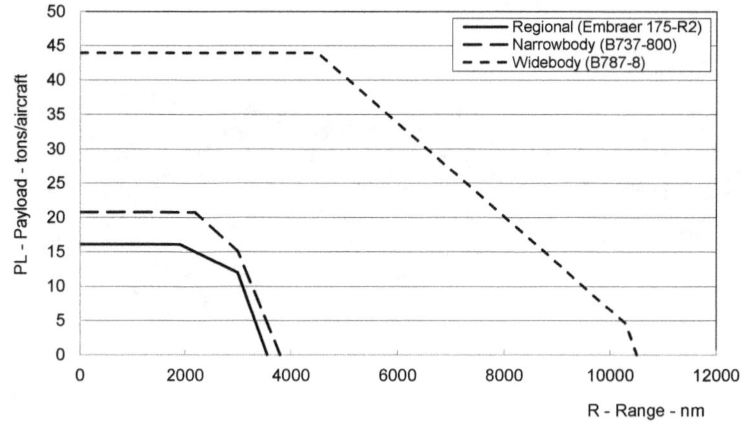

a) Payload vs range

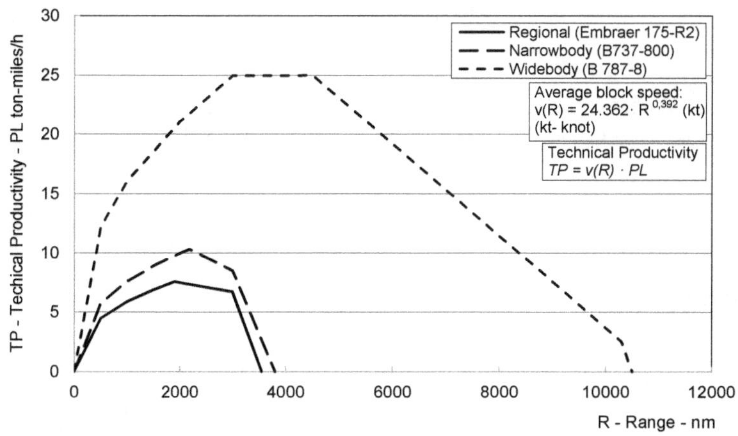

b) Productivity

Figure 3.4 Characteristics of particular aircraft categories (Boeing, 2014; Embraer, 2017; Horonjeff et al., 2010).

3.2.2 Route Networks

As mentioned above, the commercial airlines generally operate the point-to-point and hub-and-spoke networks. Figure 3.5(a, b) shows the spatial configuration of these two above-mentioned networks operated by an U.S. network airline and LCC (Low Cost Carrier), respectively.

As can be seen, Southwest Airlines operates the point-to-point network with 6 base airports and Delta airlines the hub-and-spoke network with 7 hubs, respectively (see also Table 3.2). In general, spatially, a point-to-point network covers a certain area of a country and/or a continent. It consists of airports of different size (large hubs, regional) as its nodes and direct air routes as the network links connecting them, where the aircraft perform their flights. In this case, the airports are the origins and destinations of particular flights and their passengers and freight/cargo shipments. The network size can be measured by the number of airports and air routes connecting them. For example, if the network has

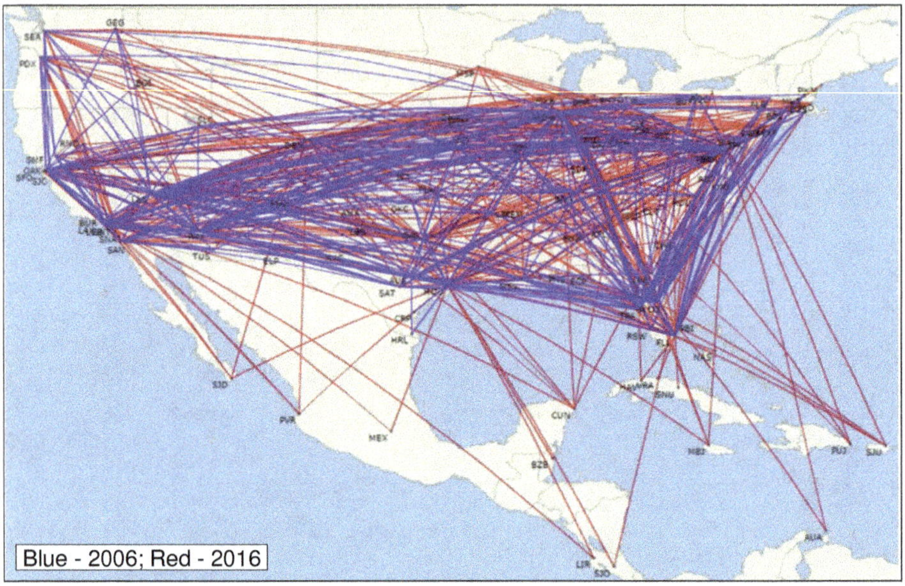

a) Point-to-point network - Case of Southwest Airlines (USA) (http://pinotglobal.com/maps/southwest- route-map/attachment/southwest-route-map-from-the-most-frequently-searched-sources-1/)

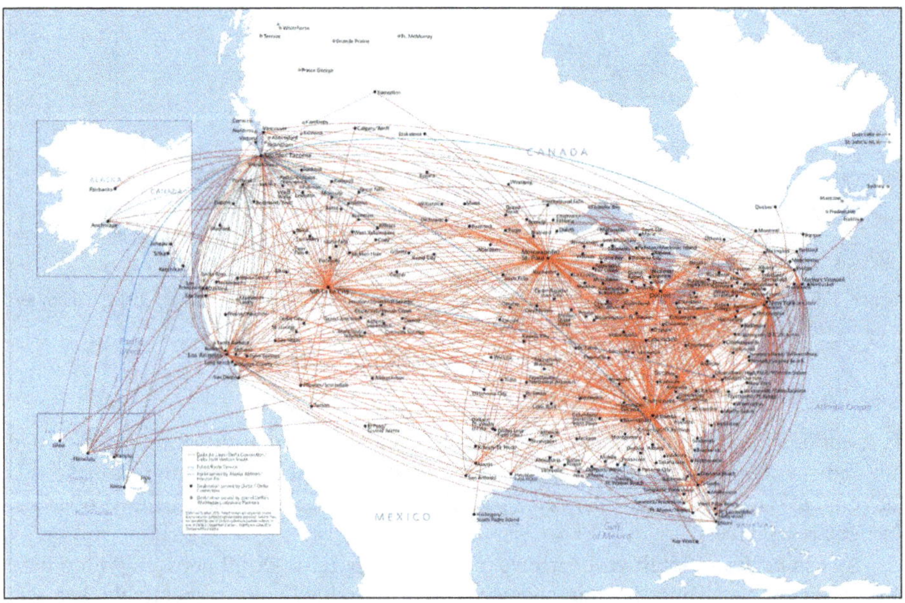

b) Hub-and-spoke network - Case of Delta Airlines (USA) (https://news.delta.com/route-map-us-canada/)

Figure 3.5 Scheme of the spatial configurations of the commercial airline networks.

(N) airports, *(N (N–1)/2)* routes fully connected them in the single direction. One or more airports of the given network are also the airline base(s). Operationally, on each route, the airline flights are scheduled with certain frequency in order to serve the expected passenger

(and freight/cargo) demand during a given period of dime (hour, day). Spatially, a hub-and-spoke network consists of airports and its nodes and air routes as links connecting them. The network nodes are one or more central (hub) and multiple peripheral (spoke) airports. The direct air routes are spread between the particular spoke and hub airports, thus indirectly connecting the spoke airports themselves. If the network consists of *(N)* airports, of which one is the hub and the rest are spokes, the number of direct routes in a single direction completely connecting the network will be *(N–1)*. Under such conditions, operationally, the passenger (and freight/cargo) demand on each route can include that to/from the hub and to/from all other spokes, thus enabling scheduling more frequent flights during a given period of time. The incoming flights from and outgoing flights to the particular spokes are scheduled at the hub airport within the relatively short time window, thus enabling relatively efficient interchange of the passengers and cargo between them on their way between particular spokes. Such clustered incoming and outgoing flights are usually called 'waves'. At some large hubs, several waves can be scheduled by the dominant airline(s) during the day (Janić, 2014).

3.2.3 Staff/Employees

The commercial airlines have also acted as relatively large staff/employers of generally two categories: on board the aircraft, i.e., pilots and cabin crews, and on the ground dealing with handling passengers and their baggage, and freight/cargo shipments at airports, as well as carrying out other commercial activities at and out of the airports. The main factors influencing the airline staff has generally proven to be the size of their fleets, represented by the number of aircraft and their seating capacity. Equation 3.1 shows the causal relationship for ten of the world's largest airlines by the annual revenues. These are: Lufthansa Group, American Airlines Group, Delta Airlines, United Airlines, Air-France – KLM, IAG (International Airlines Group), Southwest Airlines, China Southern Airlines, China Eastern Airlines, and Air China (https://www.airport-technology.com/features/worlds-biggest-airlines-2018/).

$$EMP = \beta_0 + \beta_1 \cdot N_a + \beta_2 \cdot S = 31.051 + 0.032 \cdot N_a + 0.287 \cdot S$$
$$t\text{-}stat \qquad 1.157 \qquad 2.370 \qquad 0.349$$
$$R^2 = 0.404;\ F = 2.372;\ D\text{-}W = 2.442;\ N = 10 \tag{3.1}$$

where

EMP is the number of airline employees (10^3/airline);
N_a is the number of aircraft in the fleet (-);
S is the total number of seats in the airline fleet (10^3/airline); and
N is the number of airlines (-).

As can be seen, there has been some causal relationship between the specified variables, with the most influencing variable (N_a) (number of aircraft in the fleet) in the given case. In addition, Eq. 3.2 indicates the causal relationship between the employment (dependent variable), and aircraft fleet and the number of airports served (independent variables) of the European LCC Ryanair (Period: 2009–2018) (Ryanair, 2008/2018).

$$EMP = \beta_0 + \beta_1 \cdot N_a + \beta_2 \cdot N = -3288.615 + 19.410 \cdot N_a + 38.752 \cdot N$$
$$t\text{-}stat \qquad -3.076 \qquad 4.331 \qquad 3.513$$
$$R^2 = 0.982;\ F = 192.383;\ D\text{-}W = 1.000;\ N = 10 \tag{3.2}$$

where

EMP is the average number of employees per year (-);
N_a is the number of aircraft at the end of the year (-); and
N is the number of airports served by the end of the year (-).

As can be seen, there has been a very high causal relationship between the dependent and two independent variables of the given airline. Both independent variables are of high importance. This indicates that the airline has increased its employment by increasing the aircraft fleet, which has been used to connect the increasing number of airports. Also, Figure 3.6 shows the relationship between the number of staff/employees and the size of the fleet at Lufthansa group airlines (DL AG, 2005/2018).

As can be seen, there has been the very strong causal relationship between the two above-mentioned variables during the observed period. The number of employees has increased at a decreasing rate with an increase in the number of aircraft. The above-mentioned examples indicate that, in the given context, it could generally be said that the size of airline staff/employees has mainly been positively related to the size of its fleet in terms of the number of aircraft and seat capacity, and the size of air route network in terms of the number of airports served.

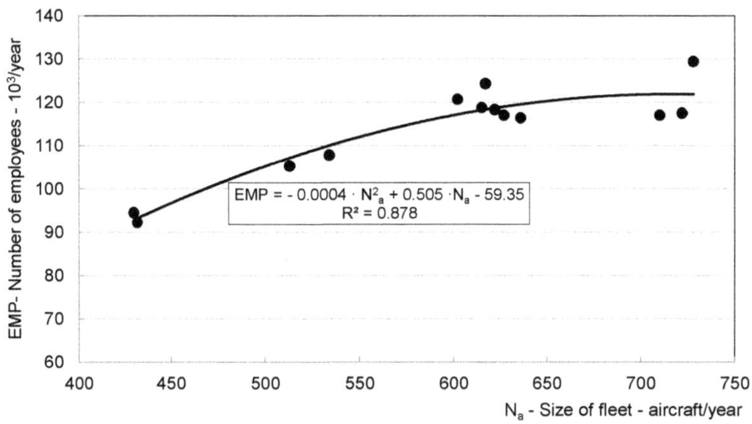

Figure 3.6 Relationship between the number of airline employees and the number of aircraft—Case of Lufthansa group (Period 2005–2018) (DL AG, 2005/2018).

3.2.4 Fuel

As mentioned in Chapter 1, the commercial airline aircraft consume Jet A fuel as necessary for carrying out the flights in their air route networks. The amount of fuel consumed is dependent on the size and structure of aircraft fleet, and their utilization. Figure 3.7 shows the average fuel consumption for two U.S. network (American) and LCC (Southwest) airlines, depending on the average daily utilization of their fleets (US BTS, 2019; http://web.mit.edu/airlinedata/www/Aircraft&Related.html/).

As can be seen, the average fuel consumption at American Airlines has varied between about 1000 and 1200 U.S. Gallons/BH (Block Hour) for the average daily aircraft utilization of U_a = 9.5–10.5 hours per day. At Southwest airlines, this consumption varied between about 750 and 800 Gallons per hour for the average daily aircraft utilization of 9.5–12.0 hours per day. In addition, the average fuel consumption of American Airlines

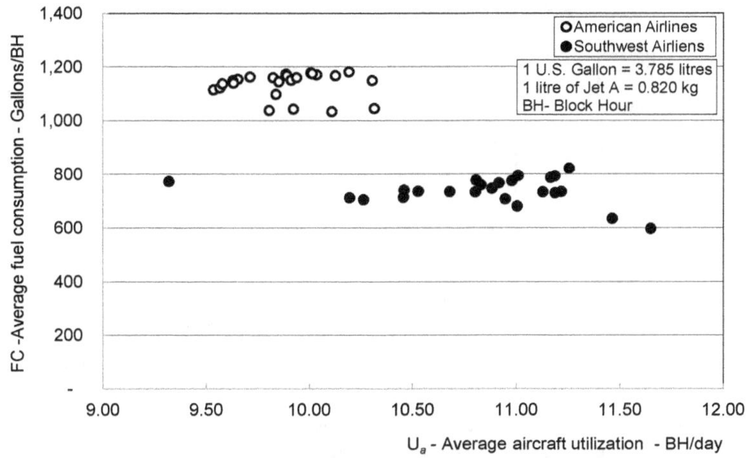

Figure 3.7 Relationship between the average fuel consumption and the average aircraft utilization—Case of the selected U.S. airlines (Period: 1995–2018) (US BTS, 2019; http://web.mit.edu/airlinedata/www/Aircraft&Related.html/).

has been higher than that of Southwest Airlines for about 40–50% mainly due to different (rather heterogeneous) fleet structure and the consequent air route network configuration (http://web.mit.edu/airlinedata/www/Aircraft&Related.html/).

3.2.5 Slots

The airline specific assets are the allocated/possessed slots at the slot-constrained airports. According to IATA (International Air Transport Association) Worldwide Airport Slots Group's guidelines, the airports worldwide are categorized as Level 1 (Non-Coordinated Airport), Level 2 (Schedules Facilitated Airport), or Level 3 (Coordinated Airport). In the summer of 2017, 123 airports in the world were Level 2 and 177 were Level 3 airports. For example, 93 airports in the EU (European Union) and 2 in the USA are slot-controlled. In the latter case, these are NY La Guardia and Reagan Washington National airport (Lenoir, 2016). At the slot-controlled airports, the airlines need to obtain slots for landings and take-offs in advance. The slot provides the airline with the right to land or take-off at an airport during a specified period of time. At the EU airports, the slots refer to accessibility to all airside and landside infrastructures necessary to operate flights on the specific date and time. At most of these airports, the slots are allocated to airlines regarding their previous use, i.e., according to the system called the "grandfather rights" including the rule "use it or lose it". For example, if an airline does not use the allocated slots (typically 80% over the period of six months), it can lose the rights to those slots. At the U.S. airports, the slots refer to using runways on the FCFS ("First Come-First Served") priority rule. As such, they enable the accommodation of the existing airline flight demand by allocation and use of the available airport runway capacity. When the demand exceeds the airport runway system capacity during particular longer periods, congestion and delays occur, which, if substantial, require the introduction of the airline flights access, i.e., slot control. Consequently, it can be said that the slot system contributes to managing flight congestions and delays at busy airports. In addition, the slots are not linked to the specific routes but allocated to airlines, which can use them

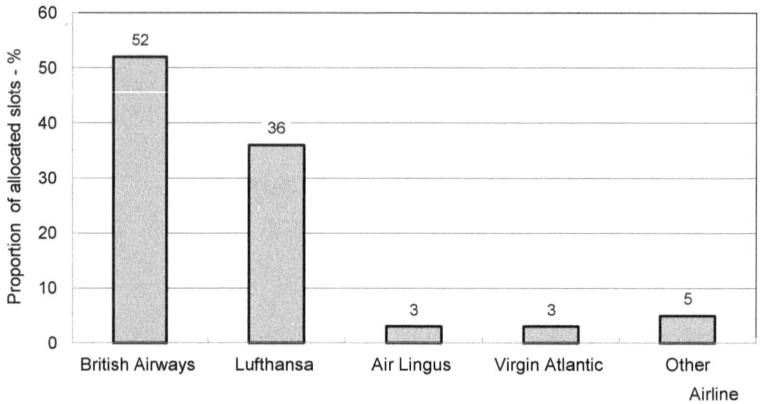

Figure 3.8 Proportion of the allocated slots to airlines—Case of London Heathrow airport (UK) (Period: Summer 2017–Peak Week) (Haylen and Butcher, 2017).

at their disposal. This implies that such allocated slots enable the airlines a freedom to design their air route networks in order to respond to the potential air passenger and freight/cargo demand. Figure 3.8 shows an example of slot allocation in the relative terms at the UK London Heathrow airport (Haylen and Butcher, 2017).

As can be seen, the UK-based BA (British Airways) has possessed the most slots also thanks to the inherited "grandfather rights", followed by German Lufthansa, Air Lingus, and Virgin Atlantic airlines.

The allocated slots are considered as airline assets because they have had commercial value on the one hand and have been the subjects of trading between airlines. Figure 3.9(a, b) shows an example for London Heathrow airport (UK) (CAPA, 2013).

Figure 9(a, b) shows that the actual commercial value of airline slots at the given airport has been in the order of millions of GBP (Great Britain Pound). Specifically, Figure 9a shows that the value of a pair of slots varied significantly during the day. That in the evening has been just about one third of that in the early morning. Figure 9b shows the very high diversity between the number of daily slots, their values, and the values of their transaction during the observed period. These figures indicate that the slots have and will continue to be valuable airline assets at the capacity constrained, i.e., slot-controlled, airports worldwide. As such, they also enable the airlines to compete both at airports and for the routes they select to operate from there. Figure 3.10 shows an example of the airline seat competition on 17–19 transatlantic routes between Amsterdam Schiphol (The Netherlands) and the corresponding number of U.S. airports.

As can be seen, the airline average market share in the given market has been proportional to the number of scheduled seats. In addition, the marginal contribution of an additional seat to increasing of the airline market share has decreased as the number of competing airlines fell, and vice versa.

3.3 Demand and Capacity

3.3.1 Demand

The above-mentioned airline sources are deployed to carry out passengers and freight/cargo shipments between their origin and destination airports during a given period

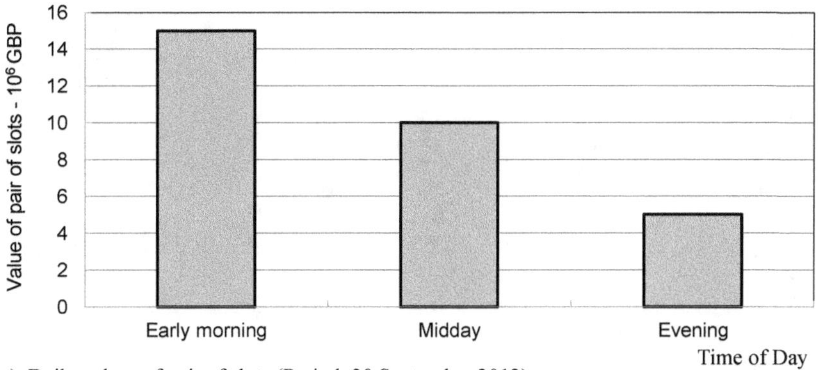

a) Daily values of pair of slots (Period: 30 September 2012)

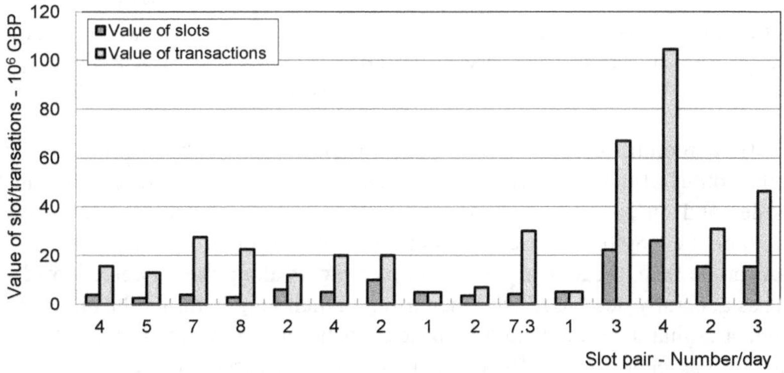

b) Value of pairs of slots and their transactions vs their number (Period: 1998-2013)

Figure 3.9 Example of the value of slots and transactions—Case of London Heathrow airport (UK) (GBP – Great Britain Pound) (CAPA, 2013).

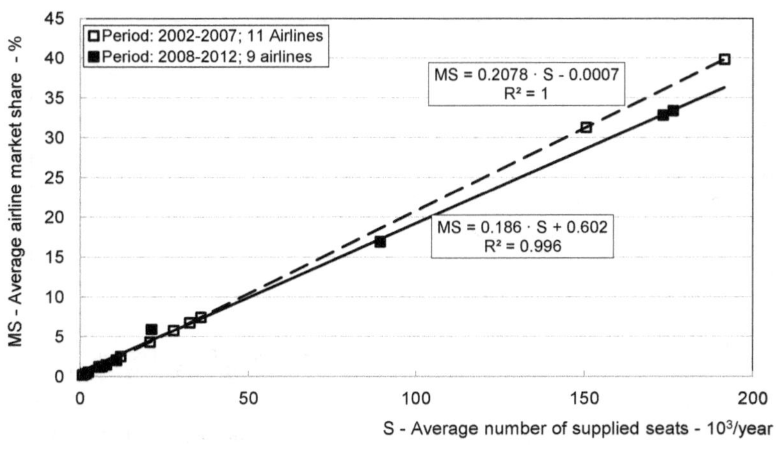

Figure 3.10 Relationship between the airline market share and the number of supplied seats—Case of Amsterdam Schiphol (The Netherlands) and U.S. airports (Kremer, 2014).

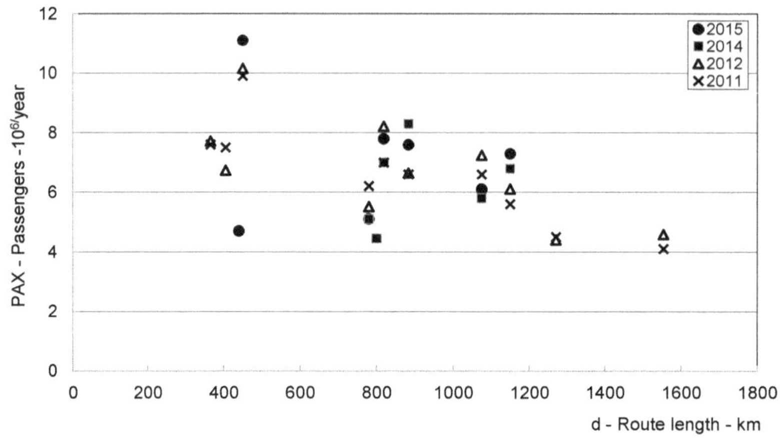

Figure 3.11 Relationship between the number of passengers and route length—Case of 12 world's busiest origin-and-destination air routes (Period: 2011–2015) (https://en.wikipedia.org/wiki/List_ofbusiest_passenger_air_routes/).

of time (day, month, season, year). As such, these represent the airline demand on the particular routes of their networks. In general, the volumes of passenger and air freight/cargos demand on particular airline routes depend on many factors. One of these has proven to be the length of particular routes, as shown in Figure 3.11.

As can be seen, the annual number of passengers along these twelve world's busiest routes has generally decreased with increasing of their length during the observed period. This indicates that the route length in these cases has acted as an "ultimate" barrier to the volumes of passenger travel. However, in some other cases the increasing numbers of passengers have and will continue to fly further and further, as shown in Figure 3.12 for the U.S. scheduled airlines.

As can be seen, the annual number of passengers has been and is also forecasted to increase almost linearly with increases in trip length during the observed period.

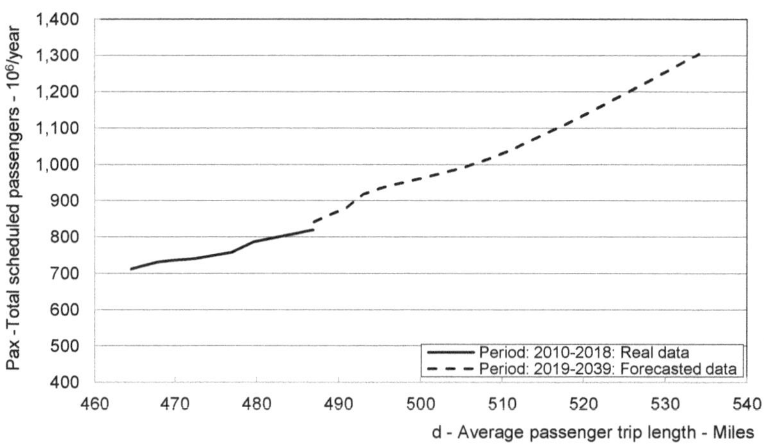

Figure 3.12 Relationship between the numbers of passengers and the average passenger trip length—Case of the total U.S. scheduled traffic (Period: 2010–2039) (US BTS, 2019).

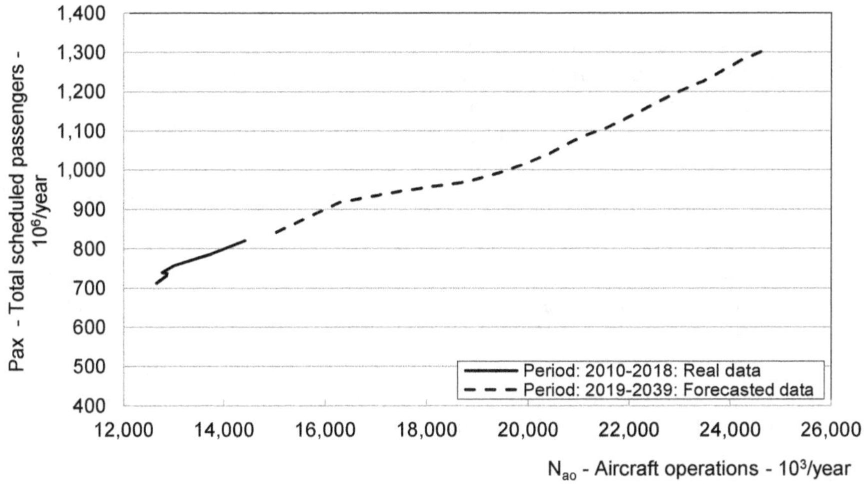

Figure 3.13 Relationship between the number of passengers and the aircraft operations—Case of the total U.S. scheduled air traffic (Period: 2010–2039) (US BTS, 2019).

In addition, these volumes of passenger demand have been and are forecasted to be supported by the numbers of airline operations as shown in Figure 3.13.

The above-mentioned figures have indicated two influencing factors of the airline passenger demand, i.e., the route length and the number of aircraft operations serving this demand. Nevertheless, the real question is about the main passenger and freight/cargo demand driving forces. In particular, the passenger demand has usually been expressed in terms of the number of *PAX* (Passengers) or *RPM* (Revenue Passenger Miles) carried out by airlines during a given period of time. The demand-driving forces have been roughly divided into the external and the airline internal ones. Regarding passenger demand, the former have frequently been the economic factors such as *GDP* (Gross Domestic Product), *PCI* (Per Capita Income), *POP* (Population), etc. The later have been the airline airfares often expressed in terms of AF (Air Fare/pax) or *Y* (Yield). The causal relationships between the above-mentioned airline passenger demand and its particular driving forces have been already under investigation for a long time at global, regional and national levels. The inclusion of particular forces has been dependent on who has been carrying out the analysis and forecasting: airlines, aircraft manufacturers, consultants, academics, etc. Equation 3.3(a, b) indicate the causal relationships between the passenger demand and its main driving forces at the U.S. airlines in their domestic market during the period 2010–2040 (the real and forecasted input data) (http://web.mit.edu/airlinedata/www/Aircraft&Related.html).

$$PAX = \alpha_0 + \alpha_1 \cdot POP + \alpha_2 \cdot GDP = -827.906 + 4.455 \cdot POP + 0.010 \cdot GDP - 14.434 \cdot Y$$

t-stat 1.311 1.727 0.769 1.480

$$R^2 = 0.988;\ F = 165.585;\ DW = 1.749;\ N = 30 \quad (3.3a)$$

and

$$RPM = \alpha_0 + \alpha_1\ POP + \alpha_2\ GDP = -599.855 + 3.071 \cdot POP + 0.027 \cdot GDP - 13.860 \cdot Y$$

t-stat 1.229 1.539 2.700 1.839

$$R^2 = 0.996;\ F = 454.860;\ DW = 1.959;\ N = 30 \quad (3.3b)$$

where

- *PAX* is the annual number of transported passengers (10^6/year);
- *RPM* is the annual volume of Revenue Passenger Miles carried out (10^9/year);
- *POP* is the annual population (inhabitants/year);
- *GDP* is the annual Gross Domestic Product (10^9 \$US/year) (2012 \$US);
- *Y* is the average annual yield reflecting the average fare paid per mile per passenger, calculated by dividing the passenger revenue by the revenue passenger miles (¢/RPM).

As can be seen, the passenger demand in terms of both (*PAX*) and (*RPM*) has been strongly dependent on the selected influencing variables—their main driving forces. As intuitively expected, it has been and will continue to be positively correlated, i.e., increasing with an increase in (*POP*) and (*GDP*) and decreasing with an increase in airline yield *Y*. In addition, Eq. 3.4 shows the relationship between the passenger demand and its main driving forces on the transatlantic route between LHR (London Heathrow) and NY JFK (New York John Fitzgerald Kennedy) airports. Figure 3.14 shows the simplified scheme.

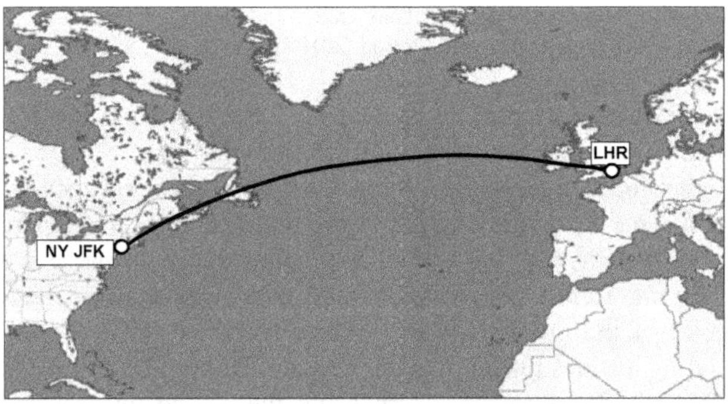

Figure 3.14 Scheme of Transatlantic route—Case between LHR (London Heathrow) and NY JFK (John Fitzgerald Kennedy) airport (https://www.youtube.com/watch?v=518QuKMZSSM/).

$$PAX_{ij} = \alpha_0 \cdot GDP_i^{\alpha 1} \cdot GDP_j^{\alpha 2} \cdot Y_{ij}^{\alpha 3} = 3.633 \cdot GDP_i^{0.219} \cdot GDP_j^{0.432} \cdot Y_{ij}^{0.508}$$
$$t\text{-}stat \qquad 5.250 \quad 0.753 \quad 4.114 \quad 1.129$$
$$R^2 = 0.866;\ F = 10.793;\ DW = 1.811;\ N_Y = 15 \qquad (3.4)$$

- PAX_{ij} is the annual number of passengers between LHR and JFK (10^6 year);
- GDP_i is Gross Domestic Product of New York area (10^9 \$US/year);
- GDP_j is Gross Domestic Product of London area (10^9 \$US/year);
- Y_{ij} is the average yield of three U.S. airlines (American, Delta, United); and
- N_Y is the number of years (N = 15) (2003–2018).

where *PAX* (Million/year), *POP* (million/year), *GDP* (billion \$US–2012/year), *Y* (average ¢/RPM – year), (*RPM* – Revenue Passenger Mile) (Period: 2010–2040 including forecasting 2019–2040) (http://web.mit.edu/airlinedata/www/Aircraft&Related.html; https://www.bea.gov/data/gdp/gross-domesticproduct/; https://www.ons.gov.uk/economy/grossdomesticproductgdp/).

As can be seen, the main driving forces between two airports have been and will continue to be *GDP* (Gross Domestic Product) in New York and London area and the airline yield (*Y*) in this case of U.S. airlines. In some way, this contradicts the influence of yield (*Y*) in Eq. 3(a, b). The relationship between the number of passengers and the airline yield on the transatlantic routes has been and will continue to be positive driving force. This implies that, despite increasing airfares, the passenger demand will continue to grow thanks to growing economics on both sides of the route. Figure 3.15 shows the relationship between the number of passengers and the airline yield on three transatlantic routes between the U.S. and Europe during the period (2010–2018).

The considered routes are: London Heathrow-New York JFK (John Fitzgerald Kennedy), London Heathrow – Los Angeles International, and Paris Charles de Gaulle – New York JFK (John Fitzgerald Kennedy). The airlines considered are American, Continental, Delta, United (2010–2018), US Airways (2010–2014). As can be seen, the number of passengers increased despite increasing of the average yield, reflecting a general increase in the average airfares.

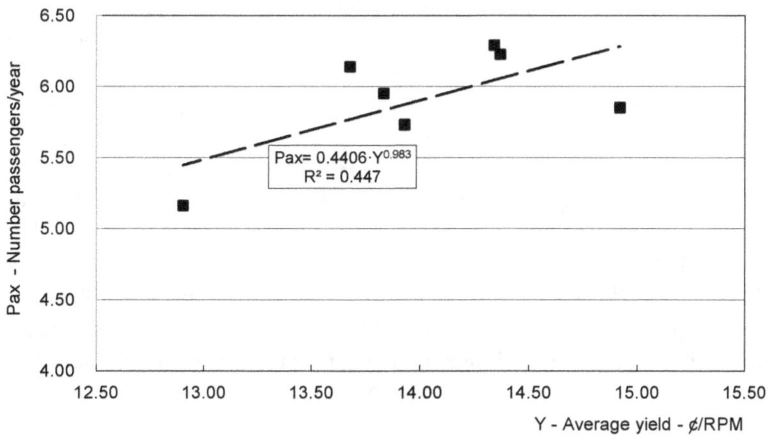

Figure 3.15 Relationship between the number of passengers and the average yield—Case of three transatlantic routes between Europe and U.S. (Period: 2010–2018) (http://web.mit.edu/airlinedata/www/Aircraft&Related.html/).

3.3.2 Capacity

3.3.2.1 Scale and scope

The elements of airline capacity have already been considered while elaborating on their means and resources used, such as aircraft fleet, air route network, staff/employees, fuel, and slots. In general, the airline capacity based on all these can be considered as "static" and "dynamic". Both capacities reflect the airline's capability to produce services for the expected air passenger and freight/cargo shipment demand during a given period (a day, week, month, a quarter year, year) under given conditions. The above-mentioned RPM (Revenue Passenger Miles) and APM (Available Passenger Miles) usually express the volume of services produced during a given period (Kilometres instead of Miles can also be used).

Similar to the other producers of services, the airlines consume the inputs for the production processes. These inputs are usually divided into the physical and non-physical. The main physical inputs as mentioned above are: capital represented by the aircraft fleet and buildings, repair/overhaul and maintenance equipment, computer and communication facilities, and airport, aircraft, passenger and baggage service facilities, labour represented by staff, and fuel/energy used. The main non-physical inputs are the above-mentioned slots as flying rights on the selected routes between particular airports included in their air route networks and "loyalty" of air passengers. Slots and consequent flying rights are mostly dependent on the market structure and policy, i.e., they have been different on the regulated and deregulated air transport. The "loyalty" of passengers has been gained either by building and maintaining the monopolistic position in the specific markets or through successful competition with other airlines and inland transport modes in the same markets (routes).

a) The airline fleet consists of the aircraft of various sizes and seat capacities. The number and capacity of particular aircraft depend on the size of the airline network, represented by the number of airports, lengths of routes between them, and the number of flights scheduled on particular routes.
b) The airline network consists of the number of airports and the lengths of routes between them. The physical/spatial size of the airline networks can be measured in order to find the geographical area of provided traffic rights.
c) The airline flights have been the base of airline operations. Each flight is defined by the aircraft type, departure and arrival time from/to the particular airports, respectively, and the scheduled (block) time between them. With scheduled airlines, the routes and the number of flights, i.e., the flight frequencies on them scheduled during a given period of time (day, week, season, and year), are announced in the airline schedule/timetable. As such, this can be considered as the airline production plan being subject to changes depending on the variation of the air passenger demand.
d) Airline outputs from the schedule/timetable are the corresponding volumes of ASM during a given period, thus, ultimately reflecting engagement of the available "static" inputs over time. Utilization of these as inputs enables the production of volumes of RPM, which can reflect the airline "dynamic" capacity. The attribute "dynamic" is applied to emphasise that the volumes of RPM carried out are the result of efforts and skills of airline management to sell the available resources (inputs) in real-time.
e) Labour is specified by the number, structure and working time of specific classes of staff operating the airline. The most common staff classification is the division into flight and non-flight, which has proven to be the most logical due to their different contributions to the airline output.
f) The airline aircraft fleet and all supporting facilities and equipment are powered by the fuel and energy, respectively, whose quantities mainly depend on the size of the airline and the efficiency of fuel and energy use.

3.3.2.2 "Static" and "dynamic" capacity

The airline "static" and "dynamic" capacity can be expressed in the quantitative terms by different indicators.

a) Indicators of "static" capacity

Some of indicators of the airline "static" capacity have been considered as follows (Janić, 2000):

- The aircraft fleet size, age, and structure in terms of different aircraft types;
- The size of the network in terms of the number of airports served;

- The average length of routes in the network; and
- The airline staff/employees.

Figure 3.16(a, b) shows the relationships between the airline fleet size, age, and the number of different aircraft types at 18 of the world's largest airlines.

Figure 3.16a shows that the average age of fleet approximately linearly increases as the fleet size increases. For example, with the airline of about 900 aircraft, the average fleet age is about 15 years. For an airline with about 250 aircraft, the fleet age is about 5 years. Figure 3.16b shows that the number of different aircraft types increase more than proportionally with an increase in the aircraft fleet size. In addition, development of the airline network in terms of the number of airports served depending on the fleet size is shown in Figure 3.17 for the European LCC Ryanair.

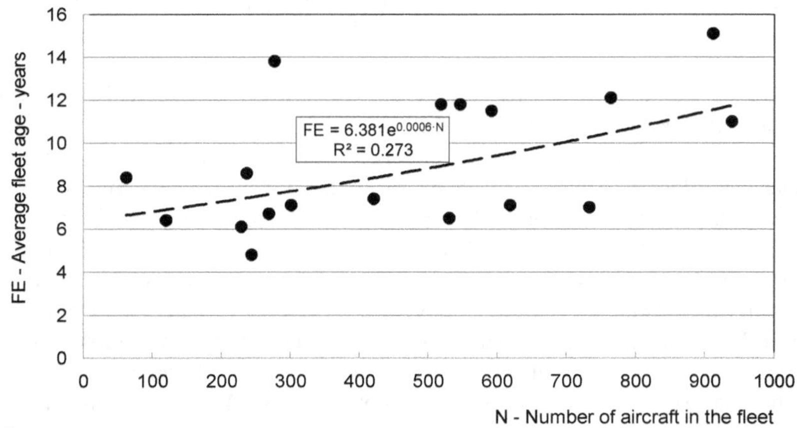

a) Average fleet age vs fleet size

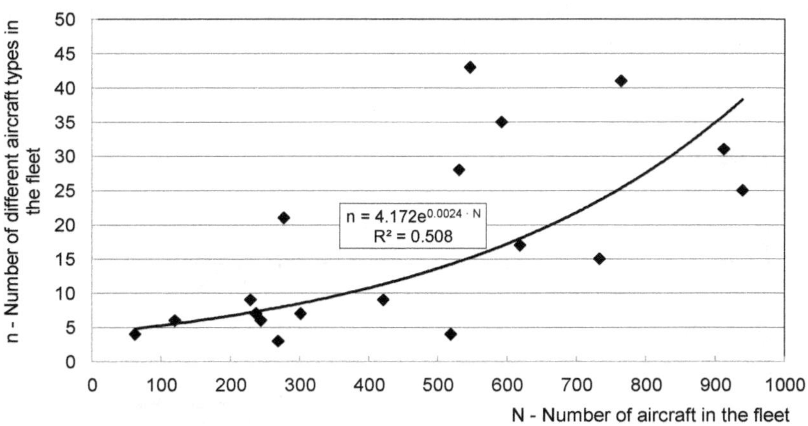

b) Structure of airline fleet vs its size

Figure 3.16 Relationship between the aircraft fleet size, age, and structure—Case of 18 of the world's largest airlines (Period: 2018) (https://www.airfleets.net/home/; https://www.planespotters.net/; https://blueswandaily.com/top-20-airline-groups-by-fleet-size/).

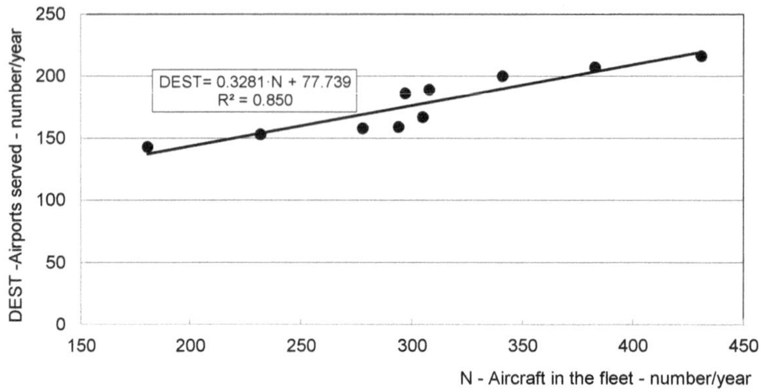

Figure 3.17 Relationship between the number of airports served and the size of airline fleet—Case of the European LCC Ryanair (Period: 2009–2018) (Ryanair, 2008/2018).

As can be seen during the observed period, the airline has expanded its network in proportion to the expansion of the aircraft fleet, and vice versa.

The length of route in an airline network is expressed indirectly by the average passenger stage length. Figure 3.18 shows examples of the development of this stage length over time at one European LCC, one U.S. LCC and one network airline.

As can be seen, the average stage length of the European Ryanair and U.S. Southwest LCC varies (increasing and decreasing) during the observed period, within the range of 600–800 mi (mi – statute mile; 1 mi = 1.609 km). That of the U.S. network Delta Airlines has generally been higher than that of both LCCs and, despite variations, generally decreased over time within a range of about 1300 to 1150 mi. In addition, Figure 3.19 shows the relationship between the number of an airline staff/employees and size of its fleet, the latter expressed in terms of the number of aircraft at the network Lufthansa Group airline.

The network Lufthansa Group airline includes Lufthansa German Airlines, SWISS, and Austrian Airlines. As can be seen, the number of staff has been increasing at a

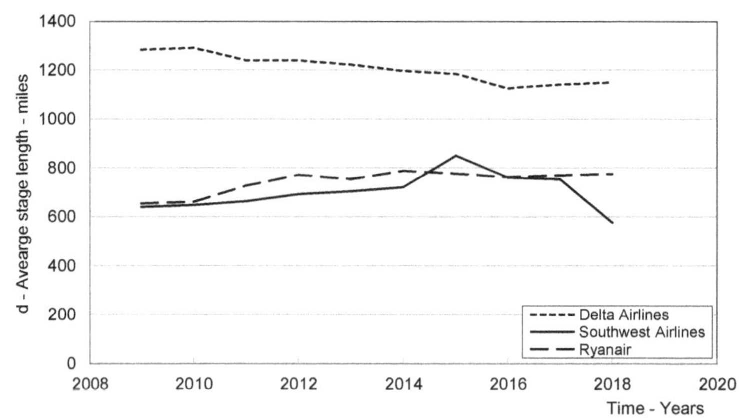

Figure 3.18 Development of the average stage length over time—Case of U.S. and European LCCs and Network Airlines (Ryanair, 2008/2018; http://web.mit.edu/airlinedata/www/Aircraft& Related.html).

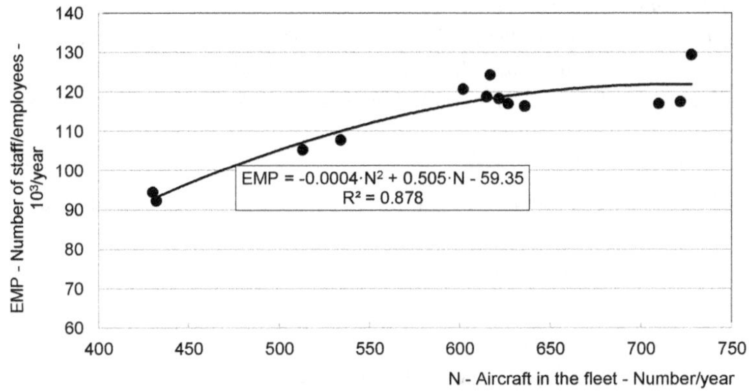

Figure 3.19 Relationship between the number of staff/employees and the number of aircraft—Case of the network airline Lufthansa Group (Period: 2005–2018) (DL AG, 2005/2018).

decreasing rate as the Group's fleet size has grown during the observed period. This indicates an increase in staff/employee work productivity over time. In addition, Eq. 3.5a gives the causal relationship between the size of staff/employees, fleet, and number of destinations served at the largest European LCC Ryanair (Ryanair, 2008/2018):

$$EMP = \beta_0 + \beta_1 \cdot N_a + \beta_2 \cdot N_d = -3288.615 + 19.410 \cdot N_a + 38.752 \cdot N_d$$
$$t\text{-}stat \quad -3.076 \quad 4.331 \quad 3.513$$
$$R^2 = 0.982;\ F = 192.383;\ D\text{-}W = 1.000;\ N = 10 \qquad (3.5a)$$

where

EMP is the staff-employees (10^3/airline);
N_a is the number of aircraft in the fleet (-);
N_d is the number of destinations served (-); and
N is the number of data series.

As can be seen, there is a very strong, almost functional, relationship between the dependent *(EMP)* and two independent variables *(N_a)* and *(N_d)*. In particular, *t*-statistics indicate the very high importance of the selected independent variables.

Also, Eq. 3.5b gives the causal relationship between the dependent variable *(EMP)* and particular dependent variables as inputs for 10 of the world's largest airlines and airline groups (Lufthansa Group, American Airlines Group, Delta Airlines, United Airlines, Air France – KLM, IAG (International Airlines Group), Southwest Airlines, China Southern Airlines, China Eastern Airlines, Air China) (https://www.airport-technology.com/features/worlds-biggest-airlines-2018/).

$$EMP = \beta_0 + \beta_1 \cdot N_a + \beta_2 \cdot S = 31.051 + 0.032 \cdot N_a + 0.287 \cdot S$$
$$t\text{-}stat \quad 1.157 \quad 2.370 \quad 0.349$$
$$R^2 = 0.404;\ F = 2.372;\ D\text{-}W = 2.442;\ N = 10 \qquad (3.5b)$$

where

EMP is the staff-employees (10^3/airline);
N_a is the number of aircraft in the fleet (-);
S is the number of fleet seats – (10^3/airline); and
N is the number of data series.

As can be seen, a certain causal relationship between the size of airline staff and the other two elements of its "static" capacity—the size of aircraft fleet and its seat capacity—has been noticed, although the relative importance of particular coefficients of the independent variables (N_a, S) remained questionable due to the low *t*-statistics.

b) Indicators of "dynamic" capacity

Some of indicators of the airline "dynamic" capacity have been considered as follows (Janić, 2000):

- Fleet productivity – load factor;
- Fleet utilization;
- Staff/employees productivity;
- Relationship between inputs, their utilization, and outputs;
- Fuel efficiency; and
- Production function.

The ratio between the airline output and input in terms of RPM and ASM is called load factor and considered as the specific indicator of the overall airline productivity. This indicates that if a greater percentage of the available/offered seats are filled, the overall productivity of the airline will be higher, and vice versa. In evaluating the productivity of particular flights, a concept of "break-even" load factor has frequently been applied in order to indicate the bottom level of their commercial feasibility (Janić, 2000). The possible relationship between the annual quantities of RPM and ASM of the U.S. airline industry during the period 1995–2018 is given as follows (http://web.mit.edu/airlinedata/www/Aircraft&Related.html):

$$RPM = -54.071 + 0.911 \cdot ASM; \; R^2 = 0.996; \; N = 23 \tag{3.6}$$

As can be seen, the average marginal load factor is about 91%. In addition, Figure 3.20 shows development of the load factor for two European and two U.S. airlines during a given period of time.

With all considered airlines, the average annual load factor has increased over time, from about 75–80% in the year 2009 to about 80–95% in the year 2018. This implies that

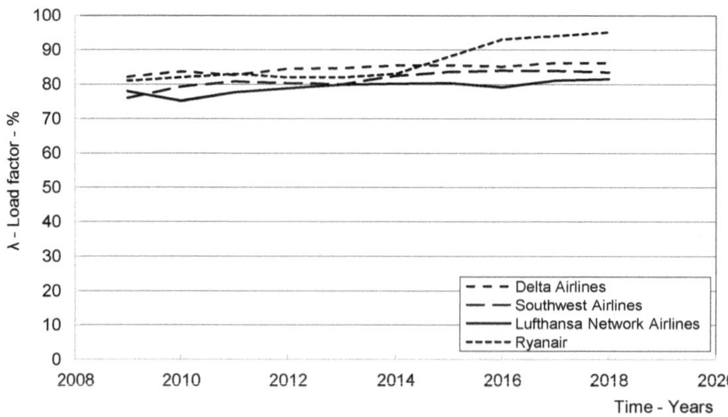

Figure 3.20 Development of airline load factor over time—Case of the selected European and U.S. airlines (DA, 2008/2018; DL AG, 2008/2018; SW, 2008/2018; Ryanair, 2008/2018).

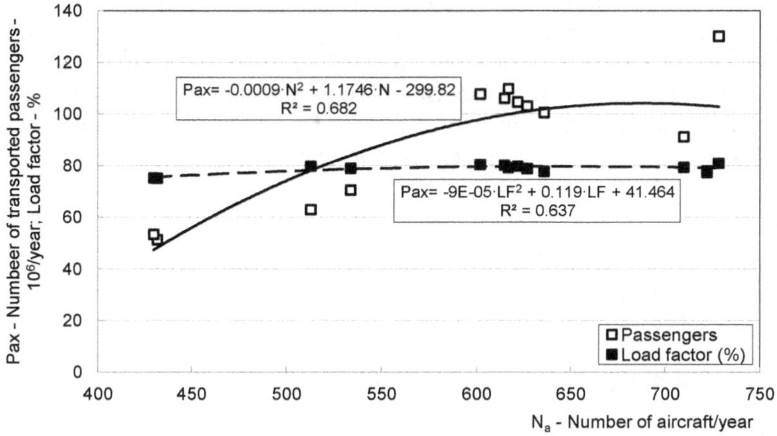

Figure 3.21 Relationship between the annual numbers of transported passengers, load factor and the annual number of aircraft—Case of Lufthansa group (Period 2005–2018) (DL AG, 2005/2018).

theses airlines have generally been continuously improving their productivity in the given context. Figure 3.21 shows the relationship between the annual number of transported passengers, operated aircraft, and load factor at Lufthansa group during the observed period.

As can be seen, the number of transported passengers increases at a decreasing rate as the number of aircraft in the fleet grows. Such development has driven the average load factor relatively constant—about 80%—during the observed period.

The fleet utilization can also be expressed by the aircraft flying time during the specified period. Figure 3.22 shows the example for two U.S. the network Delta and LCC Southwest Airlines.

As can be seen, Southwest airlines carried out 3–4 departures per day, which gives an average aircraft utilization of about 10.2–10.6 h/day. Delta airlines carried out about 5–6 departures per day, with an average aircraft utilization of about 10.2–11.6 h/day.

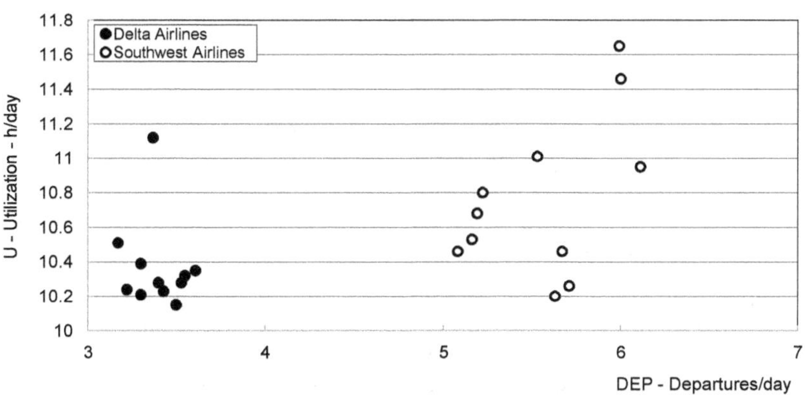

Figure 3.22 Relationship between the daily aircraft utilization and the number of flights—Case of the selected U.S. airlines (Period: 2008–2018) (DA, 2008/2018; SW, 2008/2018; https://www.transtats.bts.gov/Tables.asp? DB_ID=135; http://web.mit.edu/airlinedata/www/Aircraft&Related.html/).

At both airlines, an average flight is approximately 2 hours long (DA, 2008/2018; SW, 2008/2018; https://www.transtats.bts.gov/Tables.asp? DB_ID=135; http://web.mit.edu/airlinedata/www/Aircraft&Related.html/). In addition, the airline fleet utilization can be analysed in terms of the number of transported passengers by the engaged fleet on the given air route/airport network. Equation 3.7 gives the example for the above-mentioned 10 world's largest airlines (https://www.airport-technology.com/features/worlds-biggest-airlines-2018/):

$$PAX = \beta_0 + \beta_1 \cdot N_a + \beta_2 \cdot N_d = 40.145 + 0.091 \cdot N_a + 0.077 \cdot N_d$$
$$t\text{-}stat \qquad 1.389 \qquad 3.641 \qquad 0.692$$
$$R^2 = 0.715;\ F = 8.749;\ D\text{-}W = 2.284;\ N = 10 \tag{3.7}$$

where

PAX is the number of passengers carried – (10^6/year);
N_a is the number of aircraft in the fleet (-); and
N_d is the number of destinations served (-).

The above-mentioned causal relationship indicates that, in general, there is the rather strong dependency between the numbers of transported passengers and both influencing variables—the size of airline fleet *(AC)* and the size of air route network *(DEST)*. These numbers have increased with an increase in both these factors. It should also be mentioned that the size of airline fleet has been much more important than the size of network expressed in terms of the number of served destinations.

The airline staff/employee productivity can be expressed by the number of passengers transported during the specified period. Figure 3.23 shows an example for a large European airline.

As can be seen, the annual number of transported passengers by 3 Lufthansa Group airlines has increased more than proportionally with the growth of their staff. This indicates that the staff have permanently improved their productivity over time in the considered terms.

An additional indicator of the airline "dynamic" capacity can be also the fuel consumption of its fleet. The above-mentioned Figure 3.7 shows an example for two U.S. airlines, indicating a relatively constant average fuel consumption with increasing of the

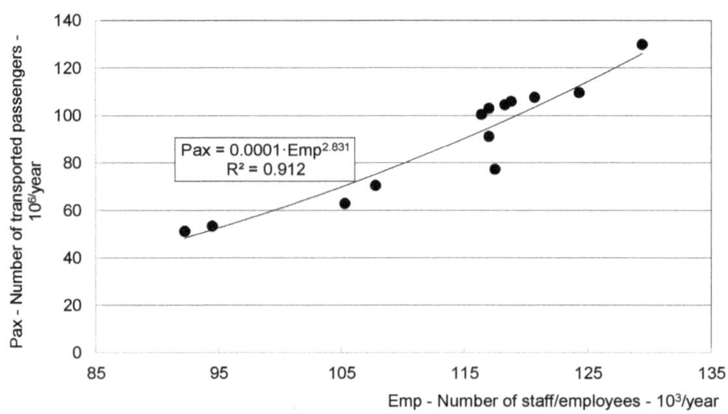

Figure 3.23 Relationship between the annual number of transported passengers and the number of staff/employees—Case of Lufthansa Group (Period: 2005–2018) (DL AG, 2005/2018).

daily utilization of their aircraft during the observed period. That at America airlines has been higher by about 57% than that of LCC Southwest airlines. This has mainly been caused by its fleet structure containing also wide-body aircraft carrying out long-haul flights, consequently consuming more fuel per unit of time.

In particular, the airlines can be considered as the firms producing transport services. As such, they are characterized by their production functions, usually expressed by the volumes of output depending on the volumes of inputs. The typical output is the transport work expressed in terms of RPK carried out during a given period. The inputs are generally labour, capital, and spatial characteristics of services, in the case of airlines, the staff/employees, the number and seat and ton capacity of the aircraft fleet, and the number of destinations/airports served, respectively. One of the generic forms of the airline production function can be as follows:

$$RPK = f(EMP; N_a; N_d) \tag{3.8}$$

where all symbols are analogous to the previous Eqs.

The airline production function is illustrated using the example of the above-mentioned $N = 10$ world's largest airlines for the year 2017. The following regression relationship is obtained (https://www.airport-technology.com/features/worlds-biggest-airlines-2018/):

$$\begin{aligned} &RPK = 4.886 \cdot EMP^{-0.165} \cdot N_a^{0.136} \cdot N_d^{0.410} \\ &t\text{-}stat \quad 5.471 \quad -0.630 \quad 0.800 \quad 1.864 \\ &R^2 = 0.507;\ F = 0.206;\ DW = 0.896;\ N = 10 \end{aligned} \tag{3.9a}$$

where RPM – 10^6/year; EMP – 10^3/year; N_a – No./year; N_d – No./year. This production function indicates that there has been a reasonable correlation between the dependent and three independent variables. At the same time, the volume of average output has decreased with the growth of the airline staff and increased with the growth of the size of fleet and the number of destinations/airports served. The elasticity of the number of destinations/airports served has been the highest. Similar production function is obtained for Lufthansa Group Airlines for the period 2005–2017 as follows:

$$\begin{aligned} &RPK = -288.742 + 4.585 \cdot EMP - 0.071 \cdot N_a \\ &t\text{-}stat \quad -5.379 \quad 4.766 \quad 0.612 \\ &R^2 = 0.935;\ F = 57.951;\ DW = 3.072;\ N = 13 \end{aligned} \tag{3.9b}$$

where RPM – 10^9/year; EMP – 10^3/year; N_a – No./year. As can be seen, this production function indicates that there has been a very strong correlation between the dependent and two independent variables. The airline output has increased with the growth of the number of staff/employees and decreased with very low elasticity as the number of aircraft in the fleet has increased. This latter feature may imply that the larger fleet was utilized at a lower rate than otherwise during the observed period. The other statistics also indicate the relevance of the chosen independent variables and significance of the entire regression.

3.3.3 Modelling Demand and Capacity

In addition to the above-mentioned analysis and quantifying of the airline demand and "static" and "dynamic" capacity based on the empirical data, the alternative approaches have contained development of the qualitative and quantitative models for estimating demand and airline operations. They relate to a route or the airline network where the flights realised by one or several aircraft types are scheduled to serve air passenger and/or fright/cargo demand under given conditions specified by the operational, economic, and/or policy constraints.

3.3.3.1 Demand

The analysis, modelling, and forecasting of the airline passenger and freight/cargo demand have been the subject of consideration by the airlines themselves, aviation agencies, industry associations and academic-research communities. The main purpose has been to alleviate, justify and support future developments. In many cases, there has been an inherent inclination towards optimism in protecting and preserving interests and objectives of the above-mentioned entities. In general, different models, such as time trend, econometric models, scenarios, ratios, market surveys, and judgments, have been used, as shown in Figure 3.24.

These models can be used for the analysis, modelling and forecasting of air passenger and freight/cargo demand on an airline individual route, set of routes and for the entire network under given conditions.

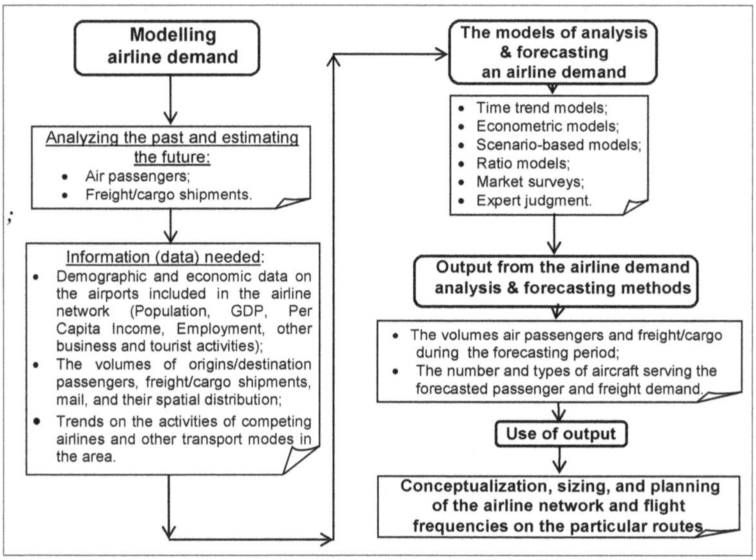

Figure 3.24 Scheme of analysis and forecasting of airline passenger and freight/cargo demand.

a) Time trend models

The time trend models are based on extrapolation of the past airline demand into the future period under the assumption that the future development (growth) will continue uninterrupted and continue development similarly as in the recent past. Figure 3.25 shows the general schemes of particular time-trend models.

In general, these models have the analytical forms containing the dependent variable *(Q)* representing the annual number of passengers on a given route, the explanatory variable *(t)* representing time (years), and the coefficients *(a)*, *(b)*, and *(c)*, whose values are estimated from past data. In general, these four types of curves can be as follows:

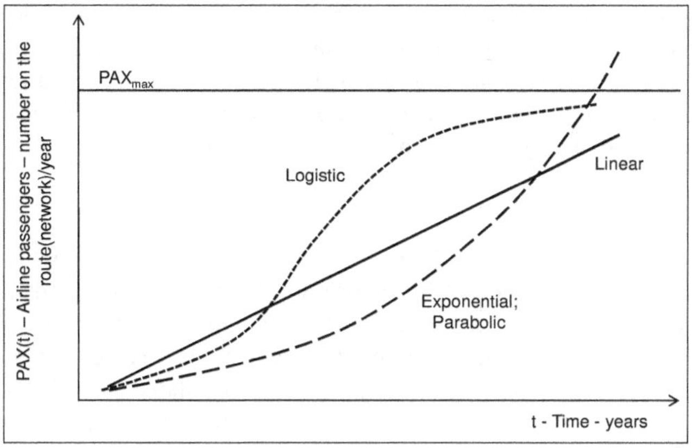

Figure 3.25 General shapes of the trend-time models.

i) Linear (or straight line)

$$PAX(t) = a + b \cdot t \tag{3.10a}$$

This implies a constant annual increment *(b)* in the airline passenger demand on a given route. Linear models indicate growth of airline demand at a constant rate over time.

ii) Exponential

$$PAX(t) = a \cdot (1 + b)^t \text{ and } \log PAX(t) = \log a + t \cdot \log(1 + b) \tag{3.10b}$$

The positive and less than one coefficient *(b)* implies that the passengers demand on a route increases at the rate *(b)* and exponentially/parabolically with time *(t)*. Taking logarithms converts the exponential to the linear model.

iii) Parabolic

$$PAX(t) = a + b \cdot t + c \cdot t^2 \tag{3.10c}$$

This curve can have different concave or convex shapes. If the coefficient *I* is greater than zero, the growth of the passenger demand on a route will increase more than proportionally over time.

iv) Gompertz

$$PAX(t) = a \cdot b^{-c \cdot t} \text{ and } \log PAX(t) = \log a - c \cdot t \cdot \log b, \quad 0 < c < 1 \tag{3.10d}$$

This curve can be used to model the airline passenger demand on a route, which will reach saturation *(a)* over a very long period of time.

v) Logistics

$$PAX(t) = \frac{PAX_{max}}{1 + b \cdot e^{-a \cdot t}} \tag{3.10e}$$

where

PAX_{max} is the absolute limit of the annual passenger demand that can be handled on a given route.

This logistics model embraces three sub-periods: in the first, it reflects relatively modest but increasing rate of growth of demand, in the second this rate is relatively constant, and in the last the rate of growth decreases when the volumes of demand approach the maximum constrained level independently of the type of constraint(s), i.e., operational, economic, policy, environmental, and/or social.

b) Econometric models

Econometric models have frequently been used to analyse (and predict) the airline passenger and freight/cargo demand. They present the mathematical/causal relationship between the overall and/or the segmented airline demand and the main driving forces, such as population, GDP, PCI, employment, the average airfare level or yield (the airline revenue per passenger kilometre), some transport service-related variables, such as travel distance, time, and the offered capacity in terms of the flight frequencies, seat capacities, etc. Based on the historical data for a given time period, the causal relationship between the airline passenger and/or freight/cargo demand as the dependent and the mentioned demand-driving forces as the independent variables can be estimated using the least-square regression technique. The main disadvantage of these models is an inherent assumption that the past relationships will very likely sustain in the future (See also Figures 3.12, 3.13 and Eqs. 3.3, 3.4). One possible analytical form of the econometric models for estimating airline demand on a given route of the network, based on the regression technique, is given as follows:

$$Q_{ij} = a_0 \cdot (GDP_i \cdot GDP_j)^{\alpha 1} \cdot (Y_{ij} \cdot d_{ij})^{\alpha 2} \cdot \tau_{ij}^{\alpha 3} \cdot f_{ij}^{\alpha 4} \cdot MS_{ij}^{\alpha 5}, \quad for\ ij \in N \tag{3.11}$$

where

Q_{ij} is the passenger demand between airports *(i)* and *(j)* (in both directions during a given time (pax/h, day, year);

$GDP_{i/j}$ is GDP (Gross Domestic Product) of the regions/around airports *(i)* and *(j)*, respectively (usually in billions \$US);

Y_{ij} is the average yield between the airports *(i)* and *(j)* (¢/RPK (Revenue Passenger Kilometre) or ¢/RPM (Revenue Passenger Mile));

d_{ij} is the two-way shortest distance between the airports *(i)* and *(j)* (km or mi) (km – kilometre; m – mile);

τ_{ij} is the total travel time between the airports *(i)* and *(j)* (h);

f_{ij} is the supply of transport capacity between the airports *(i)* and *(j)* in one or both directions during a given period) (flights/h, day, year or seats/h, day, year);

MS_{ij} is the market share of a given airline on the route between the airports *(i)* and *(j)* (proportion of the flight frequencies (seats) in the total number of flight frequencies (seats) supplied by all airlines operating on a given route (-);

α_k is the coefficient to be estimated by estimating the regression equation ($k = 1$–6).

Equation 3.11 indicates that the above-mentioned main forces drive the airline passenger demand on the route *(ij)*. On the short- and medium-haul routes (l_{ij}), the variables (τ_{ij}), (MS_{ij}), and (Y_{ij}) need to be modified in order to take into account the eventual competition from the surface transport modes, if available, and particularly with HSR (High Speed Rail).

The independent variables *(GDP_i)* and *(GDP_j)* relate to the annual GDP of the corresponding regions surrounding the airports *(i)* and *(j)*, respectively. Their product can reduce the number of variables and characterize the mutual "attractiveness" of the given regions/airports. Under conditions of the economic progress, these variables increase over time and seemingly contribute positively to the growth of airline passenger demand, and vice versa. This implies a positive elasticity, i.e., the coefficient *(a_1)*. Under some circumstances, these variables can be combined or replaced with their 'derivatives' such as PCI (Per Capita Income), population, employment, trade, investments, tourism, and exchange of other services between the airport regions. The variable *(Y_{ij})* is the average yield of a given airline operating on a route between the airports *(i)* and *(j)* during a given period of time (usually, one year). Its product with the variable *(d_{ij})* – two-way shortest travel distance—gives the average return airfare $AF_{ij} = d_{ij} \cdot Y_{hj}$. In general, when the airfare as the "travel resistance factor" increases, the number of passengers will generally decrease, and vice versa, which, at least in theory, reflects a negative elasticity, i.e., the negative coefficient *(a_2)*.

The variable *(τ_{ij})* expresses the travel time between the airports *(i)* and *(j)*. It includes the gate-to-gate aircraft time and eventually the schedule delay, i.e., the waiting time for the convenient departure flight at the airport *(i)*. Since it is based on the shortest distance, which usually does not change over time, this variable can be influential only if the aircraft technology changes (for example, if the faster regional jets replace the slower turbo-props) and/or if the flight frequencies by the given airline significantly change (Janić, 2000). Anyway, the airline passenger demand tends to increase with a decrease in route length, i.e., the shorter travel time implies negative elasticity, i.e., the coefficient *(a_3)* (see also Figure 3.10). The variable *(f_{ij})* reflects the supply of airline capacity on a given route, usually expressed by the number of seats (i.e., the flight frequency times the average aircraft size—seat capacity) or only the flight frequency. Since the supply of capacity (at reasonable airfares) intends not only to satisfy but also to stimulate demand, the latter is theoretically expected to rise with the growth of this capacity, and vice versa. This generally implies positive elasticity, i.e., the coefficient *(a_4)*. The variable (MS_{ij}) represents the competitive "power" of the given airline compared to the other airlines on a route *(ij)*. In general, the higher competitive "power" enables the attraction of a higher passenger demand, thus reflecting the positive elasticity *(a_5)*. In addition, Eq. 3.11 can also be used to analyse (and predict) the passenger demand of LCCs, but after some modifications, i.e., by keeping only the variables main driving forces, such as (GDP_{ij}) or *(PCI_{ij})* and the supply of airline capacity *(f_{ij})*. The airfares are assumed to be always acceptable for the attracted segment of passenger demand and, therefore, seemingly irrelevant for the proposed model. Also, the time-series data for particular O-D routes *(ij)* *(I, j = 1,.., N)* for the period of at least 12 to 15 years are needed for the estimation/calibration of Eq. 3.11.

c) Scenario-based models

The scenario models are usually used to demonstrate the assumed variations of the future conditions influencing the airline's either aggregated or disaggregated-passenger demand. For example, particular variables in the econometric models could be assigned a range of the future values, based on the scenarios of developing the demand-driving forces, rather than on their exact values. The disadvantage of these models is that the range between the high and the low values of particular independent variables can sometimes be so large as to cause the forecast to lose its practical value.

d) Ratio models

The ratio models usually imply the ratio between the airline passenger and freight/cargo demand at a given route and the total demand in the network. The main disadvantage of this method relates to conditions of its application—the lack of resources and expertise necessary for using some other more sophisticated methods.

e) Market surveys

The market surveys are based on collection of the information about the passenger and freight/cargo behavioural patterns—origins and destinations and the airport choice in the metropolitan area, trip purpose, and other factors useful for predicting the future behaviour. The main disadvantage of this method is its complexity, the fact that it can be very time-consuming, and the elevated financial cost of collecting the relevant data. In addition, since the data are collected for a relatively short period of time, they may not be so reflective for the medium- to longer-term planning purposes.

f) Expert judgement

The expert judgment is a component of almost all forecasting models. In this method, the assumptions on the future development and the values of particular variables are simply the matter of informal judgment. For example, the Delphi method has often been used to reach a consensus on the judgments from different experts about the values of particular variables relevant to forecasting the airline future demand. In addition, the judgment is also used for the selection of a method, years to be used as a base, analytical form of the models, and data sources.

In general, the above-mentioned models for analysis and forecasting of the airline passenger and freight/cargo demand represent the attempts of diminishing complexity of the large real systems by simplified analytical equations describing the most relevant interactions. In such context, many factors are often dropped-off due to the complexity of their analytical formalization, unnecessary increasing of the methods' complexity, lack of the relevant data, and lack of the sufficient expertise on the potential influence. Since these methods assume that the future will very much look like as the past, their products—analyses (and forecasts) of the airline passenger demand—are highly influenced by the underlying assumptions about the current and future development of the main external and internal demand-driving forces.

3.3.3.2 Capacity

In addition to the above-mentioned approach of quantifying the airline "static" and "dynamic" capacity based on calibrating the airline production function, the alternative approach has been the development of the qualitative and quantitative models of the airline operations on a route and its network. The specific class of analytical models have been dealing with the capacity of airline route, network, and size of fleet (Janić, 2000).

a) Capacity model of airline route

A route of an airline network consists of a pair of airports and a link connecting them. The link will exist if there is at least one non-stop flight scheduled between the airports in a given period (a day, week, and month). Two concepts of air route capacity have been frequently considered: the route supply capacity and the route productive capacity, both determined for some period (τ). The route supply capacity can be estimated as follows (Janić, 2000; Vuchic, 2007):

$$RSC(\tau) = f_{max}(\tau) \cdot s \tag{3.12a}$$

where

$f_{max}(\tau)$ is the maximum flight frequency, e.g., the maximum number of flights that can be carried out on a given route during time (τ) (dep/h or day); and

s is the average seat capacity of an aircraft carrying out the flights $f_{max}(\tau)$ (seats; tons of payload).

The flight frequencies, $f_{max}(\tau)$ are dependent on the capacity of the airspace where the route is stretched out. They might be determined simply as: $f_{max}(\tau) = \tau/h_{min}$, where (h_{min}) is the minimum time-based separation between successive flights scheduled on the given route applied by the ATC/ATM (Air Traffic Control/Management). The minimum separation (h_{min}) should be applied only if the given route is saturated, i.e., if there is constant demand for scheduling flights there during the time (τ). However, in the real systems, such developments have rarely taken place, so the actual number of flights scheduled on a given route during time (τ) has always been lower than the maximum number of flights $f_{max}(\tau)$. A few important influencing factors are: the insufficient passengers and/or freight/cargo demand to support the feasible frequent flights between particular airports, the traffic rights on the route(s), the overall airline business policy, etc. For example, it is known that the inter-state and inter-airline bilateral and/or multilateral agreements determine and constrain the number and availability of the landing slots and parking gates at particular airports.
The route productive capacity can be computed by formula (Vuchic, 2007):

$$PC(\tau) = RSC(\tau) \cdot v(d) = f_{max}(\tau) \cdot s \cdot v(d) \tag{3.12b}$$

where

$v(d)$ is the average aircraft gate-to-gate speed, i.e., block speed, on the route (d) (km/h; kt; kt-knots).

The product $(s \cdot v(d))$ represents the technical productivity of the aircraft performing flights $(f_{max}(\tau))$ on the route (d) (see also Figure 3.4b). This capacity appears to be very useful measure of the airline performances. It simultaneously comprises both the characteristics of route and the aircraft carrying out flights there during a given time period.

b) <u>Capacity of airline network</u>

The above-mentioned capacity concepts for the route can be extended to the airline network. The possible approach implies assuming that an airline network consists of (N) nodes/airports and $(N(N-1))$ links, each with at least one non-stop flight scheduled between any pair of airports during time (τ). The supply capacity of airline network can be estimated by summing up the supply capacities of particular non-stop routes. From Eq. 3.12a this follows:

$$NSC(\tau) = \sum_{l=1}^{N \cdot (N-1)} f_{max/l}(\tau) \cdot s_l \cdot d_l \tag{3.13a}$$

where the symbols are analogous to those in Eq. 3.12(a, b). Equation 3.13a quantifies the maximum transportation work (in terms of ASM or ASK (Available Seat Miles or

Kilometres), which can be carried out in the airline network during time *(τ)*. This can be reduced to a more real-life case by introducing the actual values of the flight frequencies.

The productive capacity of airline network can be estimated based on Eq. 13a as follows:

$$PC(\tau) = \sum_{l=1}^{N\cdot(N-1)} f_{max/l}(\tau) \cdot s_l \cdot v(d_l) \tag{3.13b}$$

where all symbols are analogous to those in Eq. 3.12b and Eq. 3.13a. As can be seen, the productive capacity of airline network represents the sum of productive capacities of particular routes. Similar to the case of supply capacity, this measure of performance may become more realistic by introducing more real-life values of the flight frequencies. The term $(v(d_l))$ can be expressed as: $v(d_l) = d_l/t(d_l)$, where $(t(d_l))$ is the average block time on the route (d_l). This time has been always estimated in advance by taking into consideration the local traffic and weather conditions on the route(s), as well as the performance of engaged aircraft. Introducing a greater number of longer routes, engagements of the larger aircraft and scheduling more flights (up to the maximum frequencies) are all the factors contributing to increasing of the productive capacity of given airline network. In addition, the delays imposed on particular flights by the ATC/ATM can prolong the scheduled block time and decrease the expected block speed. For example, if the average delay on route (d_l) is $(w(d_l))$, the actual block speed will be: $v(d_l) = d_l/[t(d_l) + w(d_l)]$. Such reduction of speed due to delays generally diminishes the productive capacity of particular flights, routes, and consequently the airline network.

c) Size of airline fleet

The size and complexity of an airline network is expressed by the number of airports and routes connecting them, the scheduled flight frequencies on particular routes, and the gate-to-gate operating times. These times are dependent on the route length and the speed of aircraft types engaged to carry out particular flights. All these are the most important factors influencing the size of an airline fleet. In order to estimate the size of this fleet, let us consider the subset of *K* routes contained in the original airline network of *L* routes $(K \subset L)$, where only one aircraft type is supposed to carry out flights during time *(τ)*. Let $(f_{k/j}(\tau))$ be the scheduled flight frequencies by the aircraft of type *j* on the route *(k)* in time (τ) $(k \in K; j \in J)$. The route length is (d_k) and the average gate-to-gate time $(t_j(d_k))$. The expected utilisation of the aircraft *(j)* on the route *(k)* during time *(τ)* is $(U_{j/k}(\tau))$. The average anticipated delay per flight *(j)* on the route *(k)* is $(w_j(d_k))$. The number of aircraft to perform the planned flights $(f_{k/j}(\tau))$ under given conditions $(AC_{j/k}(\tau))$ can be estimated as follows:

$$\sum_{k=1}^{K} f_{k/j}(\tau) \cdot [t_j(d_k) + w_j(d_k)] \le AC_{k/j}(\tau) \cdot U_{k/j}(\tau) \tag{3.14a}$$

and

$$AC_{k/j}(\tau) \ge [1/U_{k/j}(\tau)] \cdot \sum_{k=1}^{K} f_{k/j}(\tau) \cdot [t_j(d_k) + w_j(d_k)] \text{ and } t_j(d_k) = d_k/v_j(d_k) \tag{3.14b}$$

where

$v_j(d_k)$ is the average gate-to-gate speed of the aircraft *(j)* on the route (d_k).

Equation 3.14b indicates that the airline fleet size in terms of the number of aircraft will increase with an increase in the number of routes, flights scheduled there, scheduled gate-to-gate times and anticipated delays, and decrease with an increase in the aircraft average utilization during time *(τ)* (Janić, 2000; Pollack, 1982). The total size of an airline fleet can be determined by summing up the values obtained by Eq. 3.14(a, b) for all subsets of routes of the airline network where the aircraft of same and/or similar types are engaged (Janić, 2000; Pollack, 1982; Trevett, 1983).

Alternatively, since the aircraft itinerary implicitly reflects their daily (weekly) utilisation, an approach based on the queuing theory can be used to determine the airline fleet size. In order to show the characteristic of "new" procedure, the airline is assumed to operate a hub-and-spoke network, consisting of a single hub and several spokes (Janić, 2000; Sorenson, 1990). The itinerary of each aircraft carrying out flights in the network is known in advance. According to the shape of itinerary, each aircraft is scheduled to start and end its daily and/or weekly operations at the hub. Meanwhile, the aircraft fly between specific pairs of spokes connecting them through the hub. Particular flights between the same pairs of spokes are realised by the aircraft of the same or similar types. A simplified scheme of the itinerary of a couple of aircraft rotating between the spokes *(i)* and *(j)* through the hub *(h)* in time (τ) is shown in Figure 3.26.

Under the assumption that each aircraft of type *(k)* starts and finishes its itinerary at the hub *(h)*, the duration of an itinerary can be estimated as follows:

$$\tau_{ij/k} = 2 \cdot (t_{hi/k} + t_{hj/k}) + \tau_{i/a/k} + \tau_{h/a/k} + \tau_{j/a/k} \tag{3.15a}$$

where

$t_{hi/k}$, $t_{hj/k}$ are the average gate-to-gate times of the aircraft *(k)* between the hub *(h)* and the spokes *(i)* and *(j)*, respectively (it is assumed that $t_{hi/k} \equiv t_{ih/k}$ and $t_{hj/k} \equiv t_{jh/k}$) (h); and

$\tau_{i/a/k}$, $\tau_{h/a/k}$, $\tau_{j/a/k}$ are the aircraft ground handling (turnaround) times of the aircraft *(k)* at the airports *(i)*, *(h)* and *(j)*, respectively (h).

According to the queuing theory, the time $(\tau_{ij/k})$ in Eq. 3.15a can be considered as the occupancy time of a moving server (e.g., the aircraft) on a route *(h, I, h, j, h)*. When the spokes *(i)* and *(j)* are connected by the flight frequencies $(\lambda_{ij/k}(\tau))$, the number of aircraft required to sustain these connections during the time (τ) is determined by Little's formula as follows (Janić, 2000; Kleinrock, 1976):

$$AC_{ij/k}(\tau) = \lambda_{ij/k}(\tau) \cdot \tau_{ij/k} \text{ and } \lambda_{ij/k}(\tau) = 1/h_{ij/k}(\tau) \equiv f_{ij/k}(\tau)/\tau \tag{3.15b}$$

where

$f_{ij/k}(\tau)$ is the flight frequency scheduled between the spokes *(i)* and *(j)* in time (τ) (dep/h or day); and

$h_{ij/k}(\tau)$ is the average time between departing flights of class *(k)* on the route *(h, I, j, h)* in time (τ) (h, days).

From Eq. 3.15(a, b) the total number of required aircraft of type *(k)* to carry out the flights $(f_{ij/k}(\tau))$ between the airports *(i)* and *(j)* via hub *(h)* during time (τ) can be estimated as:

$$AC_k(\tau) = \sum_{ij} \left(\frac{f_{ij/k}(\tau)}{\tau} \right) \cdot \tau_{ij/k} \tag{3.15c}$$

By adding the values of $(AC_k(\tau))$ for all aircraft types, the total size of the airline fleet can be determined ($k \in K$, where K is the number of different aircraft types in the airline fleet). The procedure does not take into consideration the constrained utilisation of the specific aircraft types during time (τ). By slight modifications of Eq. 3.15a and then Eq. 3.15b, this can be easily addressed in the final estimation. Similar to Eq. 3.14(a, b),

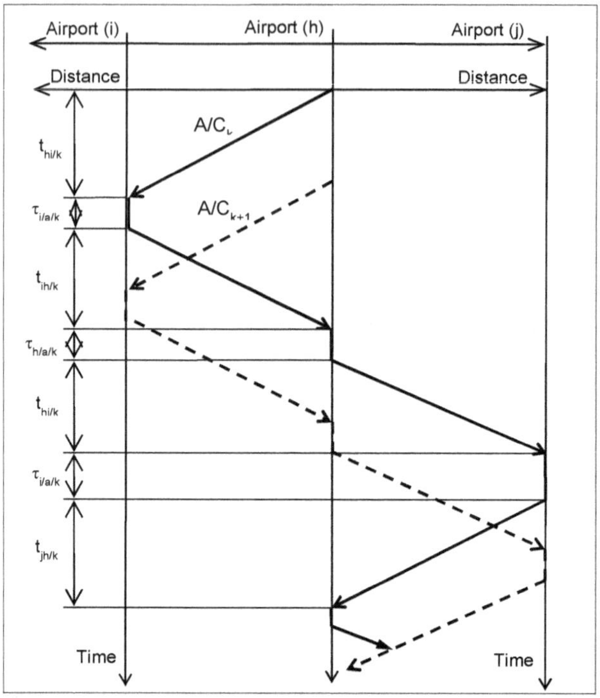

Figure 3.26 Time-space diagram of the aircraft scheduled itinerary between two spoke airports through the hub airport (Janić, 2000).

the flight delays may be also included in Eq. 3.15(a, b, c). As in the previous case, these delays will prolong the aircraft itineraries and cause a rather "artificial" need for the additional aircraft to carry out the same number of flights in the network during time *(τ)*.

In addition, the method for estimating the number of aircraft in the airline fleet was also developed (Janić, 2000). Similar to the previous case, it was assumed that the airline operates a hub-and-spokes network. Each aircraft operated between the particular spokes by passing through the hub. They started and finished their itineraries at the hub. The *(K)* types of flights scheduled from the hub might short-, medium- and long-hauls. Sorting of the particular flights into different classes could also be done according to the other criteria. It was assumed that each class of flights was assigned to the specific aircraft type (category). The period *(τ)* was divided into the peak-period (τ_p) and the non-peak period $(\tau - \tau_p)$. The number of the aircraft required in order to serve the network of N_s spokes by the type of service *(k)* was determined as follows:

$$[AC_h(\tau_p)]_k = \begin{cases} \min(\tau_{hi/k}/h_{p/k}) + 1, & \textit{if} \min \tau_{hi/k} < \tau_p \textit{ and } i \in N_S \\ \tau_p/h_{p/k} + (\tau_{hi/k} - \tau_p)/h_k + 1, & \textit{otherwise and } i \in N_S \end{cases} \tag{3.16a}$$

where

$\tau_{hi/k}$ is duration of the itinerary of the service class *(k)* equal to: $\tau_{hi/k} = t_{hi/k} + \tau_{i/k} + t_{ih/k} + \tau_{h/k}$, where $t_{hi/k}$, $t_{ih/k}$ are the gate-to-gate times of the aircraft flying from the hub *(h)* to spoke *(i)*, and vice versa, respectively (h); $\tau_{i/k}$, $\tau_{h/k}$ are the ground service time of the aircraft *(k)* at the airports *(i)* and *(h)*, respectively (min, h);

$h_{p/k}$ is the average interval between dispatching the flights of class (k) from the hub to the spokes during time (τ_p) (peak-period) (it is easy to show that: $h_{p/k} = \tau_p/f_k(\tau_p)$, where $f_k(\tau_p)$ is the number of scheduled flights of class (k) during time (τ_p)) (h, days);

h_k is the average interval between dispatching the flights of class (k) during the off-peak period $(\tau - \tau_p)$, $[h_k = (\tau - \tau_p)/[f_k(\tau - \tau_p)]$(h, days);

N_s is the number of destinations-spokes connected to the hub during time (τ_p); and

τ_p is the duration of the peak period (hours, days).

From Eq. 3.16a, the total number of aircraft required in order to perform all kinds of scheduled transport services from the hub (h) during the peak-period (τ_p) was equal to:

$$AC(\tau_p) = \sum_{k=1}^{K} [AC_h(\tau_p)]_k \tag{3.16b}$$

Equation (3.16b) implies that the aircraft, which had just finished the shortest itinerary of class k, was engaged to carry out the next itinerary of the same class during time period (τ_p). By inserting the corresponding equations linking the flight frequencies and the length of dispatching intervals in time (τ_p) and $(\tau - \tau_p)$, Eq. 3.16(a, b) would be reduced to Eq. 3.15(a, b). However, many researchers and practitioners have considered the described procedures to determine the size and structure of airline fleets as relatively crude, although they have seemed to be useful particularly when the relatively fast changes of both would need to be realised. The other more complex algorithms, including many heuristic principles, have been recommended for determining the rather "exact" size of the airline fleet. These algorithms could be used for estimating the minimum number of aircraft required in order to perform a given number of flights in the network. They have also enabled reconstruction of itineraries of the particular aircraft after they were assigned to the specific flights. The shape of each itinerary has been dependent on the sequences of successive flights allocated to the particular aircraft (Brown, 1989; Janić, 2000; Teodorović, 1988). Nevertheless, despite many efforts, distributions of the aircraft over the network and their optimal utilisation have remained a very complex, difficult, and consequently always challenging problem for the airline managers (Banfe, 1992; Janić, 2000).

3.2.2.3 Matching capacity to demand

In general, the models for matching the airline capacity to demand on a route include those on determination and optimization of the flight frequencies, which; (i) satisfy the air passenger and/or freight/cargo demand; (ii) sustain the airline market share; (iii) act as the competitive tool; (iv) minimize the total costs: (v) maximise the total profits; and (vi) maximise the quality of services. In particular, the optimization models for matching the airline capacity to demand in the airline network include the model for planning the flight frequencies to satisfy the expected demand, maximize the single airline profits, and maximize the airline profits under competitive conditions (Gordon and de Neufville, 1973; Janić, 2000; Swan and Adler, 2006; Teodorović, 1988; Teodorović and Krčmar-Nožić, 1989; Yeng, 1987).

a) Flight frequencies satisfying the passenger demand on a route

The number of flights scheduled on a route during a given time period mainly depends on the characteristics of the air passenger demand intended to be served. If the airline has a monopolistic or niche-market position on a given route, the number of flights supplied to serve the given demand during time (τ) can be estimated as follows:

$$f_l(\tau) = \frac{Q_l(\tau)}{\lambda_l(\tau) \cdot s_{k/l}} \quad \text{for } l \in L \tag{3.17}$$

where

L is the number of routes in the airline network,

$Q_l(\tau)$ is the expected number of passengers on the route (l) during (τ) (pax/day);

$\lambda_l(\tau)$ is the average load factor on the route (l) in time (τ)(-); and

$s_{k/l}$ is the seat capacity of the aircraft (k) engaged to carry out the flights on the route (l) (seats/aircraft).

a) Flight frequencies maintaining the airline market share on a route

When two airlines operate on the same route, they may either collaborate or compete with each other. If they collaborate, they will supply the flights and seats according to the negotiated bilateral contracts. Under such conditions, both airlines intend to serve the expected demand on a route. The flight frequencies for each airline can be estimated as follows (Janić, 2000):

The share $k_1(\tau)$ of the airline No. 1 in the total capacity supply on a route during time (τ) is equal to:

$$k_1(\tau) = \frac{f_1(\tau) \cdot s_1}{\sum_{i=1}^{2} f_i(\tau) \cdot s_i} \tag{3.18a}$$

where

$k_1(\tau)$ is the share in the capacity supply of the airline No. 1 during time (τ) $(0 \le k_1(\tau) \le 1.0)$;

$f_i(\tau)$ is the flight frequency of the airline I in time (τ) $(I = 1,2)$ (dep/day);

s_i is the average seating capacity of the aircraft engaged by the airline (i) to perform the flights on a given route (seats/aircraft).

Since the capacity offered by both airlines on a route should serve the total demand during time (τ), the following relation holds:

$$\sum_{i=1}^{2} f_i(\tau) \cdot s_i \cdot \lambda_i = Q(\tau) \tag{3.18b}$$

where

$Q(\tau)$ is the demand on a route during time (τ) (air passengers or tons/day); and

$\lambda_i(\tau)$ is the average route load factor of the airline (i) during time (τ).

By combining Eq. 3.18(a, b), the number of flights scheduled by the airlines No. 1 and No. 2, respectively, is equal to:

$$f_1(\tau) = \frac{Q(\tau)}{s_1 \cdot \{\lambda_1(\tau) + \lambda_2(\tau) \cdot [1/k_1(\tau) - 1]\}} \tag{3.19a}$$

and

$$f_2(\tau) = \frac{Q(\tau)}{s_2 \cdot \{[k_1(\tau)/1 - k_1(\tau)] \cdot \lambda_1(\tau) + \lambda_2(\tau)\}} \tag{3.19b}$$

Different relationships can exist between the airlines on the route. For example, if the airline No. 1 leaves the market [$k_1(\tau) = 0$], Eq. 3.19b will reduce to Eq. 3.18. This corresponds to the situation when the airline No. 2 completely serves the total demand on a route. Otherwise, in the case when [$k_1(\tau) = 1$], the airline No. 2 leaves the market and only the airline No. 1 will serve total demand. This time Eq. 3.19a will reduce to Eq. 3.18. A hypothetical example where two airlines negotiate to satisfy the expected demand on a given route together is shown in Figure 3.27.

The airline No. 1 uses the aircraft of $s_1 = 150$ seats and expects the average load factor on a route of $\lambda_1(\tau) = 0.60$. The airline No. 2 uses the aircraft of $s_2 = 250$ seats and expects the average load factor on a route of $\lambda_2(\tau) = 0.5$. Evidently, there are two opposite trends in supplying the capacities on a route. If the airline No. 1 is allowed to increase the supply of capacity on a route, the airline No. 2 should diminish its supply in order to balance the total supply to the expected demand given the above-mentioned load factors of both airlines. Since the airline No. 1 uses the smaller aircraft, the total number of flights offered on a route in time (τ) by both airlines will slightly increase. The total number of flights scheduled by both airlines during time (τ) will vary between the two extreme cases. The first case is when the airline No. 2 is the exclusive supplier of flights [$k_1(\tau) = 0$]; the other is when the airline No. 1 is the exclusive supplier of flights [$k_1(\tau) = 1$].

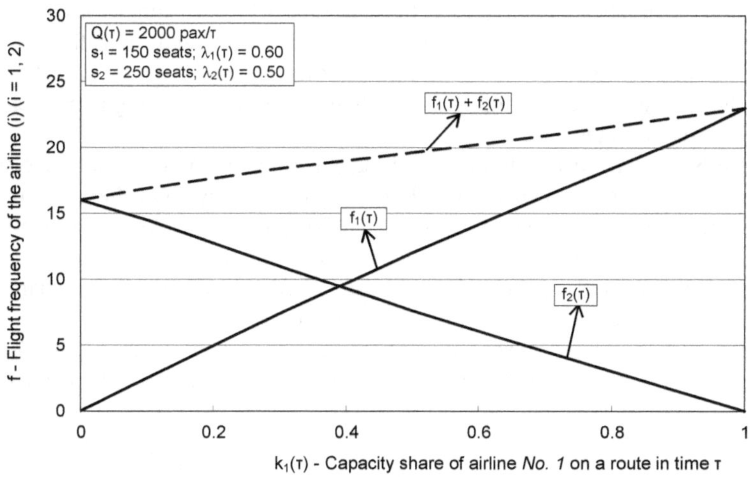

Figure 3.27 Dependence of the number of flights on the airline market share: Example of two collaborating airlines on a given route (Janić, 2000).

b) Flight frequencies as a competitive tool on a route

In partly deregulated market(s), the airlines may compete amongst each other using the flight frequencies as an exclusive competitive tool (Banfe, 1992). Thus, they can gain some market share defined by the proportion of the total demand served by each of them in a given period. For example, on the route where (K) airlines compete, the relationships between the airline market share and capacity share can be analytically expressed in the following way (Janić, 2000; Teodorović, 1983):

$$MS_{k/l} = [(f_k \cdot s_k)^a]/[\sum_{k=1}^{K}(f_k \cdot s_k)^a] \quad \text{for } l \in L \tag{3.20}$$

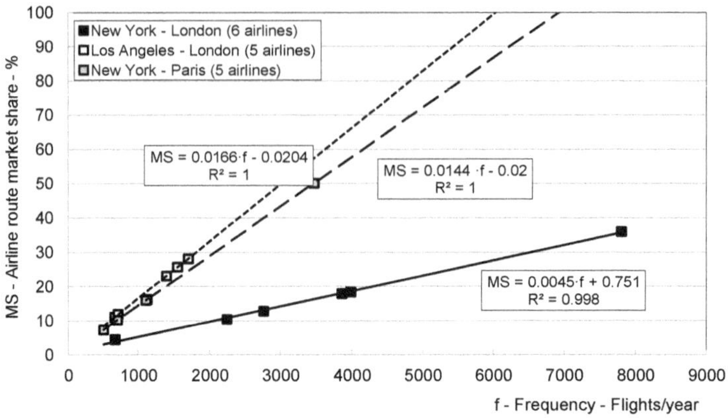

Figure 3.28 Relationships between the airline market share and the supplied capacity—Case of the selected Transatlantic routes (OAG, 2019).

where

α is an empirical constant ranging between 1 and 2.

The other symbols are analogous to those in Eqs. 3.18(a, b) and 3.19(a, b).

Equation 3.20 implies that both the aircraft seat capacity and flight frequencies are used as the airline competitive tool. Figure 3.28 shows the various empirical relationships between the market share and the capacity share in terms of the flight frequencies on different long-haul routes/markets.

On the New York to London route, six airlines operate and compete: Norwegian, Virgin Atlantic, American, Delta, United, and British Airways. On the Los Angeles to London route, five airlines compete: Norwegian, Virgin Atlantic, American, United, and British Airways. On the New York to Paris route, the competing airlines are Norwegian, Air France, Delta, United, and American. As can be seen, the airline market share at each route linearly increases as the flight frequencies increase (OAG, 2019).

Figure 3.29 shows the various theoretical relationships between the market share and the capacity share of an airline *(k)* *(k ∈ K)*. As can be seen, when the value of *(α)* is

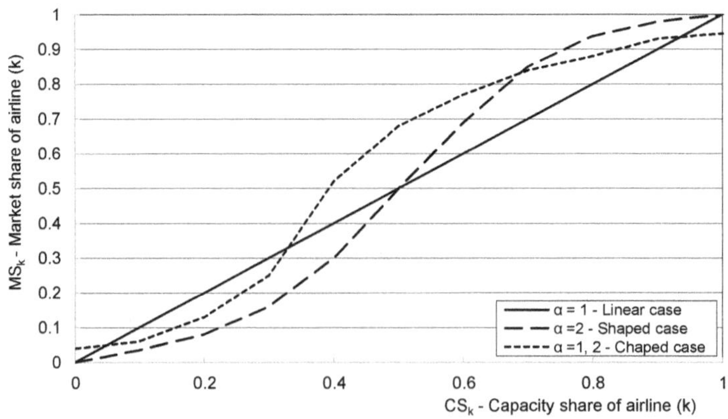

Figure 3.29 Market share vs. capacity share of the airline *(k)* on a given route—Case of the flight frequency competition (Janić, 2000).

greater than one, the curve representing the market share of corresponding airline will take an *S*-shape. Why is this so? One possible explanation relates to the expected and the actual behaviour of air passengers. First, the passengers always think about the dominant airline. Then, they make their seat-reservation at that airline expecting the best departure time (Janić, 2000).

An inherent tendency of the airlines operating in markets with free competition is to over-schedule. This may also influence the shape of the market share-capacity curve. Many airlines, especially the new entrants, have a short-term incentive to supply extra capacity in an effort to obtain an adequate market share in a relatively short time. Other airlines already operating in the same markets always answer such a challenge and expand their supply. Such a competing policy may cause an upward capacity-supply spiral. It has been common to see the airlines flying many empty seats on particular routes (markets). In order to fill these seats, these airlines have over-scheduled offering additional flights to increase their market shares. Sometimes this policy has resulted in flying more empty seats (Banfe, 1992). To quantify this behaviour, we have observed a single market (air route) where only two airlines compete between themselves. They are assumed to supply the flights on a route to sustain their temporary market shares and satisfy expected demand. They schedule their flights according to the following scenario: if airline *No. 1* increases the number of flights by, let us say, one flight, airline *No. 2* will immediately answer by supplying a new flight, and vice versa. New market shares of both airlines can be determined by Eq. 3.20. The average load factor on a route, independently of the airline, can be determined based on Eq. 3.18b as follows:

$$[f_1(\tau) \cdot s_1 + f_2(\tau) \cdot s_2] \cdot \lambda(\tau) = Q(\tau) \tag{3.21a}$$

and

$$\lambda(\tau) = Q(\tau)/[f_1(\tau) \cdot s_1 + f_2(\tau) \cdot s_2] \tag{3.21b}$$

Figure 3.30 shows an example of the dependence of the average route load factor and the airlines' market shares on the number of extra flights supplied by both competing airlines.

If both airlines simultaneously supply the same number of extra flights, their market shares will not change. The route load factor will decrease, indicating that both airlines

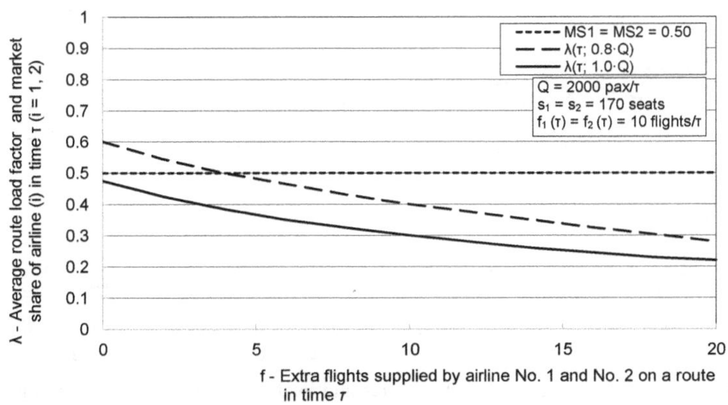

Figure 3.30 Dependence of the airline load factor and market share on the number of extra flights scheduled on a route (Janić, 2000).

fly some extra empty seats (the other factors influencing demand are assumed constant). Any decrease in demand on a route will cause an additional decrease of the load factor. Drops in demand may be caused by local and global factors, such as air traffic incidents/accidents, local crisis, regional wars and terrorists' attacks, global economic recession, etc. When demand stagnates and/or declines, the airlines should stop scheduling the extra flights. In the latter case, they should also cancel some of the existing flights in order to keep the load factor at an acceptable (at least zero-profitable) level.

What happens when one of two airlines competing on the route does not respond to the frequency challenges by the other is illustrated in Figure 3.31.

At the beginning, both airlines have equal market shares. If airline No. 1 increases the number of flights, this will attract more passengers into its system, primarily due to the passiveness of airline No. 2. Its market share will increase according to the S-shaped capacity-market share curve. However, any such increase in the capacity supply should be accompanied by the other competitive (attractive) measures, such as offering discount fares, increasing punctuality, improving cabin services, provision of more advertising, etc. Otherwise, all these may cause the flights of airline No. 1 to remain rather empty, i.e., the average load factor would decrease. The established market shares are relatively stable because the impact of market forces prevailing at the market has always been rather stable. This can be achieved primarily in the regulated and the semi-deregulated markets where there is only frequency competition between airlines. In the deregulated markets, where free entry, frequency and price competition prevail, the impact of market forces may be highly unstable, causing relatively high instability and fast changes of the airline market shares. Experience has shown that the concept of market share has become more important in the smaller rather than in the larger markets. The main reason is that the smaller markets can be controlled more easily than the larger ones. Therefore, it has appeared more desirable for particular airlines to have a greater market share in the smaller markets than a smaller market share in the larger markets. Due to decreased competition in the small markets, the average airfares may be higher (Janić, 2000).

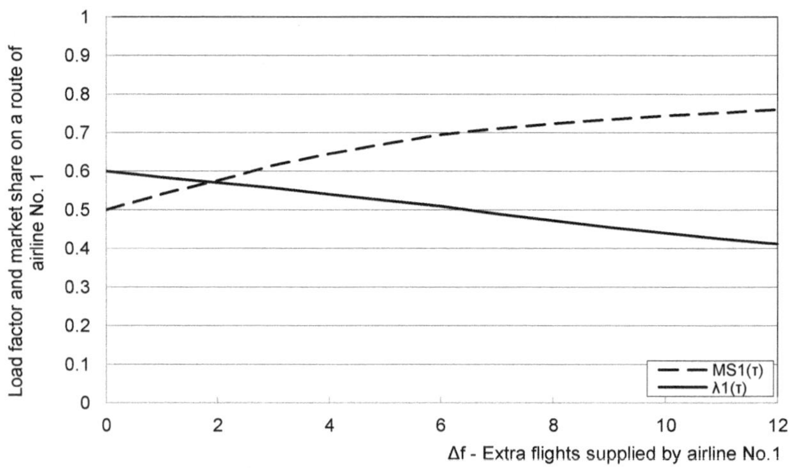

Figure 3.31 Relationship between the market shares and load factor of airline No. 1 and its supply of additional flights on a route (Janić, 2000).

c) Flight frequencies minimizing the total route cost

Various economic criteria can be applied to the determination of flight frequencies on a single route, these include revenue maximisation, cost minimisation and profit maximisation. While operating on the market(s) (route(s)) as an oligopoly, a single or a group of airlines supplies the flight frequencies to minimise the total cost, which consists of the passenger cost and the airline operating cost. This criterion is called the cost minimisation criterion. The passenger cost includes the cost of passenger time but not the cost due to airfare(s). The passenger time consists of the time waiting for the first convenient departure (e.g., the scheduled delay) and in-aircraft time. The airline operating costs comprise the cost of operating the flights on a given route in time (τ). The total route cost $C(\tau)$ can be expressed as follows (Janić, 1999; 2000; Yeng, 1987):

$$C(\tau) = Q(\tau) \cdot \left[\frac{\alpha \cdot (\tau) \cdot \tau}{2 \cdot f(\tau)} + \beta(\tau) \left(\frac{d}{v(s,d)} + w(\tau) \right) \right] + c_f(s,d) \cdot f(\tau) \tag{3.22a}$$

where

τ is the time period (h, day, month, year);

$Q(\tau)$ is the number of passengers travelling on the route in time (τ) (pax/time period);

$\alpha(\tau)$, $\beta(\tau)$ are the average cost of passenger time while waiting for departure (i.e., the schedule delay) and in-aircraft time, respectively, during time period (τ) (\$/pax-h);

d is the length of a route (km; nm);

$v(s, d)$ is the average block speed of the aircraft of seat capacity (n) on the route (d) (km/h, kts);

$f(\tau)$ is the number of flights scheduled on the route in the time period (τ) (flights/time period);

$w(\tau)$ is the average anticipated delay per a flight carried out during time (τ) (min/flight); and

$c_f(s, d)$ is the cost per flight carried out by the aircraft of seating capacity (s) on the route (d) (\$/flight).

By deriving the total cost function $(C(\tau))$ subject to continuous variable $(f(\tau))$ (all other variables are assumed to be parameters) and setting the obtained results equal to zero, the minimum of the function (3.22a) can be found as follows (Janić, 1999; 2000):

$$\frac{\partial C(\tau)}{\partial f(\tau)} = -\frac{1}{2} \frac{Q(\tau) \cdot \tau \cdot \alpha(\tau)}{|f(\tau)|^2} + c_f(s,d) = 0 \tag{3.22b}$$

The second derivative of the function $C(\tau)$ subject to the flight frequency $f(\tau)$ is positive as follows:

$$\frac{\partial^2 C(\tau)}{\partial f^2(\tau)} = \frac{Q(\tau) \cdot \tau \cdot \alpha(\tau)}{|f(\tau)|^3} = 0 \tag{3.22c}$$

From Eq. 22b, the flight frequency, which minimises the total route cost, is equal to (Janić, 1999; Janić, 2000):

$$f^*(\tau) = \left(\frac{Q(\tau) \cdot \tau \cdot \alpha(\tau)}{2 \cdot c_f(s,d)} \right)^{0.5} \tag{3.22d}$$

From Eq. 3.22d, the optimal number of flights on the route will increase at the rate of 0.5 with increasing demand consisting of more "expensive" passengers and time they are served (for example, more "expensive" can be business passengers, generally with higher valuation of their time). The optimal number of flights will decrease with an increase in the cost per flight. Combining Eq. 3.22b and Eq. 3.22d, the optimal size, i.e., seat capacity of an aircraft to be deployed on the given route in time (τ) is estimated as follows (Janić, 1996; 2000):

$$n^*(\tau) = \frac{1}{\lambda(\tau)} \cdot \left(\frac{2 \cdot Q(\tau) \cdot c_f(s,d)}{\tau \cdot \alpha(\tau)} \right)^{0.5} \tag{3.22e}$$

where all symbols are analogous to the corresponding ones in previous Eqs. If $n^*(\tau) > n_{max}$ (n_{max} is the maximum capacity of aircraft available in the airline's fleet), two or more aircraft should be deployed per single departure frequency (Janić, 1999; 2000).

Equation 3.22(d, e) is based on two assumptions: (i) the same aircraft type will carry out all flights on a given route; and (ii) the cost per flight can be calculated for any aircraft type (category). Figure 3.32 shows an example of the relationship between the optimal flight frequency, the optimal aircraft seat capacity, and the length of route.

As can be seen, the optimal frequency decreases at an increasing rate and the optimal aircraft seat capacity increases at a decreasing rate as the route length increases.

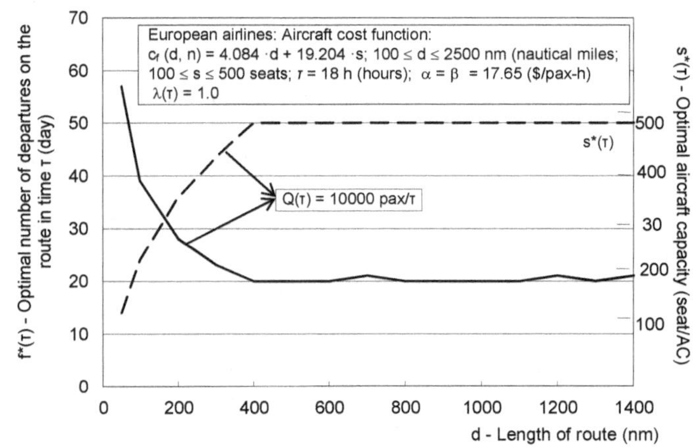

Figure 3.32 Relationship between the optimal flight frequencies, the aircraft seat capacity, and route length (Janić, 2000).

e) <u>Flight frequencies maximizing the route profits</u>

The flight frequencies which maximize the airline profits on a given route during time (τ) can be estimated as follows (Janić, 2000):

$$f(\tau) = \{[AF(\tau) + c_f] \cdot Q(\tau)\}/(c_f + c_d + c_{lf}) \tag{3.23a}$$

where

$AF(\tau)$ is the airfare on a route in time (τ)(h, day, year);

c_f, c_d is the average airline fixed and direct operating cost per passenger (not including the aircraft landing fee at the destination airport) ($/pax);

$Q(\tau)$ is the expected number of passengers on a route in time (τ) (pax/h, day, year);
c_{lf} is the landing fee at the destination airport ($/flight); and

Equation 3.23a indicates that the airline will schedule a greater number of flights on the route if it expects higher passenger demand, if it can set higher airfares, and if more passengers are expected to be attracted by the higher flight frequency. By increasing the airline operating costs, the flight frequency will be lower. In addition, the flight frequency, which maximizes the airline profits on a given route, can be estimated as follows (Janić, 2000; Teodorović, 1988):

$$f(\tau) = \left\{ \left(\frac{Q(\tau)}{AF(\tau) \cdot c_f(s,d)} \right) \cdot \left(C(\tau) - \frac{c_0(s)}{\lambda(\tau)} - c_0[Q(\tau)] \right) - \theta \cdot \alpha(\tau) \cdot \left[5.7 + 57 \cdot \left(\frac{\lambda(\tau) \cdot Q(\tau)}{s} \right)^9 \right] \right\}^{0.5} \tag{3.23b}$$

where

$C(\tau)$ is the total cost of air traffic on a given route during time (τ) ($US);
$c_0(s)$ is the minimum cost per seat on a given route ($/seat);
$c_0[Q(\tau)]$ is the cost per passenger transported on a given route during time (τ) ($US/pax); and
θ is the coefficient estimated by calibration of the model of air travel demand on a given route.

The other symbols are analogous to those in the previous Eqs.

f) Flight frequencies maximizing the quality of services on a route

The models presented above inherently include the assumption that the flight frequencies enable the airline to maximize the quality of services by minimising the cost of passenger schedule delay under given circumstances. Assuming the flight frequencies on a route as a continuous variable, the similar models for estimating the flight frequencies which minimize the total passenger scheduled delay in a given air route network were developed. In both models, the passenger schedule delay was defined as the difference between the time of intending to make a trip and the first available departing flight satisfying it. The flight frequencies scheduled on a given route of the airline network which minimize the total passengers scheduled delay is equal to (Gordon and de Neufville, 1973; Janić, 2000):

$$f_l(\tau) = \frac{Q_l(\tau)}{s} + \frac{U(\tau) - \sum_{l=1}^{L} Q_l(\tau) \cdot t_l}{s \cdot \sum_{l=1}^{L} \sqrt{Q_l(\tau) \cdot t_l}} \cdot \sqrt{\frac{Q_l(\tau)}{t_l}} \quad \text{for } l = 1,2,\cdots,L \tag{3.24a}$$

In addition, similarly, the flight frequencies minimizing the passenger schedule delays on a given route are estimated as follows (Janić, 2000; Teodorović, 1988):

$$f_l(\tau) = \frac{U(\tau) \cdot \sqrt{Q_l(\tau)}}{s \cdot \sqrt{t_l} \cdot \sum_{l=1}^{L} \sqrt{Q_l(\tau) \cdot t_l}} \quad \text{for } l = 1,2,\cdots,L \tag{3.24b}$$

where

l is the route $(l \in L)$ (L is the total number of routes in the airline network);
$Q_l(\tau)$ is the expected demand on the route (l) during time (τ) (pax/h, day, year);
t_l is the flight block time on a route (l) (h);

s is the average seat capacity of an aircraft flying in the network (seats/ac);
$U(\tau)$ is the total volume of seat-hours available to the airline fleet during time (τ) (h).

Equation 3.24(a, b) indicates again that the optimal route frequency will be higher if the smaller aircraft (with lower cost per flight) is used as well as if there is a larger number of passengers requesting service during the given time period. The airlines operating larger fleets can offer higher flight frequencies on each route of their networks and consequently provide a higher quality of services. Figure 3.33 shows the results of application of both above-mentioned models to the hypothetical example.

As can be seen, the results from the models are just slightly different to what is intuitively expected, but this does not does prevent using them for planning purposes in the given context.

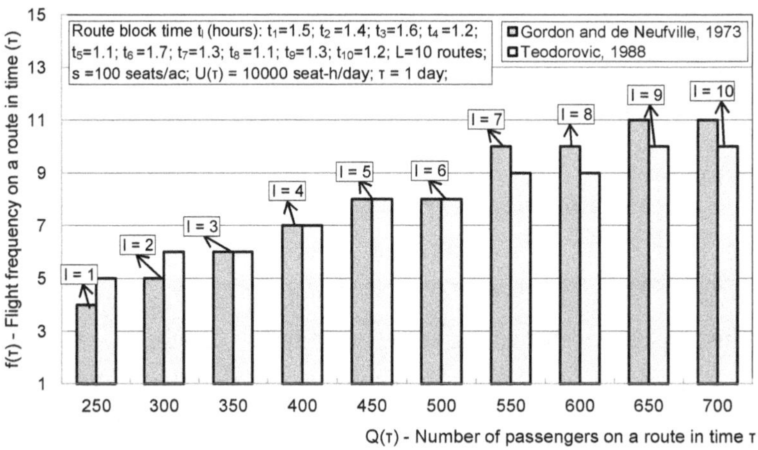

Figure 3.33 Relationship between the optimal flight frequency and the number of passengers on the route (Janić, 2000; Teodorović, 1988).

a) Maximizing profits in the airline network

Planning of the flight frequencies in the airline network is usually carried out through the schedule design. As already pointed out, the schedule may be considered the production plan of an airline; similarly to the plans of other service-producing firms, it can change over time (Etschmaier and Mathaisel, 1985; Janić, 2000). Design of the schedule usually includes the following activities: (a) determining the airports (cities) to be served, i.e., defining the potential markets (routes) and the potential (passenger and/or freight/cargo) demand there; (b) setting up the flight frequency on the particular routes to serve the expected demand; (c) determining the departure and arrival time for each flight and route by the aircraft type; (d) and estimating the potential revenues and cost (i.e., profits) from carrying out the activities (a), (b) and (c). In the actual design and planning procedure, the initial (draft) schedule has usually been checked for some time. For example, the departure times of particular flights are usually adjusted to the characteristics of demand. In addition, they may influence the planning of flight connections at particular airports. As such, they are checked regarding the operational feasibility. The flight departure times appear particularly important in the deregulated markets (routes) where the applications of available competitive tools, including adjusted departure-times, are allowed. Once the flight frequencies have been determined and their departure times checked-out, the available aircraft fleet is assigned to the specific routes and flights. After that, the rotation plan for the particular aircraft is synthesised.

Various methods for planning and designing the schedule of an airline have been developed. Most of them have included the application of sophisticated optimisation techniques developed in the operations research (Daskin and Panayotopoulos, 1989; Etschmaier and Mathaisel, 1985; Janić, 2000). In general, the complex problems called "fleet routing" and "flight and crew scheduling" have usually been handled by the specialised operational research groups engaged by the airlines. The problems have been formulated as two types of integer network flow problems, specifically pure network problems and network flow problems with side constraints. There have also been numerous objective functions and constraints. The objectives have usually been either minimisation of the fleet size (or the system costs) or maximisation of the system profits (difference between the operational revenues and costs). In addition, the numerous modifications of the schedule optimisation models can be found in the numerous references (Janić, 2000; Teodorović, 1988).

At this point, two characteristic optimisation models are described, both dealing with maximization of profits in the airline network. The first deals with the operation of a single airline on its air route network. The other considers the competition between airlines in the same air route network. Both types of models are based on the following assumptions: the air route network consists of the finite number of nodes (airports), the capacity expressed by size, structure and expected utilisation of the airline fleet during a given time period is constrained, and the capacity of air routes and airports is constrained. The objective function to be minimised is the total cost function. The function to be maximised is the total profit function (Janić, 2000; Teodorović, 1988; Teodorović and Krčmar Nožić, 1989).

i) Single airline

The problem of maximization of airline profits on the air route network is considered as the LP (Linear Programming) problem as follows (Janić, 2000; Teodorović, 1988):

$$\textit{Maximize}: \Pi(\tau) = \sum_{ij} \sum_k [p_{ij}(\tau) \cdot Q_{ij}(\tau) - c_k(s, d_{ij}) \cdot f_{ij/k}(\tau)] \quad (3.25a)$$

subject to the following constraints:

$$\sum_k \lambda_{ij/k}(\tau) \cdot s_k \cdot f_{ij/k}(\tau) \geq Q_{ij}(\tau) \quad \text{for } I \neq j,\ (I, j) \in n,\ k \in K \quad (3.25b)$$

$$\sum_k t_{ij/k} \cdot f_{ij/k}(\tau) \leq U_k(\tau) \cdot AC_k(\tau) \quad \text{for } k \in K \quad (3.25c)$$

$$\sum_j f_{ij/k}(\tau) = \sum_j f_{ji/k}(\tau) \quad \text{for } i \in N \quad (3.25d)$$

$$\sum_j f_{ij/k}(\tau) \geq [f_{ij/k}(\tau)]_{min} \quad \text{for } i \neq j,\ ij \in N;\ k \in K \quad (3.25e)$$

$$\sum_k \sum_{ij} f_{ij/k} \leq [D_i(\tau)]_{max} \quad \text{for } i \in N \quad (3.25f)$$

where

$p_{ij}\ (\tau)$ is the average airfare on the route *(ij)* during time *(τ)* ($/pax);
$Q_{ij}\ (\tau)$ is the number of passengers on the route *(ij)* during time *(τ)* (pax/h, day, year);
$c_k(s_k, d_{ij})$ is the cost per flight on the route (d_{ij}) carried out by the aircraft of seat capacity (n_k) ($/flight);
s_k is the seat capacity of the aircraft type *(k)* (seats/ac);

$\lambda_{ij/k}(\tau)$	is the average load factor of the flight carried out by the aircraft type *(k)* on the route *(ij)* during time *(τ)* (-);
$t_{ij/k}$	is the average flying time of the aircraft type (*k*) on the route (*ij*) ($t_{ij/k} = d_{ij}/v_{ij/k}$, where $v_{ij/k}$ is average block speed of aircraft type *(k)* on a route *(ij)*) (h);
$U_k(\tau)$	is the maximum utilisation of the aircraft type *(k)* during time *(τ)* (block hours/ac);
$AC_k(\tau)$	is the number of aircraft type (*k*) available during time *(τ)* (ac);
$[f_{ij}(\tau)]_{min}$	is the minimum number of flights scheduled on the route *(ij)* during time *(τ)* (dep/h, day, year);
$[D_i(\tau)]_{max}$	is the maximum number of departures that may be carried out from the airport *(i)* during time *(τ)* (dep/h, day, year); and
N	is the number of airports included in the airline network.

Constraints (3.25b–3.25f) can be explained as follows:

- Constraints (3.25b) ensure that demand on each route is satisfied by the offered flight capacity, given the average load factor.
- Constraints (3.25c) satisfy the condition that the total utilisation of the airline fleet consisting of the same aircraft type is limited by the number of aircraft in the fleet and their allowed (maximum) utilisation due to maintenance operations.
- Set of equalities (3.25d) provide the airline network conservation principle to be satisfied, i.e., the flights entering the particular network node (e.g., the airport) should also leave it during the same time period.
- Constraints (3.25e) provide that the total number of flights assigned to a given route is at least equal to or greater than the minimum number of flights scheduled on a route during a given time period.
- Constraints (3.25f) provide that the number of departures at an airport do not exceed its departure capacity during given time period.

The various sets of additional constraints may also be included in the model. However, they will increase the size of the problem. In general, the LP programming-based models produce the fractional flight frequencies, which are then rounded to the integer values. In the case of small airlines, the rounding operations of flight frequencies does not always guarantee optimal solutions. To overcome such deficiencies in the LP applications, the operational researchers proposed integer programming as a convenient operational research tool for planning the flight frequencies under such circumstances (Janić, 2000; Teodorović, 1988; Teodorović and Krčmar Nožić, 1989).

ii) Airline competition

Planning of flight frequencies can also be carried out under conditions of the frequency competition between few airlines operating in the same network of air routes. For such a case, the multi-criteria model for frequency planning and the aircraft assignment on the network was developed (Janić, 2000; Teodorović and Krčmar-Nožić, 1989). The model is based on the assumptions that the airlines competed among each other only by the flight frequencies while the airfares were fixed. The competitive airlines were assumed to have the following objectives, which should be simultaneously achieved.

They aim to determine the flight frequencies on the particular routes of the network (i.e., the number of flights), which should enable: (i) profit maximisation; (ii) maximisation of the total number of passengers attracted by the airline; and (iii) minimisation of the

total passenger scheduled delays. The analytical formulation of the corresponding optimisation problem was as follows:
Optimize the set of objective functions:

a) Airline total profit:

$$\max \Pi_k(\tau) = \sum_{ij} \{p_{ij}(\tau) \cdot Q_{ij}(\tau) \cdot \left[\frac{[f_{ij/k}(\tau)]^\alpha}{\sum_{k=1}^{K} [f_{ij/k}(\tau)]^\alpha} \right] - c_{ij/k}(s_k, d_{ij}) \cdot f_{ij/k}(\tau)\} \tag{3.26a}$$

ii) Total number of passengers attracted by the airline:

$$\max Q_k(\tau) = \sum_{ij} Q_{ij}(\tau) \cdot \left[\frac{[f_{ij/k}(\tau)]^\alpha}{\sum_{k=1}^{K} [f_{ij/k}(\tau)]^\alpha} \right] \tag{3.26b}$$

iii) Passenger schedule delays:

$$\max Z_k(\tau) = \sum_{ij} \frac{Q_{ij}(\tau) \cdot \tau}{4} \cdot \left[\frac{[f_{ij/k}(\tau)]^\alpha}{\sum_{k=1}^{K} [f_{ij/k}(\tau)]^\alpha} \right] \tag{3.26c}$$

subject to the following constraints:

$$f_{ij/k}(\tau) \leq [f_{ij/k}(\tau)]_{max} \quad \text{for } i \neq j,\ ij \in N \tag{3.26d}$$

$$\frac{[f_{ij/k}(\tau)]^\alpha}{\sum_{k=1}^{K} [f_{ij/k}(\tau)]^\alpha} \geq c_{ij/k}(s_k, d_{ij}) / [p_{ij}(\tau) \cdot Q_{ij}(\tau)] \quad for\ i \neq j,\ ij \in N \tag{3.26e}$$

$$\sum_{ij} f_{ij/k}(\tau) \cdot s_k \cdot t_{ij/k} \leq U_k(\tau) \tag{3.26f}$$

where

K is the number of competing airlines;
α is the empirically estimated constant (see Eq. 3.23b);
s_k is the seat capacity of an aircraft operated by the airline *(k)* (seat/ac); and
$U_k(\tau)$ is the total amount of available seat-hours of the fleet operated by the airline *(k)* during time period *(τ)* (seat-hours/h, day, year).

The other symbols are analogous to those in Eq. 3.25.

The objective functions (3.26a–3.26c) are non-linear in the variables which should be determined (e.g., in the flight frequencies $f_{ij/k}(\tau)$, $ij \in N$; $k \in K$). The specific sets of constraints can be explained as follows:

- The constraints (3.26d) limit the number of flights on the routes to the predetermined maximal values.
- The constraints (3.26e) provide fulfilment of the conditions for the profitable operations of the airline *(k)* along each route. According to this condition, the airline will not schedule any flight on any route if it does not expect to realise at least non-negative (zero) profits.
- The constraints (3.26f) indicate that, for each airline, the volume of transport work, which can be realised in the network is constrained by its "static" capacity, i.e., by the maximum allowed utilisation of the fleet during a given time period.

The multi-criteria integer programming method was proposed to solve this network optimisation problem. The authors made necessary modifications of the common method to make it applicable. In addition, they used heuristic approach in solving this problem (Janić, 2000; Teodorović and Krčmar-Nožić, 1989).

3.4 Quality of Services

3.4.1 Dimensions of Quality of Services

Quality of service of the airline industry can be analysed from three different aspects. The first is that of the airlines themselves. By using the quality of services as a competitive tool, they intend to maintain the current demand and attract new demand in their systems. The second aspect is that of users of the airline services—passengers and freight/cargo shippers. They choose an airline according to perception of particular attributes of its services that satisfy their temporary or permanent needs. The last one is the aspect of general public (society). On the one side, the public may indirectly be charged or benefit from the properly distributed services among the airlines and their users. On the other, similar effects may be expected by redistributing the transport services between the airlines and other transport modes.

The airline quality of service can be defined either in terms of the "passenger satisfaction" or as "continuity in satisfaction of the passenger requirements". This quality has several dimensions that can be used as criteria of choice of the air mode and the particular airline.

Generally, dimensions of the airline quality of service can be categorised as external and internal dimensions.

3.4.1.1 External dimension

The external dimension generally includes the following attributes:

a) Airline "static" capacity is defined by the number, type, and age of aircraft fleet. In addition, the number, kindness, and appearance of personnel, size of the reservation network, and easiness of access to the travel agents, operators, and all other agencies selling the airline capacities, appearance, cleanliness and convenience of the aircraft and the airport facilities intended for serving passengers reflects this capacity indirectly.

b) Size and scope of the airline network is determined by: the number of airports included in the network, the number of flights scheduled on particular routes, travel time on particular routes depending on their length, and the aircraft types engaged to perform flights. This attribute may reflect the following: diversity of destinations and possibilities for differentiation of services, convenience of the routes, departures, and flying times, the number of possible connections and the number and duration of the intermediate stops, load factor affecting the probability of finding out the empty seat at the desired flight(s), contribution to the overall congestion of air transport system, type of gains from the agreements contracted with other airlines, etc.

c) Punctuality and regularity of services reflects the airline ability to accurately perform the "promised" quality of service. It is represented by elements such as: on-time performance flights and length of delays expressing punctuality and the rate of cancellations expressed by the ratio between cancelled and carried out flights during the specified period of time, the gate assignment enabling convenient walking distance in the airport passenger terminal(s) and between different terminals, intensity of oversells, prompt baggage delivery and careful baggage handling, easiness and efficiency of procedures in a case of lost baggage, the overall efficiency, the commercial measures intended to hold existing and instigate new demand, such as the frequent flyer/loyalty programs practised by many airlines and their alliances, low (attractive) airfares becoming particularly relevant with the emergence of LCCs, and the airline financial viability expressing the overall warranty that the offered services will also be provided as promised.

3.4.1.2 Internal dimension

The internal dimension is relevant after air passengers enter the airline service system. It comprises the ground and in-vehicle (cabin) attributes of the quality of services.

a) Ground attributes are expressed by the elements such as: friendliness and kindness of the airline ground-based crew(s) (at the airline ticketing and check-in counters), continuous care for passengers, especially during long delays, readiness for re-solving overbooking problems, usually in favour of passengers, etc.
b) In-vehicle attributes depend on the size and comfort of seats in the aircraft, friendliness of cabin crew(s), cleanliness of cabin, beverage service, particularly on the long-haul flights, diversification and quality of drinks and meals, handling of children and elderly people, smoking policy, etc.

An estimation and evaluation of the airline quality of service is often a very complex task. The main complexity and even difficulty lies in the complexity of procedure and variability in the sources of relevant data. Thus, the most common way to deal with the problem has consisted of interviewing the actual and potential air passengers. The independent researchers, the consumers' protecting institutions and the airlines have mostly done this themselves. In the latest case, the airlines have provided the relevant estimates of the quality of service for themselves, their direct consumers and public. Most commonly, they have been measuring the "passenger satisfaction" by the attributes such as on-time performances, mishandled baggage, oversells, and consumer complaints. One of the typical examples of the systematic reporting of airline quality of service has been Air Travel Consumer Report, published by the U.S. Department of Transportation and its derivative AQR (Airline Quality Rating) (Bowen and Headley, 2019; http://dot.gov/airconsumer/).

3.4.1.3 Examples of quality of services

a) External attributes

Some external attributes of the airline quality of services, such as the "static" capacity and the elements of size and scope of the airline network, are given in Tables 3.1 and 3.2, and Figures 3.1 and 3.2. One specific element of the latter attribute is the number of flights scheduled on particular routes of the network. Figure 3.34 shows an example of the relationship between the flight frequencies and lengths of routes for a well-established European airline.

As can be seen, the airline scheduled a much higher number of flights (and seats) on the shorter routes than on the longer routes. This suggests that the passengers travelling

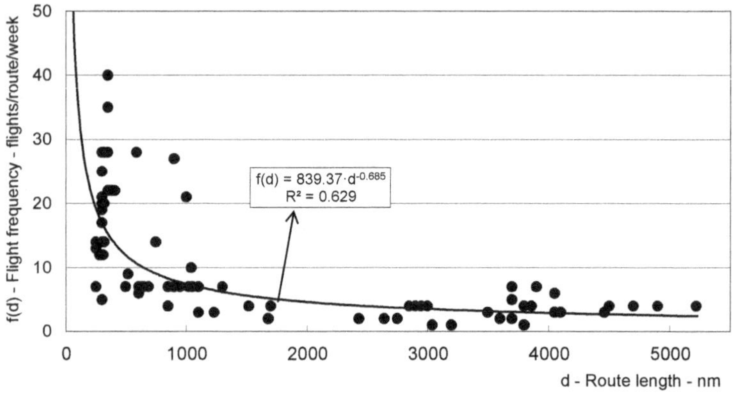

Figure 3.34 Relationship between the flight frequency per route and the route length—Case of a European airline (Janić, 2000).

on long haul-haul routes/flights had to wait longer for the first available flight than the passengers travelling on the short and medium routes/flights. This waiting time is actually defined as the passenger schedule delay, defined as the difference between the time of intending to travel until the time of the first available flight. Generally, the above-mentioned example indicates that passengers travelling on the shorter routes enjoy shorter scheduled delays due to the more frequent flights, in addition to spending a shorter time flying between origins and destinations.

The punctuality of services has been commonly expressed by the percent of on-time airline flights, i.e., on-time performance and length of delays of delayed flights. They reflect the punctuality of airline services. As such, they could be used by air passengers as criteria for choosing an airline and particular flights. In general, the frequent and business passengers can rank the airlines according to punctuality and regularity of their flights. In case of choice, they usually prefer those airlines offering high flight departure frequency, on time performance, and short eventual delays. In addition, these airlines should have high reliability of their flights, i.e., a very low cancellation rate due to the airline's reasons. Most airlines have reported these elements of service quality regularly (by month and per flight) to the public (their potential consumers), thus enabling easier evaluation of their own performances and choice of flight to their users (Bowen and Headley, 2019; EEC, 2019; US BTS, 2019; http://dot.gov/airconsumer/). Figure 3.35(a, b) shows an example of the punctuality of U.S. airlines.

Figure 3.35a shows that the average monthly on-time performance as a proportion of the number of scheduled flights departing and arriving on time has been relatively constant during the observed period. At both departing and arriving, it has varied between 60–65% and 90%. However, after an increase in the number of scheduled flights from 575 to 625 thousands/month, the corresponding on-time performance of both arriving and departing flights has been decreasing between 65% and 85%. In the given example, both on-time performances have been similar. Figure 3.35b shows that both on-time performances in the given case have been strongly correlated and that the on time arrival performance has usually been lower than its departure counterpart.

It has been mentioned that delays shorter than 15 min have not been counted. Therefore, in order to consider the airline delays as an attribute of the quality of services,

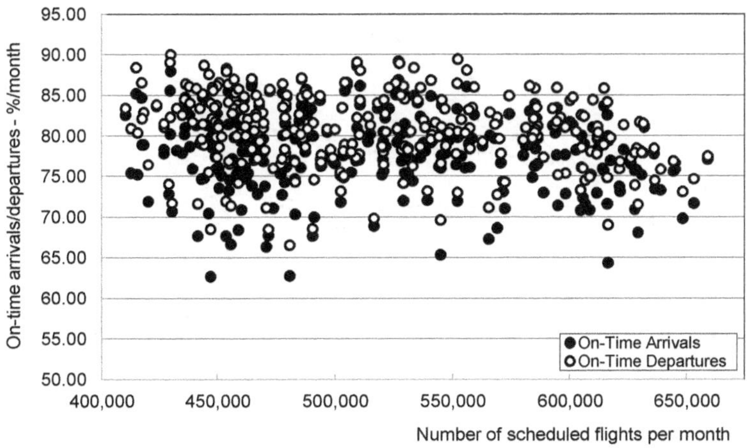

a) On-time performance vs the number of flights

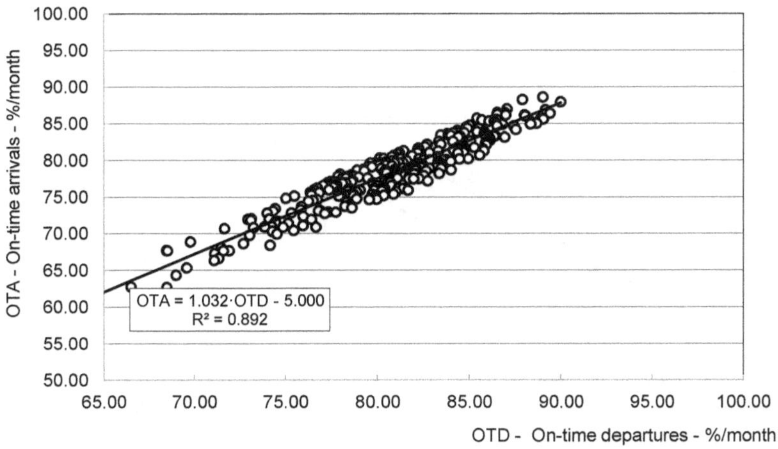

b) On-time performance: arrivals vs departures

Figure 3.35 Punctuality of services—Case of the U.S. airlines (Period: 1995–2017) (US BTS, 2019).

an example of the length of delays of delayed flights and their proportion for European airlines during the observed period is given in Figure 3.36(a, b).

Figure 3.36a shows that both average arrival and departure delays have been about 30 min/flight. Those of the arrival flights have been slightly higher than those of departures. In addition, the average proportion of the flights has increased from about 40% in the year 2015 to near 50% in the year 2019. However, this time, the proportion of delayed departure flights has been higher than that of the arrival flights. Figure 3.36b shows the rather strong linear correlation between the delayed arriving and departing flights, the former being longer than the latter (EEC, 2019).

The average load factor may also be used as the airline rather external general attribute of the airline quality of service. From the aspect of a potential marginal user, it indicates physical accessibility of the airline service (flight, route, and seat) in terms of the probability of finding an empty seat on a desired flight at a desired time. This attribute

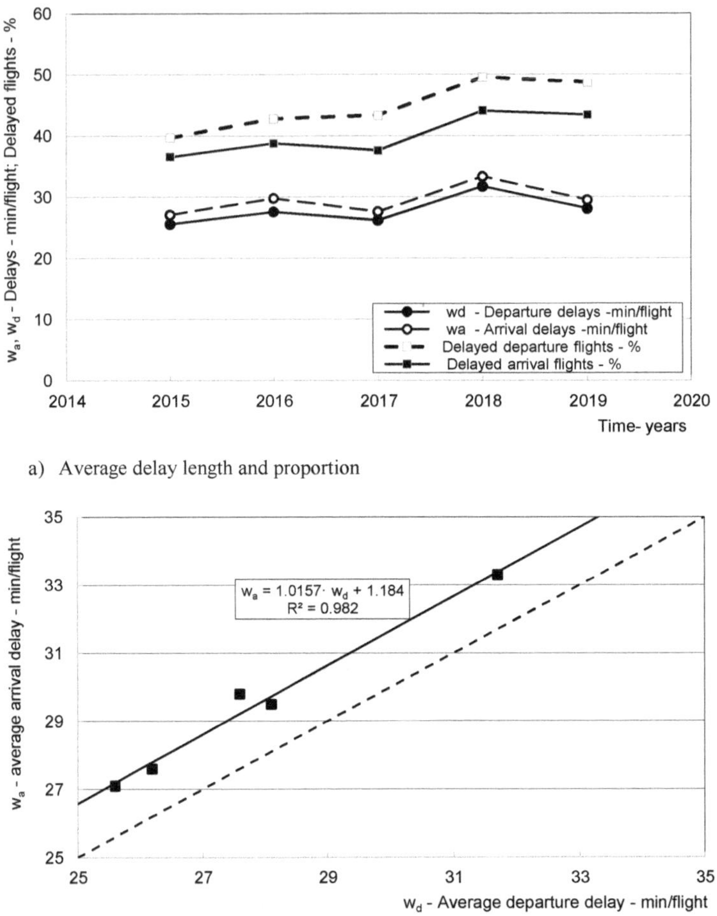

a) Average delay length and proportion

b) Relationship between arrival and departure delays

Figure 3.36 Example of the average delays per delayed flights, their proportion, and relationship—Case of European airlines (Period: April–June 2015–2019) (EEC, 2019).

inherently possesses a few disadvantages. First, it is known retrospectively, i.e., at the end of a given time period (one, three, six months, a year, etc.). Therefore, it can be applied only as an orientation to assess the future opportunity for accessing particular flights. Second, the majority of passengers do not know much about the value and meaning of an airline load factor, they are only interested in whether there will or will not be empty spaces on a route and flight where and when they want to travel. Therefore, load factor is more useful for the planners and operators of airline services than for their users/air passengers. Figure 3.37 shows an example of the relationship between the average load factor and the aircraft size, i.e., seating capacity, of the U.S. airlines (FAA, 2016).

As can be seen, the average load factor for an aircraft of an average capacity of 100–350 seats has been between 50% and 70%. This implies the rather constant corresponding probability of finding an empty seat in any of these aircraft and flights they carry out. For particular routes, the corresponding load factors can be considered

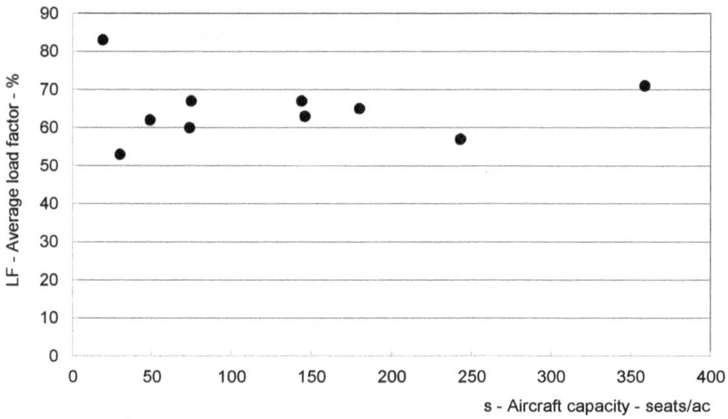

Figure 3.37 Relationship between the average load factor and the aircraft size/seat capacity—Case of U.S. airlines (Period: 2016) (FAA, 2016).

as relevant attributes of the quality of service. This approach should take into account all airlines operating on these routes by scheduling any competing or complementary services-flights.

b) Internal attributes

In addition to providing the sufficient number of flights and seats on the routes of their networks, many airlines have made efforts to provide high and uniform quality of their internal services for the passengers, both on the ground and onboard. On the ground, the airlines have mostly addressed the issue of the passenger baggage, denying passenger boarding, and general passenger complaints. In particular, passengers' complaints on the internal services of the U.S. domestic airlines have been systematically collected and made public for the long time. The official document has been published as Air Travel Consumer Report by the U.S. Department of Transportation and also used in the reports called Airline Quality Rating. They contain twelve customer complaint sub-attributes used as criteria, such as flight problems, oversales, reservations, ticketing, and boarding, fares, refunds, baggage, customer service, disability, advertising, discrimination, animals, and other (see also sub-section 3.4.2.3) (Bowen and Headley, 2019; http://dot.gov/airconsumer/). Figure 3.38 shows an example of the relative shares of these twelve attributes for the specified period of time.

As can be seen, the complaint 'flight problems' has been the most frequent, followed by the complaint 'baggage'. In addition, the total number and particular proportions of all complaints were lower in the year 2018 than in the year 2017, thus indicating their improvements. This needs always to be considered as the feedback from the airline users, i.e., air passengers.

The airlines have also provided a certain quality of services to their passengers while being onboard. For example, at present, these services differ between the short- and long-haul flights of both the conventional/network airlines and LCCs. The main attributes of these services have included categorization of in-cabin seats for business and economy regarding their comfort, i.e., size, including sub-classes of both, catering and availability of the entertainment facilities and equipment. In general, on short-haul flights, the number and diversity of services is much lower than on long-haul flights. Most airlines also standardize

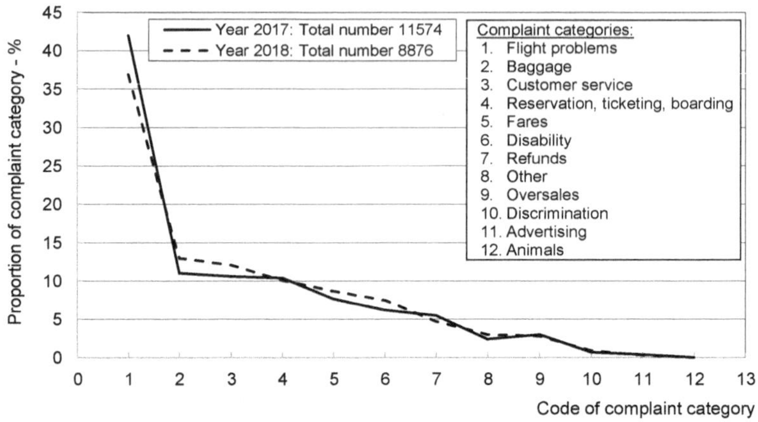

Figure 3.38 Proportion of different categories of complaints—Case of all U.S. domestic airlines (Period: 2017; 2018) (Bowen and Headly, 2019).

these services and charge for catering on short-haul flights. On long-haul flights, diversity of in-cabin services is higher.

For example, economy services provided in the long-haul flights of Virgin Airlines include: onboard WiFi, free 10 kg hand luggage, 300+hours of TV and films, complimentary drinks and snacks, a choice of 3 course meals, in-seat power for your devices, and goodie packs for kids (https://flywith.virginatlantic.com/gb/en/on-the-flight/cabin-and-seats.html/). In addition, Figure 3.39 shows one past example of the relationship between the number of different of in-cabin services and the length of flights (Janić, 2000; Lufthansa, 1995).

As can be seen, five different combinations with the same number of different types of services were offered. The number of services offered simultaneously during the flight has grown with the duration of flights. On short- and medium-haul flights (from one to four hours), various combinations of one, two and three kinds of services were offered. On the flights lasting between four and eight hours, combinations of four different

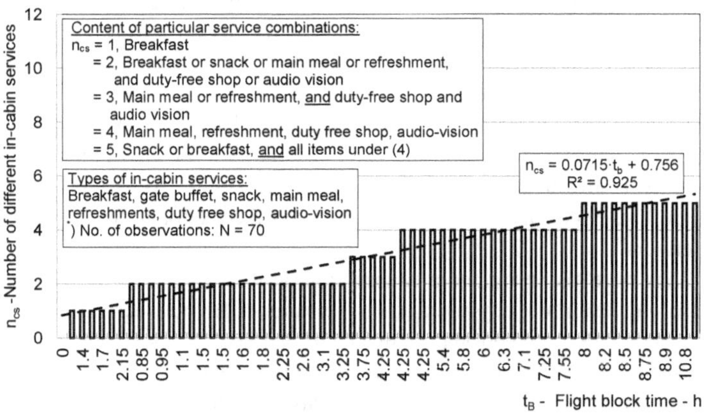

Figure 3.39 Example of the past relationship between type and quality of in-cabin services on the length of flight: Case of Lufthansa airline (Period: 1994) (Janić, 2000).

services were offered. On flights lasting between eight and eleven hours, combinations of five distinct services were offered.

3.4.2 Modelling Quality of Services

Two models of the quality of service provided by an airline are presented. The first is the model of total passenger travel time along a route. It comprises previously developed models relating to the airline operations on direct (non-stop) and indirect (one or multi-stop) routes. The model aims to point out the importance of some external time-based attributes of the quality of services, such as the schedule delay, in-aircraft time, and passenger comfort. Another model developed for the airline quality rating (AQR) provided unique criteria for comparison of the quality of services across particular airlines over time (Bowen and Headley, 2019).

3.4.2.1 Travel time on route

An airport network can be considered to consist of the set of airports and flights scheduled between them by one and/or several airlines during a given period of time. The airports may be mutually connected by direct and indirect connections (flights). A direct connection is represented by a single direct, non-stop flight scheduled between any two airports. Indirect connections contain at least two successive flights scheduled between any pair of airports via a third one (Janić, 2000; Yeng, 1987).

a) Passenger travel time on a route

Let us denote by (*ij*) the route connecting a pair of airports of an air route network. The expected travel time on this route, including the passenger schedule delay, in-aircraft time (e.g., flying time) and anticipated delay imposed on the flight, can be computed as:

$$t_{ij} = t_{s/ij} + w_{d/ij} + t_{f/ij} + w_{a/ij} \quad for \;\; i \neq j,\, ij \in N \tag{3.27a}$$

where

N is the number of airports in the air route network;

$t_{s/ij}$ is the schedule delay (i.e., "defer" time), defined as the difference between desired and available time of departure;

$w_{d/ij}$ is the delay that can be imposed on the flight before departure, defined as the difference between scheduled and actual departure time; this delay may be caused by traffic congestion at the departure airport, destination airport, and/or in the airspace between them (other mostly technical (non-traffic) reasons may also cause departure delays);

$t_{f/ij}$ is the flying time on the route (*ij*); it is the passengers' in-aircraft time; and

$w_{a/ij}$ is the flight delay on arrival that can be frequently caused by local congestion at the destination airport.

When the passengers use an indirect connection and travel between the airports *(i)* and *(j)* through the third airport *(h)*, their expected travel time can be estimated as follows:

$$t_{ij/h} = t_{ih} + t_{hj} \quad for \;\; i \neq j,\, ijh \in N \tag{3.27b}$$

where

t_{ih}, t_{hj} is the travel times on the route *(ih)* and *(h*j), respectively, each determined as in the expression (3.27a).

b) Schedule delay

The passenger schedule delay in Eq. 3.27a can be estimated with different models. These models include: the model of random incidence, the model of reduced problem of random incidence, the model(s) of schedule delay as a competitive tool, the model of schedule delay determined in dependence on the flight frequency, and the model of schedule delay in the airline network.

i) Problem of random incidence

The schedule delay can be represented by the random variable: $T_{s/ij} \equiv SD$. The main descriptor of the variable (SD) is its probability mass function $(U(SD))$ with the mean $(E(SD))$ and variance $(V(SD))$. According to the problem of random incidence, the mean $(E(T))$ can be determined as (Janić, 2000; Larson and Odoni, 1981; Turnquist and Daskin, 1982):

$$E(SD) = 1/2 \cdot [E(H) + V(H)/E(H)] \tag{3.28a}$$

where

$E(H)$, $V(H)$ is the mean and variance of the continuous random variable H, respectively, whose realisations represent the times between successive flights scheduled on a given route during time (τ) (h).

If distribution of the random variable (H), $(g(H))$ is assumed to be symmetric (or nearly so), the third moment of $(g(H))$ can be written as follows (Janić, 2000; Turnquist and Daskin, 1982):

$$E(H^3) = 3 \cdot V(H)/E(H) + [E(H)]^3 \tag{3.28b}$$

Then, from Eq. 3.28b the variance of random variable $((T))$, $(V(T))$ can be estimated as follows:

$$V(T) = V(H) + 1/3 \cdot [E(H)]^2 - \{V(H) + [E(H)]^2\}^2/4 \cdot [E(H)]^2 = V(H)/2 + [E(H)]^2/12 - [V(H)/2 \cdot E(H)]^2 \tag{3.28c}$$

The mean of random variable $((SD))$, $(E(SD))$ estimated by Eq. 3.28a can be used to indicate the time-based attribute of quality of service offered by an airline on a given route. Variance of the variable $((SD))$, $(V(SD))$ estimated by Eq. 3.28c can be used as the measure of variability of this attribute through the airline network.

The mean and variance of the flight inter-departure times on a route, $(E(H))$ and $(V(H))$ are primarily dependent on the number of flights scheduled on the route during time (τ). The number of these flights may be assumed as realisation of the continuous random variable $((F))$, with distribution $(W(F))$, its mean $(E(F))$, and variance $(V(F))$. The random variables (F) and (H) are interrelated as follows (Manheim, 1979):

$$H = \tau/F \tag{3.28d}$$

In addition, the mean and variance of random variable (H), $E(H)$ and $V(H)$ can be estimated by some suitable "tricky" methods developed in probability theory (Ang and Tang, 1975). Let the variable (H) be defined as follows:

$$H \equiv g(F) = \tau/F \tag{3.28e}$$

where (τ) is assumed to be constant. By the second and first Taylor's approximation respectively applied to $g(F)$, the $E(H)$ and $V(H)$ can be estimated in a following way:

$$E(H) \approx g[E(F)] + (g''/2!) \cdot E(F) \cdot V(F) = \tau \cdot \{1/E(F) + V(F)/[E(F)]^3\} \tag{3.28f}$$

and

$$V(H) = \{g'' [E(F)]\}^2 \cdot V(F) = [\tau^2 \cdot V(F)]/[E(F)]^2 \tag{3.28g}$$

where all symbols are analogous to those in the previous Eqs.

When the analysis of schedule delay is reduced to the specific route of an air route network, the number of flights $f(\tau)$ may be assumed to be a parameter and, consequently, the variances $(V(H))$ and $(V(F))$ in Eq. 3.28c and Eq. 3.28g are equal to zero. From Eq. 3.28a, the scheduled delay $(E(SD))$ can be estimated as follows:

$$E(SD) = 1/2 \cdot E(H) = 1/2 \cdot [\tau/f(\tau)] \tag{3.28h}$$

where all symbols are analogous to those in the previous Eqs.

ii) Reduced problem of random incidence

When the passengers requesting to travel on a given route during time (τ) are assumed to know the exact time to board their flights, they will tend to collect closer to the flight departure time. In such case, the passenger schedule delay can be estimated as follows (Janić, 2000; Teodorović and Krčmar-Nožić, 1989):

$$E(SD) = 1/4 \cdot E(H) = 1//4 \cdot [\tau/f(\tau)] \tag{3.29}$$

Equations 3.28h and Eq. 3.29 indicate that the passenger scheduled delay depends on the number of flights scheduled on a given route during time (τ), length of time (τ), and the expected (perceived) passenger behaviour.

iii) Schedule delay as the competitive tool on a route

The schedule delay has been shown to be one of the airlines' competitive tools. When they compete, for example head-to-head or collusively, by staggering the departure times in order to maximize benefits, the schedule delay of a given airline can be estimated as follows (Janić, 2000; Taneja, 1982):

$$E(SD) = \tau/[f(\tau)/I_n] \tag{3.30a}$$

where

τ is the time period (h, day);

$f(\tau)$ is the total number of flights scheduled on a specific route (market) by all competing airlines during time (τ) (dep/h or day); and

I_n is the index measuring strength of competition between the airlines operating on a route during (τ) (-).

Under such conditions, the index measuring the strength of competition between the airlines can be estimated as follows:

$$I_n = m\ 1/\sqrt{max_k MS_k} \tag{3.30b}$$

where

MS_k is the market share of the airline (k) (-).

iv) Schedule delay vs route flight frequency

According to the hypothesis that a higher number of flights can shorten the schedule delays, it was estimated as follows (Janić, 2000; Abrahams, 1983):

$$E(SD) = 5.7/f(\tau) + [Q(\tau)/S(\tau)]^9 \cdot 57/f(\tau) \tag{3.31}$$

where

$f(\tau)$ is the flight frequency on a route during time (τ) (dep/h or day);
$Q(\tau)$ is the expected passenger demand on a route during time (τ) (pax/h or day); and
$S(\tau)$ is the total number of seats supplied by an airline during time (τ) (seats/h or day).

This model was applied to estimate the flight frequencies on the specific routes of the network aiming at maximizing the airline profits under given conditions.

v) Schedule delay vs frequency in the airline network

In an airline network, each route can be considered independently as a service system. The flights and their seat capacity deployed on each route are considered as the service channels and the air passengers as customers requesting service, both during time period (τ). Under such conditions, the route load factor is analogous to the traffic load of the queuing system. The passenger scheduled delays are caused by inability of a marginal user to find an empty seat in the desired flight due to its occupation by other users. An increase in the load factor will decrease this probability and, consequently, the schedule delay will increase in proportion to the factor: $(1 - \lambda^s)^{-1/s}$, where (s) is an undetermined constant ($s \geq 1$). When the factor $(1 - \lambda^s)^{-1/s}$ is valid for all routes of the airline network, and when: $s = 1$, the total scheduled delay in an airline network during time (τ) is expressed as follows (Gordon and de Neufville, 1973; Janić, 2000):

$$E(SD) = a \cdot \sum_{i=1}^{M} \frac{Q_i(\tau)}{[1 - \lambda_i(\tau)] \cdot f_i(\tau)} \tag{3.32a}$$

where

a is the proportionality constant assumed to be equal for all routes of the network;
M is the number of routes in the airline network;
$Q_i(\tau)$ is the number of passengers expected to travel on the route (i) during time (τ) (pax/h or day); and
$f_i(\tau)$ is the number of flights scheduled on the route (i) during time (τ) (dep/h or day).

When the air passenger demand on each route of the network is completely satisfied, i.e., all passengers will be served during time (τ), the route load factor can be expressed as: $\lambda_i(\tau) = Q_i(\tau)/[F_i(\tau) \cdot n]$, where (n) is the seat capacity of an aircraft deployed on the route (i). It is assumed that all flights are carried out by aircraft with the same seat capacity.

Introducing the constraints relating to the maximum available utilisation of the airline fleet during time (τ) and determining the flight frequencies on each route in a way to minimise the route schedule delay, the total scheduled delay in the network in Eq. 3.32a can be transformed as follows (Gordon and de Neufville, 1973; Janić, 2000):

$$E(SD) = \frac{a \cdot s \cdot \left[\sum_{i=1}^{M} \sqrt{Q_i(\tau) \cdot t_i}\right]^2}{U(\tau) - \sum_{i=1}^{M} Q_i(\tau) \cdot t_i} \tag{3.32b}$$

where

s is the seat capacity of an aircraft in the airline fleet (seats/ac);
t_i is the total gate-to-gate (i.e., block) time for the flight on a route (i) (h); and
$U(\tau)$ is the number of seat-hours available to the airlines during time (τ) (seats-h).

Equation 3.32b shows that the total scheduled delay in the airline network is constant for its given configuration and aircraft fleet serving an approximately constant volume of demand during a given period of time.

The above-mentioned model enables an estimation of the minimum schedule delay to be made, depending on varying inputs. When the "point-to-point" and "hub-and-spoke" networks were compared, the results, as expected, indicated that the hub-and-spokes network is preferable for minimising the total schedule delays. At the same time, this network appeared to be less attractive in comparison to the "point-to-point" network because of its lower connectivity. In addition, introduction of larger aircraft justified by the volume of demand can compromise the passenger schedule delays. Namely, when larger aircraft are used, the number of flights will decrease and the schedule delays will increase on the particular routes (Gordon and de Neufville, 1973; Janić, 2000).

c) Flight Time

Estimating the flying time on a given airline route is based on the typical relationship between distance, speed and time, known from kinematics. According to this relationship, the route block time ($t_{b/ij}$) in Eq. 27(a, b) can be estimated as follows:

$$t_{b/ij} = d_{ij}/[v_{b/ij}(d_{ij})], \quad for\ i \neq j;\ ij \in N \tag{3.33a}$$

where

d_{ij} is the length of the route connecting airports *(i)* and *(j)* (km or miles); and
$v_{b/ij}(d_{ij})$ is the average block speed assumed to be constant on route (d_{ij}) (km/h or knots).

The average block speed is always dependent on the length of the route, which depends on the deployed aircraft type, cruising altitude, traffic conditions along the route (i.e., the delay that may be imposed on flight), and complexity of departing and arriving procedures at both origin and destination airport(s). These factors may reduce the average speed along the given route and consequently prolong the scheduled flying time. If this frequently occurs, the affected airlines consider this while planning the block times of their flights. Some relationships between the average block speed and route length are shown in Figure 3.40 (Janić, 2000).

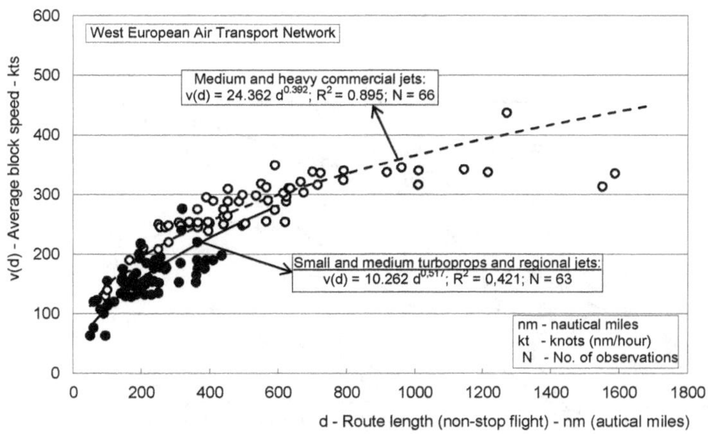

Figure 3.40 Relationship between block speed and route length—Case of West European air transport network (Janić, 2000).

As can be seen, the upper regression curve corresponds to the medium and large jets with seat capacities between 100 and 450 seats. The lower curve corresponds to the small and medium turboprops and jets with seat capacities from 10 to 100 seats, used mostly for regional transport. Both curves indicate that the average speed increases as the non-stop flying distances increase at a decreasing rate. The regional aircraft always have a lower average speed and travel at lower altitudes, up to FL250 (FL250 = 25000 ft; 1 ft ~ 0.305 m). Combining the relationships on Figure 3.39 and Eq. 3.33a, the average block time on the route *(ij)* can be expressed as follows:

$$t_{b/ij}(d_{ij}) = d_{ij}/(a_0 \cdot d_{ij}^{a_1}), \quad for\ i \neq j;\ ij \in N \tag{3.33b}$$

where

a_0, a_1 are the coefficients of the regression equation estimated by the least-square method (for the cases shown in Figure 3.39, $a_1 \leq 1.0$).

Some other examples have indicated the linear relationship between the route block time and route length (Janić, 2000).

d) Take-off and landing delays

The airline take-off and landing delay, $(w_{d/ij})$ and $(w_{a/ij})$, respectively, in Eq. 3.27a, as well as en-route delays, have been elaborated in Chapters 2 and 4, where the quality of services provided to the aircraft at airports and in the ATC/ATM (Air Traffic Control/Management) system have been elaborated.

e) Travel time on direct and indirect routes

In general, the airline users/air passengers can be offered different routes to travel between a given pair of airports in the airline network. For example, at least two kinds of air connections may be established between a given pair of airports. The first one is represented by direct (non-stop) flights. The other one is represented by indirect (one or multi-stop) flights scheduled between any two airports via a third one. Figure 3.41 shows the simplified scheme.

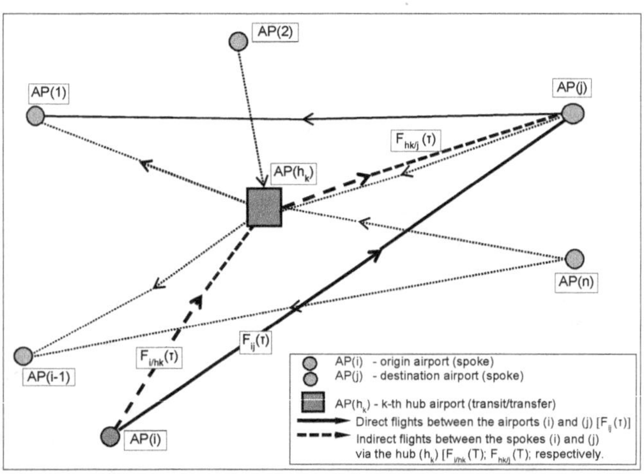

Figure 3.41 Scheme of connecting pairs of airports by direct and indirect connections/flights (Janić, 2000).

The time that air passengers spend in the airline system while using a direct air connection/flight between the airports *(i)* and *(j)* can be estimated based on Eqs. 3.27, 3.28–3.33 as follows:

$$t_{ij}(\tau) = (1/4) \cdot [\tau/f_{ij}(\tau)] + w_{d/ij} + d_{ij}/v_{ij}(d_{ij}) + w_{a/ij}, \quad for\ i \neq j;\ ij \in N \tag{3.34a}$$

When the route indirect air connection between the airports *(i)* and *(j)* is used through the third airport h_k (k = *1,2...*), the time spent in the airline system can be estimated as follows:

$$t_{ij/h}(\tau) = t_{ih}(\tau) + t_{hj}(\tau) = (1/4) \cdot [\tau/f_{ih}(\tau)] + w_{d/ih} + d_{ih}/v_{ih}(d_{ih}) + w_{a/ih} + (T_{d/hj} - T_{a/ih}) + w_{hj} + d_{hj}/v_{hj}(d_{hj}) + w_{a/hj}, \quad for\ i \neq h \neq j;\ ihj \in N \tag{3.34b}$$

where

$T_{a/ih}$ is the arrival time of a flight at airport *(h)* from the airport *(i)*; and
$T_{d/hj}$ is the departure time of a flight from the airport *(h)* to airport *(j)*.

When the airline schedule at the hub is coordinated, the air passengers from the incoming will spend the minimum time at the hub until the next outgoing flights ($T_{d/hj}$ – $T_{d/hj}$ = T_{hc}, where (T_{hc}) is the minimum connection time).

Equation 3.34(a, b) indicates that the indirect connections are always longer than their direct counterparts in terms of both travel distance and time. This can be explained as follows: the direct route between the airports *(i)* and *(j)* is always shorter than the total length of a two-segment route connecting these airports through the third one; along two shorter sub-routes *(ih)* and *(hj),* the aircraft fly at lower block speeds (see Figure 3.39); the departure and arrival delays are more likely to happen on the shorter and denser incoming and outgoing routes at the airport *(h)* than on the direct route connecting the airports *(i)* and *(j)*; and at the hub airport, the passengers spend the connecting time between the incoming and outgoing flights. The number of flights serving denser demand on the indirect routes *(ih)* and *(hj)* during time *(τ)* can contribute to shortening the corresponding passenger schedule delays. However, this is not always sufficient to compensate for the time advantage and convenience of the direct but less frequent flights connecting the same pair of airports (Janić, 2000; Yeng, 1987).

3.4.2.2 Load factor

The air passengers can consider the aircraft/flight load factor as an indicator of the probability for finding an empty seat at the time of wishing to travel. Usually, most of them are not familiar with the term and its real meaning. They only consider if there is an empty seat for booking. The airlines always aim to keep the load factor of their flights at least at the aero-profitable level. When the average cost, airfare, flight frequencies and air passenger demand for the route *(ij)* in time *(τ)* are known, the load factor enabling zero-profits can be estimated as follows:

$$f_{ij}(\tau) \cdot C_{ij}(s_{ij}, d_{ij}) \leq AF_{ij}(\tau) \cdot f_{ij}(\tau) \cdot s_{ij}(\tau) \cdot \lambda_{ij}, \quad for\ i \neq j;\ ij \in N \tag{3.35a}$$

and

$$\lambda_{ij} \geq \frac{c_{ij}(s_{ij}, d_{ij})}{AF_{ij}(\tau) \cdot s_{ij}(\tau)}, \quad for\ i \neq j;\ ij \in N \tag{3.35b}$$

where

$f_{ij}(\tau)$	is the number of flights on the route *(ij)* during time (τ) (flights/day);
$C_{ij}(s_{ij}, d_{ij})$	is the cost per flight on the route *(ij)* (d_{ij} is the route length (km, nm) ($US/flight);
$AF_{ij}(\tau)$	is the average airfare per passenger on the route *(ij)* during time (τ) ($US/pax);
$s_{ij}(\tau)$	n_{ij} is the average aircraft/flight seat capacity on the route *(ij)* during time (τ) (seats/ac);
$\lambda_{ij}(\tau)$	is the average load factor of a flight scheduled on the route *(ij)* during time (τ) (-); and
N	is the number of airports in the airline network.

The cost of flight $(c_{ij}(s_{ij}, d_{ij}))$ can include direct or total, direct and indirect operational costs. The load factor $(\lambda_{ij}(\tau))$ varies between zero and one, i.e., $0 \leq \lambda_{ij}(\tau) \leq 1.0$. If the airfares are set higher, the load factor will be lower. For example, when the load factor is relatively high, say between 0.8 and 0.9, the probability of finding an empty seat will significantly decrease, so the quality of service expressed by accessibility to the airline free capacity will diminish. This may be particularly important for last-minute passengers, i.e., those intending to book the seat close to the flight departures. In addition, the quality of services for air passengers already boarded on the flight could deteriorate in terms of individual privacy and comfort.

3.4.2.3 AQR (Airline Quality Rating)

a) Structure

The AQR (Airline Quality Rating) was developed as a rather comprehensive method/tool for the systematic assessing and evaluating of the quality of services of U.S. airlines. Actually, it was designed and for the first time announced in the year 1991, with the intended purpose of assessing and evaluating the overall quality of airlines. As such, AQR was assumed to be an objective method/tool based on using publicly available data on the standardised set of quality attributes/criteria of airline services (Bowen and Headley, 2019). These attributes/criteria had to be of relevance for the users/air passengers that were asked to evaluate the quality of airline services and also be obtainable from the published data as well. The main attributes/criteria have been: (i) On-time arrival; (ii) Involuntary denied boarding; (iii) Mishandled baggage; and (iv) Customer complaints, with = twelve sub-attributes/criteria. These all have been considered as important for air passengers. The monthly data for all attributes/criteria have been provided by the U.S. Department of Transportation in its Air Travel Consumer Report (http://dot.gov/airconsumer/).

i) On-Time Performance (OT) (Weight: +8.63)

The attribute/criterion "on-time performance" relates to the airline flights counted to be "on time", i.e., if operated within 15 minutes of the scheduled time shown in the airlines' CRSs (Computerized Reservations System(s)). In addition, the cancelled and diverted flights are counted to be late including those delayed due to the different mechanical problems. The AQR method/tool uses the percent of on-time arriving flights during a month (see also Figure 3.34(a, b)).

ii) <u>Involuntary Denied Boardings (DB) (Weight: –8.03)</u>

The attribute/criterion involuntary "denied boarding" relates to passengers with confirmed reservations who are involuntarily denied boarding because their flight(s) have been sold out. The AQR method/tool uses the ratio of involuntary denied boarding/10000 pax/month.

iii) <u>Mishandled baggage (MB) (Weight: –7.92)</u>

The attribute/criterion "mishandled baggage" relates to claims for lost, damaged, delayed, or pilfered baggage. The AQR method/tool uses the ratio of the number of lost, damaged, delayed, and/or pilfered baggage/1000 passengers.

iv) <u>Consumer complaints (Weight: –7.17)</u>

The attributes/criteria of "consumer complaints", including twelve specific complaint categories, are reported by air passengers in writing, by telephone, via e-mail, or in person. The AQR uses these complaints of the various categories by clustering them into the larger customer complaint attribute/criteria. Then it is expressed as the consumer complaint ratio in terms of the number of complaints/100000 passengers.

The relative importance of the above-mentioned attributes/criteria has been represented by the weights characterized by their value and direction of impact on the airline quality. While the above-mentioned attributes/criteria differ among the airlines and over time, the weights and their impacts are constant. The model/tool AQR combining both attributes/criteria and their weights for assessing the quality of a given airline can be expressed as follows:

$$AQR = \sum_{i=1}^{4} w_i \cdot A_i / \sum_{i=}^{4} w_i \qquad (3.36a)$$

w_i is the weight of *(i)*-th attribute/criteria; and
A_i is the value of *(i)*-th attribute/criteria of quality of service.

In addition, the AQR used for estimating the quality of a given U.S. airline has been as follows (Bowen and Headley, 2019):

$$AQR = \frac{(+8.63 \cdot OT) + (-8.03 \cdot DB) + (-7.92 \cdot MB) + (-7.17 \cdot CC)}{(8.63 + 8.03 + 7.92 + 7.17)} \qquad (3.36b)$$

b) <u>Examples</u>

The average AQR for the U.S. airline industry, including nine airlines, during the observed period (2011–2018) is shown in Figure 3.42. These are: Alaska, American, Delta, Frontier, Hawaiian, JetBlue, Southwest, Spirit, and United.

As can be seen, the AQR score of the U.S. airline industry has been negative during the observed period. This implies that the attribute/criterion 'on-time performance' with the highest positive weight has not compensated for the other three with the negative weights. In addition, the industry quality has gradually increased during the observed period.

In addition, Figure 3.43(a, b) shows the AQR score and ranking of Delta and Southwest Airlines among the above-mentioned nine airlines during the observed period (2011–2018).

Figure 3.43a shows that, similar to at the entire industry, the AQR scores of Delta and Southwest airlines have been negative during the observed period. At the same time, while both have changed, the AQR of Delta was higher than that of Southwest airlines.

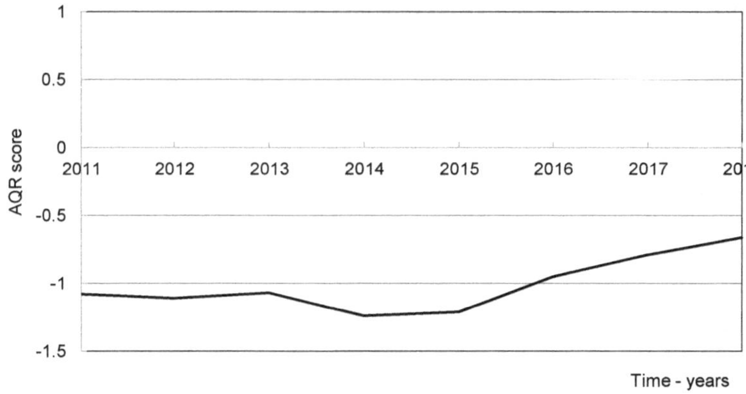

Figure 3.42 AQR (Air Quality Ranking) score—Case of the U.S. airline industry (Period: 2011–2018) (Bowen and Headley, 2019).

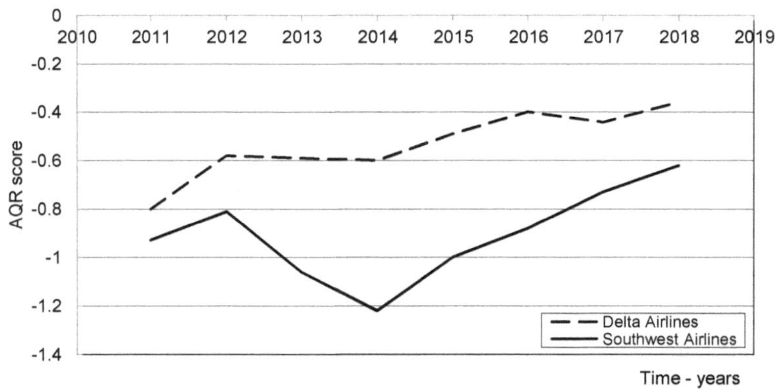

a) AQR score

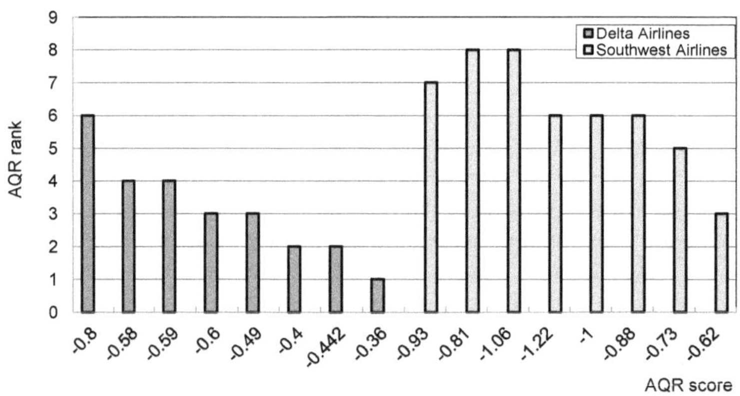

b) AQR rank vs AQR score

Figure 3.43 AQR (Airline Quality Ranking)—Case of the U.S. Delta and Southwest airlines (Period: 2011–2018) (Bowen and Headley, 2019).

Figure 3.43b shows that the AQR rank of both airlines has decreased with the drop in AQR score at both airlines. The scores of Delta have been lower than those of Southwest airlines, thus confirming its continuously higher rank during the observed period. The proposed AQR method/tool has been used for about 28 years, starting in the year 1991. As such, it has enabled a rather consistent and systematic way of monitoring the quality of services useful for both U.S. airlines and their users/air passengers.

3.5 Economics

3.5.1 Components

The airline economics generally relates to the output and the related revenues, costs, and their difference in terms of profits (positive) and losses (negative).

As already mentioned, the airline operations generate different outputs. These are: Available Seat Kilometres (Miles) (ASKs, ASMs), Revenue Passenger Kilometres (Miles) (RPKs, RPMs), and freight/cargo Available and Revenue Ton Kilometres (Miles) (ATKs, RTKs). They are all available in the airline annual reports, together with other operational/traffic and economic/financial data. In some cases, they can be given per month, or even per day. Since these outputs are highly aggregated, they can be used as a unique measure for fair comparison between different airlines operating in the same or different regions, and the aircraft of the similar or different types operated by particular airlines.

The airline revenues are obtained by charging users, i.e., air passengers, freight/cargo shippers, and postal services. The airline costs are imposed by buying the inputs for carrying out flights, such as aircraft fleets, their insurance and maintenance, fuel, air crews and ground staff, and others. For example, in the U.S. airline industry, about 75% revenues have been from air passengers, 15% from freight/cargo shippers, and 10% from the other transport-related services. In particular, the main passenger airlines get up to about 10% revenues from belly freight/cargo (https://www.avjobs.com/history/airline-economics.asp/).

The airline costs are generally classified as direct and indirect operating costs. The direct operating costs include: the costs of flight operations (crew, fuel, maintenance and overhaul, and aircraft ownership, i.e., depreciation and amortisation). The indirect operating costs are categorized as: (i) ground operating costs including servicing of passengers and aircraft at airport stations, aircraft landing fees, and reservations/sales charges; and (ii) system operating costs including marketing, administrative and general overhead items, in-flight services, and ground equipment ownership (ICAO, 2017). Figure 3.44 shows the structure of total costs of the U.S. airline industry, shown to be rather stable over a relatively long time period.

As can be seen, the share of direct and indirect operating costs has been the same (50:50). The difference between the revenues and operating costs represents the airline profits. In addition to the annual reports, the relevant data on the airline revenues, costs, and profits are also presented in the standardized form in the reports of the national (for example U.S. FAA, EC) and international organizations, such as IATA (International Air Transport Association) and ICAO (International Civil Aviation Organization) (IATA, 2018; Janić, 2000). More detailed elaboration of the above-mentioned components of the airline economics in the given context follows and includes the aircraft and airline costs, airline economies of scale, and airline profits.

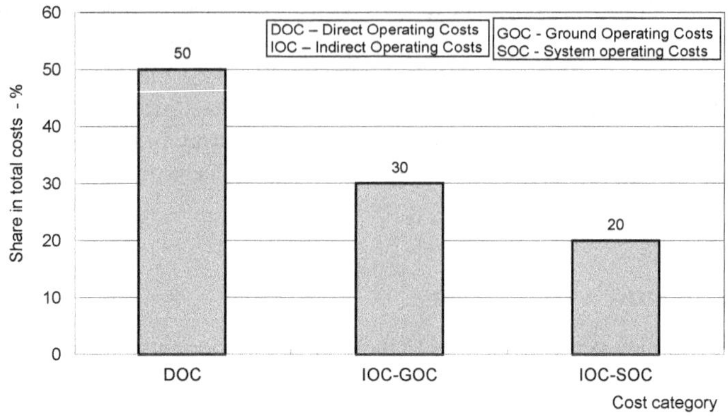

Figure 3.44 Structure of the total costs—Case of the U.S. airline industry (ICAO, 2017).

3.5.2 Aircraft Costs

As mentioned above, in terms of their length, routes, flights, and the aircraft operating them are generally classified as short-, medium-, and long-haul. For example, in Europe, Eurocontrol defines the medium-haul routes of length from 1500–4000 km (930–2490 mi; 810–2160 nmi), the short-haul as shorter and the long-haul routes as longer than that. The U.S. American Airlines define the short-/medium-haul routes of length less than 4800 km 3000 mi (3000 mi; 2600 nm) and the long-haul as longer than that. The United Airlines defines the short-haul routes and corresponding flights of length less than 1100 km (700 mi; 610 nm), and the long-haul flights routes as longer than 4,800 km (3000 mi; 2,600 nm) (1 mi = 1.609 km; 1 nm = 1.852 km). As far as the aircraft types are concerned, the U.S. Delta Airlines defines Boeing 717, MD-88 and MD-90 as the short-haul domestic aircraft, Boeing 757, 737, Airbus A319, A321 as the long-haul domestic and Boeing 777, 767, 747, Airbus A330 and Boeing 757 as the long-haul. Lufthansa considers its wide-body aircraft Airbus A330/A340, A350, A380, and Boeing B747 as the long-haul, the narrow-body aircraft A320 and B737 as the medium-haul, and the regional jets Embraer E-Jets and Bombardier CRJ-900 as the short-haul aircraft (https://en.wikipedia.org/wiki/Flight_length/).

The operating cost of a flight carried out by the assigned aircraft type on a given route generally depends on the route length and the aircraft size, i.e., seat or payload capacity. One example for the cost of medium- to short-haul flights operated by the narrow-body and regional jets is estimated as follows (Swan and Adler, 2006):

$$C_f(s, d) = -105.564 + 10.372 \cdot s + 1.699 \cdot d \quad (\$US/flight)$$
$$t \qquad -0.150 \qquad 10.509 \qquad 0.944$$
$$R^2 = 0.970;\ F = 114.981;\ DW = 1.271;\ N = 10$$
$$(s - (-);\ d - mi;\ 1\ mi = 1.609\ km) \tag{3.37a}$$

where all symbols are the same as in the previous Eqs. As intuitively expected, the flight costs linearly increase with an increase in the aircraft size, i.e., seat capacity and route length. By dividing the flight cost by the number of seats and route length, the average cost per seat-km (seat-mile) can be obtained, as shown in Figure 3.45(a, b).

Figure 3.45(a, b) shows that the average aircraft cost decreases more than proportionally with increasing of the route length and/or seat capacity, thus indicating

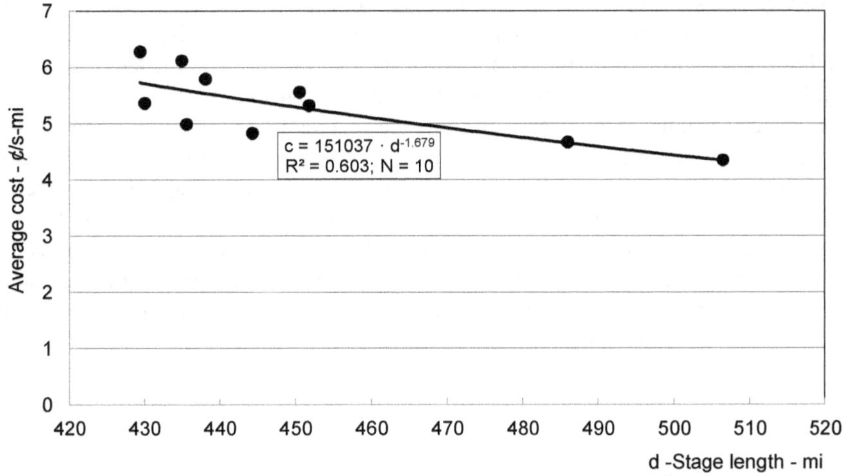

a) Cost vs stage length

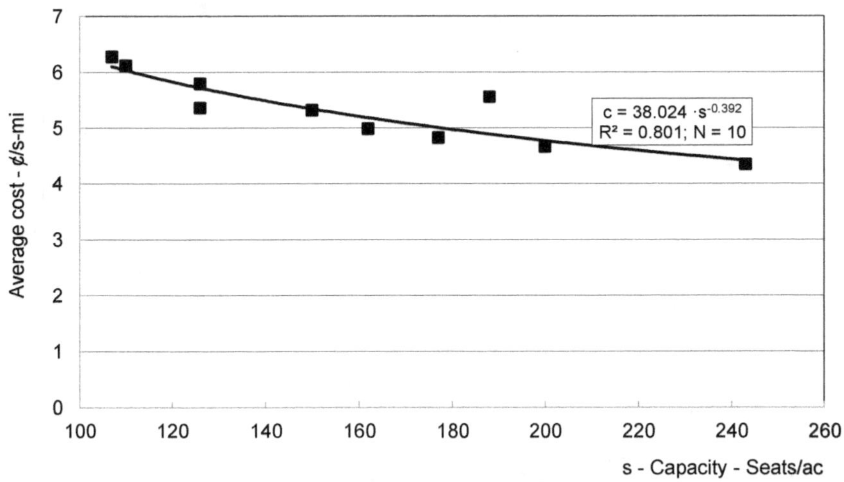

b) Cost vs capacity (seats/ac)

Figure 3.45 The average operating cost of narrow-body aircraft (Swan and Adler, 2006).

some existence of economies of scale. This means that the flights carried out by larger aircraft on the longer routes can have lower cost per unit of output than otherwise. A similar example of the relationship between the flight operating cost, aircraft seat capacity, and route length for the wide body aircraft operating on long-haul routes is estimated as follows (Hong et al., 2015):

$$C_f(s, d) = -44704.220 + 244.362 \cdot s + 7.745 \cdot d \quad (\$US/flight)$$
$$t \quad -2.099 \quad 2.732 \quad 1.430$$
$$R^2 = 0.875;\ F = 21.083;\ DW = 1.523;\ N = 9$$
$$(s - (-);\ d - \text{mi};\ 1\ \text{mi} = 1.609\ \text{km}) \quad (3.37b)$$

where all symbols are the same as in Eq. 3.37a. As can be seen, again, the flight costs linearly increase with an increase in the aircraft size, i.e., seat capacity and route length. By dividing the flight cost by the number of seats and route length, the average cost per seat-mile (seat-km) can be obtained as shown in Figure 3.46(a, b).

As can be seen, in the given case, the average cost per unit of output (¢/s-mi) initially decreases with increasing aircraft seat capacity and/or route length and then increases with a further increase of both. This indicates that the economies of scale exist on certain long-haul routes and aircraft types deployed there, but also may not exist on the very long-haul routes operated by very high capacity aircraft. A different impression can be had if considering the relationship between the average unit costs per aircraft output and the volume of output during a given period of time. Figure 3.47 shows the corresponding example for 45 wide-body aircraft types operated by the U.S. airlines.

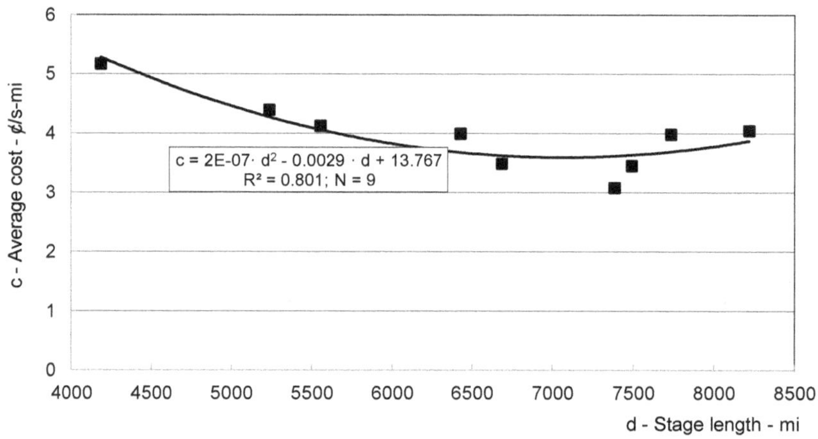

a) Cost vs stage length

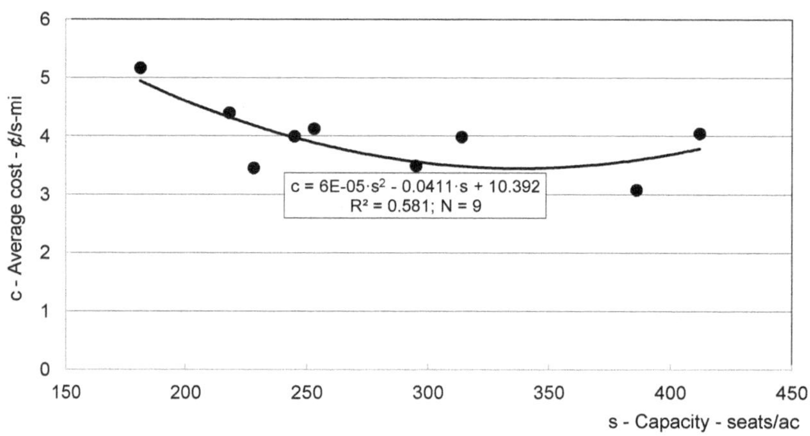

b) Cost vs capacity (seats/ac)

Figure 3.46 Relationship between the average direct operating costs and the aircraft seat capacity and stage length—Case of selected wide body aircraft (Hong et al., 2015).

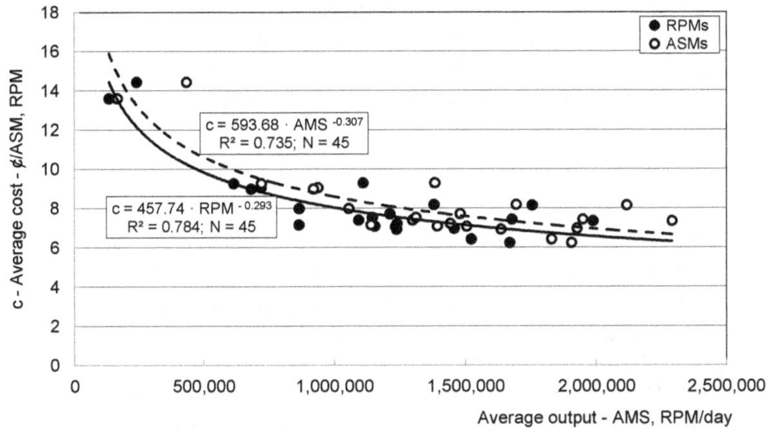

Figure 3.47 Relationship between the average aircraft cost and the quantity of output—Case of wide-bodies operated by U.S. airlines (Period: 2014) (US BTS, 2019; www.Plane Stats.com/).

As can be seen, the average cost per output in terms of both ¢/ASM and ¢/RPM has decreased more than proportionally with an increase in aircraft daily output, thus indicating the existence of economies of scale in the given case. As expected, the average cost per RPM has been higher than that per ASM (The ratio RPM/ASM reflects the average load factor).

The structure of the above-mentioned aircraft cost per flight has been particularly interesting and relevant. Figure 3.48 shows the example for three aircraft wide-body aircraft of the U.S. airlines operating long-haul routes/flights. This time, the structure of direct operating cost is given for block hour (the time between the engines switching on at the gate of departing and switching-off at the gate of destination airport).

As can be seen, in the given case, the average share of Fuel costs in the total costs was the highest (13.8%), followed by Crew cost (63.2%), Maintenance cost (9%), and AC (Aircraft cost) (about 9%). The share of Insurance and other costs were the lowest

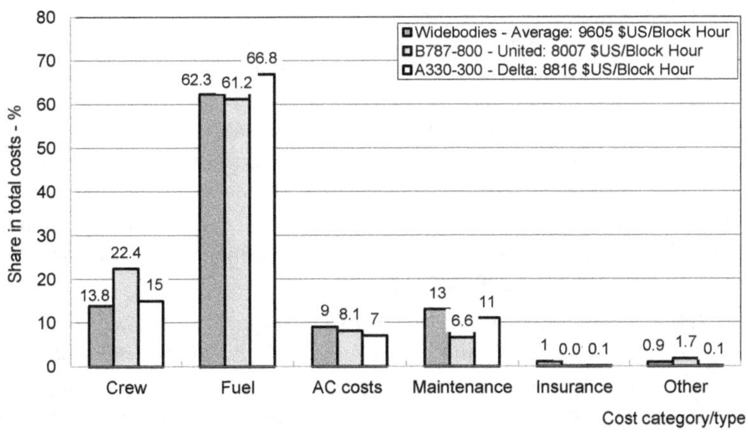

Figure 3.48 Example of the structure of aircraft direct operating costs—US airlines (Period: 2014) (US BTS, 2019; www.PlaneStats.com/).

(1% and 0.9%, respectively). The shares of particular direct operating cost components also differed for the mentioned aircraft types. For example, the share of Crew cost was higher for B787-800 while that of Fuel costs was higher for A330-300 aircraft. The share of AC cost was higher for B787-800 and the share of Maintenance cost was higher for A330-300 aircraft.

3.5.3 *Airline Costs*

In addition to the costs related to particular aircraft types, the airlines and airline groups have published their costs in more or less absolute terms, i.e., the total amounts/year. The international airline associations and aviation organizations also present these costs in different forms and criteria of aggregation, for example, the totals for airline industries of particular regions of the world, etc. One of the most detailed reports of airline costs is the U.S. FAA's Form 41, which requires the country's airlines to report the aircraft operating costs at two levels of details—airline (network, LCC) and aircraft category (small narrow-bodies, large narrow-bodies, wide bodies, total fleets) (https://www.transtats.bts.gov/Tables.asp?DBID=135; http://web.mit.edu/airlinedata/www/Aircraft&Related.html).

3.5.3.1 Cost structure

The airline costs are categorized as direct and indirect operating costs with the particular components analogous to those of the aircraft.

a) Crew/staff costs are those attributed to the pilots on aircraft, involved in testing, training, being in reserve, and being different instructors. In addition to this, there are the costs of the cabin crew.

b) Fuel costs are the expenses the airlines pay for the fuel powering their aircraft. In the absolute values, these costs are generally influenced by the external and internal factors. The former factors are usually the prices of crude oil at the global scale and the specific airline arrangements with the local fuel suppliers at airports (https://inflationdata.com/articles/inflation-adjusted-prices/historical-crude-oil-prices-table/). The internal factors are the composition and fuel efficiency of the airline fleets, influenced by the aircraft design and pattern of operations, i.e., carrying out flights, and the number and characteristics of flights (short, medium, and long-haul). Similarly to the other cost categories, the fuel costs are reported in the absolute terms combined with the quantities of fuel bought/consumed during the specified time period (usually one year). If these costs are divided by the annual volumes of the given airline output in terms of ASK (Available Seat Kilometres) or RPK (Revenue Passenger Kilometres), the corresponding averages can be obtained. Figure 3.49(a, b) shows the example of the average fuel costs for the large European airline Lufthansa Group, including the Network airlines (Lufthansa German Airlines, SWISS and Austrian Airlines) and Eurowings (Germanwings, Eurowings Europe, Brussels Airlines) (DL AG, 2008/2018).

   Figure 3.49a shows two distinct periods of developing the average fuel cost in the given case. The first is when they have increased and the second when they have decreased mainly influence, among other factors, also by the changes of prices of crude oil in the global markets. At the same time, the average cost per RPK has always been higher than their counterparts per ASK. Figure 3.49b again shows two segments of changing the average fuel cost per unit of the airline output. In the first

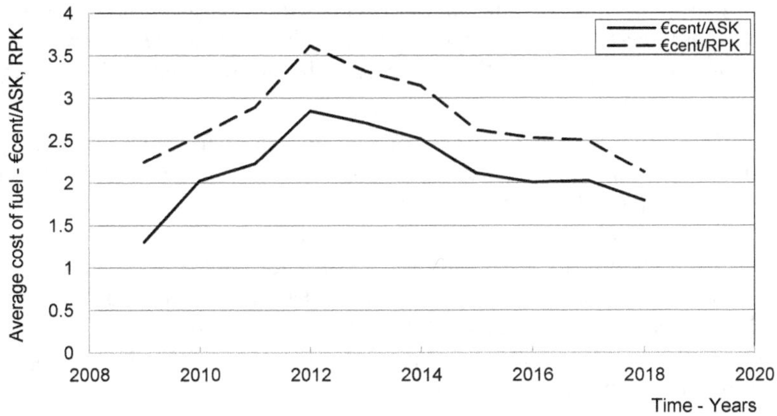

a) Average fuel cost vs time

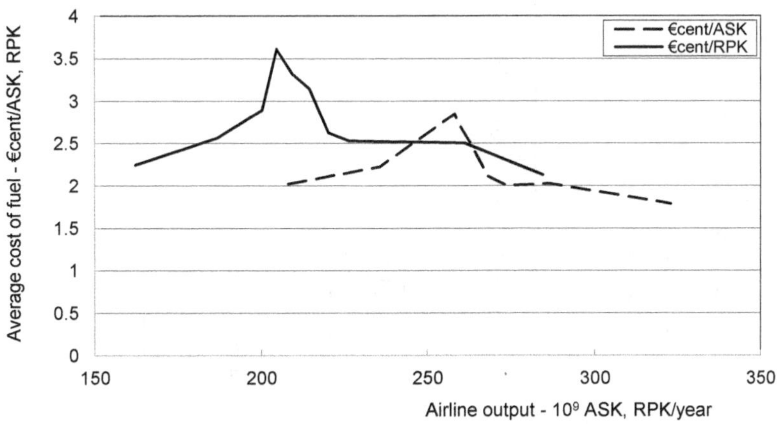

b) Average fuel cost vs the airline output

Figure 3.49 Relationship between the average fuel cost, time, and the volume of airline output—Case of Lufthansa Group (Germany) (Period: 2009–2018) (DL AG, 2008/2018).

it increased and in the second decreased with an increase in the volumes of both ASK and RPK. This again indicates the stronger influence of the external factors, such as the price of crude oil in the global markets, than increasing of the airline output.

c) Ownership/depreciation costs depend on the depreciation rate and assumed residual value of the aircraft fleet. As such, these costs represent the capital costs of aircraft ownership. The airline depreciation policies are different at particular airlines, mainly depending on the aircraft salvage value and depreciation period. The common depreciation period for aircraft at most airlines has been 20–25 years.

d) Maintenance costs include those for consumed materials and labour for the routine and capital repair of aircraft and engines. Labour costs usually represent the larger cost component for checks, removal, and installation of aircraft components, and operational checks. Material and part costs relate to all consumable and non-consumable materials, use of components and other materials used in the maintenance processes.

e) Other costs include those of the aircraft fleet insurance, costs of navigation charges, and all other operating costs. Figure 3.50(a, b) shows the share of particular direct operating cost components at two distinct—the network and LCCs – U.S. and European airlines.

Figure 3.50a shows that the shares of the crew/staff, maintenance, and aircraft ownership are higher and that of fuel/tax/insurance and others are lower for Southwest than for Delta Airlines. The greatest relative difference appears to be in the shares of maintenance and other costs. Figure 3.50b shows that the shares of crew/staff, maintenance, and other costs are higher at Lufthansa group than at Ryanair. However, those of the fuel/tax/insurance and aircraft ownership are substantially higher at Ryanair. In general, the above-mentioned shares of the particular direct operating cost components indicate that the same airline business model–network or LCC—in different regions can also differ in terms of the relative cost structure.

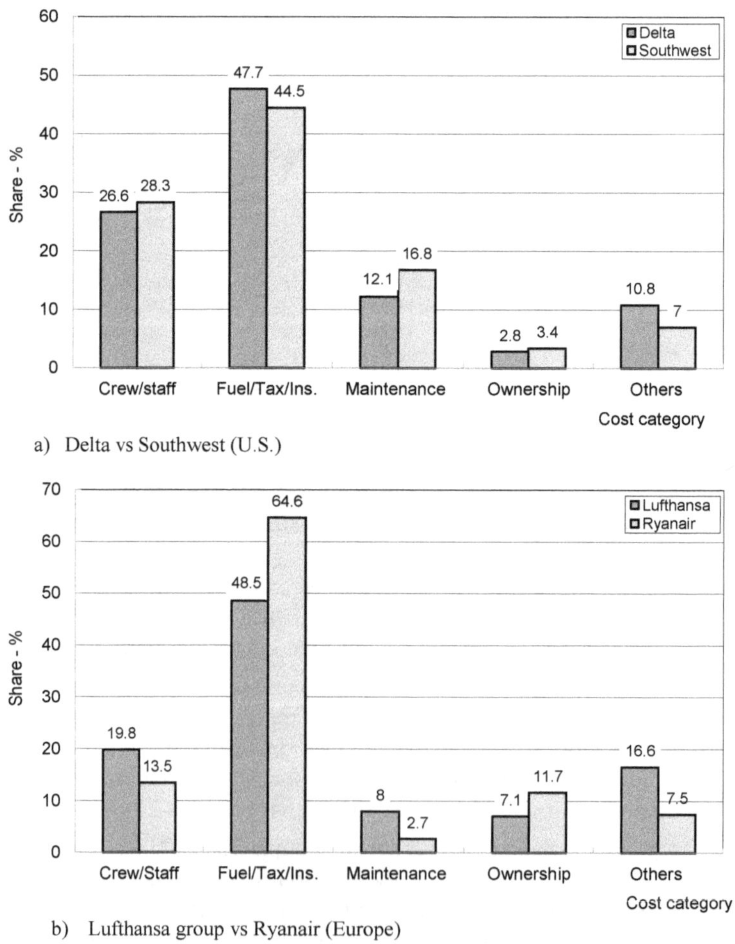

Figure 3.50 Structure of the direct operating costs of the selected U.S. and European airlines (Period: 2018) (DA, 2018; DL AG, 2018; Ryanair, 2018; SW, 2018).

3.5.3.2 Economies and diseconomies of scale

Economies of scale imply that the average cost per a given firm's output decreases as the volumes of its output increase. With airlines, this implies that the average operating costs decrease as their output increases in terms of ASK (ASM) and RPK (RPM) during the specified period. Diseconomies of scale imply the opposite. The question is if economies or diseconomies of scale exist in the airline industries as mentioned in this context. Figure 3.51 shows the example of economies of scale for the U.S. airline industry, including 7 network airlines (American, Continental, Delta, Northwest, United, US Airways, and America West) and 5 LCCs (Southwest, jetBlue, AirTran, Frontier, and Virgin America), during the period 1995–2018 (https://www.transtats.bts.gov/Tables.asp?DBID=135; http://web.mit.edu/airlinedata/www/Aircraft&Related.html/).

As can be seen, during the observed period the volumes of output and the corresponding average operating costs of the network airlines have been generally higher than that of their LCC counterparts. For network airlines, the average costs have varied from about 9.5 ¢/ASM to 15 ¢/ASM for the volumes of output of about 600–800 billion ASM/year, respectively. At the LCCs, the average costs have increased from about 5 ¢/ASM to 10 ¢/ASM for increasing of the volumes of output between 40 and 200 billion ASM/year, respectively. Such developments indicate that economies of scale did not exist for network airlines and that diseconomies of scale have been strongly present in the LCCS in the given case. Figure 3.52 confirms the absence of economies of scale at U.S. network Delta and LCC Southwest Airlines.

As can be seen, the average costs and the corresponding volumes of output were higher for the network Delta than for the LCC Southwest Airlines during the observed period, as intuitively expected. At Southwest Airlines the average cost has been between about 11 ¢ASM and 13 ¢ASM for the volumes of output of 100–160 billion ASK/year, respectively. At Delta Airlines, the average cost has been between 13¢/ASM and 15¢/ASM for the volumes of outputs between 200 and 240 billion ASM/year, respectively. Figure 3.53 shows that economies of scale at the above-mentioned two airlines do not exist regarding the average stage length.

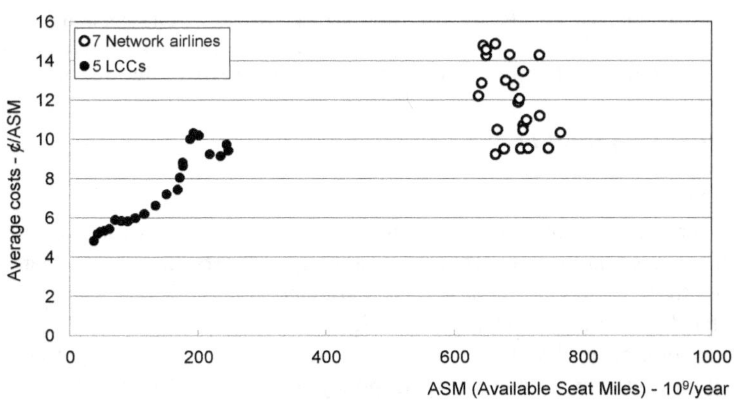

Figure 3.51 Relationship between the average airline unit cost and the volume of output (Period: 1995–2018) (https://www.transtats.bts.gov/Tables.asp?DBID=135; http://web.mit.edu/airlinedata/www/Aircraft&Related.html/).

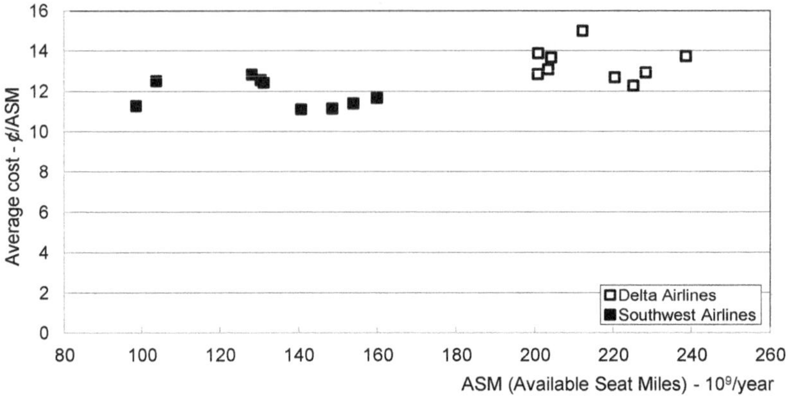

Figure 3.52 Relationship between the average costs and the available capacity of the selected U.S. airlines (Period: 2010–2018) (SW, 2008/2018; DA, 2008/2018; https://www.transtats.bts.gov/Tables.asp?DB_ID=135; http://web.mit.edu/airlinedata/www/Aircraft&Related.html).

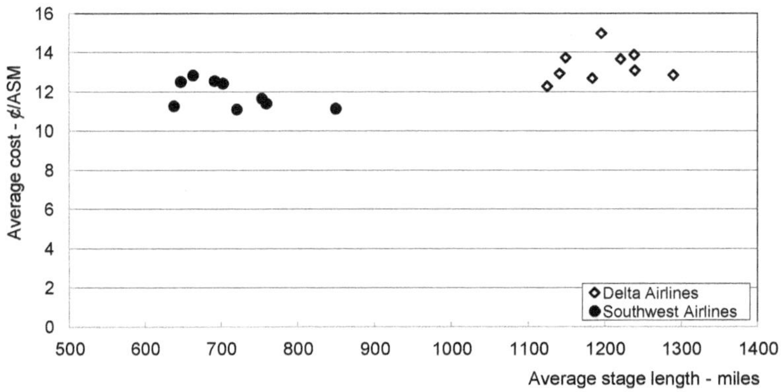

Figure 3.53 Relationship between the average costs and the average stage length at the selected U.S. airlines (Period: 2010–2018) (SW, 2008/2018; DA, 2008/2018; https://www.transtats.bts.gov/Tables.asp?DB_ID=135; http://web.mit.edu/airlinedata/www/Aircraft &Related.html).

3.5.4 Airline Profitability

3.5.4.1 Airfares

Airlines cover their operational direct and indirect costs with revenues obtained by charging their users—air passengers and freight/cargo shippers. In addition, many airlines get revenues from some other commercial non-aviation activities. In general, charging users is carried out through airfares and freight/cargo fees. They are set up according to the airline pricing policy, which is usually designed at both strategic and tactical (operational) levels. The strategic pricing policy aims to fulfil two airline objectives: (i) covering direct and indirect operating costs, and (ii) gaining some profits for expansion of scale and scope of capacity and operations under conditions of sufficient demand. The tactical pricing policy aims to set up airfares to stimulate the air travel demand for recently established and low-density routes, sustain the existing demand during economic crises, attract low-fare passengers, and fill seats that are expected to be empty otherwise. This also implies that the airfares are dependent on the characteristics of demand in terms of time of travel, trip purpose, social-economic characteristics of potential users, etc. Figure 3.54(a, b) shows how the airfares depend on the time of booking.

Figure 3.54a shows that the LCC Transavia (Europe) charges exponentially higher airfares if one is booking the given flight closer to its departure date (time). Actually, these airfares are close to that of common for LCC flights (for example, less than 100 €) only if booked about 25–30 days before the flight departure. Figure 3.54b shows that over a period of 5 years, the actual average airfare in the U.S. domestic market was lower than the average fare if the flight had been booked between 150 and 10 days before its departure. During these last 10 days, this actual airfare sharply increases and

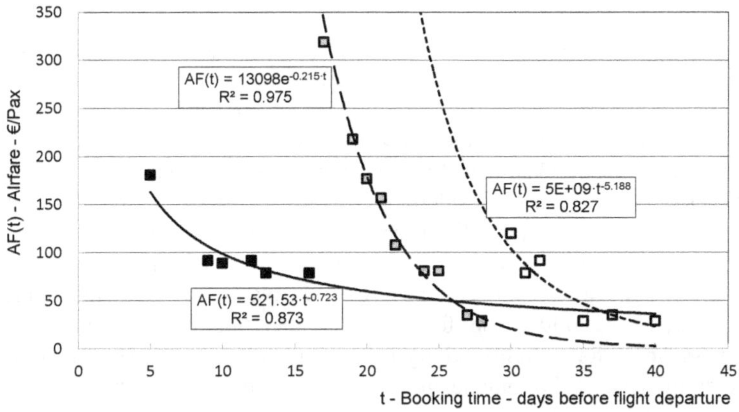

a) Transavia flight – Case of Amsterdam (The Netherlands) - Ljubljana (Slovenia) (Period: 3 December 2019 - 12 January 2020)

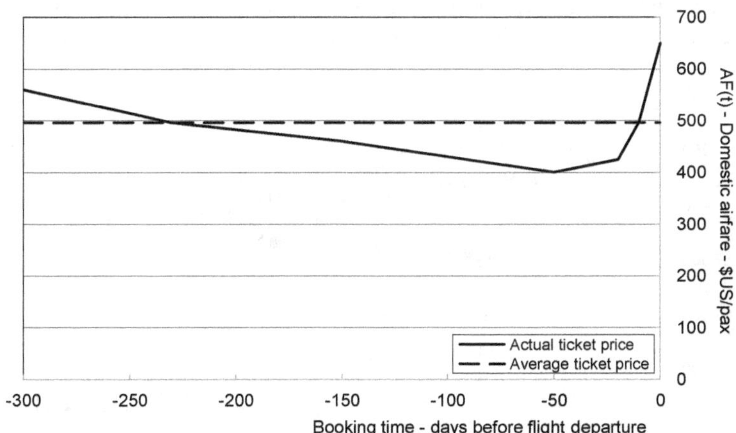

b) Case of flights in North America

Figure 3.54 Example of the airline pricing depending on time before the flight departure (https://www.cheapair.com/blog/cheapair-coms-5th-annual-airfare-study-reveals-the-best-time-to-buy-airline-tickets/; https://www.cnbc.com/2017/05/11/whens-the-best-time-to-book-a-flight.html/; https://www.transavia.com/nl-NL/home/).

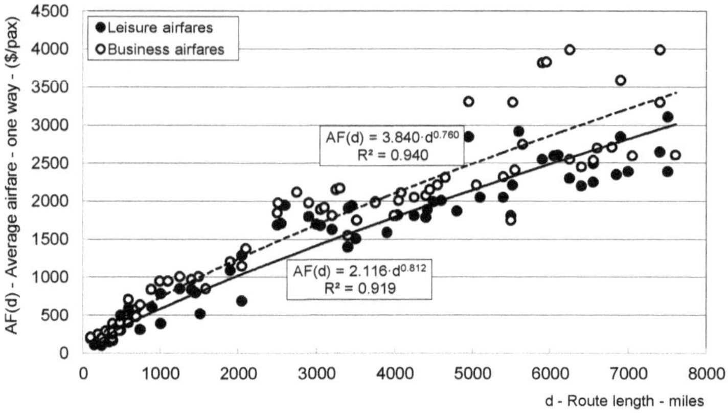

Figure 3.55 Relationship between the airfare, route length, and trip purpose—Case of KLM routes scheduled from Schiphol Airport (Amsterdam) (Janić, 2000; KLM, 1995).

becomes about 30% higher than the average airfare. In addition, the airfares for trips carried out during the weekends and/or during peak and off-peak periods have proven to be different, in some cases considerably. Regarding the trip purpose and the social-economic characteristics of potential users-air passengers, the airfares have also been quite different. In general, both have been categorised as business and leisure. Business trips on a given route are almost always more expensive than leisure trips (Button, 1993; Doganis, 1992; Janić, 2000). Also, following the operational costs, the airfares paid by users-air passengers for traveling along the given route also depend on its length. In general, they increase as the route length increases, as shown in Figure 3.55.

Two general types of airfares are presented—those offered to business and those offered to leisure air passengers. Both classes of airfares increased at a decreasing rate as the route length increased. Furthermore, as intuitively expected, the business airfares were always higher than the leisure ones. For a long ago these relationships can be considered as a rather generic in the given context. Last but not least, Figure 3.56 shows

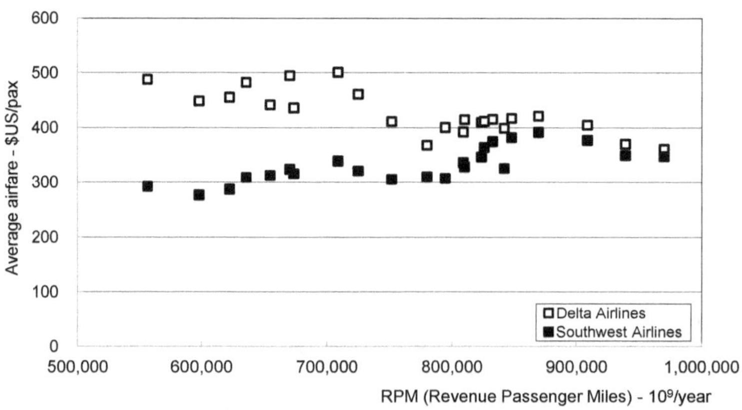

Figure 3.56 Relationship between the average airfare per passenger and the quantity of airline revenue output—Case of two U.S. airlines (Period: 1995–2018) (https://www.transtats.bts.gov/Tables.asp?DB_ID=135; http://web.mit.edu/airlinedata/; www/Aircraft&Re lated.html).

the relationship between the average airfare per passenger and the airline revenue output of the selected U.S. airlines.

As can be seen, at the lower volumes of revenue output the differences in the average airfares were higher, i.e., substantially greater at Delta than at Southwest airlines. However, with an increase in the volumes of output, these differences have been decreasing and have almost equalized at an equivalent annual volume. The main reason has been the rapid growth of the LCC Southwest in the U.S. domestic market and the rise of its airfare competition with the network Delta airlines in this case. This has forced Delta to also reduce its airfares and make them competitive to those of Southwest.

The airline airfares have also generally been influenced by the regulatory rules and procedures in the air transport markets. Generally, these can be regulated, semi-regulated or deregulated markets. In the regulated and semi-regulated markets, the airfares are regulated at the national and international level and agreed among the airlines and respected aviation authorities. To this end, various bilateral and/or multi-lateral agreements have been contracted between airlines. They always regulate the level and type of airfares, possible ways of covering the costs, sharing the revenues, etc. They also specify how to supply the transport capacity (the number of flights scheduled on particular routes during a given time period, aircraft types, the flight schedules, etc.). In the deregulated markets, the airlines are free to set up the competitive airfares, enter and/or leave the particular routes, and schedule the flights by chosen aircraft types (Janić, 2000; OECD, 1988).

3.5.4.2 Yield

The specific characteristic in the domain of airlines charging their users, i.e., air passengers and air freight/cargo shippers, is yield expressed in terms of revenues per revenue passenger mile/kilometre. Figure 3.57 shows the example for the selected European and U.S. network airlines and LCCs.

As can be seen, during the observed period, two sub-periods of change in the average yield over time can be distinguished for all considered airlines. The first was until about

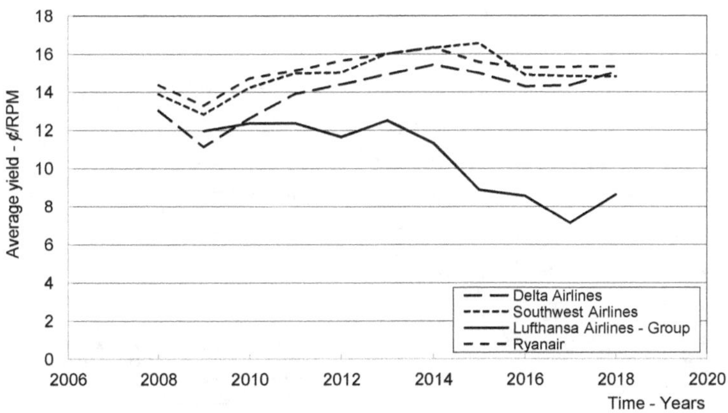

Figure 3.57 Development of the average yield over time—Case of the selected U.S. and European airlines (RPM – Revenue Passenger Mile) (DA, 2008/2018; DL AG, 2008/2018; Ryanir, 2008/2018; SW, 2008/2018; https://www.transtats.bts.gov/Tables.asp?DB_ID=135; http://web.mit.edu/airlinedata/; www/Aircraft&Related.html).

the year 2013/2015 when the average yield generally was increasing, and the other when it was decreasing, staying rather constant, and then it started to increase again. During most of the observed period, the yield of Lufthansa was decreasing and staying much lower than that of other airlines. It was, in particular, much lower than that of its LCC counterpart Ryanair. Despite the lower average airfares (see Figure 3.56), the average yield of LCC Southwest was very similar but higher than that of the network Delta airlines. The yield of Ryanair was comparable and even a bit higher than that of Southwest. This indicates the similarities of outcomes from the LCC business model of the airlines of similar size, scale and scope of operations although operating in different market environments (Europe and U.S.).

3.5.4.3 Profits

The airline profits are defined as the difference between their revenues and costs. When this difference is positive, the given airline is profitable, otherwise it suffers losses. In the latter case, it can simply be forced to stop its business and leave the market. In cases when particular segments of operations (passenger, freight/cargo segment, particular routes, and/or flights) are failing, the affected airline can simply abandon them. In some cases, the third-party subsidies can cover these losses and consequently sustain the corresponding segments under given conditions. Nevertheless, the airline industry in particular regions of the world and most individual airlines are generally profitable on the annual scale. Figure 3.58 shows the example of the average revenues and costs of the selected U.S. airlines over time.

As can be seen, both average revenues and costs have been higher at 7 network airlines that at their 5 LCC counterparts during the observed period. In particular, starting from the year 2003, the positive difference between the revenues and costs of 7 network airlines has generally been increasing, thus indicating a strengthening of their profitability. The difference between revenues and costs of 5 LCCs has been always positive, thus indicating their constant profitability during the entire observed period. Starting from the year 2015, this difference was higher than previously, but always remained less than that of its counterpart of 7 network airlines.

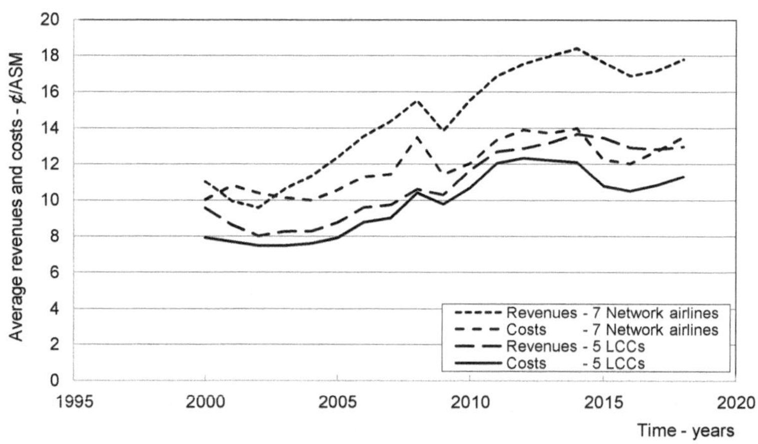

Figure 3.58 Development of the average revenues and costs—Case of the selected U.S. airlines over time (US BTS, 2019; https://www.transtats.bts.gov/Data_Elements_Financial.aspx? Data=7/).

3.5.5 Modelling Airline Economics

3.5.5.1 Route cost

One of the main economic characteristics of an airline network is the total and average costs of particular routes. Once the total route costs are determined, the average cost per unit of airline output (per passenger or p-km or p-mile) can be estimated as follows:

$$\overline{c}_{ij}(\tau) = \frac{c_{ij}(s_{ij}, d_{ij}) \cdot f_{ij}(\tau)}{d_{ij} \cdot Q_{ij}(\tau)}, \quad \textit{for } i \neq j;\, i, j \in N \tag{3.38}$$

where all symbols are analogous to those in Eq. 3.33a. Figure 3.59 shows a generic example of the relationship between the average route cost and density of passenger demand served by an increased number of flights during a given period of time.

As can be seen, the average unit cost decreases more than proportionally as the number of passengers, corresponding flights, and length of route. This illustrates that there may be economies of scale and density due to increases in passenger density and route length.

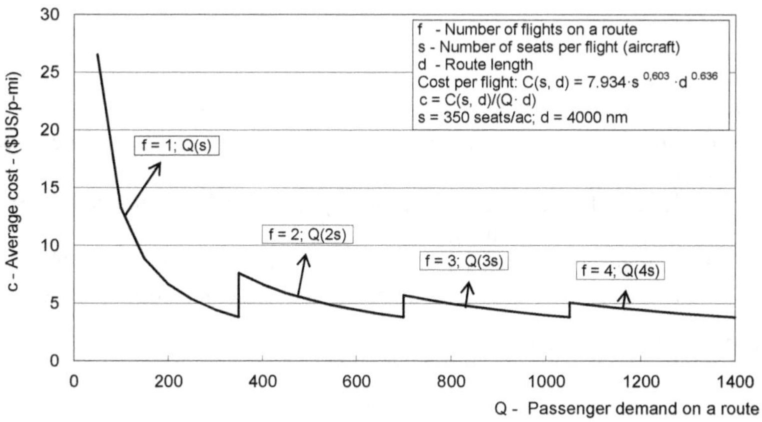

Figure 3.59 Relationship between the average cost and passenger demand on a given route (Janić, 2000).

3.5.5.2 Economies of network configurations

After deregulation and liberalization of air transport markets, some airlines have developed hub-and-spoke networks and some others, including new entrants, have developed point-to-point networks. In particular, the U.S. network airlines previously operating point-to-point networks established strong hub-and-spoke networks after the deregulation of the U.S. airline industry in 1978. The interesting questions have been and are still: "Why did most of the U.S. airlines change their network structures?" and "Why did large European airlines keep and further consolidate their already existing hub-and-spoke networks after liberalisation of the EU (European Union) aviation market in 1993?" The answer was relatively simple. The hub-and-spoke networks were expected to be relatively cheaper than the point-to-point networks. The following hypothetical example was to indicate if these expectations had been "true" or "false".

An airline is assumed to schedule its flights in its air route network consisting of *(N)* airports as the network's nodes. Each route and flight scheduled is assumed as a link of the network. If the airline strictly operates in a point-to-point network, *(N(N–1))* one-way flights need to completely connect these airports. The strict "hub-and-spoke" networks consisting of *(N)* nodes can be fully connected by *(2(N–1))* one-way flights through the central node, i.e., the hub airport ($N > 2$). Figure 3.60 shows the simplified scheme of two network configurations (see also Figure 3.5).

The total operating cost of the point-to-point network can be estimated as follows (Janić, 2000; O'Kelly, 1986). Let (Q_{ij}), (d_{ij}), and (c_{ij}) be the air passenger demand (pax/unit of time), route length (km, miles), and the average unit cost ($US/p-km), respectively, between the airports *(i)* and *(j)* $(I, j \in N)$. The total costs of operating the network are equal to:

$$C_1 = \sum_{i=1}^{N} \sum_{j=1}^{N} Q_{ij} \cdot d_{ij} \cdot c_{ij} + c_1 \cdot (N \cdot (N-1))/2 \quad for\ (i \neq j) \tag{3.39a}$$

where

c_1 is the average cost of maintaining each route of the network ($/route).

Similarly, the total operating cost of the hub-and-spoke network is equal to:

$$C_2 = \sum_{i=1}^{N} \sum_{j=1}^{N} [(q_{ih} \cdot d_{ih} \cdot c_{ih} + q_{hj} \cdot d_{hj} \cdot c_{hj}) + (Q_{ij} \cdot d_{ij} \cdot c_{ij})] + c_2 \cdot (N-1)$$
$$\text{for } (i,j,h \in N) \tag{3.39b}$$

where

i, h, j is the origin, hub, and destination airport, respectively;
q_{ih}, q_{hj} is the air passenger demand on the routes (d_{ih}) and (d_{hj}), respectively (pax/unit of time);
d_{ij}, d_{hj} is the length of routes connecting the airports *(i)* and *(h)*, and the airport *(h)* and *(j)*, respectively (km, miles);
c_{ih}, c_{hj} is the average unit cost on the routes (d_{ih}) and (d_{hj}), respectively ($US/p-km); and
c_2 is the average cost of maintaining each route of the network ($/route).

The other symbols are analogous to those in Eq. 3.39a.

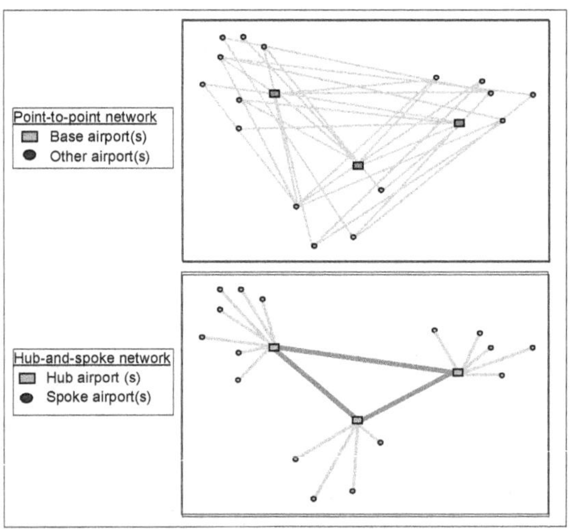

Figure 3.60 Schemes of configurations of the airline air route network.

Equations 3.39(a, b) indicate that the cost of indirect connection is always higher than the corresponding cost of direct connection, i.e., $(d_{ih}c_{ih} + d_{hj}c_{hj}) > d_{ij}c_{ij}$ for any (I, j, h). Thus, the total cost of a hub-and-spoke network will be higher than the cost of equivalent point-to-point networks. This may turn an airline away from operating a hub-and-spoke network. Nevertheless, the total savings due to reducing the number of routes of the hub-and-spoke network have been always greater than the total incremental costs. Therefore, the hub-and-spoke network can be a preferable configuration in terms of the total operating costs. Since there is a concentration of passenger flows on the smaller number of routes, three options of supplying the transport capacities can be applied. First, the flight frequencies can be increased by use of the same aircraft fleet. This will reduce the schedule delay and partially alleviate the travel inconveniences caused by the detours and longer flying time(s) along the indirect routes. Second, the airline may use the larger aircraft and keep the same flight frequency as in the point-to-point network to serve the denser air passenger demand. This results in an increase in the seat density on the particular routes that may produce the economies of scale and economies of density and, thus, create conditions for reducing airfares. Last, concentration of air passengers on the indirect routes enables economies of scale and economies of density.

As mentioned above, the airline can also schedule flights to minimise the total network costs, which include its operational costs and the costs of passenger time. As mentioned above, the costs of passenger time consist of the cost of schedule delay and the cost of in-aircraft (route) time. Under such conditions, the route-cost function (see Eq. 3.38) can be applied in order to illustrate the minimisation of the cost on direct route (ij) of a given airline network as follows:

$$C_{ij/min}(\tau) = [2 \cdot Q_{ij}(\tau) \cdot \alpha(\tau) \cdot \tau \cdot c_{ij}(s_{ij}, d_{ij})]^{0.5} + Q_{ij}(\tau) \cdot \beta(\tau)(d_{ij}/v_{ij} + w_{ij}) + c_0 \cdot Q_{ij}(\tau) \quad (3.40a)$$

From Eq. 3.40a, the minimum average airline cost per passenger is equal to:

$$\bar{c}_{ij/min}(\tau) = \left[\frac{2 \cdot \alpha(\tau) \cdot \tau \cdot c_{ij}(s_{ij}, d_{ij})}{Q_{ij}(\tau)}\right]^{0.5} + \beta(\tau) \cdot (d_{ij}/v_{ij} + w_{ij}) + c_0 \quad (3.40b)$$

The minimum marginal airline cost per passenger is equal to:

$$mc_{ij/min} = \partial C_{ij/min}(\tau)/\partial Q_{ij}(\tau) = \left[\frac{\alpha(\tau) \cdot \tau \cdot c_{ij}(s_{ij}, d_{ij})}{2 \cdot Q_{ij}(\tau)}\right]^{0.5} + c_0 \quad (3.40c)$$

Equations 3.40(b, c) indicate that both average and marginal cost per passenger increases at a rate of (0.5) as the value of passenger time $(\alpha(\tau))$ and cost per flight $(c_{ij}(s_{ij}, d_{ij}))$ increase, the latter depending on the aircraft capacity (n_{ij}) and route length (d_{ij}). At the same time, this cost decreases at a rate of $(1/Q_{ij})^{0.5}$ as the passenger density on the route increases.

After changing the network routing scheme, the passenger flows will be redistributed from $N(N–1)/2$ direct routes of the point-to-point network to $(N–1)$ indirect routes of the hub-and-spoke network. Such redistribution will increase the concentration of passenger demand and flights serving the demand on each route of the hub-and-spoke network as follows:

$$Q_{ih}(\tau) = \sum_{j=1}^{N} [q_{ih}(\tau) + Q_{ij}(\tau)] \text{ and } Q_{hj}(\tau) = \sum_{i=1}^{N} [q_{hj}(\tau) + Q_{ij}(\tau)] \quad (3.40d)$$

As can be seen, both concentrated demand flows are always greater than basic O/D demand flow $(Q_{ij}(T))$. This will reduce the average and marginal cost on both indirect routes *(ih)* and *(hj)* for the factors $[Q_{ij}(T)/Q_{ih}(T)]^{0.5}$ and $[Q_{ij}(T)/Q_{hj}(T)]^{0.5}$, respectively, compared to the corresponding cost on the direct route. In addition, the cost per flight on both routes will be lower due to shorter flying distances (d_{ih}) and (d_{hj}). When the number of spoke airports in the hub and spoke network is relatively large and the passenger demand high, the significant concentration of demand on the particular routes can be achieved, thus enabling fulfilment of the following conditions:

$$\bar{c}_{ih/min}(\tau) + \bar{c}_{hj/min}(\tau) < \bar{c}_{ij/min}(\tau) \quad for\ (i,h,j \in N) \tag{3.40e}$$

The similar relationship is valid for the marginal costs. The above condition is an exclusive consequence of the existence of economies of scale and density. Thanks to them, the unit cost of flying along longer (indirect) routes may become lower than the cost of flying on direct (shorter) routes. In many cases, these unit costs have been used as the basis for setting airfares. Following such logic, the airline may set lower airfares on indirect than on direct routes.

3.5.5.3 *Pricing policy*

As mentioned above, airlines apply different pricing policies for their services depending on their business model (network, LCC), market conditions (regulated, semi-deregulated and deregulated), characteristics of demand and own operating costs. Taking all these factors into account has resulted in the wide diversity of airfares among airlines and their networks. In the regulated markets, both the flight frequencies and airfares are bilaterally or multilaterally agreed by airlines, their governments, and the international aviation associations. In the semi-deregulated markets, the airfares are agreed upon and the flight frequencies are freely scheduled. Under such conditions, the flight frequencies are applied as an exclusive competitive tool for achieving profitable operations on the particular routes. In order to illustrate some characteristics of the semi-deregulated markets, let $(AF_{ij/r})$ be the agreed/regulated airfare on the route *(ij)* operated by a given airline. If each flight scheduled on this route should be at least zero profitable, the following condition has to be satisfied:

$$Q^{*}_{ij/r} \cdot AF_{ij/r} = c_0 \cdot Q^{*}_{ij/r} + C_{ij}(s_{ij}, d_{ij}) \tag{3.41a}$$

where

$Q^{*}_{ij/r}$ is the demand per flight on the route *(ij)* (pax/flight); and
c_0 is the average fixed cost per passenger ($US/pax).

The other symbols are analogous to those in previous Eqs. Let (Q_{ij}) and (f_{ij}) be the total air passenger demand and the corresponding total number of flights on route *(ij)*, respectively. If the flights are assumed to behave as natural monopolists, the demand (Q^{*}_{ijr}) can be estimated as: $Q^{*}_{ij/r} = Q_{ij}/f_{ij}$. From Eq. 3.41a, it follows that: $Q^{*}_{ij/r} = c_{ij}(s_{ij}, d_{ij})/(AF_{ij/r} - c_o)$. By combining both Eqs. for estimating the flight demand $(Q^{*}_{ij/r})$, the number of flights, which provides the zero-profits on the route *(ij)*, is equal as follows:

$$f_{ij} = [(AF_{ij/r} - c_0) \cdot Q_{ij}]/C_{ij}(s_{ij}, d_{ij}) \quad (flights) \tag{3.41b}$$

where $[(\$US/pax - \$US/pax) \cdot pax]/(\$US/flight) = flights$.

Equation 3.41b indicates the influence of regulated airfare on the supply of flight frequencies on a given route. Because of avoiding eventual losses, the airline will schedule the flight frequencies just up to the level zero-profitable level. Therefore, if some local, regional, and/or central authorities and/or governmental authorities request more flights on the given route(s), they should be prepared to cover (subsidise) losses of the non-profitable (additional) flights. Because regulated airfares do not always relate to the specific airline cost and the characteristics of potential users-air passengers, they reflect the inherent defects of semi-deregulated market(s).

In deregulated markets, the airlines set up airfares and flight frequencies freely according to the following main criteria: profitability of services and maintaining and attracting new air passenger demand. One of the earliest models for setting up the average airfare according to these criteria was as follows (Janić, 2000; Teodorović, 1983):

$$AF_{ij} \geq C_{ij}(s_{ij}, d_{ij})/(\lambda_{ij} \cdot s_{ij}) + c_0 \tag{3.41c}$$

where

λ_{ij} is the average load factor on the route (ij) $(0 < \lambda_{ij} \leq 1)$.

The other symbols are analogous to those in the previous Eqs.

Figure 3.61 shows a hypothetical example of the relationships between the airline output, airfares and costs.

The airlines operating in deregulated markets can freely set up airfares that are adjusted to the characteristics of potential air passenger demand. This is called the demand differentiation in terms of pricing and costing (Janić, 2000; Maddala and Miller, 1989). As can be seen, three types of airfares are offered to three classes of potential air passengers. The relationships between the volume of demand and airfare are represented by down-sloped demand curves, one for each class of potential users and each corresponding to the homogenous population of potential users. That means that they possess approximately the same social-economic characteristics and preferences relating to the cost, quality of service, and willingness to accept the airfare for air trip. In addition, the users of particular classes are sensitive to changes in airfares, i.e., if the airfares increase the demand will decrease, and vice versa. In the fully competitive markets, the demand curve is equivalent to the marginal revenue curve since only the passengers who accept the offered airfares will actually appear on the market. In this example, the prices are assumed to attract the

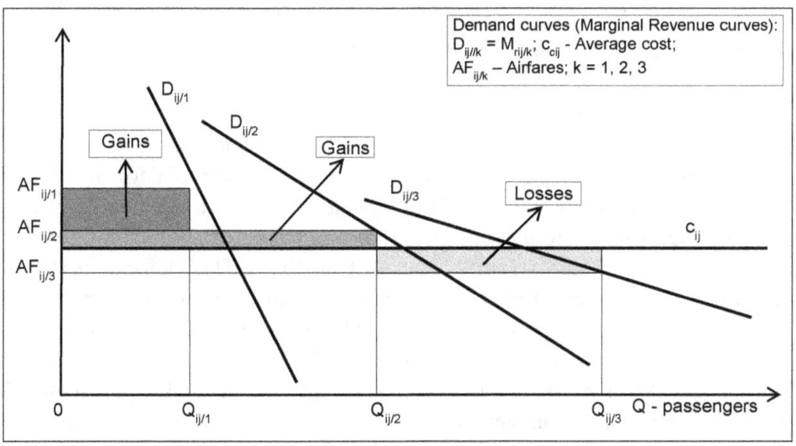

Figure 3.61 Example of the airline demand differentiation: pricing and costing (Janić, 2000).

widest range of potential users. If set up at the level $(AF_{ij/k})$, they will attract $(Q_{ij/k})$ users $(k = 1, 2, 3)$. In order to simultaneously maximise the revenues and passenger welfare, the airline should allocate the capacity (seats) for $(Q_{ij/1})$ passengers of Class *1*, $[Q_{ij/2} - Q_{ij/1}]$ passengers of Class *2*, and $[Q_{ij/3} - Q_{ij/2}]$ passengers of Class *3*. The total revenue gained by application of such capacity allocation policy is equal to:

$$R_{ij} = \sum_{k=1}^{3} AF_{ij/k} \cdot (Q_{ij/k} - Q_{ij/k-1}) \tag{3.41d}$$

It appears quite clear that the airline will always assign the highest priority and most of the available capacity to the most expensive users—air passengers. The space for the lowest-charged air passengers will be strictly limited. The passengers of this Class are frequently offered to fill in the potentially empty seats. Once the aircraft type has been determined, the average cost per passenger is assumed to be fixed and not dependent on the volume of demand (horizontal line c_{ij} in Figure 3.60). Then, the total profits on the route *(ij)* can be estimated as follows:

$$\Pi_{ij} = \sum_{k=1}^{3} (AF_{ij/k} - \bar{c}_{ij}) \cdot (Q_{ij/k} - Q_{ij/k-1}) \tag{3.41e}$$

The relationships between particular airfares and the average cost $(\bar{c}_{ij})$ indicate that the airline may gain profits and suffer from losses depending on the attracted Classes of air passengers. For example, if the airline applies the "first-best" pricing rule, the airfares charged to the low-charged user will be equal to the marginal cost (Button, 1993; Janić, 2000). Since, in this case, the marginal cost is always lower or equal than the average cost, the airline will suffer losses, but the users' surplus will be maximal. The airfares charged to the high-charged passengers can be set either to the marginal or average cost, aiming at covering losses from the low-charged passengers (see the shaded areas in Figure 3.60). The described pricing policy is always designated in advance, i.e., before starting the business cycle. Since the results, either profits or losses, are known only at the end of period, it seems to be risky. Particularly, the losses may be caused by mistakes in the setting up of airfares at the beginning of a business period. Then, the significant discount of airfares due to the eventual price competition (i.e., "price war") and an unexpected variability of demand, which cannot be accurately predicted at the beginning of the period, may also cause (severe) losses. In addition, changing the airline cost due to significant changes in the airfares of inputs may cause losses at the end of period. Nevertheless, despite these disadvantages, this charging policy has been applied by most airlines operating in the deregulated markets.

3.5.5.4 Yield management

The above-mentioned proportions of particular Classes of air passengers in the total passenger demand on the specific route(s) can be used as an input for division of the available airline seat capacity into the different categories. Such allocated seat capacity and corresponding airfares are used as an initial input for operation of the airline CRS (Computer Reservation System) (Janić, 2000; OECD, 1988). The CRS enables the airline to sell its seats (service) in real time. It can create a mixture of air passengers of different Classes for the particular flights and routes, which ensures the maximum revenues ("yield") under the given conditions (Belobaba, 1987; 1989; Glover et al., 1982; Janić, 2000; Smith et al., 1992). As such, the process is usually called "yield management". The main premise of the "yield management" process has been "selling the right seat to the right customer at the right price" aiming at maximizing the airline revenues. Based on the above premise, "yield management" has been considered as the tactical decision-making process consisting of the following phases (Button, 1993; Janić, 2000):

- Setting up the flight capacity;
- Estimating the demand for full-fare passengers;
- Setting the optimal overbooking levels under given conditions;
- Estimating the number of discount seats;
- Setting up the airfares and restrictions on the particular seat Classes; and
- Up-dating the booking limits of particular airfare Classes until closing down the booking procedure (i.e., until the flight departure).

In order to successfully perform the above-mentioned particular phases of the "yield management" process, the airlines have had use the sophisticated CRSs, including large and fast computers and communication links and sophisticated software based on the mathematical programming and operation research techniques for controlling such complex processes. The software supporting the successful "yield management" has had to fulfil three functions: overbooking, discount-seat allocation (i.e., imposing the limitations on the available seats at given prices), and traffic management (i.e., controlling the passenger itineraries) (Belobaba, 1987; 1989; Janić, 2000; Lee and Hersh, 1993; Smith et al., 1992).

a) Overbooking

Overbooking function of the "yield management" process is performed in order to maximize the net revenues associated with overbooking decisions. This implies accepting more bookings than the flight's seating capacity. The objective is to provide an opportunity for compensation of the revenue losses due to the cancelled reservations and the non-show air passengers. Setting a proper booking limit is a relatively complex task. For example, if the booking limit is set too high, and if the air passengers coming to board the flight physically confirm all requests, some of them will not be boarded. Consequently, the airline has to pay compensations to the involuntary denied passengers as the over-sale. Otherwise, if the booking limit is set too low and if more air passengers have cancelled or not shown up, the flight will depart with empty seats that could be occupied by the turned-away demand. The revenues and costs as the outcome from the overbooking function can generally have different relationships. In such cases, the "optimal" overbooking level has to optimize the net profits as the differences between the expected revenues and costs of over-sales. This "optimal" overbooking limit can be achieved by continuously balancing the additional (marginal) revenues gained by selling a reservation and the costs of an additional over-sale (Janić, 2000; Smith et al., 1992). Figure 3.62 shows a simplified scheme of the possible relationships between the marginal revenues and costs of over-sales.

The potential revenue (vertical axis) will increase in line with the overbooking level (horizontal axis). However, despite the revenue growth, the net benefit will decrease due to increasing the costs of over-sale. As can be seen, the "optimal" overbooking level will be reached at the point where the marginal revenue earned by an additional reservation becomes equal to the marginal cost of an additional over-sale. Detailed mathematical models based on probability theory have been developed in order to determine the "optimal" overbooking level under the varying conditions related to the characteristics of demand, flight seat capacity, and time (Janić, 2000; Teodorović, 1988).

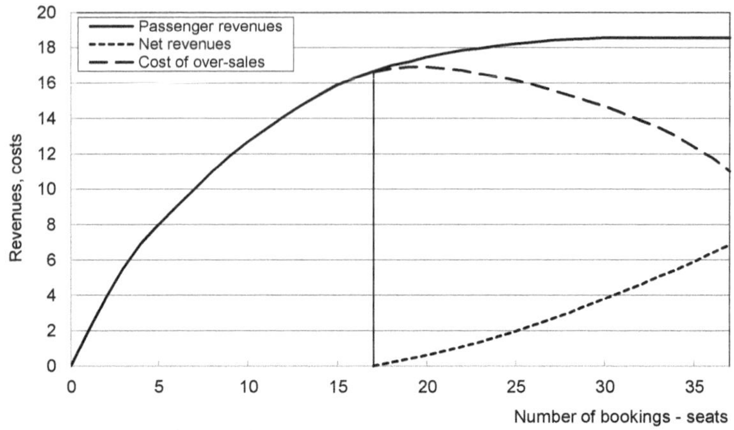

Figure 3.62 Relationship between the marginal revenues and the costs of over-sales (Janić, 2000; Smith et al., 1992).

b) Discount seat allocation

The overbooking function provides the complete "yield management" solution when all passengers pay the same airfares on a given flight, i.e., when all available seats are offered and sold at the same airfare. However, when the airline competes by the airfares, the need for control availability of the various fare-type seats will be needed. The easiest way to clarify the problem of discount-seat allocation has been to analyse an example of allocation of two fare Classes—discount and full airfares. In general, two methods for seat allocation have been developed: the non-nested method, where the airfare-class buckets have been considered as strictly independent, and the nested-seat allocation (booking) method, where the discount airfares (seats) have been nested within the full-fare seats (Curry, 1990; Janić, 2000; Lee and Hersh, 1993).

The non-nested seat-allocation method allocates seats to the various airfares (booking classes) independently. The sum of all allocations is equal to the overbooking capacity. The size of particular booking classes may change over time and characteristics of passenger demand. The main inherent defect of this method is the possible rejecting of the request from the highest value booking class, although the total booking limit has not been reached. The nested seat-allocation (booking) method is based on nesting the booking classes according to their revenue values. The limits on the particular classes may change over time and depending on the characteristics of the expected air passenger demand. The simplified scheme of the non-nested seat-allocation method is shown in Figure 3.63(a, b).

One of the earliest developed rules for the optimal allocation of seats was for the conditions of two airfare classes and "nested" discount within the full airfare class. According to this rule, the flight revenues can be maximised if the low airfare class is closed for additional booking under the following conditions: the expected revenue from selling the other low-fare seat exceeds the expected revenue that could be obtained by selling the same seat at a full fare. Mathematically, the discount fare passenger request (AF_2) should be accepted under the following condition:

$$AF_2 \geq p(q_1 > s_1) \cdot AF_1 \tag{3.42}$$

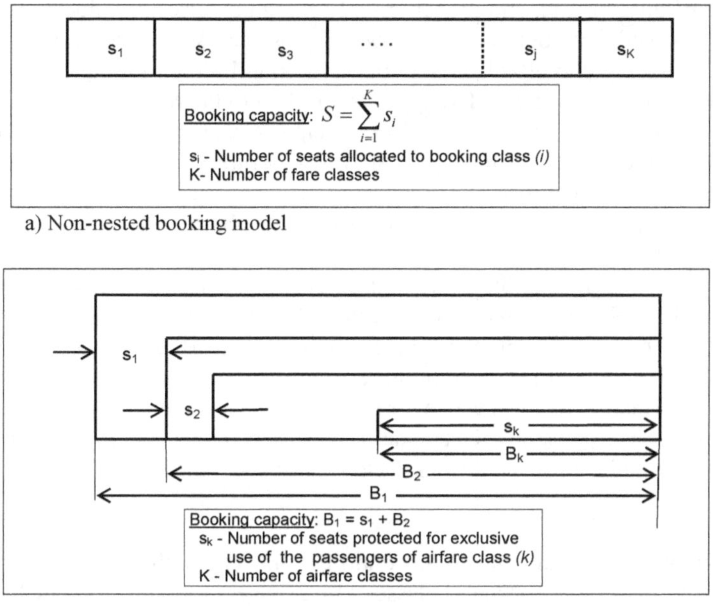

Figure 3.63 Simplified schemes of two basic seat allocation methods (Janić, 2000).

where

AF_1 is the full airfare ($US/pax);
AF_2 is the discount airfare ($US/pax);
p is the probability operator;
q_1 is the number of requests for the full fare Class 1 (-); and
s_1 is the number of seats protected for the full fare Class 1 (-).

The smallest value of (q_1) that satisfies condition (3.41a) is the booking limit maximising the revenues from the bookings of airfare Class 1. Figure 3.64 shows a scheme of the decision-making dilemma faced by a booking agent who has to decide to accept or reject a discount seat request (Janić, 2000; Smith et al., 1992).

According to the decision tree in Figure 3.64, if $(p \cdot AF_1 < AF_2)$, the discount request will be rejected. Otherwise, the request will be accepted. The probability $p \equiv P(q_1 > s_1)$ is dependent on factors such as the expected demand of full fare passengers, accuracy of

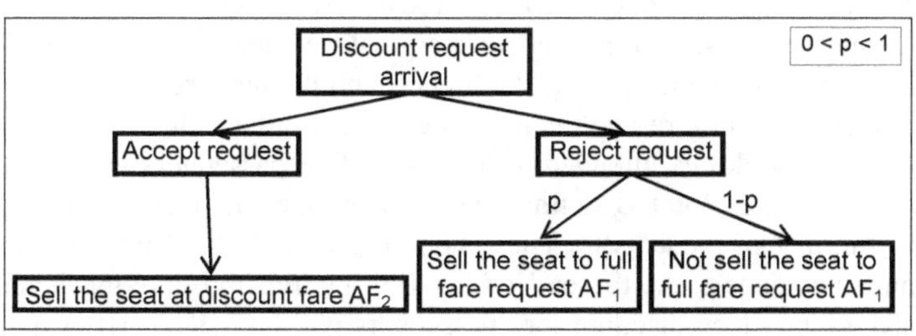

Figure 3.64 Typical decision tree that a booking agent might face (Janić, 2000).

forecasting of this demand, and probability of selling up. Discount allocation decisions should be up-dated from time to time, depending on the actual variations of passenger demand. The principle to establish an optimal booking limit on multiple fare classes is the same as in the case with two fare types. According to the rule defined by Eq. 42, the marginal value of the airfare request is always weighted against the marginal values of all other airfares at the time when the request arrives. As a rule, as the number of reservations allocated to the particular airfare classes increases, the marginal expected value of an additional reservation of that fare type decreases (Belobaba, 1987; 1989; Janić, 2000; Smith et al., 1992).

Although the airfare is the same, the probability of an additional sale decreases with each additional seat offered. If the number of offered seats is much higher than the expected demand, the marginal revenue will only slightly increase with an additional seat offered (i.e., the marginal returns diminishes with increasing of the capacity). Let (R_i) be the expected marginal seat revenue for airfare Class *(i)* when the number of available seats of that class has just increased by one. The expected revenue of (S_i)-th seat of airfare class *(i)*, $R_i(S_i)$ will be equal as follows (Belobaba, 1989; Janić, 2000):

$$R_i(s_i) = AF_i \cdot p(q_i > s_i) \tag{3.43a}$$

where all symbols are analogous to those in Eq. 3.42. The optimal inventories of seats that should be sold to the particular classes of passengers can be determined according to the rule:

$$R_i(s_{ji}) = AF_j, \quad for\ i < j, j = 1,2,..,K \tag{3.43b}$$

where

s_{ji} is the number of seats protected from the users of class *(j)*; they are available exclusively to the users of class *i (i < j)*.

Equation 3.43a indicates that, for *(K)* airfare classes nested within each other, the required number of comparisons of the marginal revenues is equal to *[K(K–1)/2]*. The decision criteria *(R)* can also be applied to the dynamic control, in which booking limits may be revised as the flight departure nears (Belobaba, 1989; Janić, 2000). The booking limits on each airfare class *(j)* can be determined as:

$$BL_j = s - \sum_{i<j} sji \tag{3.43c}$$

where

s is the flight seating capacity (seats/flight).

Figure 3.65 shows the simplified scheme of an analysis of the marginal revenue aiming at estimating discount allocation for three airfare classes.

For each airfare class, an independent marginal revenue curve is defined. The full airfare class marginal revenue curve is the highest and the most squeezed. The marginal revenue curve of discount fare-class is the lowest one (on the right). Point A defines the optimal number of bookings intended to the full airfare passengers and withheld from any discount passenger. Point B defines the total number of bookings to allocate the medium and full airfare passengers. These bookings are withheld from deep (the third class) discount passengers. At the end of booking period, the booking levels for full airfare, moderate and deep discount fare passengers will be: $(BL_1 - BL_2) \equiv s_{21}$; $(BL_2 - BL_3) \equiv s_{21}$, and BL_3, respectively. The optimal booking limits for particular fare classes

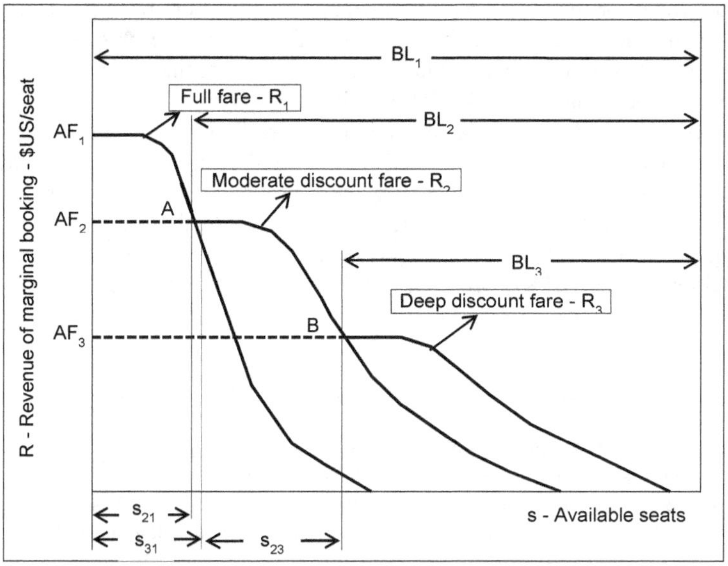

Figure 3.65 Relationship between the revenues of marginal reservation and the number of seats of particular airfare classes (Belobaba, 1989; Janić, 2000).

are (BL_1), (BL_2), and (BL_3), respectively (Belobaba, 1987; 1989; Brumelle, 1990; Janić, 2000; Smith et al., 1992).

c) Traffic Management

As mentioned above, after many airlines have started to operate the hub-and-spoke networks, the flights scheduled on the specific routes have been boarded by air passengers travelling from a single origin(s) (spoke(s)) to many destinations (other spokes) of the airline network by passing through the hub. Thus, the number of different passenger classes, trip purposes and airfares has substantively. Under such conditions, the "classic" yield management process has become very complex, inefficient and difficult to solve. Consequently, two fields have been additionally elaborated: the air passenger itinerary control and integration of the yield management process with pricing and scheduling. This resulted in the concept of "virtual nesting", which was first applied by American Airlines (Smith et al., 1992). The concept was based on defining the particular itinerary/fare classes of potential users, determining their characteristics and assigning the appropriate revenue values to them. The characteristics of particular itinerary/fare classes were not explicitly stored in the CRS. Their availability and characteristics were determined independently for each request by combining indexing with the flight/bucket availability. The objective was to minimise the variability of particular itinerary/airfare classes within the particular buckets and maximise the variability between different buckets. The airline had to determine the value of each bucket (Smith et al., 1992). This revenue value reflected the potential contribution of the itinerary/airfare class to the total expected revenue over the entire network, not only to the revenue of its own class. In order to evaluate the potential contribution of the specific itinerary/airfare class, it was necessary to establish the minimum acceptable revenue contributions. The limit represented the lower requirement for the air passenger itinerary to qualify for space on each flight leg in the connecting network. Then, the particular itinerary/airfare classes clustered into several buckets according to their values were nested so that the sale was first restricted for the low-level bookings, regardless of their itinerary/airfare class.

The overbooking, discount seat allocation, and traffic management, as components of the yield management process, were based on the short-run forecasting of air passenger demand. Therefore, yield management was expected to be more efficient if the forecasts of demand were more accurate. In other words, the accurate forecasting of air passenger demand became fundamental for running an efficient yield management system (Belobaba, 1987; 1989; Brumelle, 1990; Janić, 2000; Smith et al., 1992).

References

Abrahams, M. (1983). A service quality model of air travel demand: an empirical study. Transportation Research-A, 17A(5): 385–393.

Ang, H. S. A. and Tang, H. W. (1975). Probability Concept in Engineering Planning and Design: Volume I—Basic Principles. John Wiley and Sons, Inc., New York, USA.

Banfe, F. C. (1992). Airline Management. Prentice Hall, Englewood Cliffs, New Jersey, USA.

Belobaba, P. P. (1987). Airline yield management: an overview of seat inventory control. Transportation Science, 21(1): 63–73.

Belobaba, P. P. (1989). Application of probabilistic decision model to airline seat inventory control. Operations Research, 37(2): 183–797.

Boeing. (2014). 787 Airplane Characteristics for Airport Planning, Boeing Commercial Airplanes, Seattle, Washington, U.S.A.

Bowen, D. B. and Headley, E. D. (2019). Airline Quality Rating 2019, EMBRY-RIDDLE-Aeronautical University, Scholarly Commons College of Aviation, Prescott, Arizona, USA.

Brumelle, L. S. (1990). Allocation of airline seats between stochastically dependent demands. Transportation Science, 24(3): 183–192.

Button, J. K. (1993). Transport Economics. 2nd Ed., Edward Elgar, P. L., Grower House, England, UK.

CAPA. (2013). Heathrow airport's slot machine: hitting the jackpot again? Centre for Aviation, https://centreforaviation.com/analysis/reports/heathrow-airports-slot-machine-hitting-the-jackpot-again-108646/.

Curry, E. R. (1990). Optimal airline seat allocation with fare classes nested by origins and destinations. Transportation Science, 24(3): 193–204.

Daskin, S. M. and Panayotopoulos, D. N. (1989). Lagrangian relaxation approach to assigning aircraft to routes in hub and spoke networks. Transportation Science, 23(2): 91–99.

DA. (2008/2018). Annual Report, Delta Airlines, Atlanta, USA.

DL AG. (2005/2018). Lufthansa Group – Annual Report, Deutsche Lufthansa AG, Investor Relations, Lufthansa Aviation Centre, Frankfurt/Main, Germany, Investor.relations@dlh.de.

Doganis, R. (1992). The Airport Business. Routeledge, London and New York, UK.

DVB. (2019). An Overview of Commercial Aircraft 2018–2019, DV Bank SE Aviation Research (AR), Schiphol, Netherlands.

EEC. (2019). CODA Digest All-causes Delay and Cancellations to Air Transport in Europe-Q2 2019, The European Organisation for the Safety of Air Navigation (EUROCONTROL), Brussels, Belgium.

Embraer. (2017). Airport Planning Manual - E JETS E2, EMBRAER S.A., Sao Jose dos Campos—S.P., Brazil.

Etschmaier, M. M. and Mathaisel, D. F. X. (1985). Airline scheduling: an overview. Transportation Science, 19(2): 127–138.

FAA. (2016). Economic Values for FAA Investment and Regulatory Decisions: Aircraft Capacity and Utilization Factors, A Guide Final Report, Federal Aviation Administration, Washington D.C., USA.

FG. (2018). World Airline Rankings 2018, Flight Global, London, UK.

Glover, F., Glover, R., Lorenzo, J. and McMillan, C. (1982). The passenger mix problem in the scheduled airlines. Interfaces, 12(3): 73–79.

Gordon, S. and de Neufville, R. (1973). Design of air transportation networks. Transportation Research, 7(3): 207–222.
Haylen, A. and Butcher, L. (2017). Airport Slots, Briefing paper, Number CBP 488, 12 June, House of Commons Library, London, UK, www.parliament.uk/commons-library.
Hong, S., Ganxiang, H., Jiesong, Z. and Yu, W. (2015). Wide-body aircraft trip direct operating cost model based on trans-log function. Proceedings of the International Conference on Logistics, Engineering, Management and Computer Science, Shenyang, China, pp. 726–729.
Horonjeff, R., McKelvey, X. F., Sproule, J. W. and Young, B. S. (2010). Planning and Design of Airports, Fifth Edition, McGraw-Hill Education, Mc Grow hill Book Company, New York, USA.
http://web.mit.edu/airlinedata/www/Aircraft&Related.html
http://www.airportspotting.com/worlds-largest-airlines/
http://pinotglobal.com/maps/southwest-route-map/attachment/southwest-route-map-from-the-most-frequently-searched-sources-1
http://dot.gov/airconsumer/
https://www.cheapair.com/blog/cheapair-coms-5th-annual-airfare-study-reveals-the-best-time-to-buy-airline-tickets/)
https://www.cnbc.com/2017/05/11/whens-the-best-time-to-book-a-flight.html/
https://www.transtats.bts.gov/Data_Elements_Financial.aspx?Data=7/
https://www.bea.gov/data/gdp/gross-domestic-product
https://www.ons.gov.uk/economy/grossdomesticproductgdp
https://en.wikipedia.org/wiki/World%27s_largest_airlines
https://www.airport-technology.com/features/worlds-biggest-airlines-2018/
https://atwonline.com/datasheet/2019-world-airline-report-world-airline-fleets
https://news.delta.com/route-map-us-canada/
https://www.statista.com/statistics/262971/aircraft-fleets-by-region-worldwide/
https://contentzone.eurocontrol.int/aircraftperformance/details.aspx?ICAO=E190&/
https://inflationdata.com/articles/inflation-adjusted-prices/historical-crude-oil-prices-table/
https://en.wikipedia.org/wiki/List_of_busiest_passenger_air_routes
https://www.youtube.com/watch?v=518QuKMZSSM/
https://www.airfleets.net/home/
https://www.planespotters.net/
https:// blueswandaily.com/top-20-airline-groups-by-fleet-size/
https://www.avjobs.com/history/airline-economics.asp/
https://www.transtats.bts.gov/Tables.asp?DB_ID=135/
https://flywith.virgin atlantic.com/gb/en/on-the-flight/cabin-and-seats.html/
https://en.wikipedia.org/wiki/Flight_length/
https://www.transavia.com/nl-NL/home/
IATA. (2018). The Economics of Airline Financial Performance and Wider Economic Benefits, International Air Transport Association, Paris, France, www.iata.org/economics.
ICAO. (2017). Airline Operating Costs and Productivity. International Civil Aviation Organization, Economic Development, Teheran, Iran, https://login.icao.int/my.policy.
Janić, M. (1999). The trans european air transport network: an analysis of quality of service and spatial performance. pp. 241–268. *In*: Aura Reggiani and Deniele Fabri (eds.). Networks and Developments in Economics Spatial Systems: New Perspectives. Avebury, UK.
Janić. (2000). Air Transport System Analysis and Modelling: Capacity, Quality of Services and Economics, Gordon and Breach Science Publishers, Volume 16, Amsterdam, The Netherlands.
Kleinrock, L. (1976). Queuing Systems, Volume II: Computer Applications. A Wiley Interscience Publications, John Wiley and Sons, New York, USA.
KLM. (1995). Timetable, Amsterdam, The Netherlands.
Kremer, S. (2014). The Developments in the Competition in the Transatlantic Air Transport Market in the Periods before and after the Implementation of the US-EU Open Skies Agreement, Bachelor Thesis, European Public Administration – Programme European Studies, University of Twente, Enschede, The Netherlands.
Larson, C. R. and Odoni, R. A. (1981). Urban Operations Research. Prentice Hall, Inc. Englewood Cliffs, New Jersey, USA.

Lee, C. T. and Hersh, M. (1993). A model for dynamic airline seat inventory control with multiple seat bookings. Transportation Science, 27(3): 252–265.

Lenoir, N. (2016). Research for TRAN Committee: Airport Slots and Aircraft Size at EU Airports: In-Depth Analysis, Directorate-General for Internal Policies Policy Department B: Structural and Cohesion Policies, Transport and Tourism Research, European Parliament, Brussels, Belgium.

Lufthansa. (1995). Timetable, Frankfurt, Germany.

Maddala, S. G. and Miller, E. (1989). Microeconomics: Theory and Application. McGraw Book Hill Book Company, New York, USA.

Manheim, L. M. (1979). Fundamentals of Transport System Analysis: Volume 1: Basic Concepts. The MIT Press, Cambridge, Massachusetts, London, England, New York, USA.

OAG. (2019). Busiest Routes 2019, Luton, UK, https://www.oag.com/.

OECD. (1988). Deregulation and Airline Competition, OECD Publications, Paris, France.

O'Kelly, E. M. (1986). The location of interacting hub facilities. Transportation Science, 20(2): 92–106.

Pollack, M. (1982). Airline route-frequency planning: some design trade-offs. Transportation Research-A, 16A(2): 149–159.

Ryanair. (2008/2018). Annual Report, Dublin Airport, Ireland, www.ryanair.com/.

Smith, C. B., Leimkuhler, F. J. and Darrow, M. R. (1992). Yield management at American airlines. Interfaces, 22(1): 8–31.

Sorenson, N. (1990). Airlines Competitive Strategy: A Spatial Perspective. PhD Thesis, University of Washington, Washington DC, USA.

SW. (2008/2018). Annual Report, Southwest Airlines CO., Dallas, Texas, USA.

Swan, M. W. and Adler, N. (2006). Aircraft trip cost parameters: a function of stage length and seat capacity. Transportation Research Part E, 42(2): 105–115.

Taneja, K. J. (1982). Airline Planning: Corporate, Financial and Marketing, Lexington Books, D. C. Heath and Company, Lexington, Massachusetts, Toronto, Canada.

Teodorović, D. (1983). Flight frequency determination. Journal of Transport Engineering, 109(5): 747–757.

Teodorović, D. (1988). Airline Operations Research, Gordon and Breach Science Publishers, London, UK.

Teodorović, D. and Krčmar-Nožić, E. (1989). Multicriteria model to determine flight frequencies on an airline network under competitive conditions. Transportation Science, 23(1): 14–25.

Trevett, C. J. (1983). Fleet planning models. The Aeronautical Journal, 87(863): 88–94.

Turnquist, A. M. and Daskin, S. M. (1982). Queuing models of classification and connection delay in railyards. Transportation Science, 16(2): 207–230.

US BTS. (2019). DOT Form 41, Bureau of Transportation Statistics, U.S. Department of Transportation, Washington, D.C., USA.

Vuchic, V. R. (2007). Urban Transit Systems and Technology. John Wiley & Sons, ISBN: 978-0-471-75823-5, p. 624.

Yeng, J. C. (1987). Routing Strategies for an Idealised Airline Network. PhD Thesis, University of California, Berkeley, USA.

www.PlaneStats.com/

Chapter 4

ATC/ATM (Air Traffic Control/Management)

4.1 Introduction

The ATC/ATM (Air Traffic Control/Management)[1] is considered as the infrastructure component of the air transport system. It consists of the controlled airspace established over particular countries, continents and oceans, technical/technological components, such as radio-navigational facilities and equipment located on the ground and in space (satellites), and their complements in terms of the aircraft and the operating staff/employees (the ATC controllers) on the ground and the aircraft crews. The ATC controllers apply the operating rules and procedures in order to control and manage the aircraft/flight flows safely, efficiently, and effectively. This implies that the aircraft/flights, as the system's users, have to be served without potential mutual conflicts and the with the surrounding terrain (safety). In addition, they have to be provided with the intended time-fuel optimal trajectories between their origin and destination airports (effectiveness) and without imposing the substantial costly delays caused by the ATC (effectiveness, efficiency). This is achieved by providing the system with a sufficient capacity for it to appropriately meet the expected demand.

4.2 The System

4.2.1 Airspace

In order to fulfil the above-mentioned objectives, the ATC large controlled airspace is divided into the smaller parts at two levels. In the first, the airspace is divided into AZ (Airport Zone) with the airport airside area (runways, taxiways, and the apron/gate complex), TMA (Terminal Manoeuvring Area) with the approach and departure trajectories towards a single and/or different runways of a given airport(s), LAA (Low Altitude Airspace) enabling the aircraft/flight climbing to and descending from the cruising altitudes, and HAA (High Altitude Airspace) with the fixed air route network or FRA (Free[2] Route Airspace) where the cruising phase of flights takes place on the

---

[1] In the further text the term ATC instead of "ATC/ATM" will be used with the same meaning.

[2] FRA (Free Route Airspace) is the specified airspace where the aircraft/flights may freely plan a route between a defined entry and exit point. For example, at the end of the year 2019, most European airspace is expected to implement FRA, with all airspace having this type of operation by 2021/2022 (http://www.eurocontrol.int/articles/free-route-airspace/).

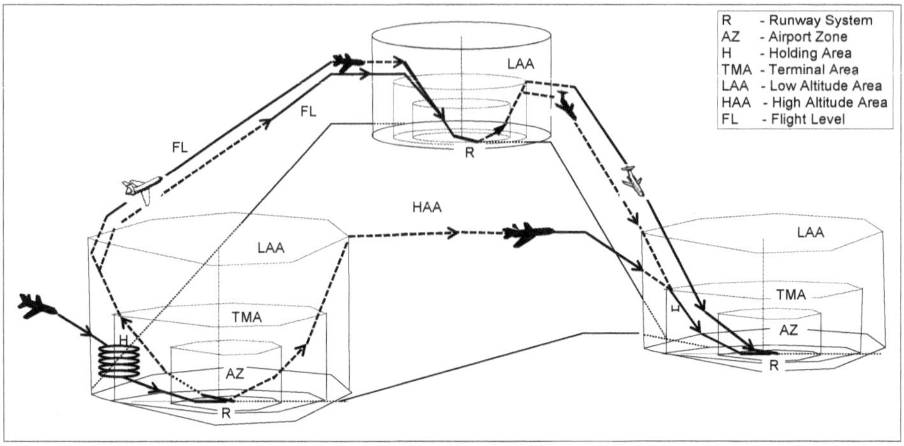

Figure 4.1 Scheme of division of the ATC airspace (Janić, 2000).

assigned FLs (Flight Levels). Figure 4.1 shows the simplified scheme of division of an ATC airspace (Janić, 2000).

In particular, in the LAA (Low Altitude Area) and HAA (High Altitude Area) of a given airspace, shown in Figure 4.1, the ARTCC (Air Route Traffic Control Centre) (U.S.) or ATCC/ATMC (Air Traffic Control/Air Traffic Management Centre) (Europe) manages the air traffic flows and control movement of individual aircraft/flights. At the second level, each of the above-mentioned airspaces is divided into smaller parts, called the ATC sectors; these are assigned to one or a team(s) of the ATC controllers (ICAO, 2016; 2018). In addition, while moving through an ATC sector, an aircraft/flight can operate according to IFR (Instrument Flight Rules) or VFR (Visual Flight Rules) (This latter mainly in the U.S.). According to this, the IFR aircraft/flight is primarily responsible for maintaining the assigned flight trajectory approved by the ATC controller, also guaranteeing the safe separation from other traffic. The VFR aircraft/flight is responsible for performing the primary navigation along the chosen trajectory and maintaining a safe separation from other VFR traffic. In the airspace where both IFR and VFR aircraft/flights operate, division of responsibility between pilots and ATC controllers for safe separation is the same as in the case of the IFR aircraft/flights.

Regarding the pattern of operations, each ATCC/ATMC and its sectors can generally handle four categories of aircraft/flight flows[3] as follows (Song et al., 2006):

- Transit flights, i.e., overflights;
- Inside flights, between the origin and destination airports in the airspace under jurisdiction of a given ATCC/ATMC and/or of its particular sectors;
- Outgoing flights between the origin airports under the Centre's and/or its sectors' jurisdiction and the destination airports outside it/them; and
- Incoming flights between the origin airports outside the Centre's and/or its sectors' jurisdiction and destination airports within it/them.

[3] Air traffic flow consists of the relatively closely-spaced aircraft/flights moving approximately in the same direction(s) along the specified air route(s). These aircraft/flights usually cruise at the same and/or different FLs.

4.2.2 Technical/Technological Components

The main technical/technological components of an ATC system are CNS (Communication Navigation Surveillance) systems, including communications, navigation, and surveillance systems. These are based on digital technologies, including space satellites designed at the various levels of automation. In addition, there are various tools supporting the ATC tactical and strategic control and management of the individual aircraft/flights and the air traffic flows, respectively.

a) Communications include the channels for transmission of information between (i) pilots and ATC controllers, such as CPDLC (Controller Pilot Data-Link Communications), which enables exchange of digital messages without the need to use R/T (Radio Telephony); (ii) between particular ATC control units; and (iii) between the ATC staff and its outside environment. The forthcoming communications under development in the scope of EU (European) SESAR (Single European Sky ATM Research) and U.S. NextGen (Next Generation Air Transportation System) research programs mostly relate to the further increase in automation of the non-voice transmissions of information throughout the ATC system (http://www.sesarju.eu/; http://www.faa.gov/nextgen/).
b) Navigation is based on the ground aids and space satellites, both used by the aircraft/flights for primary navigation. The ground aids are: VOR (VHF Omnidirectional Range), DME (Distance Measuring Equipment), VOR/DME, and RNAV (Area Navigation) system (use information about the distance and azimuth by the VORTAC stations as input), the over-water en-route OMEGA, LORAN-C, DNS (Doppler Navigation System), INS (Inertial Navigation System), the terminal area/airport ILS (Instrument Landing System), and the airport navigational equipment, including the approach lighting systems, slope indicators, surface detection equipment, and others. The forthcoming navigational procedures are to be primarily based on RNAV and/or RNP (Required Navigation Performance), enabling flying along the most preferred 4D (Four-Dimensional) gate-to-gate trajectories/routes expected to contribute to an increased flexibility of airspace usage, and consequently a higher efficiency and effectiveness of flights while maintaining the required level of safety (Janić, 2000).
c) Surveillance is enabled by the various types of primary and secondary (beacon) radars. They enable the ATC controllers to monitor separation between aircraft, and consequently make the appropriate decisions. In particular, Mode S secondary radars possessing functionality as DAPs (Downlink of Aircraft Parameter(s)) enable the direct (automatic) exchange of information between the ATC system and FMS (Flight Management System) via A/G (Air/Ground) communication data link, which excludes verbal R/T (Radio/Telephone) communication. Also, the ATC controller's alerting/informing tools are: STCA (Short-Term Conflict Alert), which checks possible conflicting trajectories and gives an alert prior to the loss of minimum separation between the aircraft/flights, MSAW (Minimum Safe Altitude Warning), which gives an alert if an aircraft/flight is in danger of colliding with the terrain, and the APW (Area Penetration Warning) system, which informs that an aircraft/flight will penetrate a restricted area. Last, but not least, the aircraft pilots can carry out the self-monitoring of their mutual separation by means of ADS-B (Automatic Dependent Surveillance) in combination with the CDTI (Cockpit Display Traffic Information) system, which may additionally relieve the workload of ATC controller(s) (http://www.sesarju.eu/; http://www.faa.gov/nextgen/).

d) The ATM (Air Traffic Management) enables the ATC control to perform tasks/functions at two-levels by: (a) providing safe separation between aircraft/flights according to the prescribed rules; and (b) enabling safe, efficient, and effective movement of air traffic flows in the given airspace under given conditions. The ATM system's sub-functions are: AM (Airspace Management), ATS (Air Traffic Services), and ATFM (Air Traffic Flow Management). AM focuses on the availability of airspace in order to provide optimal aircraft/flight trajectories. The ATS as a core component of the ATC control function continuously provides conditions for safe, efficient, and effective movement of air traffic flows. The ATFM balances the air traffic flows as the demand for service and the available ATC system's capacity, handling this demand at both the tactical and strategic level, aiming at preventing the rather long ground/airport and air/airspace aircraft/flight delays, and eventual compromising of the required level of safety (EEC, 2007; 2008; Janić, 2000).

e) Tools supporting the above-mentioned ATC system's operations are: CARD (Conflict and Risk Display), which helps the ATC controller(s) in detecting the potential conflicts within a given time frame, SYSCO (System Supported Coordination), enabling the ATC controller(s) screen-to-screen or electronic coordination, aiming at coordinating and transferring flights between the sectors, DMAN (Departure Manager) and AMAN (Arrival Manager), contributing to the optimization of airport runway system departure and arrival capacity usage, respectively; FDP (Flight Data Processing) systems (one per ATC center) processing all information related to flights scheduled in the given time horizon (from gate-to-gate (airport departure/arrival gates)) and distributing such processed information to all stakeholders involved, such as the ATC controllers, neighboring ATC centers, airports, etc. In addition, the system of electronic flight strips replacing the paper strips reduces the need for manual functions and consequently the related ATC controller(s) workload (http://www.sesarju.eu/; http://www.faa.gov/nextgen/).

4.2.3 Staff/Employees

The ATC staff/employees include the ATC controllers as well as the aircraft pilots. The former are responsible for monitoring, controlling, and managing the aircraft/flights in the airspace of their jurisdiction. At the level of ATC sectors, they are usually organized as teams of two or three persons with the precise division of control tasks and responsibilities. The aircraft pilots participate in the ATC processes by following, in addition to their own, instructions and guidance obtained by the ATC controllers. These mainly relate to maintaining their assigned trajectories and separation from the other aircraft/flights.

4.3 Demand and Capacity

4.3.1 Demand

The demand at a given ATC unit managing/controlling the air traffic in the airspace/sector under its jurisdiction is usually expressed by the number of aircraft/flights requested to be handled during a given period of time, which can be an hour, day or year. For the operational purposes and the short-term prediction of the ATC controller's workload, the hourly number of aircraft/flights is usually taken as relevant. For the long-term strategy of handling demand, the annual number of aircraft/flights to be handled is usually relevant. This latter number is also important for planning the long-term development of the ATC capacity. For such purpose, the forecasting of long-term demand in terms of the annual number of aircraft/flights expected to use the given airspace has to be carried out. In most cases, this number should include all categories of flights, such as those by the commercial air carriers, air taxi/commuters, general aviation, and military (FAA, 2005).

4.3.2 Capacity

The ATC airspace capacity is usually expressed by the maximum number of aircraft/flights allowed to enter the given sector during a given period under given conditions. Similar to the case of demand, the period can be an hour, day, month, or year. In particular, in the European ATC system, the declared hourly capacity is expressed by the number of aircraft/flight entries assessed by the ASNP (Air Navigation Service Provider) and declared to the CFMU (Central Flow Management Unit), aiming at protecting the given sector from overloading (EEC, 2003). The given conditions can be specified regarding: (i) configuration of the aircraft/flight trajectories; (ii) facilities and equipment used by the ATC unit and the aircraft pilots, and (iii) the intensity, volume, structure, time and space distribution, and continuity of the aircraft/flight demand during the specified period of time. The airspace capacity can be estimated for the terminal area, the individual air route(s)/corridor(s), and/or the network of air routes, the latter two cases established in the wider (LAA and/or HAA) airspace. The ATC capacity based on the ATC controller's workload is expressed by the number of aircraft/flights that can be simultaneously controlled while in the sector, within the predefined workload threshold during a given period, as given in Table 4.1.

In general, the required number, duration, and order of execution of the particular control tasks in handling the aircraft/flights in the sector inherently influence the above-mentioned ATC controller's workload and the corresponding capacity (Janić, 1997; 2000; Majumdar and Plack, 2001; Welch, 2015).

Table 4.1 Example of the ATC controller's workload thresholds (EEC, 2003).

Workload threshold (%)	**Explanation**	**Working time** (min/h)
0–17	Very light load	0–10
18–29	Light load	11–17
30–53	Medium load	18–31
54–69	Heavy load	32–41
≥ 70	Overload	≥ 42

4.3.3 Modelling Demand and Capacity

4.3.3.1 Demand

As mentioned above, the ATC demand is expressed by the number of flights requested to be handled in the given airspace during a given period of time under predefined conditions. In most cases, modelling of this demand implies analysing the past and current developments over time using trend-based method(s) and then extrapolating them over the specified future period of time, as shown in Figure 4.2(a, b).

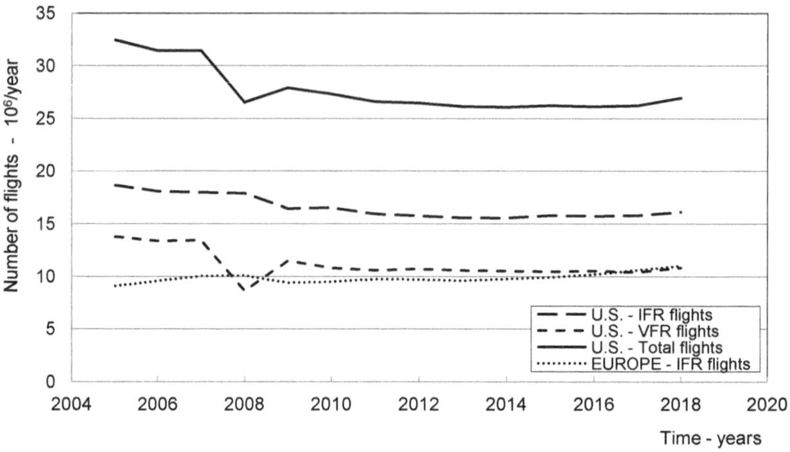

a) The ATC demand development over time

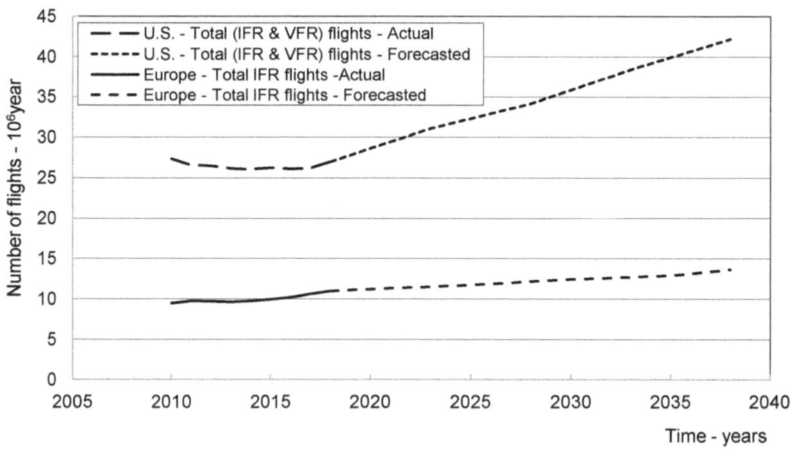

b) The ATC demand development and forecast over time

Figure 4.2 Example of development and forecasting of the ATC demand—Case of the annual number of IFR and VFR flights handled by U.S. ATO (Air Traffic Organization) and European EUROCONTROL (EEC, 2018; FAA, 2018; 2019).

Extrapolation of these trends is based on the average annual growth rates of this demand, inherently depending on the main driving forces of the air transport demand, such as population, GDP (Gross Domestic Product), PCI (Per Capita Income), globalization of trade and tourism, the generalized air travel costs/airfares, and, in some specific cases, competition from the ground transport modes (for example, HSR (High Speed Rail) in Europe, Japan, and China on the short- and medium-haul routes) (Albalate et al., 2014; Janić, 1993).

Figure 4.2a shows the variation of the total ATC demand over time in terms of the annual number of flights handled by the U.S. ATC system caused by the variations in the numbers of both IFR and VFR flights. While during more than half of the period these both decreased, during the remaining part, they gradually recovered, but did not reach the level(s) from the beginning of the period. In the European ATC system, the annual number of handled flights generally gradually increased after the drop in the year 2009 caused by the overall economic crisis in the year 2008. In addition, the number of IFR flights handled by the U.S. ATC system was about 50% higher than that in Europe, the latter being comparable to their U.S. VFR counterparts. Figure 4.2b shows the annual ATC flight demand expected to be handled by the ATC system in both regions during the future period 2018–2038. As can be seen, the almost linear and rather stable growth has been forecasted implicitly suggesting the stability of the main above-mentioned general air transport demand-driving forces. In particular, at the end of the period, the U.S. ATC flight demand is expected to be about three times higher than its European counterpart (EEC, 2018; FAA, 2018; 2019).

4.3.3.2 Capacity of TMA (Terminal Manoeuvring Area)

a) System

The TMA (Terminal Manoeuvring Area) established above and around an airport zone ensures the safe, efficient, and effective handling of arriving and departing aircraft flows under given conditions. In either analytical or simulation modelling of its capacity, the terminal area is usually represented as a prism with a polygon-shaped base and corners as the entry/exit gates. In such case, these entry/exit gates are the connecting/transfer locations between the terminal area and LAA (Low Altitude Airspace). They are usually defined by the navigational aids, which enable the arriving aircraft to perform, if necessary, the holding pattern before entering the terminal area. After entering the terminal area, these aircraft follow the assigned arrival trajectories to the landing runway. These trajectories can be defined by: (i) the navigational aids located inside the terminal; (ii) two- (2D), three- (3D) or four- (4D) aerial navigation system; (iii) the ATC controller's radar vectoring, including the aircraft headings, heights, and speeds. In any case, when all aircraft have to follow such trajectories, they are known as STARs (Standard Terminal Arrival Route (s)). Similarly, the departure routes can be defined. If all aircraft are required to follow them, they are known as SID (Standard Instrument Departure) routes or DPs departure procedures (DPs). Figure 4.3 shows the simplified scheme of the set of STARs in TMA.

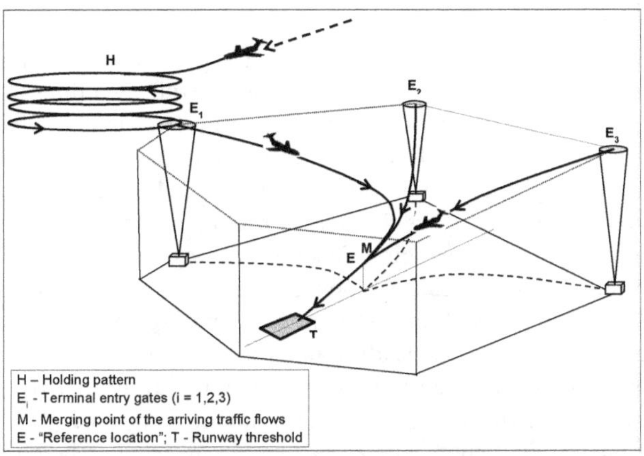

Figure 4.3 Scheme of the aircraft arriving trajectories in TMA (Terminal Manoeuvring Area) (Janić, 2000).

b) Assumptions

Development of the models of the terminal area capacity has been usually based on the range of assumptions. Some of them introduced in development of the analytical model of the terminal area capacity are as follows (Janić, 2000):

- The ATC controller applies the radar vectoring of the aircraft arriving in a given TMA;
- The arriving aircraft trajectories consist of the three straight-line segments converging through TMA to a common point located vertically above the extended runway centerline. This implies that each trajectory has a common last segment, which is actually the final approach path (see Figure 4.3);
- The arriving aircraft generally decrease their speeds while approaching the landing runway. The speeds are changed along the arriving trajectories each time the aircraft leaves one and passes to another segment of the trajectory, i.e., when it changes its heading;
- The horizontal separation rules are exclusively applied between the arriving aircraft;
- The radar coverage ensures monitoring of the horizontal separation between the arriving aircraft;
- The arriving and departing aircraft are usually separated by the vertical separation rules;
- The model is based on the concept of "ultimate" or "saturation" capacity, implying that there is the constant demand for service by arriving traffic during a given period of time. Under such conditions, in order to estimate the "ultimate" capacity of TMA, it is necessary to estimate the average minimum time between the successive aircraft passing through the "reference location";
- The "reference location" is adopted to be the runway entry gate (point E on Figure 4.3); and
- The average minimum time as an average of the minimum inter-arrival times of the different combinations of arriving aircraft on different trajectories is based on the fact that the minimum ATC separation rules must not be violated at any time, i.e., neither before nor after the aircraft passing through the runway entry gate.

c) Structure of the model

The minimum inter-arrival time at the runway entry gate is estimated by considering different combinations of sequences of the arriving aircraft regarding their trajectories and speeds. Each sequence contains a pair of aircraft. The time between the first-in-sequence (leading) and the second-in-sequence (trailing) aircraft is denoted by $(t_{kl/ij})$ where: (k) is the leading aircraft trajectory, (l) is the trailing aircraft trajectory, (i) is the type of leading aircraft, and (j) is the type of trailing aircraft. The three straight-line segments are used to approximate their trajectories as shown in Figure 4.4.

When each terminal entry gate is connected to the runway entry gate with only one trajectory (i.e., when the STARs exist), the symbols (E_k) and (E_l) can represent the terminal entry gates from which the leading and trailing aircraft appear, respectively.

The probability of occurrence of the sequence of aircraft types (ij) at the entry gate when the leading aircraft (i) arrives from the terminal entry gate (k) and the trailing aircraft (j) from the terminal entry gate (l) is denoted by $(p_{kl/ij})$. The appearance of the leading aircraft (i) at the entry gate (k) and the trailing aircraft at the entry gate (l) are

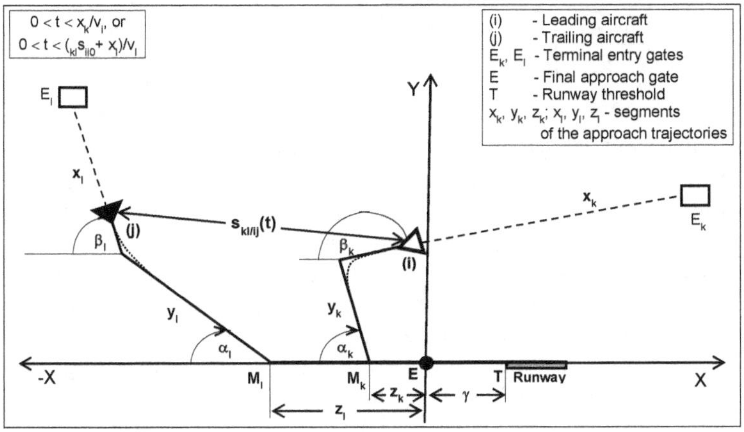

Figure 4.4 A simplified scheme of the arriving trajectories *(k)* and *(l)* in the terminal area with the horizontal distance $s_{klij}(t)$ between a pair of aircraft *(i)* and *(j)* (Janić, 2000).

independent events. In addition, the mix of arriving aircraft types can differ at particular terminal entry gates. For example, at some entry gates more short- and/or medium-haul and at some others more long-haul aircraft can enter TMA. The mix of arriving aircraft can also change over time, i.e., during different time periods of the day. By taking all of the above-mentioned into account, the probability of occurrence of the aircraft sequence *(ij)* arriving from the entry gates *(k)* and *(l)*, respectively, is equal to:

$$p_{kl/ij} = p_k \cdot p_{i/k} \cdot p_l \cdot p_{j/l} \tag{4.1}$$

where

- p_k is the proportion of arriving traffic entering TMA at the entry gate *(k)*;
- p_l is the proportion of arriving traffic entering TMA at the entry gate *(l);*
- $p_{i/k}$ is the probability that the leading aircraft of type *(i)* enters TMA at the entry gate *(k)*; and
- $p_{j/l}$ is the probability that the trailing aircraft of type *(j)* enters TMA at the entry gate *(l)*.

From Eq. 4.1, the average inter-arrival time at the runway entry gate *(E)* of different combinations of the arriving trajectories and the aircraft types is equal to:

$$\bar{t}_a = \sum_{ijkl} p_{kl/ij} \cdot t_{kl/ij} \tag{4.2}$$

From Eq. 4.2, the corresponding arriving capacity is equal as follows:

$$\mu_a = 1/\bar{t}_a \tag{4.3}$$

The distribution of arriving traffic between the terminal entry gates and the mix of aircraft types at particular gates, as the inputs for the model, are constant during a given period of time. In order to estimate the capacity at the "counting location", the aircraft inter-arrival times $(t_{kl/ij})$ must first be minimized for all *(i)*, *(j)*, *(k)*, *(l)* with respect to the applied ATC separation rules, and then used to compute the values $(\bar{t}_a)$ and (μ_a).

For each pair of the aircraft *(i)* and *(j)* entering TMA through the entry gates *(k)* and *(l)*, respectively, the inter-arrival times at the runway entry gate $(t_{kl/ij})$ in Eq. 4.2 are estimated as follows:

$$t_{kl/ij} = \left(\frac{s_{0/kl/ij} + x_l}{v_j} + \frac{y_l + z_l}{u_j} \right) - \left(\frac{x_k}{v_i} + \frac{y_k + z_k}{u_{ji}} \right) \tag{4.4}$$

where

$s_{0/kl/ij}$ is the "initial separation" between the aircraft *(i)* and *(j)*, i.e., the distance between the aircraft *(j)* and the terminal entry gate (E_l) measured along the trajectory *(l)* of the aircraft *(j)* at the time: $t = 0$, when the aircraft *(i)* enters the terminal at the entry gate (E_k); this "initial separation" can be negative when the aircraft *(j)* enters the terminal area before the aircraft (i), and vice versa;

x_k, y_k, z_k is the length of three segments of the trajectory *(k)* spreading between the terminal entry gate (E_k) and the runway entry gate *(E)*;

x_l, y_l, z_l is the length of three segments of the trajectory *(l)* spreading between the terminal entry gate (E_l) and the runway entry gate *(E)*;

v_i, u_i is the speed of aircraft type *(i)* along the segments (x_k, y_k) and (z_k) respectively; and

v_j, u_j is the speed of aircraft type *(j)* along the segments $({}_{kl}s_{ij0}, x_l, y_l)$ and (z_l), respectively.

Equation 4.4 indicates that the inter-arrival time $(t_{kl/ij})$ for a given aircraft pair is an increasing function of the initial separation $(s_{0/kl/ij})$. In order to maximise the aircraft flow through the runway entry gate, i.e., achieve the TMA capacity, the inter-arrival time $(t_{kl/ij})$ for each pair of aircraft *(ij)* entering the terminal at the entry gates (E_k) and (E_l), respectively, must be minimised, while all the time preventing violation of the ATC minimum separation rules. This can be written as follows:

$$\min s_{0/kl/ij} \tag{4.5a}$$

subject to

$$s_{kl/ij} \geq \delta_{ij} \quad for\ \forall t \in \left(0; \frac{x_k}{v_i} + \frac{y_k + z_k}{u_i} + \frac{\gamma}{w_i} \right) \tag{4.5b}$$

where

$s_{kl/ij}(t)$ is the horizontal separation between the aircraft *(i)* and *(j)* at some time *(t)* after entered the terminal at the gates (E_k) and (E_l), respectively;

δ_{ij} is the minimum horizontal separation rule applied between the aircraft type *(i)* and type *(j)*;

γ is the length of the final approach path connecting the runway entry gate *(E)* and the runway threshold *(T)* common for all arriving aircraft;

w_i is the average speed of aircraft type *(i)* along the final approach path.

The minimum separation $(s_{kl/ij})$ in Eq. 4.5 can be achieved when the "initial separation" $(s_{0/kl/ij})$ guarantees that the separation between two aircraft is: $s_{kl/ij}(t) = \delta_{ij}$ at the moment (t^*) when the two aircraft approach closest to each other. Consequently, the horizontal distance between two aircraft *(i)* and *(j)* entering the terminal area at the entry gates (E_k) and (E_l) and following the trajectories *(k)* and *(l)*, respectively, can be presented by the complex trigonometric function whose form changes for the different possible positions of two aircraft on three segments of their trajectories. This function can take thirteen different forms, depending on the mutual positions of the aircraft *(i)* and *(j)* on the particular segments of the trajectories *(k)* and *(l)*. One of these forms, corresponding to the case shown in Figure 4.4, is as follows (Janić, 2000):

$$s_{kl/ij}(t) = \sqrt{At^2 + 2Bt + C} \tag{4.6}$$

where

$$A = v_i^2 - 2v_i v_j \cos(\beta_l - \beta_k) + v_j^2$$

$$B = [x_k - s_{0/kl/ij}(v_i v_j)] \cdot [v_j \cos(\beta_l - \beta_k) - v_i] - x_l \cdot [v_j - v_i \cos(\beta_l - \beta_k)] - y_l \cdot [v_j \cos(\beta_l - \alpha_l) - \\ - v_i \cos(\beta_k - \alpha_l)] + y_k \cdot [v_j \cos(\beta_l - \alpha_k) - v_i \cos(\beta_k - \alpha_k)] - (z_l - z_k) \cdot (v_j \cos\beta_l - v_i \cos\beta_k)$$

$$C = [x_k - s_{0/kl/ij}(v_i/v_j)]^2 - 2 \cdot [x_k - s_{0/kl/ij}(v_i/v_j)] \cdot [x_l \cos(\beta_l - \beta_k) + y_l \cos(\beta_k - \alpha_l) - \\ - y_k \cos(\beta_k - \alpha_k) + (z_l - z_k) \cos\beta_k] + x_l^2 + 2 \cdot x_l \cdot [y_l \cos(\beta_l - \alpha_l) + (z_l - z_k) \cos\beta_l - y_k \cos(\beta_l \\ - \alpha_k)] + y_l^2 + y_k^2 - 2 \cdot y_k y_l \cos(\alpha_l - \alpha_k) + (z_l - z_k)^2 - 2 \cdot (z_l - z_k) \cdot (y_k \cos\alpha_k - y_l \cos\alpha_l)$$

For the chosen trajectories and the mix of aircraft types, all variables in Eq. 4.6, except $(s_{0/kl/ij})$ and *(t)*, can be considered as parameters, thus enabling the minimum horizontal distance between the aircraft *(i)* and *(j)*, $min\ s_{kl/ij} = s_{kl/ij}(t^*)$ occurring in moment (t^*) to be found. In addition, for: $s_{kl/ij}(t^*) = \delta_{ij}$ the corresponding $(s_{0/kl/ij})$ and $(t_{0/kl/ij})$ can also be computed. Then, given the inter-arrival times $(t_{0/kl/ij})$ and the probabilities (p_k), (p_l), $(p_{i/k})$ and $(p_{j/l})$, the average inter-arrival time $(\bar{t}_a)$ and the corresponding terminal area capacity (μ_a) can be computed using Eq. 4.2 and Eq. 4.3, respectively. Under these conditions, the set of "initial separations" $(s_{0/kl/ij})$ represents the necessary initial separation conditions for achieving the arriving traffic flow at the TMA capacity level.

d) <u>Application of the model</u>

The above-mentioned model is applied by using the inputs from the case of terminal area established around Belgrade Airport (Serbia). The arriving aircraft, categorized in three categories regarding the wake-vortex characteristics as Light, Medium, Large, and Heavy, are assumed to enter the terminal area at three entry gates. The ATC applies exclusively horizontal separation rules between these aircraft while in the terminal area and during the final approach and landing, depending on the mentioned wake-vortex categories (ICAO, 2016). The results are shown in Figure 4.5(a, b).

The results indicate some intuitive conclusions. Figure 4.5a shows that when the most intensive traffic arrives from the terminal entry gates which are at the opposite side from the landing runway, so that the trajectories that are less direct and which have a bigger curvature are more often in use, the capacity will decrease. In the given example, the approach trajectory from the entry gate No. 1 has the biggest curvature. Figure 4.5b shows that the TMA capacity first decreases and then increases with an increase in the proportion of heavy aircraft in the mix. The former is caused by applying the larger separation rules between these and other aircraft. The latter happens thanks to an increase in the average speed of aircraft flows thanks to the increased presence of heavy aircraft (with higher speeds), despite a simultaneous increase of the corresponding ATC separation rules (Janić, 2000).

4.3.3.3 Capacity of air route

a) <u>System</u>

The long straight air route defined by the navigational aids or RNAV (Area Navigation) system with several available flight levels represents the system for which the capacity is estimated under given conditions. The simplified scheme of an air route such as this is shown in Figure 4.6.

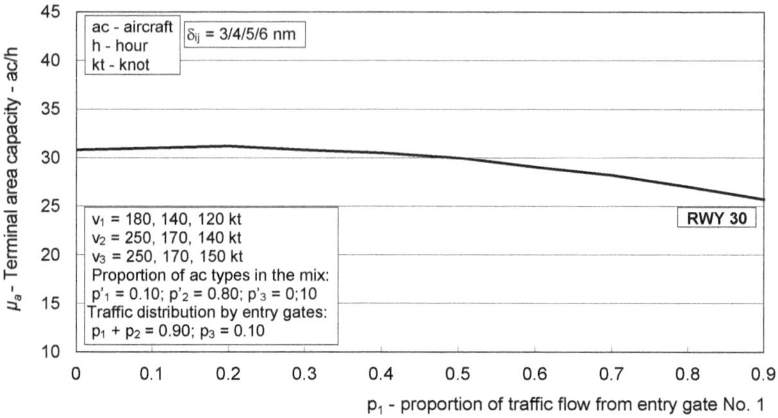

a) Capacity vs distribution of arriving traffic between the TMA entry gates

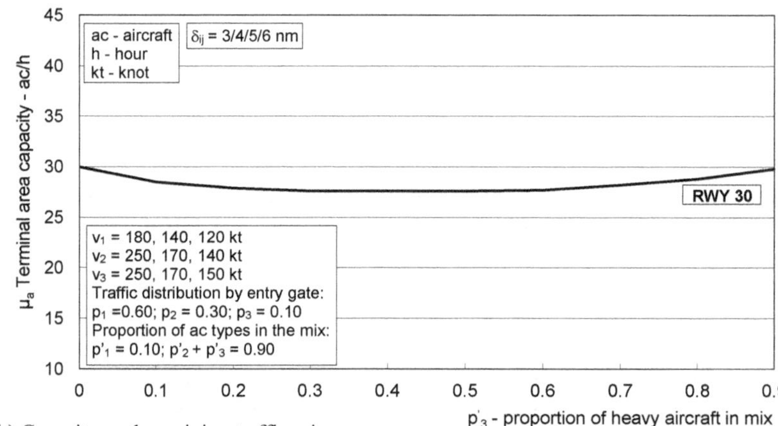

b) Capacity vs the arriving traffic mix

Figure 4.5 Relationships between the TMA capacity and the pattern and characteristics of arriving traffic (Janić, 2000).

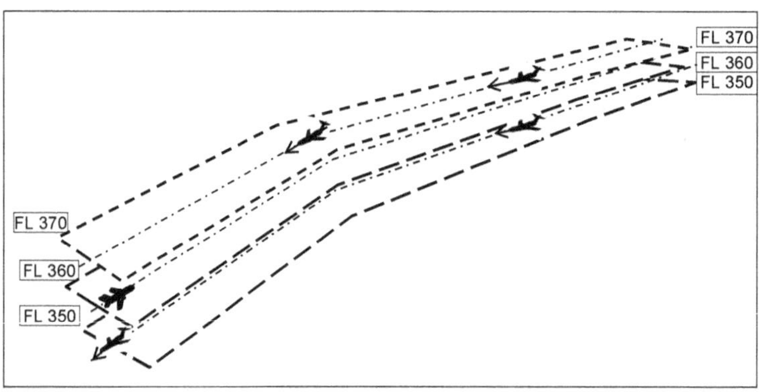

Figure 4.6 Scheme of an air route for estimating its capacity—Case of HAA (High Altitude Airspace).

The aircraft either fly along the above-mentioned air route between any two air-route intersections or pass through an intersection of different air routes while maintaining constant altitudes and cruising speeds. Nevertheless, since small differences between the speeds of aircraft flying in the same direction on the same altitude may exist, the potential overtaking conflicts can happen including the overtaking between the aircraft flying along the same route, the latter in the very rare cases.

The ATC controller resolves these conflicts by instructing the corresponding aircraft to adjust their speeds or change flight level. In the latter case, this implies some kind of "sifting" of the original aircraft flows requesting to use the same air route and fly at the same flight level(s). The "sifting" and consequent homogenization of the original aircraft flow is completed after the "sifted" aircraft arrive at the new assigned flight levels. Such procedure actually results in creating the rather homogenous aircraft flows at each flight level of a given air route regarding their cruising speeds. In addition, the aircraft expected to enter the overtaking conflicts while continuing along the route can be separated from the original flow by one of the following actions: (i) diverting to the other alternative air route(s), enabling them to fly in the same direction; (ii) assigning the other free flight level on the original route; (iii) performing the holding pattern some time before entering the air route and requested flight level. This can be carried out either on the ground before departure (ground holding procedure) or somewhere in the LAA or HAA airspace nearby the required air route. Using the RNAV system provides an increasing flexibility of using the available airspace and options for resolving both potential overtaking and crossing conflicts.

b) Assumptions

The analytical model for estimating the capacity of an air route where the air traffic takes place according to the specified "what-if" scenario is based on the following assumptions:

- Similar to the case of the TMA capacity model, the influence of the "human factor" (i.e., the ATC controller workload) on the air route capacity is not considered; it is implicitly assumed that the ATC controllers can manage the air traffic flows at the air route capacity level.
- If the ATC can establish the minimum separation rules between the aircraft representing the constant demand for entering the given air route during a given period of time, the corresponding output flow at the "saturation" or "ultimate" capacity level through the "reference location" will be realized. In this case, the "reference location" for counting the aircraft passing through can usually be an arbitrarily chosen point/location along the air route. When estimating the capacity of intersection of the air routes, the "reference location" is usually chosen to be the intersection itself (Janić, 2000). However, the above-mentioned "saturation" conditions relatively rarely happen, particularly along the straight-line air routes. This is because, as mentioned above, the ATC controller(s) primarily act to prevent the potential overtaking conflicts by homogenizing the air traffic sub-flows at the particular flight levels subject to the aircraft cruising speeds. Consequently, the aircraft with uniform speeds operate at the particular flight levels of the give air route.
- The ATC applies the minimum distance- or time-based horizontal separation rules between the aircraft flying on the same flight level(s).
- Each of the above flow "sifting" actions compromises the original intensity of the total and consequently the sub-flows at different flight levels of a given air route.

c) Structure of the model

In order to model the capacity of a given air route operating under the above-mentioned traffic scenario, it has been assumed that the "sifted" aircraft flow occurs at a given flight level according to the Poisson process. This implies that the aircraft inter-arrival times at the entrance of a given air route are the realizations of the stochastic variable, which can be represented by the shifted exponential density function as follows (Janic, 2000):

$$f(t) = \begin{cases} \lambda_{ar} e^{-\lambda_{ar}(t-\tau_{ar/min})}, & for\ t > 0 \\ 0, & otherwise \end{cases} \tag{4.7}$$

where

λ_{ar} is the constant intensity of the original (non "sifted") air traffic flow (ac/unit of time);

$\tau_{ar/min}$ is the ATC minimum time-based horizontal separation rules between the aircraft in "sifted" flow (time units/ac); and

t is time as the continuous variable (time units).

The average inter-arrival time of the aircraft forming the "sifted" flow at the "reference location" (i.e., also the entry point of a given air route) is determined as follows (Ang and Tang, 1975):

$$E(T_S) = \int_0^\infty t \cdot \lambda_{ar} \cdot e^{-\lambda ar(t - \tau ar/min)} dt = 1/\tau_{ar/min} \tag{4.8a}$$

From Eq. 8a, the intensity of "sifted" traffic flow, i.e., the actually achievable capacity of the flight level of a given air route on the single is equal to:

$$\mu_{ar} \equiv \lambda_{s/ar} = 1/E(T_s) = 1/(1/\lambda_{ar} + \tau_{ar/min}) = \lambda_{ar}/(1 + \lambda_{ar} \cdot \tau_{ar/min}) \tag{4.8b}$$

The minimum separation rule (τ_{min}) in Eq. 4.8 (a, b) can either be taken as the constant value (i.e., parameter) or realisation of the stochastic variable dependent on the skills of the ATC controllers in order to establish the minimum separation between each pair of aircraft in the "sifted" aircraft flow on a given flight level. When the ATC distance-based separation rules are applied, the average value of (τ_{min}) can be estimated as follows (Siddiqee, 1973):

$$\tau_{ar/min} = \delta_{ar}/v_{ar/f} \tag{4.8c}$$

where

δ_{ar} is the ATC minimum longitudinal distance-based separation rules applied to the given route; and

$v_{ar/f}$ is the average speed of "sifted" air traffic flow on the distance δ_{ar}.

If the different "sifted" aircraft flows take place on (N) flight levels of the given route, the total air route capacity can be estimated as follows:

$$\Lambda_{ar} = \sum_{i=1}^{N} \lambda_{s/ar/i} \equiv M_{ar} = \sum_{i=1}^{N} \mu_{ar/i} \tag{4.8d}$$

where

$\lambda_{s/ar/i}$ is the constant intensity of the "sifted" aircraft flow on the flight level (i) of the given air route (ac/unit of time).

Some additional characteristics of aircraft flows can be obtained for a given air route and its particular flight levels indirectly related to their capacity. For example, the average traffic density on a flight level of the given route can be estimated as follows:

$$D_{ar/i} = (\lambda_{s/ar/i} \cdot \tau_{ar/i})/L_{ar/i} \tag{4.8e}$$

where

$L_{ar/i}$ is the length of air route on the flight level *(i)* (km; nm); and

$\tau_{ar/i}$ is the average aircraft flying time on the flight level *(i)* of the given route (time units) ($\tau_{ar/i} = L_{ar/i}/v_{ar/i}$).

The other symbols are analogous to those in the previous Eqs.

From Eq. 4.8e, the total traffic density on the given air route with *(N)* flight levels is equal to:

$$D_{ar} = \sum_{i=1}^{N} D_{ar/i} \tag{4.8f}$$

where all symbols are analogous to those in the previous Eqs.

d) Application of the model

A hypothetical example of the relationship between the intensity of "sifted" air traffic flow, i.e., the capacity of a flight level of the given air route, and the intensity of original air traffic flow is shown in Figure 4.7. Two types of the ATC minimum separation rules between the aircraft flying on the same flight level of the given route are considered: the past δ_{ar} = 10 nm and the present δ_{ar} = 5 nm (ICAO, 2016). The average speed of aircraft flow is assumed to be 560 kt, implying that this flow consists of the heavy commercial jets flying along the HAA (High Altitude Area) route (nm – nautical mile; kt – knot).

As can be seen, the capacity of the flight level will increase at a decreasing rate as the intensity of original flow increases and the ATC minimum separation rules decrease, as was intuitively expected.

As already mentioned, the particular intersections of air routes (sometimes defined as the "way points") may come more frequently into saturation. In order to model their capacity under such conditions, it is assumed that the arriving aircraft flows converge before and diverge after passing through the intersections at the same flight levels. All

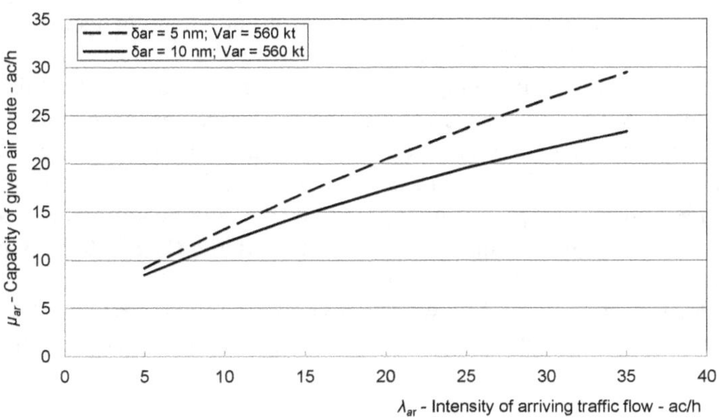

Figure 4.7 Relationship between the air route capacity and the arriving air traffic flow—Case of HAA (High Altitude Area) airspace.

the while, these aircraft are separated by appropriate ATC separation rules. The basic structure of the capacity model of an intersection of air routes was presented in the previous literature (Janić, 2000).

In addition, let us consider a route in HAA airspace of length: L = 3000 nm (1 nm = 1.852 km) where the aircraft fly at: N = 9 flight levels (FL 310–FL 390) (1 nm = 1.852 km; FL 310/390 = 31000/39000 ft; 1 ft = 0.305 m). The aircraft enter the route at each flight level longitudinally separated by the ATC minimum time-based separation rules of: $\tau_{ar/i/min}$ = 10 min/ac. This gives the maximum arrival rate on each flight level of the route of: $\lambda_{ar} \equiv \mu_{ar} = 60/\tau_{ar/i/min} = 60/10 = 6$ ac/h (ac – aircraft; h – hour). The average aircraft speed along the route on each flight level is constant and equal: $v_{ar/i}$ = 496 kt or 560 kt). Then, the average flying time of the aircraft along the route on each flight level is estimated to be: $\tau_{ar/i} = L_{ar/i}/v_{ar/i} = 3000/496 \approx 6.0$ h and $3000/560 \approx 5.4$ h (i = 1,2,.., N). In addition, the maximum number of aircraft simultaneously flying on any flight level of the given route is estimated as: $n_{ar/i} = \mu_{ar/i} \cdot \tau_{ar/i} = 6 \cdot 6.0 = 36$ ac/FL and $6.0 \cdot 5.4 = 32$ ac/FL. This implies that they fly separated by the average distance of: $3000/36 \approx 83$ nm and $3000/32 \approx 94$ nm, respectively. From Eq. 4.8e, the average aircraft density per flight level is estimated to be: $D_{ar/i} = 36/3 = 12$ ac/1000 nm and $32/3 \approx 11$ ac/1000 nm, respectively. From Eq. 4.8f, the air route density across all nine flight levels is equal to: $D_{ar} = N \cdot D_{ar/i} = 9 \cdot 36/3 = 324/3 = 108$ ac/1000 nm and $9 \cdot 32/3 = 288/3 = 96$ ac/1000 nm. In this case, the average distance between these aircraft in the horizontal plane across all flight levels is equal to: $3000/324 \approx 9$ nm and $3000/288 \approx 10$ nm, respectively.

4.3.3.4 Capacity of air route network

a) System

For the purpose of modelling capacity, an air route network, established in the airspace between two continents, i.e., over the ocean, is considered as the system. The airspace is assumed to be without ground-based navigational facilities and equipment, including radar coverage, thus preventing provision of the direct ATC monitoring and control of air traffic flows there. Under such conditions, the aircraft has to perform RNAV by using the traditional compass and/or the satellite navigation systems, such as GPS (Global Position System) (USDD, 2008). During that phase of flight, the aircraft follow the routes called the "tracks", defined by the series of WPs (Way Point(s))[4] indicating locations along the tracks where the course, speed, and/or altitude can change. In most cases, these routes/tracks coincide with the great-circles, i.e., the shortest distances between any two points on the globe, implying orthodrome-based navigation. In addition, before entering and after leaving these routes/tracks, the aircraft can also use the ground-based navigational facilities and equipment such as VOR and DME on the range of about 240 Nm (440 km). Also, the radar coverage, and consequently the radar monitoring, can also be provided during the phases of flights over the continental parts of the flights. Figure 4.8(a, b) shows an example of such traffic patterns (Dief, 2018; ICAO, 2017; NATS, 2015; https://www.flightradar24.com/).

The managerial/control step follows by delivering an initial info to the airline aircraft about the "optimal" route/track that can be followed under the expected conditions based

[4] In the given case, WP (Way Point) is a predetermined geographical position defined by the latitude/longitude coordinates (the altitude is not included) (ICAO, 2017).

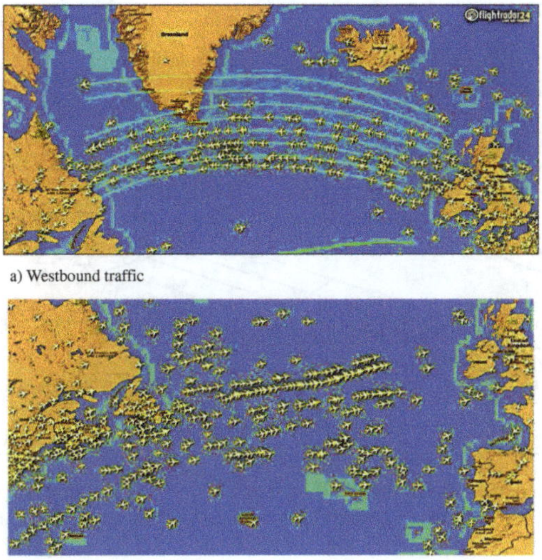

a) Westbound traffic

b) Eastbound traffic

Figure 4.8 En-route airspace—Case of Transatlantic air traffic pattern (https://www.flightradar24.com/).

on the flight destination, type, take-off weight, prevailing weather/winds, and the ATC route charges. Before entering the oceanic airspace, the aircraft contacts the ATC Oceanic Center requesting the already given track, including the estimated time of arrival at the entry gate of the track. This enables the ATC Oceanic controllers to estimate the required separation between the aircraft and consequently issue clearances to the pilots. The assigned route/track can coincide or be different from the initial one, but the aircraft has to follow this latest anyway. After entering the oceanic airspace, the aircraft are obliged to report their position when crossing the WPs along their routes/tracks, which includes predicting the time of crossing the next and the successive WPs ahead. These enable the ATC Oceanic controllers to maintain a safe separation between aircraft (ICAO, 2017; NATS, 2015).

For reporting their positions, the aircraft use the satellite communication CPDLC, HF (High Frequency) link and/or alternatively ADS-C & ADS-B systems. In the latter case, the controller-pilot-controller voice communication is replaced by the automatic downlink transfer of the position reports and the other flight information if necessary (ICAO, 2017). Figure 4.9 shows the simplified scheme of the above-mentioned air route network connecting Europe and America in Transatlantic airspace used as the case for modelling and estimating capacity.

As can be seen, the network consists of the set of air routes/tracks spreading between two continents. The air traffic in the network is managed and controlled by OATCC (Oceanic Air Traffic Control Centers). The managerial step is to set up the network, i.e., the set of routes/tracks, for the next day, i.e., 24 h in advance. This is carried out based on the expected weather conditions (to reduce the impacts of headwinds where possible and to use benefits from tailwinds) and the airline preferences regarding the routes submitted in advance. As a result, 6–7 routes/tracks are commonly designed each day, which always start and finish at the same 5 or 6 oceanic entry points at both sides of the Atlantic. The tracks for the following day are published at 22:00 h. The same and or different sets of routes/tracks can be used for the flights in either westbound or eastbound direction.

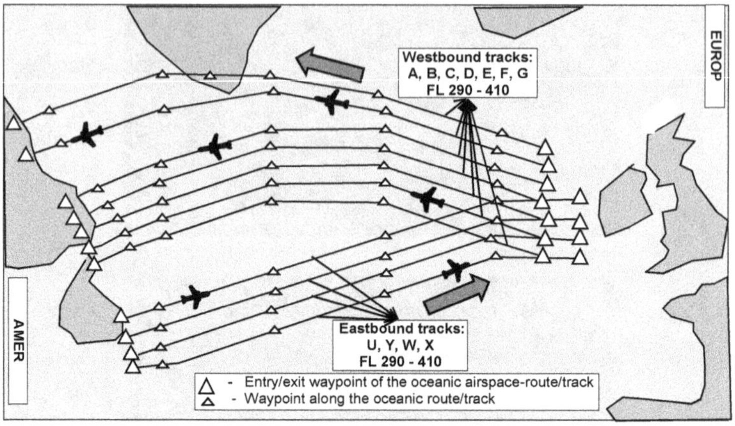

Figure 4.9 Simplified scheme of Transatlantic airspace and corresponding air route network (ICAO, 2017; NATS, 2015; http://d-maps.com/carte.php?num_car=3211&lang=en/).

In addition, regarding daily scheduling, the westbound flights in Figure 4.9 typically depart from Europe during the daylight between early morning and late afternoon (11:30 h–19:30 h UTC (Coordinated Universal Time) or GMT (Greenwich Mean Time) in order to arrive in North America between early afternoon and late evening, i.e., during daylight. The opposite eastbound flights are scheduled to depart from North America to Europe in the evenings (01:00 h UTC to 08:00 h UTC on the North American side), thus enabling passengers to arrive at their destination in the mornings.

b) Assumptions

Modelling capacity of a given air route network operating according to the specified "what-if" scenarios is based on the following assumptions:

- The air route network consists of the set of long-haul routes connecting the given set of origin and destination airports of the aircraft/flight flows;
- The network has a fixed configuration in terms of the number of long-haul routes/tracks and the available flight levels on each of them during a given period of time;
- Each route consists of three successive segments: the first between the origin airport and the entry of long-haul route/track, the second as the long-haul route/track where the longest portion of the cruising phase of flights is performed, and the last between the exit of long-haul route/track and the destination airport.
- The long-haul routes/tracks where the aircraft fly in the same direction are approximately parallel with each other;
- The aircraft remain all the time at the assigned flight levels of the given long-haul route(s)/track(s);
- The aircraft in the flow(s) on the same flight level(s) maintain approximately the same speed, thus eliminating the overtaking conflicts and need for their resolving by changing the flight level/altitude; this can however be carried out if the aircraft have not immediately reached the "optimal" flight level and if there is a free (requested) flight level;

- Each flight level of particular long-haul routes/tracks is considered as the "service channel" with the constant service rate/capacity of handling the aircraft flows;
- The aircraft flows arrive to the "service channel(s)" according to the Poisson processes of the constant intensity during a given period of time;
- The intensity of arriving aircraft flows does not exceed the capacity of the particular "service channels", in which case, the average queue lengths and related delays remain finite;
- Despite the intensity of aircraft/customer flows being lower than the capacity of the particular "service channel(s)", delays in entering these "channels" can occur due to the stochasticity of individual arrivals and deviations from the ATC minimum time-based separation rules; these delays are supposed to be realized at the origin airports before departures in the scope of the "ground holding" procedures or in the air before entering the routes/track(s);
- The fixed routing procedure based on the assignment of the network's total capacity to the total arriving flows in proportion to the "capacity of particular channel(s)" is applied during the given period of time; this enables almost minimization of the aircraft flows' total time of passing through the network; and
- The "capacity of particular channel(s)" relevant for assignment of air traffic flow demand is proportional to the sum of the reciprocal of the average flying time along its first and the third segment, and the service rate/capacity of the second segment.

c) Structure of the model

The capacity model of the given long-haul air route network of the above-mentioned spatial configuration is based on an analogy with the queuing networks (Kleinrock, 1975; 1976). This also implies that the concepts and definitions of its "ultimate' and "practical" capacity are applicable in the given context similarly as in the case of airport runway system, TMA, and air route. Consequently, by respecting the above-mentioned assumptions, the capacity models are summarized as follows:

i) "Ultimate" capacity

Let the given air route network be characterized by $(N(\Delta t))$ routes/tracks and $(Mi(\Delta t))$ flight levels per single track *(i)* during some time (Δt) for which the capacity is estimated. Under conditions of the constant demand for service, the "ultimate capacity" of a flight level *(j)* of a given route/track (i) of the network can be estimated as follows:

$$\mu_{ij}(\Delta t) = 1/\tau_{ij/min}(\Delta t) \tag{4.9a}$$

where

$\tau_{ij/min}(\Delta t)$ is the ATC minimum time-based separation rule applied during time (Δt) between aircraft flying on the flight level *(j)* of the route/track *(i)* (min).

Under rather hypothetical conditions implying the continuous demand of aircraft "perfectly packed" to enter each route/track and flight level according to the ATC minimum separation rules, the "ultimate" capacity of the network can be estimated from Eq. 4.9a as follows:

$$\mu(\Delta t) = \sum_{i=1}^{N(\Delta t)} \sum_{j=1}^{M_i(\Delta t)} \mu_{ij}(\Delta t) \tag{4.9b}$$

where all symbols are analogous to those in the previous Eqs.

From Eq. 4.9b, the number of aircraft simultaneously in the network under given conditions can be estimated as follows:

$$n(\Delta t) = \sum_{i=1}^{N(\Delta t)} \sum_{j=1}^{M_i(\Delta t)} \mu_{ij}(\Delta t) \cdot \tau_{ij}(\Delta t) \tag{4.9c}$$

where

$\tau_{ij}(\Delta t)$ is the average aircraft flying time on the flight level *(j)* of the route/track *(i)* during time *(Δt)* (h).

The other symbols are analogous to those in the previous Eqs.

ii) "Practical" capacity

The "practical" capacity of a given air route network can be expressed by the maximum number of aircraft handled during a given period of time under conditions of imposing the average maximum delay on each of them before entering the network. Consequently, if the network is considered as the steady state multi-channel queuing system of type *M/G/1* (i.e., the entire network is assumed to operate as the single service channel system), the average delay of an aircraft before entering the network can be estimated as follows (Kleinrock, 1975; 1976):

$$\bar{W}(\Delta t) = \frac{\lambda(\Delta t) \cdot \{1/[\mu(\Delta t)]^2 + \sigma_s^2\}}{2 \cdot [1 - \lambda(\Delta t)/\mu(\Delta t)]} \tag{4.10a}$$

where

$\lambda(\Delta t)$ is the intensity of aircraft flows requesting entering the network during time *(Δt)* (ac/unit of time).

The other symbols are analogous to those in the previous Eqs.

After specifying the maximum average delay as: $\bar{W}(\Delta t) \equiv W^*(\Delta t)$, the maximum intensity of aircraft flows representing the network "practical" capacity under given conditions can be estimated from Eq. 4.10a as follows:

$$\lambda(\Delta t) \equiv \mu^*(\Delta t) = \frac{2 \cdot W^*(\Delta t) \cdot [\mu(\Delta t)]^2}{1 + 2 \cdot W^*(\Delta t) \cdot \mu(\Delta t) + [\mu(\Delta t)]^2 \cdot \sigma_s^2} \tag{4.10b}$$

where

$W^*(\Delta t)$ is the maximum average delay imposed on each aircraft before entering the network during time *(Δt)* (min/ac);

σ_S^2 is the standard deviation of the service time independent of the time *(Δt)* (min/ac).

The other symbols are analogous to those in the previous Eqs. In Eq. 4.10(a, b), the "ultimate" capacity of the network *(μ(Δt))* can be estimated using Eq. 4.9b. A similar model can be used to estimate the "practical" capacity of a given route/track of a network with the flight levels as the "service channels" as well as the "practical" capacity of a given flight level operating as the "service channel" itself.

iii) Matching capacity to demand

Matching capacity to demand in the abovementioned air route network implies assigning the aircraft flows entering the network in proportion to the "ultimate" and "practical" capacity of its particular "service channels", i.e., routes/tracks and their particular flight levels. This is justifiable in cases when the capacities of particular "service channels" are different due to any reasons, in the given context, in most cases due to weather. Namely, flying time and length along some routes/tracks can be unexpectedly extended, some of their flight levels can be unavailable, or they could simply be closed. Under such conditions, let the intensity of aircraft flows demanding to enter and pass through the given air route network during the time period *(Δt)* be equal to:

$$\gamma(\Delta t) = \sum_{k=1}^{K} \sum_{l=1}^{L} \gamma_{kl}(\Delta t) \tag{4.11a}$$

where

K, L is the number of origin and destination airports of the aircraft flows; and

$\gamma_{k,l}(\Delta t)$ is the intensity of aircraft flows between airports *(k)* and *(l)* during the time *(Δt)* (ac/unit of time).

The intensity of aircraft flows assigned to the "service channel" *(ij)* of the network during time *(Δt)* according to the fixed routing procedure can be estimated as follows (Kleinrock, 1975; 1976):

$$\lambda_{ij}(\Delta t) = \left[\frac{1/\tau_{ki} + \mu_{ij}(\Delta t) + 1/\tau_{ij}(\Delta t) + 1/\tau_{jl}}{\Sigma_{k=1}^{K}\Sigma_{i=1}^{N} 1/\tau_{ki} + \Sigma_{i=1}^{N}\Sigma_{j=1}^{M_i}[\mu_{ij}(\Delta t) + 1/\tau_{ij}(\Delta t)] + \Sigma_{j=1}^{M_i}\Sigma_{l=1}^{L} 1/\tau_{jl}} \right] \cdot \gamma(\Delta t) \tag{4.11b}$$

where

N is the number of long-haul parallel routes/tracks in the network (-);

M_i is the number of flight levels on the long-haul route/track *(i)* (-);

τ_{ki}, τ_{jl} is the average aircraft flying time along the first *(ki)* and the third *(jl)* segment, respectively, of the route/track *(kl)* (units of time/ac);

$\tau_{ij}(\Delta t)$ is the average aircraft flying time through the "channel", i.e., the cruising segment *(ij)* of the route/track *(kl)* (units of time/ac); and

$\mu_{ij}(\Delta t)$ is the service rate, i.e., capacity, of the "channel", i.e., the cruising segment *(ij)* of the route/track *(kl)* (ac/unit of time).

The other symbols are analogous to those in Eq. 4.11a. The capacity $(\mu_{ij}(\Delta t))$ in Eq. 4.11b is estimated as follows:

$$\mu_{ij}(\Delta t) = 1/[\delta_{ij}/V_{ij}(\Delta t)] \text{ or } \mu_{ij}(\Delta t) = 1/\bar{t}_{min/ij}(\Delta t) \tag{4.11c}$$

where

$\delta_{ij}(\Delta t)$ is the minimum longitudinal separation between successive aircraft through the "channel", i.e., the cruising segment *(ij)* of the route/track *(kl)* during time *(Δt)* (nm/ac; km/ac);

$V_{ij}(\Delta t)$ is the average speed of aircraft flow(s) along the "channel", i.e., the cruising segment *(ij)* of the route/track *(kl)* during time *(Δt)* (kt; km/h); and

$\bar{t}_{min/ij}(\Delta t)$ is the minimum time separation between the successive aircraft entries and along the "channel", i.e., the cruising segment *(ij)* of the route/track *(kl)* during time *(Δt)* (time units/ac).

The average aircraft flying time $\tau_{ij}(\Delta t)$ in Eq. 4.11b is estimated as follows:

$$\tau_{ij}(\Delta t) = d_{ij}/V_{ij}(\Delta t) \tag{4.11d}$$

where

d_{ij} is the length of "channel", i.e., the cruising segment *(ij)* of the route/track *(kl)*.

The average speed $V_{ij}(\Delta t)$ of the aircraft flow(s) in Eq. 4.11d is estimated as follows:

$$V_{ij}(\Delta t) = v_{ij}(\Delta t) \pm w_{ij}(\Delta t) \tag{4.11e}$$

where

$v_{ij}(\Delta t)$ is the average indicated speed of the aircraft flow(s) along the "channel", i.e., the cruising segment *(ij)* of the route/track *(kl)* during time (Δt) (kt; km/h); and

$w_{ij}(\Delta t)$ is the average wind speed along the "channel", i.e., the cruising segment *(ij)* of the route/track *(kl)* during time (Δt) (kt; km/h) ("+" indicates the tail wind and "–" the head wind).

In addition, the average delay of an aircraft before entering the "channel", i.e., the cruising segment *(ij)* of the route/track *(kl)* can be estimated as follows:

$$\bar{W}_{ij}(\Delta t) = \frac{\lambda_{ij}(\Delta t) \cdot \{1/[\mu_{ij}(\Delta t)]^2 + \sigma^2_{s/ij}\}}{2 \cdot [1 - \lambda_{ij}(\Delta t)/\mu_{ij}(\Delta t)]} \tag{4.11f}$$

where

$\sigma_{s/ij}$ is the standard deviation of the service time of the "channel", i.e., of the cruising segment *(ij)* of the route/track *(kl)* (time units/ac).

The other symbols are analogous to those in the previous Eqs.

The average flying time of the aircraft flow $(\gamma(kl))$ between the origin airport *(k)* and the destination airport *(l)* is estimated as follows:

$$\bar{\tau}_{kl}(\Delta t) = \bar{W}_{ij}(\Delta t) + \tau_{ki} + 1/\mu_{ij}(\Delta t) + \tau_{ij}(\Delta t) + \tau_{jl} \tag{4.11g}$$

The average time of an aircraft from the flow passing through the network during time (Δt) is estimated as follows:

$$\bar{\tau}(\Delta t) = \frac{1}{\gamma(\Delta t)} \cdot [\Sigma_{k=1}^{K}\Sigma_{i=1}^{N}\tau_{ki} + \Sigma_{i=1}^{N}\Sigma_{j=1}^{M_i}[\bar{W}_{ij}(\Delta t) + 1/\mu_{ij}(\Delta t) + \tau_{ij}(\Delta t)] + \Sigma_{j=1}^{M_i}\Sigma_{l=1}^{L}\tau_{jl}] \tag{4.11h}$$

where all symbols are analogous to those in the previous Eqs.

d) Application of the model

i) The network case

The three models mentioned above are applied by using a part of data related to the air route (Transatlantic) network, shown in Figure 4.9, handling the westbound traffic between Europe and North America. This implies that the network consists of: $N = 7$ routes/tracks (A, B, C, D, E, F, G), each with $M_i = 11$ most preferred flight levels (FL 310 – FL 410). The almost parallel routes/tracks are laterally separated by the standard lateral distance of: $S = 30$ nm, called RLatSM (Reduced Lateral Separation Minima), instead of the previous: $S = 60$ nm, starting from the year 2015 (i.e., from 1 to 1/2 degree of latitude). Supported by SLOP (Strategic Lateral Offset Procedure), this separation still guarantees safe aircraft deviation around the route/track centerline of about or one or two nm (nm – nautical mile) (Dief, 2018; NATS, 2015). The aircraft flying on the same FL of the given/route track are longitudinally separated by the ATC minimum time-based separation rules of: $\tau_{ij/min}(\Delta t) = 10$ min. The ATC minimum vertical separation rules between successive flight levels are: $h = 1000$ ft thanks to the RVSM (Reduced Vertical

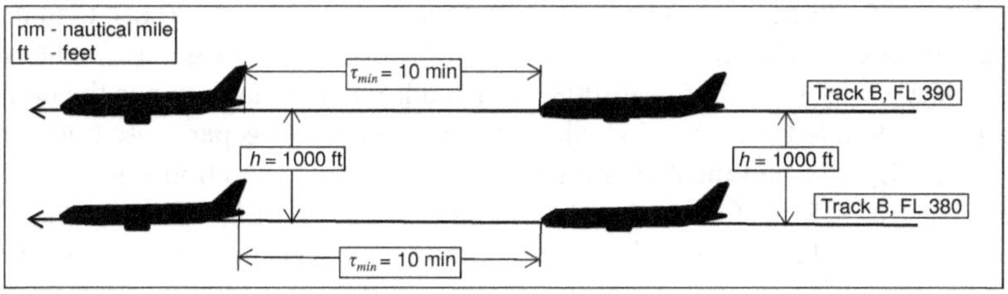

a) Lateral and longitudinal separation rules

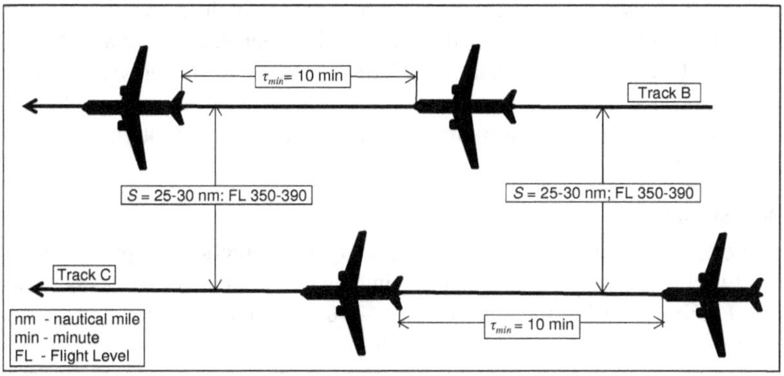

b) Vertical and longitudinal rules

Figure 4.10 Scheme of the ATC minimum separation rules applied to the aircraft flying in the oceanic (Transatlantic) airspace in the given example (1 nm = 1.852 km; 1 ft = 0.305 m).

Separation Minima) program implemented in the year 2004. Figure 4.10(a, b) shows the simplified scheme (ICAO, 2016; 2017).

In addition, recently, the new ATC longitudinal separation rules of: $\tau_{ij/min}(\Delta t)$ = 5 min have been introduced between the adequately equipped aircraft (with ADS-B system) operating on the same flight levels over the Gander and Shanwick North Atlantic oceanic airspace (http://www.ainonline.com/aviation-news/air-transport/2011-05-09/longitudinal-airspace-separations-reduced-over-north-atlantic/).

ii) "Ultimate" capacity

Based on the ATC minimum longitudinal separation distance of: $\tau_{ij/min}(\Delta t) = 10$ min, the capacity of the given flight level *(j)* of the route/track *(i)* during the period of $(\Delta t) = 1$ h is equal to: $\mu_{ij}(\Delta t) = 1/\tau_{ij/min}(\Delta t) = 60/10 = 6$ *ac/h*. In addition, if the average aircraft cruising speed, independently of the route/track and flight level, is: $V = 490$ kt, then the longitudinal distance-based separation between any two aircraft will be about: $\delta \approx 82$ nm. If this flight level capacity is equal for all *(N)* routes/tracks and their equal number of available flight levels ($M_i = 11$, for $i = 1, 2, \ldots, 7$), the total network "ultimate" capacity will be: $\mu(\Delta t) = N \cdot M_i \cdot \mu_{ij}(\Delta t) = 7 \cdot 11 \cdot 6 = 462$ *ac/h*. If the air traffic is developing at a constant rate in the westbound direction during the period of: $\Delta\tau = 8$ h (for example continuously between 11:30–19:30 UTC or GMT), the total "ultimate" capacity of the network under given conditions will be: $\mu(\Delta t = 8) = 462 \cdot 8 = 3696$ ac/period. In addition, if the aircraft average flying time along any flight level of any route/track is about: $\tau_{ij}(\Delta t) \equiv \tau(\Delta t) = 6.50$ h (The average length of a route/track independently of the flight level(s): $L = 3000$ nm and the average aircraft cruising speed: $V = 490$ kt), then the total number of aircraft simultaneously present in the network under the given conditions will be: $n(\Delta t) = n(\Delta t) \cdot \tau(\Delta t) = 462 \cdot 6.5 = 3003$. As an additional scenario, when the number of routes/tracks is doubled (i.e., $N = 14$), the number of flight levels per route/track remains the same (i.e., $M_i = 11$ and the ATC minimum longitudinal separation rule decreases by half (i.e., $\tau_{ij/min}(\Delta t) = 5$ min), the network "ultimate" capacity will double, i.e., $\mu(\Delta t) = N \cdot M_i \cdot \mu_{ij}(\Delta t) = 14 \cdot 11 \cdot 12 = 1848$ *ac/h*. For the period of: $\Delta\tau = 8$ h, this capacity will be: $\mu(\Delta t = 8) = 1848 \cdot 8 = 14784$ ac/period. This is four times higher than in the previous case. The number of aircraft simultaneously present in the network will be: $n(\Delta t) = n(\Delta t) \cdot \tau(\Delta t) = 1848 \cdot 6.50 = 12012$, which is again four times higher than in the previous scenario. Also, the average longitudinal distance-based separation between successive aircraft flying on the same flight level(s) will be shorter by half compared to the previous scenario as: $\delta = 5/60 \cdot 490 \approx 41$ nm. Similarly, the "ultimate" capacity of the network handling the eastbound traffic under given conditions of given/constant configuration and demand for service can be estimated.

iii) "Practical" capacity

Based on the above-mentioned "ultimate" capacity, the "practical" capacity of the given air route network can be estimated by specifying the maximum average delay imposed on each aircraft before entering the network (Eq. 4.10b). Figure 4.11 shows dependence of the network's "practical" capacity on the maximum imposed average delays, the standard deviation of average service time, i.e., deviations from the ATC minimum time-based separation rules, and the "ultimate" capacity.

As can be seen, if the average aircraft service time, i.e., the ATC minimum time-based separation rules, are almost perfectly adjusted (i.e., without any deviations), then, independently of the length of imposed delays on the aircraft entering the network, the corresponding "practical" capacity will remain very close to its "ultimate" counterpart under the given conditions. If the deviations from the aircraft service time increase even by a couple of minutes, the "practical" capacity will substantially decrease. At the same time, it will increase at a decreasing rate as the average imposed delays increase, and consequently come closer to its "practical" counterpart. Since, in reality, the perfect

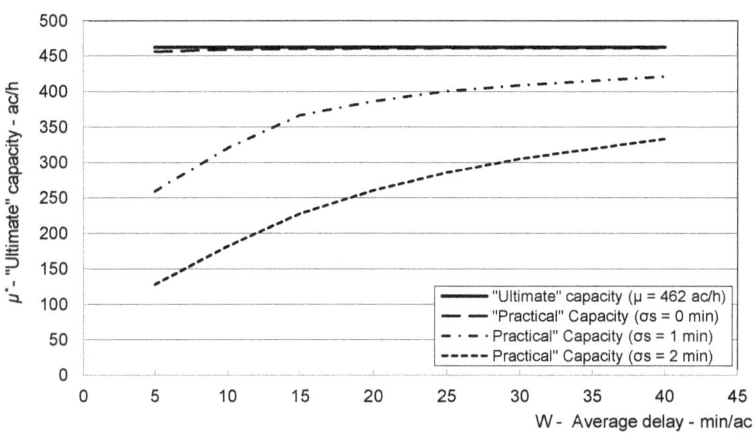

Figure 4.11 Relationship between the "practical" capacity of the given air route network, the average delay imposed on aircraft before entering it, and deviations of the aircraft service time.

conditions for getting the "ultimate" capacity are almost impossible to achieve, its "practical" counterparts can be used for a range of practical purposes.

iv) Matching capacity to demand

The three models mentioned above are applied by using a part of data related to the air route (Transatlantic) network shown in Figure 4.8(a, b). To this end, the existing characteristics are modified for a given day in terms the number of air available routes/tracks, flight levels, the "ultimate" hourly and daily capacity per flight level, and the total average flying time as given in Table 4.2.

As can be seen, the total available "ultimate" capacity of the given network is: $\mu_a(\Delta t) = 3120$ ac/24 h. The total fully utilized "ultimate" capacity of the given network is: $\mu(\Delta t) = N \cdot M_i \cdot (1/\tau_{ij/min}) = 5 \cdot 7 \cdot (60/10) = 210$ (ac/h) $\cdot$ 24 h = 5040 ac/24 h. This implies that the share of total "ultimate" capacity, which can be allocated to the aircraft/flight demand of: $\gamma(\Delta t) = 1500$ ac/24 h requesting service in the network is: $U = \mu_a(\Delta t)/\mu(\Delta t) = 3120/5040 = 0.619$ or $\approx 62\%$. In addition, the utilization of the network full and available "ultimate" capacity is: $U_a = \gamma(\Delta t)/\mu_a(\Delta t) = 1500/3120 = 0.48$ or $\approx 48\%$. Utilization of the fully available "ultimate" capacity under given conditions would be: $U = \gamma(\Delta t)/\mu(\Delta t) = 1500/5040 = 0.297$ or $\approx 30\%$. By taking into account the available "ultimate" capacity and the total average flying time through the particular "service channels" and routes of the network, the above-mentioned traffic demand is assigned by Eq. 4.11b as given in Table 4.3.

Table 4.2 Characteristics of the air route network in the given example.

FL *(i/j)*[1)] (10³ ft)	350	360	370	380	390	400	410	
Route track	**Available "ultimate" capacity (ac/h; ac/Δt = 24 h)**							**Total O-D air route flying time (h)**
1	3/72	5/120	6/144	6/144	6/144	6/144	6/144	8.05
2			6/144	6/144	6/144	6/144	6/144	7.85
3			5/120	5/120		6/144	6/144	7.55
4				6/144	6/144	6/144	6/144	7.39
5			4/96			6/144	6/144	7.10

[1)] This time also includes the average aircraft/flight delay before entering the network.

Table 4.3 Assignment of the aircraft/flight daily demand to particular air routes/tracks and their flight levels regarding the available "ultimate" capacity in the given example.

FL *(i/j)* (10³ ft)	350	360	370	380	390	400	410		
Route track	**Assigned demand** (ac/24 h)							**Total** (ac/24 h)	**Share** (%/track)
1	35	58	69	69	69	69	69	438	29.2
2			69	69	69	69	69	345	23.0
3			59	59		69	69	256	17.1
4				69	69	69	69	276	18.4
5			44			69	69	185	12.3

In addition, Figure 4.12 shows the relationship between the number of allocated aircraft/flights to particular routes/tracks and their corresponding available "ultimate" capacities.

As can be seen, this confirms that the larger number of aircraft/flights has been assigned to the particular routes/tracks approximately in proportion to their available "ultimate" capacity under given conditions.

From Table 4.2, it can be estimated the average aircraft flying time through the network is: $\tau_0(\Delta t)$ = 7.59 h. This is carried out by taking into account the average delay before entering the network estimated by Eq. 4.11f (applied to the network level), and the total time along the average O-D routes estimated from Eq. 4.11h as: $\overline{\tau}(\Delta t) = \tau_0(\Delta t) + \overline{W}(\Delta t) = 7.57 + 0.020 = 7.59$ h/flight. Regarding the average length of O-D routes through the network assumed to be: L = 3500 nm, the average aircraft/flight speed will be: $V \approx L/\overline{\tau}(\Delta t) = 3500/7.59 \approx 461$ kt. In addition, the number of aircraft simultaneously present in the network from Eq. 3 will be equal to: $n = \gamma \cdot n = \gamma \cdot \tau_0(\Delta t) = (1500/24) \cdot 7.57 \approx 473$.

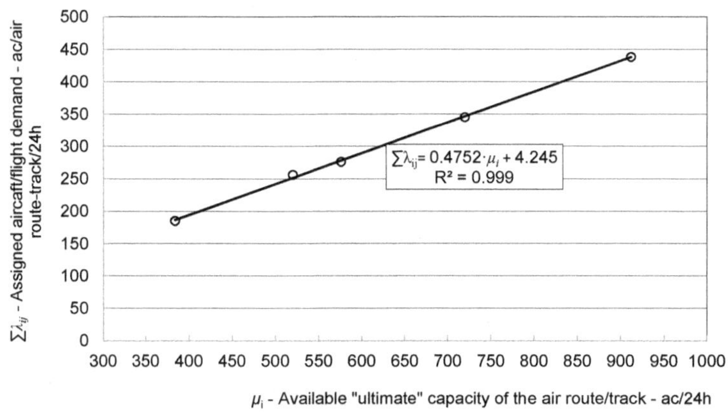

Figure 4.12 Relationship between the assigned aircraft/flight demand and the "ultimate" capacity of air routes/tracks of the network in the given example.

4.3.3.5 Capacity based on the ATC controller workload

a) <u>General</u>

In the given system/operating environment, the workload of an (human) operator may be defined as the rate of engagement of his/her available capabilities to carry out the assigned tasks. In engaging particular capabilities, the operator is assumed to use selective attention.[5] This directly influences the operator's capacity expressed by the amount/number of the successive tasks, which can be carried out during a given period of time under the given conditions. Frequently, the time-consuming tasks should be carried out within the available (limited) time. Consequently, the ratio between the required and the available time for performing a given set of tasks represents the operator's workload

[5] In general, the selective attention is based on the human's ability to focus on what is important while blocking out the rest. Therefore, it is the process of paying attention to the relevant stimuli while ignoring the irrelevant stimuli. Since the capacity for paying attention is limited, the corresponding flow of received information is also limited (Treisman, 1964; McLeod, 2018).

factor. In the case of the ATC controller as an operator, this is the factor of his/her mental workload (Carlson and Jensen, 1982; Janić, 1997; Lehrl, 1988; Loft et al., 2007; Spiegel and Bryant, 1978).

b) Main tasks of ATC controller

In general, an ATC controller carries out two types of control tasks in order to enable safe, efficient, and effective control/management of the air traffic flows passing through the sector of his/her jurisdiction. These are: (i) monitoring the radar display and other screens for getting some additional information, both visually; and (ii) Communications—A/G with the aircraft crews and G/G (Ground/Ground) with other ATC controllers—(both when necessary) (Janić, 1997).

i) Monitoring task(s)

The ATC controller monitors the radar display and the other screens by actively observing the current traffic situation from time-to-time, i.e., in discrete time intervals and passively almost continuously. The outcome from the active monitoring which takes some time is the "mental picture" of the current traffic situation stored in his/her short-term memory. In addition, the ATC controller also monitors the other screens containing the flight data and updates the changes of air traffic situation if they are not done automatically.

ii) Communications tasks

During the rather passive monitoring of air traffic in the sector of his/her jurisdiction, the ATC controller can actively carry out the voice A/G (Air/Ground) communications with the aircraft crews if necessary. While carrying out these tasks, he/she cannot actively carry out the monitoring task(s), and vice versa. The voice communications tasks are based on the previous monitoring task(s) creating the "mental picture" of traffic situation. Once created, they are delivered as instructions and assumed to be implemented when to the corresponding aircraft start to follow them. The implementation of the specific control tasks changes the traffic situation, requiring a new round of its active monitoring, forming a new "mental picture" and storing it again in the ATC controller's short term-memory. In the forthcoming ATC system, the A/G voice communications are expected to be replaced by the digitalized end-to-end-communications containing the necessary messages exchanged through CPDLC link. In addition, the end-to-end communication system will enable exclusion of the aircraft crew from the A/G communication, particularly during handling between different ATC sectors. Nevertheless, any of the systems will still impose the workload on ATC controller(s) due to designing and exchanging the related messages for coordinating air traffic in the sector (EEC, 2009). Anyway, the A/G voice communications seem to remain necessary in cases of resolving the potential conflicts.

c) Assumptions

The models of an ATC sector capacity based on the ATC controller's workload are based on the following assumptions:

- The capacity of an ATC sector can be expressed by the number of aircraft simultaneously under control and/or by the number of aircraft served during a given period under the given conditions. In the latter case, the "given conditions" are defined by the constant demand for service (i.e., the "ultimate" capacity) or by the average delay imposed on each aircraft before being served (i.e., the "practical" capacity). This average delay can be also used as an indicator of the quality of service provided by the ATC controllers to aircraft while taking over them in the sector;

- The incoming aircraft generate monitoring and other communication control tasks for the ATC controller's in the sector, requiring their execution through monitoring on the radar and display and communicating means by the above-mentioned A/G and G/G communication channels; the intensity and processing time of arriving aircraft and related tasks are assumed to be constant during a given period; this implies that he/she operates as the service channel processing the incoming control task demand under given conditions;
- Carrying out different categories of tasks (monitoring and communication) is ultimately mutually exclusive, i.e., during the active monitoring, the ATC controller cannot actively carry out other communication tasks, and vice versa;
- Different categories of the above-mentioned control tasks are carried out without pre-emption, i.e., once one starts, it is performed until completion despite the others may appear in the meantime; and
- The traffic scenario in the sector is known in advance, i.e., the number of aircraft, their flying times through the sector, and the related number of control tasks per aircraft are approximately known in advance, i.e., before they enter the sector.

d) Structure of the Model(s)

i) Capacity based on the number of aircraft simultaneously under control

The number of aircraft simultaneously under control by an ATC controller in the sector of his/her jurisdiction can be expressed as follows:

$$N = \Delta\tau/(\tau_r + \tau_{c/d} \cdot q_{c/d}) \quad \text{if } p_c = 0 \tag{4.12a}$$

$$N \cdot (\tau_r + q_{c/d} \cdot \tau_{c/d}) + p_c \cdot \frac{N \cdot (N-1)}{2} \cdot \tau_c \leq \Delta\tau \quad \text{if } p_c > 0 \tag{4.12b}$$

where

N is the number of aircraft simultaneously monitored in the sector (-);

$q_{c/d}$ is the proportion of aircraft requesting climbing and/or descending in the sector (-);

τ_r is the total time of taking an aircraft over from the neighbouring sector and monitoring it on the radar screen while in the sector (sec);

$\tau_{c/d}$ is the total time of monitoring an aircraft climbing and/or descending while in the sector (sec);

τ_c is the total time of resolving and monitoring the potential conflicts between any pair of aircraft in the sector (sec);

$\Delta\tau$ is the ATC controller's available time for controlling aircraft in the sector (s/h; min/h) (s-second; min – minute; h – hour); and

p_c is the probability of potential conflict between a pair of aircraft in the sector (-);

Equation 4.12(a, b) implies the total time for executing particular tasks per an aircraft and not the time per an individual task because a few same tasks can be executed for the same aircraft while in the sector.

After transforming Eq. 4.12a into an explicit quadratic form, the number of aircraft *(N)* simultaneously monitored in the sector can be estimated as follows:

$$N^2 + N \cdot \left[\frac{2 \cdot (\tau_r + q_{c/d} \cdot \tau_{c/d})}{p_c \cdot \tau_c} - 1 \right] - \frac{2 \cdot \Delta\tau}{p_c \cdot \tau_c} \leq 0 \tag{4.13a}$$

where

$$A = 1.0; B = \left[\frac{2 \cdot (\tau_r + q_{c/d} \cdot \tau_{c/d})}{p_c \cdot \tau_c} - 1 \right]; C = -\frac{2 \cdot \Delta\tau}{p_c \cdot \tau_c}; \text{ and } N^*_{1,2} = \frac{-B \pm \sqrt{B^2 - 4AC}}{2A} \tag{4.13b}$$

The times (τ_r), $(\tau_{c/d})$, and (τ_c) in Eqs. 4.12(a, b) and 4.13(a, b) are usually estimated by the empirical measurements (EEC, 2003; Welch, 2015).

ii) Capacity based on the number of aircraft served during given time period

• "Ultimate" capacity

The "ultimate" capacity of an ATC sector expressed by the maximum number of aircraft served under conditions of constant demand for service during a given time period can be estimated as follows (ac/h) (Kleinrock, 1975; 1976):

$$\mu_{ULT} = N/\tau_s \tag{4.14}$$

where

N is the maximum number of aircraft simultaneously under control in the sector (ac), which can be estimated from Eq. 4.12(a, b); and

τ_s is the average aircraft flying time through the sector (min).

From Eq. 4.14, the "ultimate" capacity of the ATC sector increases with an increase in the number of aircraft simultaneously handled and decrease of their flying time there.

• "Practical" capacity

The "practical" capacity of an ATC sector expressed by the maximum number of aircraft served during a given period of time under conditions of imposing an average delay on each of them can be estimated as follows (Kleinrock, 1975; 1976):

$$\overline{W}^*_q = \frac{(\lambda^*/\mu^2_{ULT}) \cdot (1 + C_s^2)}{2 \cdot (1 - \lambda^*/\mu_{ULT})} = \frac{\lambda^*(1 + C_s^2)}{2 \cdot \mu^2_{ULT} \cdot (1 - \lambda^*/\mu_{ULT})} = \frac{\lambda^*(1 + C_s^2)}{2 \cdot (\mu^2_{ULT} - \lambda^* \cdot \mu_{ULT})} \tag{4.15a}$$

where

$\overline{W}^*_q$ is the maximum average delay imposed on an aircraft before entering the given ATC sector (min/ac);

λ^* is the intensity of arriving aircraft in the given ATC sector (ac/min);

μ_{ULT} is the service rate, i.e., the "ultimate" capacity of the given ATC sector (ac/min); and

C_s is the coefficient of variation of the aircraft/flight service time in the sector (-).

From Eq. 4.15a, the "practical" capacity of the given ATC sector guaranteeing imposing the specified average delay on each arriving aircraft is equal to:

$$\lambda^* \leq \frac{2 \cdot \overline{W}^*_q \cdot \mu^2_{ULT}}{(1 + C_s^2 + 2 \cdot W^*_q \cdot \mu_{ULT})} \tag{4.15b}$$

where all symbols are analogous to those in the previous Eqs.

Equation 4.15b indicates that the "practical" capacity of the given ATC sector generally increases with increasing of its "ultimate" capacity and the average imposed delays.

iii) Capacity based on the ATC controller internal "mental" model

The ATC controller performs the above-mentioned control tasks by his/her internal "mental" model. This implies that the information of these tasks are stored into his/her short-term memory. This memorized picture is then used for creating particular control tasks in combination with information obtained through A/G communications with aircraft/pilots and G/G communications with ATC controllers from the neighbouring sectors. These control tasks are the products of decision making by the ATC controller. Despite the fact that it looks possible, the ATC controller is not able to execute monitoring and communications activities/tasks, which both actually generate workload. This has been one of the essential assumptions in developing the ATC controller's "ultimate" capacity model. The additional assumption was that the human operator (man) behaves as the single server system with the given service rate (capacity) to process various information and corresponding tasks during a given time period under conditions of constant demand for service. Under such circumstances, the ATC controller's workload has been expressed by the workload coefficient *(ρ)* as follows (Janić, 2000; Kleinrock, 1976):

$$\rho_w = \tau_s/\tau_a \tag{4.16a}$$

where

τ_t is the average service time of the single control task (s, min); and,
τ_a is the average inter-arrival time of the control tasks (s, min).

Equation 4.16a indicates that *(ρ_w)* can be considered as the server's load factor, defined analogously as the "utilisation rate" in the theory of queuing systems. Generally, the workload will increase if the time for executing the control task, the intensity of arriving control tasks, or both simultaneously increase.

The above-mentioned concept is elaborated in more detail as follows: The time needed for an ATC controller to make one observation of the traffic situation on the radar display can be quantified by the formula:

$$\tau_1^* = \lambda_c \cdot \tau_f \cdot p_c \cdot H \cdot \tau_{01} + [(\lambda_c \cdot \tau_f \cdot p_c)^2 - (\lambda_c \cdot \tau_f \cdot p_c)] \cdot \tau_{02}/2 \tag{4.16b}$$

where

λ_c is the intensity of aircraft flow entering the airspace (sector) (ac/h);
τ_f is the average aircraft flying time through the sector (min);
H is the total entropy the ATC radar controller is facing while observing the radar display to identify the current and prospective status of the specific aircraft (bit);
τ_{01} is the time needed for processing the unit of information obtained by observation of the radar display (b/s);
τ_{02} is the time required in order to establish the appropriate relationship between any pair of blips on the radar display (this time is primarily dependent on the quality of displays; primary radar, beacon and alphanumeric displays can be available) (s); and
p_c is the proportion of blips (aircraft) under observation on the radar display.

Since the traffic situation in the given sector changes continuously, the ATC controller should resume observation of the radar display after some time to update the traffic picture. Observations of the radar will be more frequent (with shorter time gaps in between) if the controller expects that the traffic situation is going to change considerably and rapidly (TMA is typical example where these traffic characteristics are frequently observed). The time between any two successive observations of traffic situation on radar display is estimated to be equal as follows:

$$\tau_2^* = [3 \cdot \varepsilon_0 \cdot m \cdot (\tau_{0z} + \tau_{wz})]/[\Phi^{-1}(0.5 - p_0/2] \quad (4.16c)$$

where

- m is the number of implemented control tasks that produce a considerable change in the actual traffic situation in comparison to its remembered picture (-);
- τ_{oz} is the average time of creating and executing a control task (s);
- τ_{wz} is the time in which the control task can be saved in the ATC controller's short-term memory (after time (τ_{wz}) the task will be forgotten, and consequently the new creating process should start) (s);
- ε_o is the maximum tolerable error that can emerge during a comparison of two pictures of traffic situation formed by two successive observations of radar display (this error is the random variable with Gauss (normal) probability distribution) (-);
- p_o is the probability that the error (ε_o) is going to surpass the tolerable value (ε_o) set up in advance (-); and
- Φ^{-1} is the inverse Laplace's function.

Under conditions when the ATC radar controller successively observes the radar display and executes the control tasks, his (her) workload due to carrying out both generally stabilizes and the corresponding workload coefficient satisfies the condition as follows:

$$\rho_w = \rho_m + \rho_z = 1 \quad (4.16d)$$

where

- ρ_m is the coefficient of workload caused by monitoring of the radar display: $\rho_m = \tau_1^* \cdot (\tau_1^* + \tau_2^*)^{-1}$; and
- ρ_z is the coefficient of workload caused by creating and executing the control tasks for the aircraft in the sector $\rho_z = \lambda_c \cdot n \cdot \tau_{0z}$, where (n) is the number of control tasks performed for one aircraft in the sector.

The other symbols are analogous those in the previous Eqs. From Eq. 4.16(b, c, d), the variable (λ_c) is treated as the unknown variable in the third degree polynomial equation as follows:

$$a_0 + \sum_{k=3}^{3} a_k \cdot \lambda_c^n = 0 \quad (4.16e)$$

where

$$a_0 = -[6 \cdot \varepsilon_0 \cdot m \cdot (\tau_{0z} + \tau_{wz})]/\{n \cdot \tau_{0z} \cdot \tau_f^2 \cdot p_c^2 \cdot \tau_{02}^2 \cdot [\Phi^{-1}(0.5 - p_0/2)]\}$$

$$a_1 = [6 \cdot \varepsilon_0 \cdot m \cdot (\tau_{0z} + \tau_{wz})]/\{\tau_f^2 \cdot p_c^2 \cdot \tau_{02}^2 \cdot [\Phi^{-1}(0.5 - p_0/2)]\}$$

$$a_2 = [2 \cdot (H \cdot \tau_{01} - 0.5 \cdot p_c \cdot \tau_{02})]/[p_c^2 \cdot \tau_{02} \cdot \tau_f]$$

$$a_3 = 1.0$$

A real root of the polynomial Eq. (4.16e) represents the "ultimate" capacity of the ATC radar controller under the given operating scenario.

iv) Application of the models

The models mentioned above are applied to the hypothetical ATC sector by using the input data as follows: $\tau_r = 43$ sec (routine task); $\tau_{c/d} = 15$ sec (monitoring task during climbing/

descending): τ_c = 70 sec (task of resolving and monitoring crossing conflict(s)); 10 sec (task of resolving and monitoring along and opposite track conflict(s)). The maximum ATC controller's workload is assumed to be $\Delta\tau = 0.7\ \text{h} = 0.7 \cdot 60\ \text{min} = 42$ min/h (EEC, 2003). Some results are shown in Figure 4.13(a, b, c).

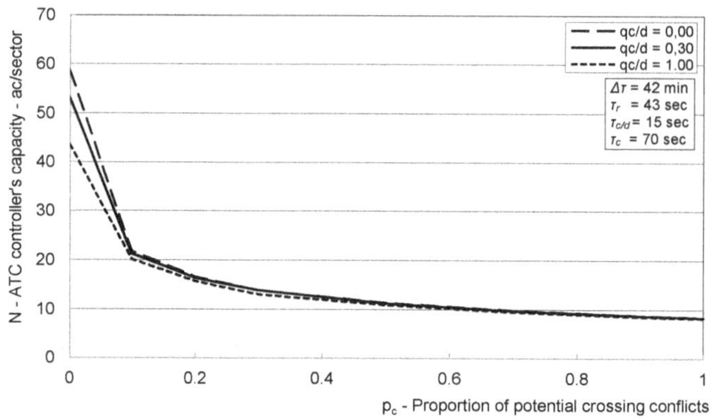

a) The ATC controller's capacity vs proportion of the potential crossing conflicts

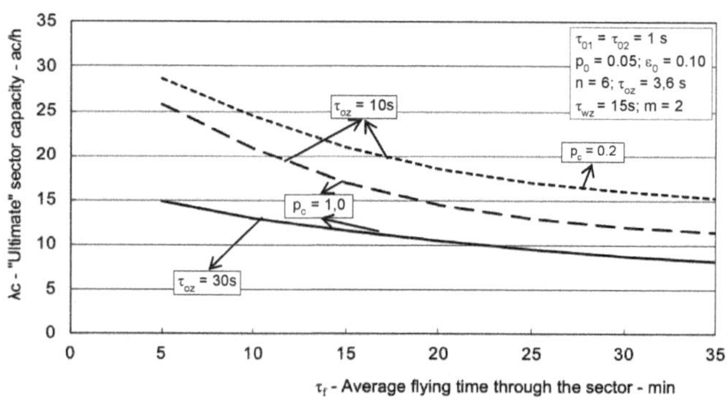

b) The ATC controller's capacity vs the average aircraft flying time through the sector

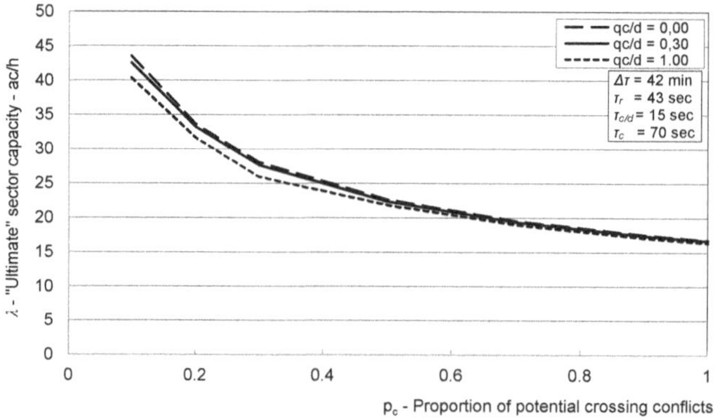

c) "Ultimate" sector capacity vs proportion of the potential crossing conflicts

Figure 4.13 Relationships between the ATC sector capacities based on the ATC controller's workload, flying time, and proportion of the potential crossing conflicts in the sector.

Figure 4.13a shows that the ATC controller's capacity decreases more than proportionally as the proportion of potential crossing conflicts increases, requiring a relatively long time to be resolved in the given context. Such development appears reasonable regarding increasing complexity of traffic in the sector imposing a higher workload on the ATC controller within the limited time budget. At the same time, given the proportion of potential crossing conflicts, the proportion of climbing or descending aircraft/flights in the sector does not significantly influence on the ATC controller's workload and consequently the sector's capacity. Figure 4.13b shows that the sector "ultimate" capacity decreases more than proportionally as the aircraft flying time through the sector increases. It also decreases with an increase in the average time of executing the particular control task and tactics of observing specific blips (aircraft) on the radar display. The results also indicate that there can be some reserve for increasing the sector capacity in the tactics of monitoring the radar display. The model implicitly indicates that the partial or complete automation of particular ATC functions will inevitably influence the structure and content of the ATC controller internal "mental" model, workload, and consequently the corresponding capacity.

Figure 4.13c shows that the "ultimate" sector capacity behaves similarly to its counterpart based on the ATC controller's workload. It decreases more than proportionally as the proportion of potential crossing conflicts increases and remains mostly unaffected by an increase in the proportion of climbing and descending aircraft/flights in the sector under the given conditions. In addition, Figure 4.14 shows the relationship between the ATC "practical" sector capacity and the average delay imposed on each aircraft/flight before entering the sector.

As can be seen, the "practical" capacity increases at a decreasing rate as the average delays imposed on the aircraft/flights before entering the sector increase. For the specified average delay, this capacity increases while increasing its "ultimate" counterpart, as intuitively expected. Anyway, these imposed delays frequently manifest as ground-hold delays before aircraft/flight departures from their origin airports.

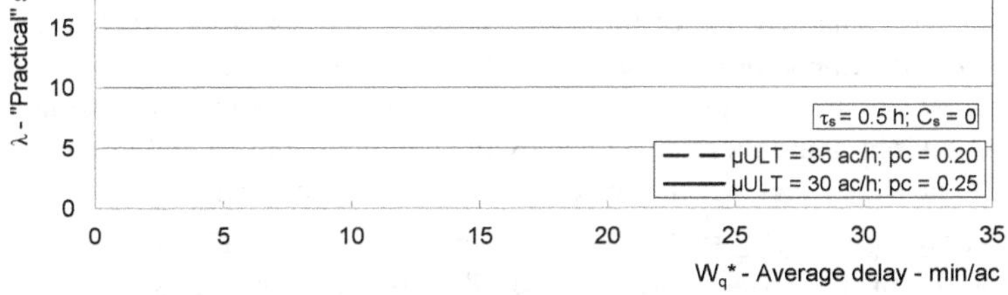

Figure 4.14 Relationships between the ATC sector "practical" capacity based on the ATC controller's workload and the aircraft/flight delay before entering the given sector.

4.4 Quality of Services

4.4.1 Description

The quality of services provided by an ATC system or ANSP to the users, i.e., aircraft/flights in the airspace of its jurisdiction, can be expressed by different indicators of performances such as: (i) the aircraft/flight delays and related costs caused by the ATFM operations at and around airports and during en-route phases of flights; (ii) the en-route flight efficiency; and (iii) safety reflecting the actually occurred air accidents and incidents due to direct and indirect involvement/contribution of the ATC system and the potential risk of their occurrence as well (EEC, 2019).

4.4.2 Delays

4.4.2.1 Causes

As described in Chapter 2, congestion and delays occur whenever the demand exceeds the capacity of the service facility. Delays of the flights managed by a given ATC system are defined as the deviation of their actual time pattern from the planned ones. The planned flight time between origin and destination airport (gate-to-gate) is the time it takes without the influence of other traffic/flights, failures of the ATC system's particular components and/or shortage of its staff, and/or the impact of bad weather. In other words, similarly to the airport arrival and departure delays, the flight delays in the rest of controlled airspace, such as LAA and HAA airspace, are defined as the differences between the actual and planned time of passing through the "reference locations", which, in most cases, are the borders of particular ATC sectors. In general, they are caused by the temporal relationships between the demand and the system's capacity. These relationships can be influenced by many factors. The most important causes of flight delays are bad weather, ATC system disruptions and staff, system capacity, and the continuous growth of flight demand.

a) Bad weather materializes as fog and low cloudiness, thunderstorms, snow showers, strong winds, etc. In general, it may temporarily restrict the use of a particular airspace for some time. Such conditions require an increase in the separation rules between the aircraft, which may result in a reduction of the airspace capacity, consequently causing delays of the affected flights. In some cases, the affected airspace can be completely compromised, requiring the affected aircraft to bypass it, which actually prolongs the corresponding flights and causes them to be delayed. Despite the long-term efforts to improve technology on board the aircraft, airports, and ATC system in combination for increased sophistication of the ATFM procedures, bad weather has remained one of the main causes of delays in the ATC system. Even complete implementation of the new technologies, for example in the scope of the U.S. NextGen and European SESAR research and development programs, will not completely eliminate these delays, but will eventually reduce and make them more predictable (https://www.faa.gov/nextgen/; https://www.sesarju.eu/).
b) Disruptions of the ATC system and staff can also cause flights delays. In general, disruptions in the given context happen due to the technical failures of the ATC system's particular components and/or temporal shortage of the ATC staff due to any reason, including industrial actions. In many cases, these conditions require reorganization of the controlled airspace in terms of the number of ATC sectors, and consequently a reduction of the system's declared capacity. Under such conditions,

guaranteeing safe and efficient operations according to the prescribed separation rules and procedures requires constraining of the aircraft/flight demand by imposing the rather long delays or even refusing services to the affected flights, resulting in their cancellation.

c) The ATC system capacity as the maximum number of flights to be handled during a given period of time under given conditions is balanced with the flight demand structured in the airline schedules. The latter can also be a cause of flight delays. Namely, by adjusting the departure and arrival times of their flights to user/air passenger demand as much as possible, the airlines have frequently bunched their flights into relatively short periods (i.e., "peaks" lasting quarter, half and/or one hour) (The "waves" consisting of tens of the arriving and departing flights at the airline hub airports scheduled in an hour or so are typical examples of such bunched flights). Since such concentrated flight demand often intentionally exceeds the capacity of airports, airspaces and ATC systems, congestions and delays occur. One of the options for reducing such congestions and delays is known as "spreading the peaks", implying the introduction of different operational and economic measures, the latter particularly at airports in the form of congestion charges.

d) Continuous growth of the flight demand has also caused an increase in delays of particular flights. This has happened due to the increasing demand on the one hand, and the limited opportunities for expansion of the existing ATC and airport capacity on the other.

4.2.2.2 Indicators

The flight delays aiming at being used for assessment of the quality of services provided by a given ATC system can be indirectly or directly quantified by a range of indicators. Some of them are the on-time arrival performance, i.e., punctuality, shares by particular cause, duration, and related costs. In the given case, these indicators can specified and estimated for the entire and/or parts of LAA and HAA airspace in both absolute and relative terms, depending on the time period or the averages per unit of the system's output, i.e., the number of flights handled during the specified time period.

a) On-time performance, i.e., arrival flight punctuality

The on-time performance, i.e., punctuality of flights, has been an indicator of the quality of services used (and relevant) by airlines, ATC system(s), airports, and their users, i.e., air passengers, as well. Figure 4.15 shows examples of the on-time performance, i.e., punctuality of the arriving flights in Europe and U.S. during the specified period of time. For the above-mentioned stakeholders involved, this is generally defined as the proportion of flights delayed at arrivals less than 15 minutes compared to the published airline schedule (EEC, 2019; https://www.transtats.bts.gov/OT_Delay/otdelaycause1.asp?type=5&pn=1).

As can be seen, the on-time performances, i.e., the punctuality of arriving flights at 34 main airports have fluctuated in both regions during the observed period of time (2008–2017). Three characteristic sub-periods can be noticed. In the first sub-period (2010–2011), that in the USA was higher than in Europe. In the second one (2013–2015), the arrival punctuality was higher in Europe than in the USA. During the last sub-period (2016–2017), the arrival flight punctuality in Europe fell below its counterpart in the USA.

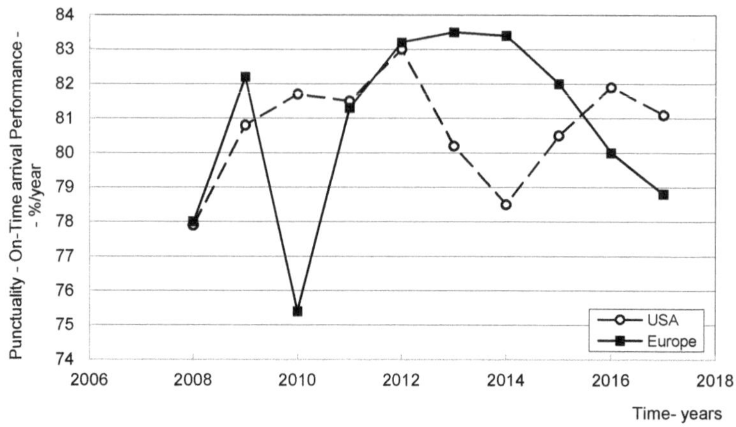

Figure 4.15 Example of the on-time performances, i.e., arrival flight punctuality over time—Case of 34 main airports in Europe and U.S. (EEC/FAA, 2017; https://www.transtats.bts.gov/HomeDrillChart.asp/).

b) <u>Share of flight delays by cause</u>

Figure 4.16 shows some examples of the flight delays by cause in Europe and U.S. (EEC/FAA, 2017; https://www.transtats.bts.gov/OT_Delay/ot_delaycause1.asp?type=5& pn=1).

As can be seen, the most frequent cause of flight delays in Europe was the "ATC system capacity". The less frequent cause was "weather", followed by "ATC staff". "ATC disruptions" was the least frequent cause. In the USA, the most frequent cause of flight delays was "air carrier delays", followed by "ATC system" cause. "Weather" was a much less frequent cause, while "security" appeared to be negligible.

a) <u>Duration of flight delays</u>

Duration of the flight delays is usually expressed in the time units (minutes) per flight and per delayed flight. Figure 4.17(a, b) shows the examples of duration of the en-route

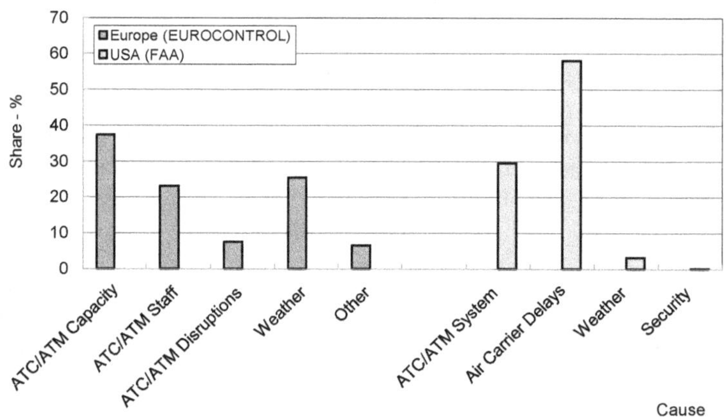

Figure 4.16 Example of the en-route flight delays by cause—Case of Europe and U.S. (Period: 2018) (EEC/FAA, 2017; https://www.transtats.bts.gov/OT_Delay/otdelaycause1.asp?type=5&pn=1).

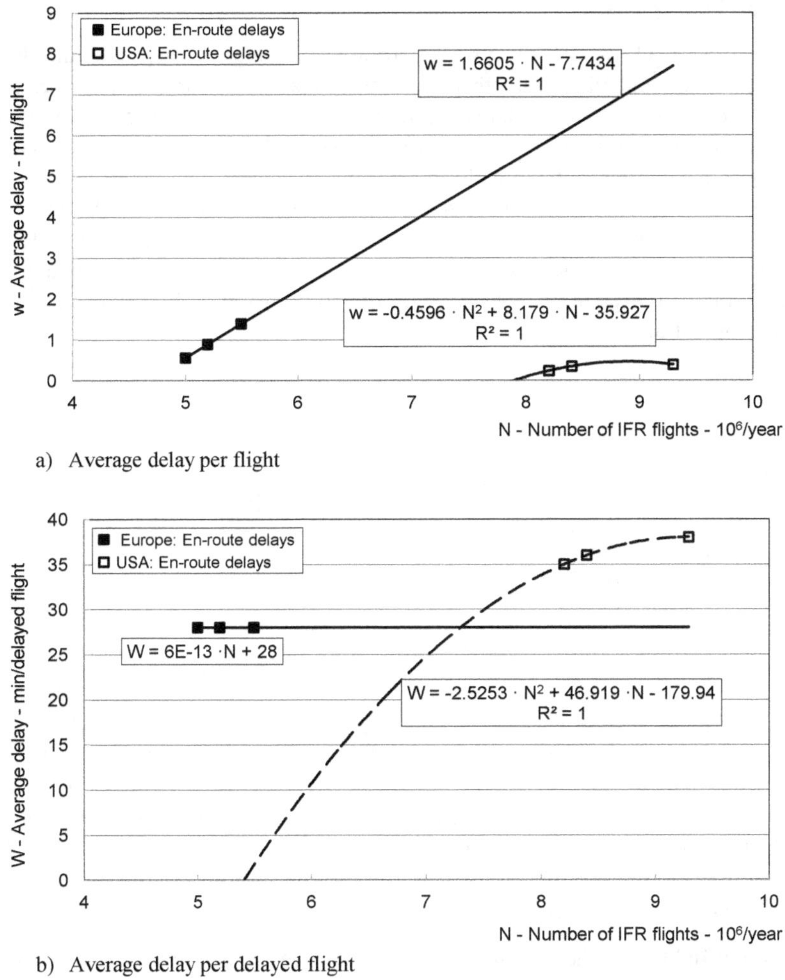

Figure 4.17 Relationship between the en-route flight delays and the number of flights handled—Case of Europe and U.S. (flights between 34 larges airports) (Period 2008, 2015, 2017) (EEC/FAA, 2017).

flight delays per flight and per delayed flight handled during the specified period of time in Europe and U.S., respectively (EEC/FAA, 2017).

Figure 4.17a shows that both in Europe and U.S., the average en-route delay per flight has tended to increase with an increase in the annual number of handled flights, at a much higher rate in Europe than in U.S. At the same time, in Europe, these delays have been much more significant than in the U.S., despite the number of the handled flights in Europe being lower by about 60%. Figure 4.17b shows that, in Europe, the average en-route delay per delayed flight has been relatively constant, while in the USA, it has been increasing as the number of handled flights grows. In addition, the U.S. delays have been higher than in Europe by about 68%.

b) Costs of flight delays

The costs of the ATC en-route delays relate to the direct flight costs, including the fuel, crew, maintenance, and other costs, then the costs of reactionary/propagated delays as the network effects, and the airline related passenger costs, including the costs of rebooking, compensation, etc. Figure 4.18 shows an example of the ATC en-route average cost depending on the number of handled flights during the specified period of time.

As can be seen, in the European airspace managed by EUROCONTROL, the average en-route cost per flight increased at first and then decreased at a decreasing rate as the number of handled flights increased. In particular, the cost decreasing as the number of handled flights increases indicates growing efficiency of the ATC system in the given segment of operations.

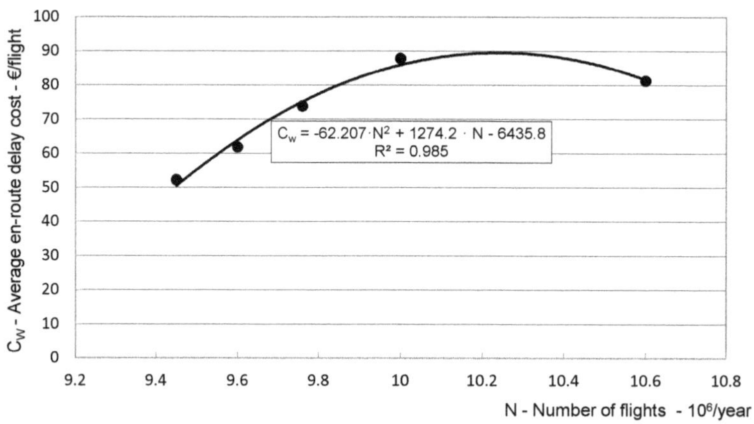

Figure 4.18 Relationship between the average en-route delay cost per flight and the number of handled flights by an ATC system—Case of EUROCONTROL area (Period: 2014–2018) EEC, 2016).

4.4.3 En-Route Flight Efficiency

The en-route flight efficiency is an indicator of service quality reflecting the ability of a given ATC system to handle flights according to their plans under given conditions. As such, the en-route flight efficiency has horizontal (distance) and vertical (altitude) components.

The horizontal efficiency is expressed as the ratio of actual and planned distances of the en-route phase of flights. As such, it actually indicates the percentage of extension of the en-route phase of flights relative to that in the planned ones (EEC/FAA, 2017). Figure 4.19 shows the example of this indicator as en-route "inefficiency" in Europe and USA over time.

As can be seen in both ATC systems, the level of en-route flight "inefficiency" was around 2.7% and 3%, i.e., quite comparable. During the entire observed period, except in the last year, the "inefficiency" was lower in the U.S. than in the European ATC system. In Europe in the year 2017 it was 2.81%, while in the U.S. it was 2.86%. At the same time, the average horizontal en-route flight extension route extension in the year 2017 was 12.5 Nm in USA and 7.1 Nm in Europe (1 Nm = 1.852 km). This indicates that, in absolute terms, the average additional distance was longer in the U.S. due to the longer flights while, at the same time, the flight "inefficiency" was generally lower

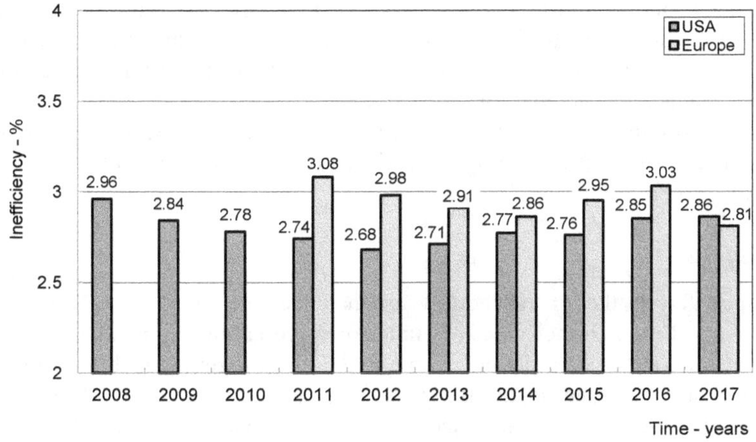

Figure 4.19 Development of the horizontal en-route flight "inefficiency" over time—Case of Europe and U.S. flights between the 34 main airports (EEC/FAA, 2017).

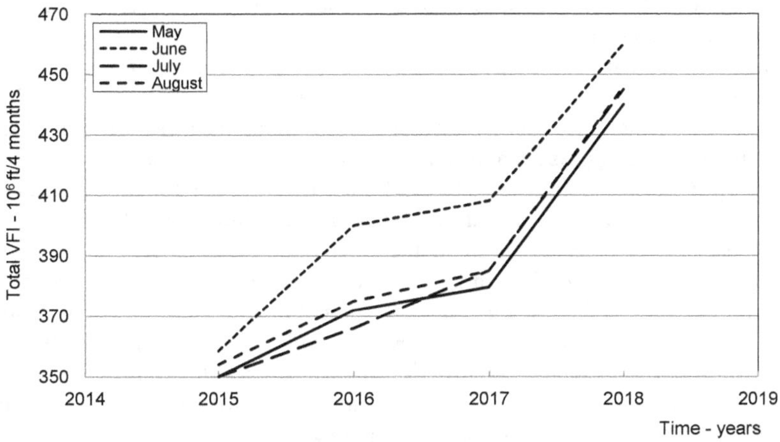

Figure 4.20 Examples of development of the total VFI (Vertical Flight Inefficiency) over time—Case of Europe (Period: 2018) (EEC, 2019).

(EEC/FAA, 2017). Such horizontal "inefficiency", independently of the proportion and actual extensions of the en-route cruising phases of flights, affected their economics by increasing the operating costs.

The vertical en-route flight "inefficiency" is defined as the total amount of constrained flight en-route cruising altitudes in the absolute terms. Figure 4.20 shows an example for Europe for four months of each year during the observed period (EEC, 2019).

As can be seen, the vertical flight inefficiency in the European ATC system (EUROCONTROL) increased during the specified period of time (2015–2018). In each year of the period, it was the highest in June due to the high seasonality of air traffic and the lowest but variable between May and July. That in August was somewhere between the previous two. In addition, the number of airport pairs with constrained altitude increased over time. For example, it was about 80 in the year 2015 and during the first half of 2016, about 125 during the last half of 2016 and the first half of 2018, and more than 250 during the second half of 2018 (EEC, 2019).This vertical flight inefficiency also affected the economics of flights due to the additional fuel consumption linked to flying at altitudes that are non-optimal for fuel efficiency.

The airport related ATC inefficiencies, such as those in TMA, during landings and take-offs, and the aircraft taxiing-in and taxiing-out, have been elaborated in Chapter 2 (EEC/FAA, 2017).

4.4.4 Safety

In general, safety in the air transport system has been expressed in both absolute and relative terms. In the former case, an indicator representing the number of accidents and incidents during a given period of time in the given region (country, continent, and worldwide) has been used. In the latter case, the indicator, which has been used, has expressed the number of air traffic accidents and/or incidents per unit of the system's output, such as the number of flights, flight hours, flight and passenger kilometres, etc., all during a given time period in the given region, as mentioned above. Under such conditions, the system's external and internal causes and eventual contribution of the particular stakeholder involved in the particular accidents and incidents have not been explicitly considered (Janić, 2000a).

In addition to airlines and airports, the ATC has been one of the main operational stakeholders dealing with and being responsible for maintaining safety of air traffic in the airspace of its jurisdiction. Consequently, safety of operations of an ATC unit has been one of its main objectives. As such, it can also be considered as an indicator of the quality of services provided to users, i.e., aircraft/flights, under given conditions. Under such circumstances, it is usually expressed by the number of air traffic accidents and incidents occurred during a given period of time in the given region with the ATC unit's direct and or indirect influence/contribution. In the given context, "accidents" are considered to be events with or without fatalities and/or directly and/or indirectly damaged properties. The "incidents" are considered to be the events of violation of some operational rules and procedures, which could potentially turn into accidents, but have not. The term "ATC contribution" implies events to which the ATC unit has contributed directly or indirectly. Figure 4.21(a, b) shows some statistics of the characteristics of the above-mentioned safety indicators of the European ANSP EUROCONTROL (EEC, 2019).

Figure 4.21a shows the relationship between the number of fatal accidents and their relative share and the annual total of both fatal and non-fatal accidents with both direct and indirect ATC contribution. As can be seen, during the observed period of time, both the number of events and the share of fatal accidents in the total number of accidents were between about 1–2 and 1–2.5%. These figures indicate that contribution of the ATC to air traffic accidents in the given case was extremely low, thus reflecting its rather high level of safety in providing ANSs.

Figure 4.21b shows the relationship between the number of incidents with ATC contribution and the number of processed TSUs (Transport Service Unit(s)) during the specified period of time. The incidents in this context generally relate to the events such as: separation rules infringements, unauthorised penetration of airspace, runway incursions, and the ATC specific events (EEC, 2019). As can be seen, during the observed period, this number of incidents increased more than proportionally with an increase in the number of processed en-route TSUs, i.e., the volumes of handled traffic. This can

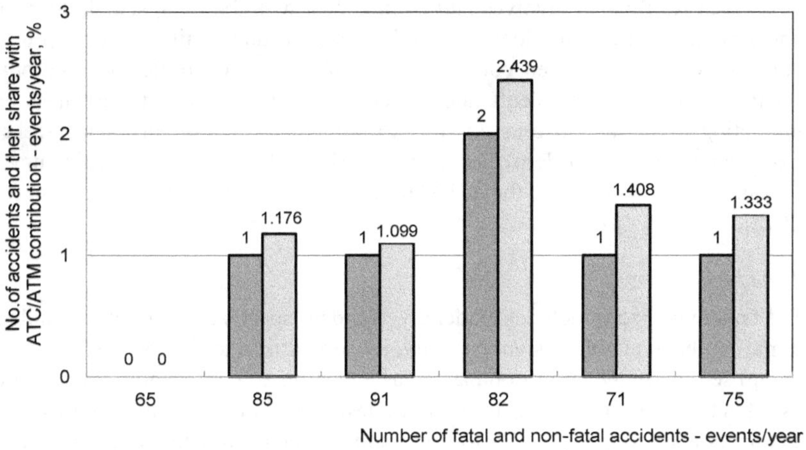

a) Accidents with the ATC contribution and their share vs the total number of accidents (Period: 2013-2018)

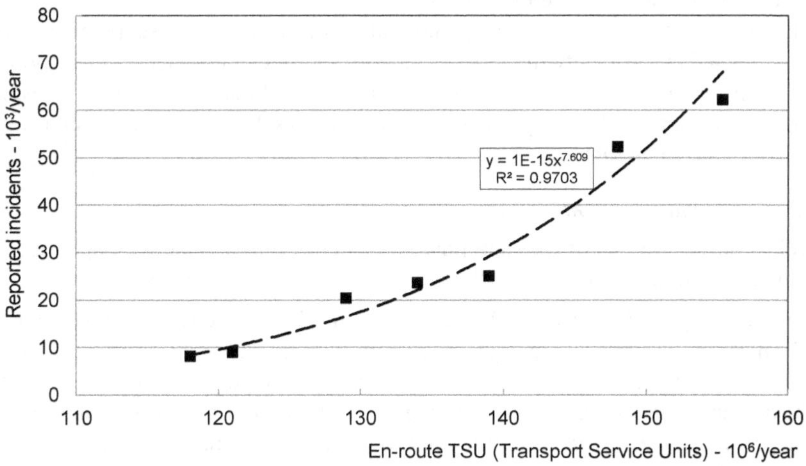

b) Reported incidents vs the processed en-route TSUs (Period: 2012-2018)

Figure 4.21 Some indicators of the ATC safety—Case of the European ANSP – EUROCONTROL (EEC, 2019).

be explained in two ways. First, the incident reporting regularity and, consequently, the reporting system have been improving over time. Second, the system has tended towards an increasing instability, containing an inherent risk of occurrence of air traffic incidents, under conditions of handling increasing volumes of air traffic.

4.4.5 Measures for Improving Quality of Services

As mentioned above, the capacity of a given ATC unit is constrained. In order to balance such constrained capacity to the current and future flight demand safely, efficiently, and effectively, the ATFM process is used, for example, in Europe by ANSP EUROCONTROL and in the USA by ANSP FAA NAS (National Airspace System). In carrying out the process, a given ANSP also involves and collaborates with the main stakeholders, such as airport operators, the individual ATC units, users-airlines, military authorities and other stakeholders. In the given context, safety of air traffic implies accommodating flights without the risk of incidents and accidents. Efficiency and effectiveness imply accommodating flights as close as possible to their plans. These particularly relate to minimizing their arrival and departure delays, horizontal and vertical inefficiency, and related costs. The main phases of the ATFM process are planning and management (EEC/FAA, 2017; ICAO, 2018).

4.4.5.1 ATFM planning

The ATFM planning phase includes, under given and prospective conditions, establishing, i.e., estimating the available system capacity, its comparison to the forecasted air traffic demand and setting up the targets on particular indicators of performance, i.e., quality of services. This relates to reviewing the airspace design, technical facilities and equipment aiming at investigating opportunities for increasing capacity, updating the ATC operation rules and procedures, availability of the staff (ATC controllers), in order to provide the adequate capacity on time to serve the forecasted demand, and training of the staff to be delivered to the particular stakeholders involved.

The outcome from this phase is quantification of the scale and scope of imbalance between demand and capacity under given conditions, and consequently identifying the actions for mitigating such imbalance. In this case, the estimated capacity is considered as declared and submitted to the particular stakeholders as the basic input for carrying out the ATFM management phase.

4.4.5.2 ATFM management

The ATFM phase consists of four sub-phases: strategic, pre- and tactical, operating, i.e., fine-tuning or optimization of traffic flows, and post-operations analysis.

a) The strategic ATFM sub-phase includes the activities planned a long time in advance (from several months to one day). These activities are directed towards resolving the major differentials between demand and capacity. For example, months in advance, the demand data are assembled from the airline flight plan data. By using these data, the preliminary estimates of traffic loads over any critical navigation point or in any ATC sector are made in cooperation with the other stakeholders involved, such as the airlines, airports, and the local ATC authorities. As the time progresses, the initial database is updated by introduction of the additional flight information received from the airlines (for example, cancellation of the already announced flight and/or planning of additional flights). The outcome from this sub-phase is identification of the possible imbalances between demand and capacity and definition of possible solution for their mitigation, i.e., optimization of the available capacity to facilitate with the expected demand in a way to achieve the specified targets on the indicators of performances representing the quality of services.
b) The pre-tactical ATFM sub-phase includes the measures carried out one day in advance. This relates to analysing and comparing the expected demand and capacity, usually resulting in adjusting the related figures made in the strategic sub-phase. In this sub-phase, optimization of the capacity is carried out by organization of the available resources in terms of management of the ATC sector configuration, using of the alternative flight procedures, etc. The outcome from this process is the plan

period containing the capacity resources and measures for managing the expected traffic during a 24-hour period.

c) The tactical ATFM sub-phase contains the measures undertaken on the day of operation. The aim is to define eventual disturbances, including staffing, meteorological conditions, eventual failures of the system components, including the ground and air infrastructure, and then select the measures to mitigate their impact.

d) The ATFM operating, i.e., fine-tuning or optimization of traffic flows, sub-phase includes measures influencing the traffic flows before departures and during en-route phases of flights. These measures include the aircraft/flight sequencing and metering in combination with imposing some delays on particular flights, resulting in fine-tuning departure flows. This fine-tuning results in the cleared take-off times within the specified time windows of, for example, –5/+5, –5/+10, –15/+30 min. The latter measures include prediction of the possible high congestion and delays 5 to 10 minutes in advance, which are expected to last for a relatively short time, i.e., from a few minutes to a quarter of an hour. These congestions and delays are mitigated or even prevented according to the fixed plan, including re-directing the affected aircraft/flights to less loaded ATC sectors and air routes. To this end, direct instructions are given to these aircraft/flights, such as: adjusting speeds in order to absorb the expected delays at high altitudes; radar vectoring to avoid temporarily congested areas; performing en-route holding patterns; increasing the longitudinal separation rules at the entrance points of particular ATC sectors in order to reduce the high intensities of demand, etc. The ATC applies these actions in advance, i.e., just before the events occur, i.e., in the time span less than 2 minutes (EEC, 2000; 2003; EEC/FAA, 2017).

e) The post-operations analysis is the last sub-phase of the ATFM management phase. This includes analysis of application of the previous sub-phases and their evaluation regarding whether the specified targets of performances, i.e., particular indicators of the quality of services, have been met. The reports are further used for development of innovative and best practices, including undertaking the necessary improvements in the future. These improvements would mainly relate to increasing of the ATC system capacity by: improvement of utilization of the existing facilities and building of new facilities, installation of new equipment in the airports, airspace and ATC, increasing the sophistication of the ATC operations, and the overall reduction of aircraft separation rules in combination with application of the advanced procedures thanks to implementation of the new sophisticated technologies, such as those mentioned above in the scope of the U.S. NextGen and European SESAR research and development program (https://www.faa.gov/nextgen/; https://www.sesarju.eu/).

4.4.6 Modelling Quality of Services

4.4.6.1 General

Modelling quality of services of an ATC unit has usually dealt with the flight delays in the particular system's components, such as airports, terminal areas, LAA and HAA airspace. In particular, modelling the aircraft/flight delays at airports during their landings and take-offs has been most frequently under consideration. However, those delays imposed on the flights while in other parts of the airspace have rarely been under consideration.

These delays can generally happen due to performing high-altitude holding patterns before passing from one to the other ATC sector(s), limiting the cruising speed due to congestion of the ATC sector(s) ahead, and the above-mentioned extension of the planned trajectories and constraining the cruising altitudes. The former inefficiency represents the actual flight delays. The latter one reflects some "hidden delays", such as waiting at the current altitudes for the safe passage to the desired/requested ones. With the exception of delays imposed on the aircraft/flights at airports before take-off as the "ground holding" delays carried out at the departure gates with the engines off, all other delays cause additional fuel consumption and related emissions of GHGs.

The airport delays imposed on the aircraft/flights before landings and take-offs are modelled in Chapter 2 as the indicators of quality of services provided to aircraft/flights in the airport airside area. At this place, the flight horizontal and vertical flight inefficiency is modelled.

4.4.6.2 Model of the horizontal flight "inefficiency"

a) System

The trajectories of flights are spread through the ATC controlled airspace between the origin and destination airports. In most cases, they are planned in advance in order to be as optimal as possible regarding length, trip time, fuel consumption, and related costs. However, during their realization, such optimality cannot be always achieved due to various reasons. Consequently, they may become longer in terms of distance, trip time, or both, causing the additional fuel consumption and related emissions of GHGs, and the related costs as well. These flights are considered as horizontally "inefficient". In the opposite cases, some flights can improve their planned horizontal efficiency thanks to the weather conditions. This frequently happens when they fly down the jet streams with strong tail winds increasing their ground speed and consequently shortening the scheduled trip time.

b) Assumptions

Modelling the horizontal "inefficiency" of the above-mentioned flights is based on the assumption that their actual trajectories are always different (longer or shorter) to the planned ones, thus causing additional or shorter trip time, additional or lower fuel consumption and related emissions of GHGs, and consequently higher or lower corresponding costs (fuel, crew, users/passengers).

c) Structure of the model

i) Horizontally "inefficient" flights

The delay of a horizontally inefficient flight can be estimated as follows:

$$w_{hfe} = L_a/v_a - L_p/v_p \tag{4.17}$$

where

L_a, L_p is the actual and planned length, respectively, of the flight trajectory (nm, km); and

v_a, v_p is the average aircraft/flight speed along the actual and planned trajectory, respectively (kt, km/h).

According to the above-mentioned assumption, in Eq. 4.17 it is assumed that $L_a \geq L_p$ and $v_a < v_p$. From Eq. 4.17, the flight additional fuel consumption and GHG emissions are estimated as follows:

$$FC_{hfe} = w_{hfe} \cdot f_c = (L_d/v_a - L_p/v_p) \cdot f_c \tag{4.18a}$$

and

$$EM_{hfe} = w_{hfe} \cdot f_c \cdot e_m = (L_d/v_a - L_p/v_p) \cdot f_c \cdot e_m \tag{4.18b}$$

where

f_c is the average flight fuel consumption per unit of time (Tons of fuel//h); and
e_m is the emission rate of GHG from burning Jet A fuel (ton of CO_{2e}/ton of fuel) (CO_{2e} – Carbon Dioxide equivalents).

Summing up the above-mentioned quantities for the flights affected during the given period of time, the corresponding totals can be estimated.

ii) Horizontally eventually efficient flights

An aircraft/flight is assumed to cruise at the assigned FL *(i)* of the long air route, such as, for example, Transatlantic Westbound track, of the length *(L)*. The strong jet stream is reported to take place on the lower FL *(j)* and the aircraft is given the ATC clearance to descend and continue at that FL *(j)*, aiming at shortening the trip time by increased speed thanks to the strong tailwind. At the same time, the aircraft can be imposed with higher fuel consumption and related emissions of GHG due to flying at the lower FL *(j)*. The shortening of the trip time can be estimated similarly as in Eq. 4.17. The potential improvement of the flight horizontal efficiency in terms of the eventual savings in fuel consumption and related emissions of GHG by such FL change can be estimated as follows:

$$FC_i = f_{c/i} \cdot \left(\frac{L}{v_i}\right) \text{ and } FC_j = f_{c/j} \cdot \left(\frac{L}{v_j}\right) \tag{4.19a}$$

and

$$EM_i = f_{c/i} \cdot \left(\frac{L}{v_i}\right) \cdot e_m \text{ and } EM_j = f_{c/j} \cdot \left(\frac{L}{v_j}\right) \cdot e_m \tag{4.19b}$$

where

$f_{c/i}$, $f_{c/j}$ is the average fuel consumption on FL *(i)* and FL *(j)*, respectively (kg/min; tons/h)
L is the length of air route (nm; km); and
v_i, v_j is the aircraft/flight speed on FL *(i)* and FL *(j)*, respectively (kts; km/h).

The other symbols are analogous to those in Eq. 4.18(a, b).

In Eq. 4.19(a, b), the planned horizontal efficiency of the given flight is considered as improved if the fuel consumption at FL *(j)* is lower than that at FL *(i)*, i.e., if $FC_j < FC_i$. This should happen under conditions when: $(f_{c/j} > f_{c/i})$, $(v_j > v_i)$, and $(v_j = v_i + v_w)$, where (v_w) is the speed of jet stream, i.e., the tail wind (kt; km/h). In addition, the aircraft/flight operating within the jet stream should not exceed its maximum speed regarding its design/constructive constraints. Similarly, as in the case of horizontal flight inefficiency, summing up the above-mentioned quantities for the flights flying within the jet stream in the given ATC airspace during the given period of time, the corresponding totals can be estimated.

d) Application of the models

i) Horizontally "inefficient" flight(s)

The model of flight horizontal "inefficiency" is applied to the hypothetical flight carried out by B737-800 aircraft along the air route with a length of: $L_p = 1000$ nm. The results are shown in Figure 4.22(a, b).

Figure 4.22a shows the linear relationship between the flight's extended cruising time and its horizontal "inefficiency". For example, with increasing of the horizontal inefficiency, i.e., extension of the cruising distance/length by 1%, the flight's cruising time will increase by about 1.3%. Figure 4.22b again shows the linear relationship between the flight's additional fuel consumption and its horizontal inefficiency. For example, with increasing of the horizontal inefficiency, i.e., extension of the cruising

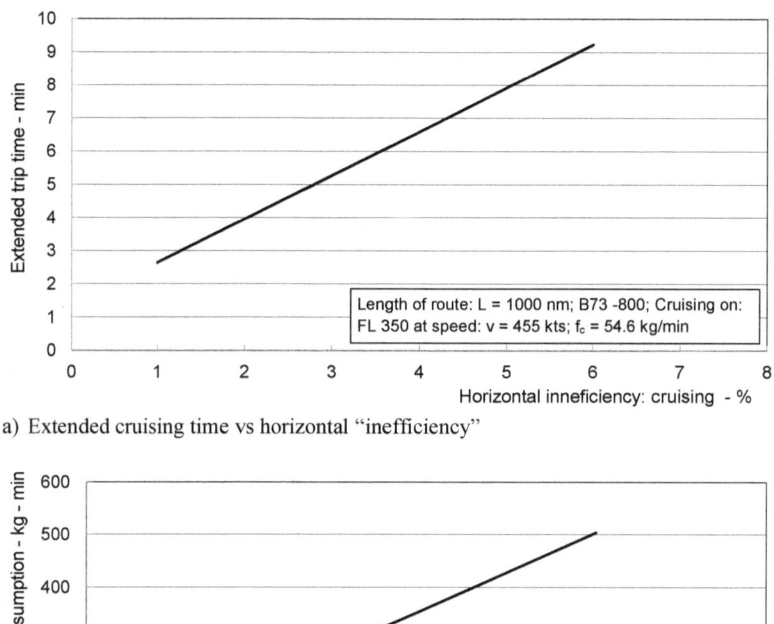

a) Extended cruising time vs horizontal "inefficiency"

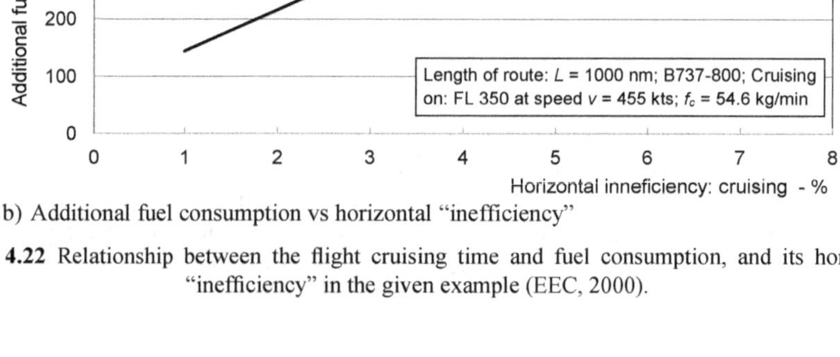

b) Additional fuel consumption vs horizontal "inefficiency"

Figure 4.22 Relationship between the flight cruising time and fuel consumption, and its horizontal "inefficiency" in the given example (EEC, 2000).

distance/length by 1%, the flight's additional fuel consumption will increase by about 72 kg/min (EEC, 2000). By multiplying these amounts by the factor: $e_m = 4.42358$ (Masiol and Harrison, 2014), the corresponding additional emissions of GHG due to the flight horizontal inefficiency can be obtained.

ii) Horizontally eventually efficient flight(s)

The model of the horizontally eventually efficient flight is applied to the hypothetical flight carried out by the aircraft A330-300 and B777-300 on the initial FL *(i)* 370, which is changed for FL *(j)* 330 due to expecting potential savings in the trip time and fuel consumption from the jet stream with speed of: $v_w = 100$ kts (tail wind) along the changing distance (https://www. skybrary.aero/index.php/Jet_Stream#Related_Articles). The other inputs are given in Table 4.4.

Table 4.4 Characteristics of the aircraft/flight speed and fuel consumption in the given example (EEC, 2000).

FL[1)]	Aircraft type	v - Cruising speed (TAS[2)]) (kts)	$f_{c/.}$ - Fuel consumption (nominal) (kg/min)
	A330-300		
330		465	96.0
370		459	87.4
	B777-300		
330		474	133.3
370		482	129.1

[1)] FL – Flight level (10^3 ft); [2)] True air speed.

The results are shown in Figure 4.23(a, b) as the relationships between savings in the trip time and fuel consumption and the distance of operating within the jet stream.

Figure 4.23a shows that the savings in the trip time for both aircraft/flight types linearly increase with an increase in the length of route when flying within the jet stream. These savings are higher for A 330 aircraft/flight due to the higher differences in the ground speed supported by the jet stream (465 + 100 – 459 = 106 kts) compared to that of its counterpart B777 (474 + 100 – 482 = 9 kts). Consequently, these time differences increase with the length of route when flying within the jet stream. Figure 4.23b shows that the fuel consumption also linearly increases with an increase in the length of the route when flying within the jet stream. Despite the lower savings in trip time, these fuel consumption savings are higher for the B777 aircraft/flight than for the A330, mainly because of its higher fuel consumption rates (FL 330 – 39% and FL 370 – 48%), as given in Table 4.4.

4.4.6.3 Model of the vertical flight inefficiency

a) System

The cruising phase of flights, depending on their planned length, takes place partially in the LAA and mostly in the HAA. Both areas are divided into the three-dimensional ATC sectors of different prismatic and/or (rarely) cylindrical shapes. In the HAA sectors, the cruising phase of flights in the horizontal plane is carried out along the assigned air

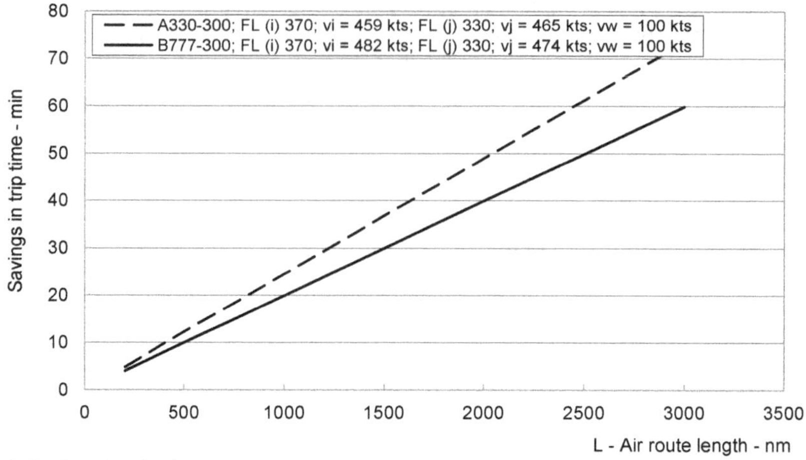

a) Savings in trip time

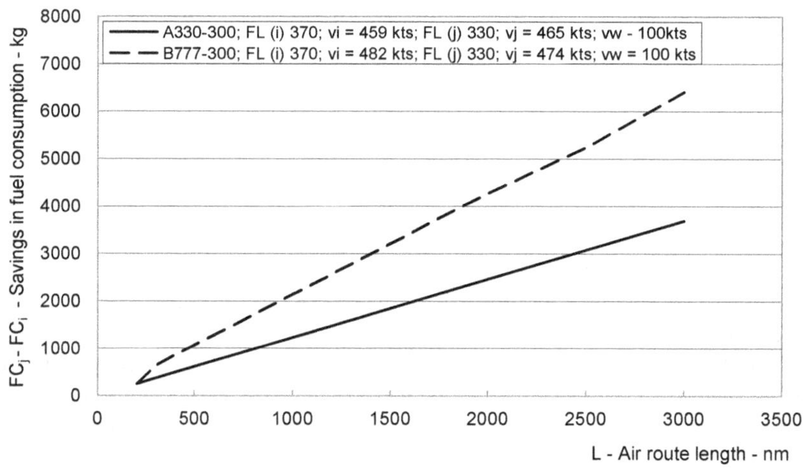

b) Savings in fuel consumption

Figure 4.23 The aircraft/flight savings in trip time and fuel consumption thanks to operating within the jet stream in the given example.

routes. These can be a part of the ATC route network with a fixed layout defined by the radio-navigational facilities and equipment or chosen by the aircraft/flights themselves. In the latter case, the aircraft/flights freely choose air routes between the entry and exit points of the sector independently on the existing ATC route network, if there is a portion of the FRA (Free Route Airspace). In such case, the configuration of the air route network is flexible in terms of time and existing traffic conditions, which, in turn, enables more flexibility regarding the utilization of the available airspace (EEC, 2016; 2019). Figure 4.24(a, b) shows the schemes in the horizontal plane.

In the vertical plane of both air route-flexibility cases shown in Figure 4.24(a, b), the aircraft/flights follow the assigned FLs. This however does not exclude that they can

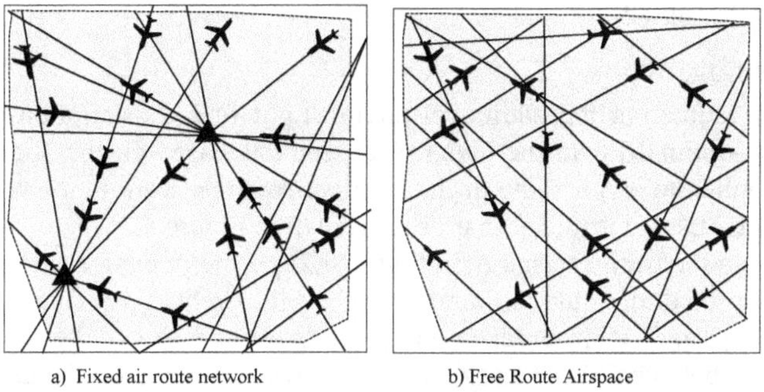

Figure 4.24 Scheme of a HAA sector with the air route networks of different flexibility in the horizontal plane.

request to change FLs while being in the sector. The ATC applies the minimum horizontal and vertical distance-based separation rules between the aircraft/flights on the same and different FLs, respectively. In addition, the horizontal distance-based separation rules are applied between the aircraft/flights changing FLs and between these and the others maintaining the unchanged/constant FLs.

b) Assumptions

Modelling the vertical inefficiency of flights includes development of the analytical model for estimating the aircraft/flight additional fuel consumption and related emissions of GHGs due to cruising at FLs that are non-optimal for fuel consumption. The model is based on the following assumptions:

- The HAA sector of any prismatic or a cylindrical shape with the fixed and flexible air route configurations is considered; as such, the sector is characterized with its volume of airspace;
- The aircraft/flight flows entering the sector at the specified entry points are of constant intensity during their average time of passing through; particular aircraft/flights are assigned the same or different FLs while entering the sector;
- The flights on the particular air routes are assumed to maintain constant or request a change of their FLs while in the sector;
- The aircraft/flights request the FL change immediately or sometime after entering the sector due to the FLs that are non-optimal for fuel consumption that were previously assigned upon entry;
- The ATC applies the minimum horizontal separation rules between the aircraft/flights changing FLs and the others maintaining the assigned FLs;
- The ATC does not influence the aircraft/flights maintaining constant FLs in order to provide the safe separation to those requesting changing FLs; and
- The aircraft/flights are assumed to be completely randomly distributed on the air routes and their flight levels in the sector according to the time-spatial homogenous Poisson processes (Homogeneity is assumed based on the assumed constant intensity of aircraft/flight flows entering the sector during their average flying times along its air routes and their flight levels) (Larson and Odoni, 2007).

c) Structure of the model

i) Fixed air route network

The model for estimating the additional aircraft/light fuel consumption due to flying on the fuel non-optimal FL in the given ATC sector is based on the above-mentioned assumptions applied to an air route of the fixed air route network in a HAA sector with several available FLs and traffic scenarios, as shown in Figure 4.25.

According to the traffic scenario on Figure 4.25, sometime after entering the sector on FL *(i)* of the fixed air route, an aircraft/flight (AC_c) requests a change to the more fuel-optimal FL *(j)* $(j > i)$. In order to achieve this change, the (AC_c) should cross the intermediate FLs between FL *(i)* and *(j)* while always remaining on the air route. At the time of crossing each of these FLs, the (AC_c) should be separated from the aircraft/flights there by the ATC distance-based minimum separation rules. Consequently, an imaginary prism as the safe separation "buffer" is associated with the (AC_c). The basis of this prism is determined by the width of air route and the ATC minimum distance-based separation rules between the (AC_c) and the aircraft/flight in front and behind it, as shown in Figure 4.25. The height of the prism is defined as the distance between the current and requested FL. Due to the already existing traffic flows on the intermediate and requested FL, the (AC_c) will be able to pass safely through with a certain probability. According to the above-mentioned assumption of the time-spatial Poisson distribution of aircraft/flights on the air routes and their FLs, this probability can be estimated for occurrence of the set of successive events as follows; the (AC_c) is separated from the aircraft/flights on each intermediate, including the required FL, at least by the ATC minimum distance-based separation rules at times when it crosses them. Under such conditions, this probability is equal to:

$$P_{ij}[X(V,t)=0] \equiv P_{ij} = \prod_{k=i+1}^{j} p_k(t_k) = \prod_{k=i+1}^{j} e^{-D_k(t_k)\cdot(a\cdot b)} \quad (4.20a)$$

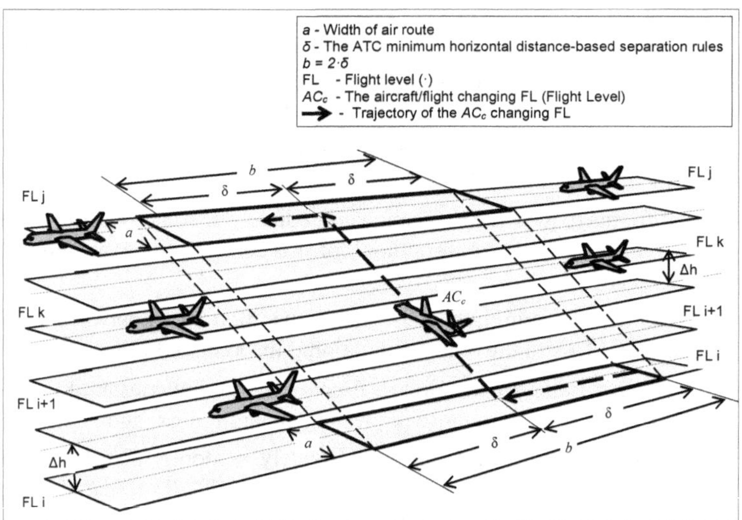

Figure 4.25 Scheme of an air route of the fixed network in a HAA sector and traffic scenario used in the model.

where

V is the volume of imaginary prism associated with the (AC_c) during changing FL (km^3; nm^3);
t is time (min; h);
i, j is the current and requested FL;
t_k is the moment when the (AC_c) crosses FL (k) (min; h);
$p_k(t_k)$ is the probability of a space gap occurring at the moment (t_k) for safe crossing FL (k);
$D_k(t_k)$ is the average spatial density of aircraft/flight flows on FL (k) (ac/km^2; ac/nm^2);
a, b is the width and length, respectively, of the imaginary prism ($b = 2\delta$) (km; nm); and
δ is the ATC minimum distance-based separation rules applied to the aircraft/flights operating on the same FLs or while changing them (km; nm).

The aircraft/flight density $D_k(t_k)$ in Eq. 4.19a can be estimated as follows:

$$D_k(t_k) = n_k(t_k)/L_k = [\lambda_k(t_k) \cdot \tau_k]/L_k \tag{4.20b}$$

where

$n_k(t_k)$ is the number of aircraft/flights on FL (k) of the given air route (-);
$\lambda_k(t_k)$ is the average intensity of aircraft/flight flows entering the given air route on FL (k) (ac/min; ac/h);
τ_k is the average aircraft/flight flows time on FL (k) of the given route (min; h); and
L_k is the length of air route measured on FL (k) (km; nm).

The intensity of aircraft/flight flows $(\lambda_k(t_k))$ in Eq. 4.20b is assumed to be constant during time (τ_k). In general, it is equal to:

$$\lambda_k(t_k) = \gamma(\tau_k) \cdot q_k(\tau_k) \text{ and } \sum_{k=i}^{j} q_k(\tau_k) = 1 \tag{4.20c}$$

where

$\gamma(\tau_k)$ is the intensity of aircraft/flight flows in the sector during time (τ_k) (ac/min; ac/h); and
$q_k(\tau_k)$ is the proportion of the aircraft/flight flows entering the sector on FL (k) (-).

The intensity of aircraft/flight flows $(\gamma(\tau_k))$ in Eq. 4.20c is assumed to be constant during time (τ_k). From Eq. 4.20(a, b, c), the average fuel consumption of the (AC_c) while passing through the sector under given conditions is equal to:

$$FC_{vfe/ij} = P_{ij} \cdot \{f_{c/i} \cdot \tau_{0/i} + f_{c/ij} \cdot (j-i) \cdot \Delta\tau + f_{c/j} \cdot [\tau_j - (j-i) \cdot \Delta\tau - \tau_{0/i}]\} + (1 - P_{ij}) \cdot (f_{c/i} \cdot \tau_i) \tag{4.20d}$$

where

$f_{c/i}, f_{c/j}$ is the average fuel consumption by the *(ACc)* on FL (i) and (j), respectively (ton/h);
$\tau_{0/i}$ is the time after entering the sector, air route and FL (i) when the (AC_c) requests changing FL (i) for FL (j) (min; h);
$\Delta\tau$ is the average time of passing between any two successive FLs (min); and
τ_i, τ_j is the average time of (AC_c) on FL (i) and FL (j), respectively, of the given route (min; h).

The other symbols are analogous to those in the previous Eqs.

In Eq. 4.19d, the average time $(\Delta\tau)$ can be estimated as follows:

$$\Delta\tau = \Delta h/w \tag{4.20e}$$

where

Δh is the vertical distance between any two successive FLs (ft; m); and

w is the vertical (climbing/descending) speed of the (AC_c) while changing FL (ft/min; m/min).

The first term in brackets in Eq. 4.20d represents the fuel consumption of the (AC_c) before requesting the FL change. The second term represents its fuel consumption while changing FL. The last term represents its fuel consumption between arriving at the requested FL and leaving it, i.e., air route and the sector. The last term in Eq. 4.20d represents the fuel consumption of the (AC_c) if not managing to change FL with the given probability. From Eq. 4.20d, the corresponding emissions of GHG can be estimated as follows:

$$EM_{vfe/ij} = EC_{vfe/ij} \cdot e_m \tag{4.20f}$$

where all symbols are analogous to those in Eqs. 4.18b and 4.19d.

ii) <u>Flexible air route network - FRA (Free Route Airspace)</u>

The model for estimating the additional aircraft/light fuel consumption due to flying on the fuel non-optimal FL in the given ATC sector is based on the above-mentioned assumptions applied to the flexible air route network in a HAA FRA sector with several available FLs and traffic scenarios shown in Figure 4.26. The possible shape of an idealized cylindrical FRA sector is shown just as an illustration because otherwise it can be of any prismatic configuration.

Similarly to the HAA sector with the fixed air route network, sometime after entering the sector on FL (i) of the self-specifying air route, an aircraft/flight (AC_c) requests a change from the fuel non-optimal to the more fuel-optimal FL (j) $(j > i)$. In order to achieve such change, the (AC_c) should cross the intermediate FLs between FL (i) and FL (j) while all the time remaining on the same air route. At the time of crossing each of these FLs, the (AC_c) should be separated from the aircraft/flights there by the ATC distance-based minimum-separation rules. Consequently, an imaginary cylinder as the safe separation "buffer" is associated with the (AC_c), as shown in Figure 4.26. Under an assumption that the aircraft/flights on each flight level between FL(i) and FL(j) are distributed according to the spatial homogenous Poisson process, the probability of changing FL(i) for FL(j) can be estimated similarly to Eq. 4.20a as follows:

$$P_{ij}[X(V, t) = 0] \equiv P_{ij} = \prod_{k=i+1}^{j} p_k(t_k) = \prod_{k=i+1}^{j} e^{-D_k(t_k)\cdot(\pi\cdot\delta^2)} \tag{4.20g}$$

where all symbols are analogous to those in Eq. 4.20a.

The aircraft/flight density on FL (k) in Eq. 4.20e, $D_k(t_k)$, can be estimated as follows:

$$D_k(t_k) = n_k(t_k)/A_k = [\lambda_k(t_k) \cdot \tau_k]/A_k \tag{4.20h}$$

A_k is the area of the HAA sector at FL (k) (km^2; nm^2).

The other symbols are analogous to those in Eq. 4.20b.

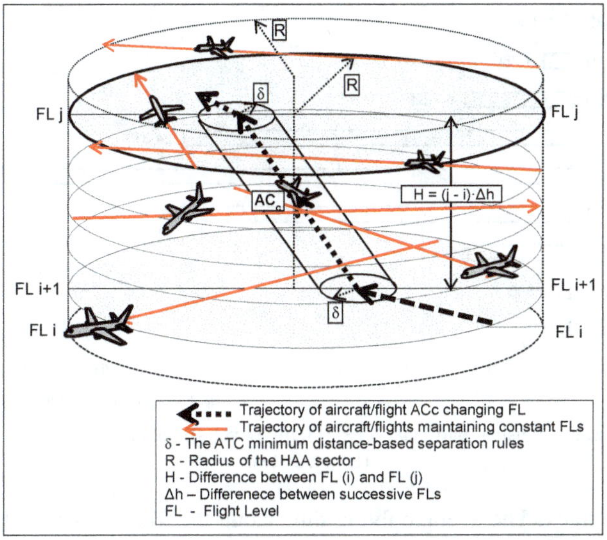

Figure 4.26 Scheme of a HAA sector with flexible air route network and traffic scenario used in the model.

The intensity of aircraft/flight flows $(\lambda_k(t_k))$ can be estimated analogously to Eq. 4.20b, and the additional fuel consumption $(FC_{vfe/ij})$ and related emissions of GHG $(EM_{vfe/ij})$ as in Eq. 4.20d and Eq. 4.20f, respectively.

d) Application of the models

i) Fixed air route network

Application of the model of the additional fuel consumption of an aircraft/flight due to operating on the fuel non-optimal in a HAA sector with fixed air route network is illustrated by using the inputs from the case of transatlantic air traffic (Section 4.4.2). The given HAA "sector" contains the fixed air route network as the day-time westbound Organized Track System consisting of (A–G) tracks, each with the available FLs 270-390 (270 – 39 · 10^3 ft). The ATC issues the so-called oceanic clearance to each aircraft/flight intending to pass through this airspace with three elements: route, FL, and speed, aiming at providing the aircraft/flight safe lateral, vertical, and longitudinal separation. In particular, depending on the current traffic situation, the aircraft/flights can be allowed to change the current FLs by the cruise climb. This can be requested either due to operating at the fuel non-optimal FL or decreasing the aircraft weight due to the fuel being consumed, thus enabling operating at higher FLs, or bad weather. In some specific cases, an aircraft/flight can request "a block of flight levels" in order to operate with a "flexible" vertical profile and gradually climb while staying on the track (ICAO, 2017).

The considered aircraft is A330-300, which has carried out about 25% of flights in the given airspace (https://www.anna.aero/2018/04/18/transatlantic-tracks-europe-americas-us-majors-lead-way-s18-london-heathrow-tops-airport-table/). The length of track on each FL is assumed to be: L = 2816 Nm (5215 km) which is equivalent to the distance between London and New York: L_0 = 3016 Nm (5585 km), decreased by approximately 100 Nm (185.2 km) at both ends due to the distance required for entering

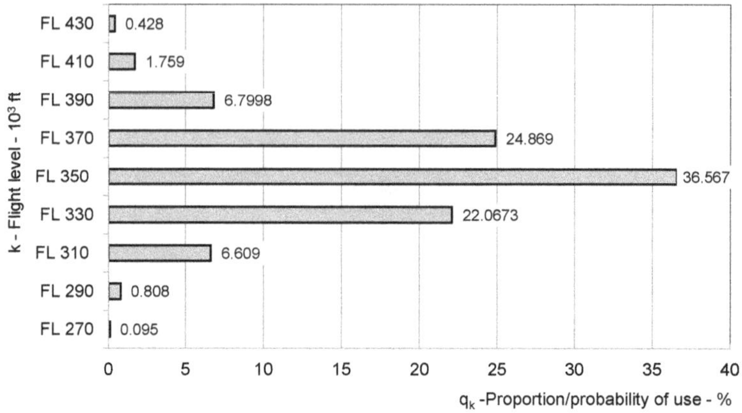

Figure 4.27 Example of using particular FLs (Flight Level(s)) along transatlantic air routes/tracks (Dhief, 2018).

and leaving the track. The average flying time along the track on each FL is adopted to be: $\tau_k = 6.0$ h ($k = 1, 2, \ldots, 9$), which corresponds to the average cruising speed of: $v = 469$ kt (869 km/h) (A330-300) (EEC, 2000). The ATC minimum longitudinal separation rules between aircraft/flights on the same FL of the particular tracks is assumed to be: τ_{min} = 10 min, which, regarding the average speed of: $v = 469$ kt, gives the minimum horizontal distance-based separation rules of: $\delta = 469 \cdot (10/60) = 78.17$ Nm. In addition, the relative use of FLs of the particular tracks is shown in Figure 4.27.

As can be seen, the most commonly used was FL 350, followed by FL 370 and FL 350. The least used were FL 270, FL 430, FL 410, and FL 290. Under such conditions, it is assumed that the aircraft/flight *(ACc)* enters the air route/track at FL *(i)* ≡ FL 310, FL 330, or FL 350, and later on, after time: $\tau_{0/i} = 2.0$ h, requires a change from these to FL *(j)* ≡ FL 330, FL 350, or FL 370, respectively. If it manages to follow the given combination of FLs, the aircraft/flight *(ACc)* stays at the FL it switched to until it leaves the air route/track. The average rates of fuel consumption on particular FLs are $f_{c/310}$ = 101.4, $f_{c/330} = 96.0$, $f_{c/350} = 91.3$, and $f_{c/370} = 87.4$ km/min. These rates while changing flight levels are $f_{c/310-330} = 161.3$, $f_{c/330-350} = 150.2$, and $f_{c/350-370} = 139.7$ kg/min. The average climbing speeds are $w = 1690$, 1520, and 1360 ft/min, respectively (EEC, 2000). The aircraft/flight flows are assumed to continuously enter the air route/track during the average time of staying there, i.e., $\tau_k = 6.0$ h ($k = 1, 2,\ldots,9$). This gives the average density of aircraft/flights on particular FLs as: $D_k = (q_k \cdot \lambda \cdot \tau_k)/L = (q_k \cdot \lambda \cdot 6)/2816$ (ac/nm), where *(λ)* is the intensity of aircraft/flight flows entering the air route/track (ac/h) and *(qk)* is the probability of using FL *(k)* given in Figure 4.27. The results, consisting of the probabilities of successful change of FL and related fuel consumption, are shown in Figure 4.28(a, b).

Figure 4.28a shows that the probability of changing FL decreases more than proportionally with increasing of the intensity and density of aircraft/flows entering the air route/track and on its FLs. This implies that it decreases if the density of aircraft/flights on the requested FLs is higher than what was intuitively expected. If the difference between the current and requested FL is increasing, the probability of their successful and safe exchange will be almost negligible, tending to zero. For example, passing safely from FL 270 to FL 370 will be almost impossible even under conditions of moderate

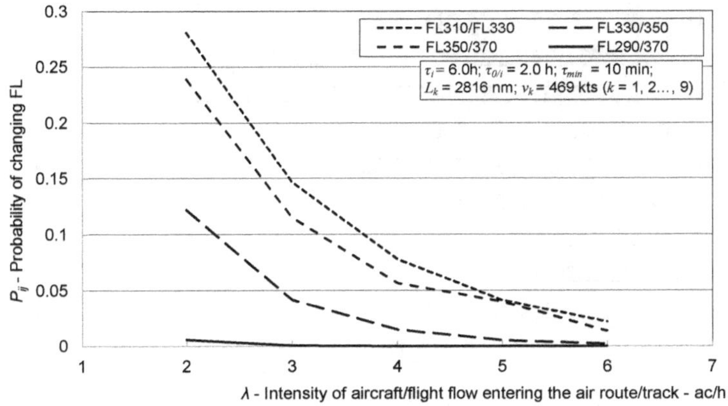

a) Probability of changing given combination of FLs

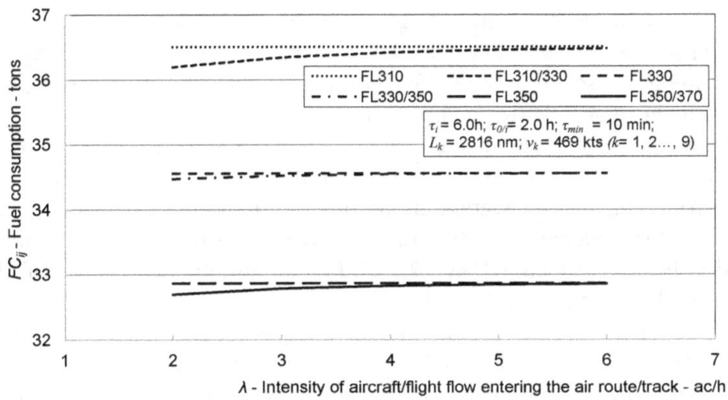

b) Fuel consumption due to changing given combination of FLs

Figure 4.28 Relationships between the probability of changing FL (Flight Level), related fuel consumption and the intensity of aircraft/flight flows entering the air route/track in the given example.

intensities and densities of aircraft/flows on the intermediate FLs. Figure 4.28b shows that the average fuel consumption due to FL change is generally lower than that of remaining at the current FL. In any combination of FL change, it decreases at a decreasing rate as the intensity and density of aircraft/flight flows on the requested FLs increase, mainly influenced by decreasing probability of successful safe FL change. In the absolute terms, these differences are also dependent on the fuel consumption rates at particular current and requested FLs, i.e., higher at combinations of the lower FLs than at the higher FLs.

ii) Flexible air route network - FRA (Free Route Airspace)

Application of the model of the additional fuel consumption of an aircraft/flight due to operating on the fuel non-optimal in a HAA sector with the flexible air route network, i.e., FRA (Free Route Airspace) is illustrated by using the inputs from the European case. The given HAA "sector" contains the flexible air route network with the available FLs 310–390 (310–39 · 10^3 ft) (EEC, 2017). Similarly to the case of the fixed route network,

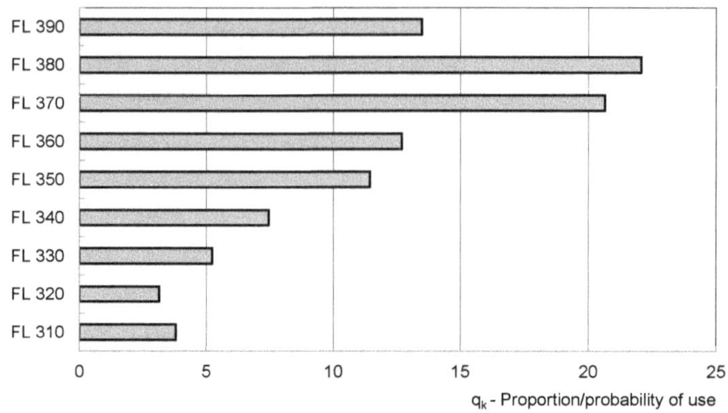

Figure 4.29 Example of distribution of flight levels used in the given example—Case of ECAC (European Civil Aviation Conference) area (June 2016–week 25) (EEC, 2016).

ATC gives a clearance to the aircraft/flight requested routes after checking if they satisfy the minimum separation rules. The considered aircraft requesting to changing its current FL while in the given sector is B 737–800, which has about a 10% share in the fleets of traditional scheduled and charter, and 40% in the fleets LCCs (EEC, 2016).

The area of the sector is assumed to be: $A_k = 41000$ nm^2 with an average route length through it of: $L_k = 230$ nm ($k = 1, 2,.., 9$). Given the average aircraft speed of: $v_k = 453$ kts (839 kmh), this gives the average flying time through the sector of: $\tau_i = 0.5$ h. The ATC minimum longitudinal separation rules between the aircraft/flights on the same FL is assumed to be: $\delta_{min} = 5$ nm (ICAO, 2016). In addition, the relative use of FLs in the given sector is shown in Figure 4.29.

As can be seen, the most commonly used was FL 380, followed by FL 370 and FL 390. The least used were FL 320, FL 330, and FL 310. Under such conditions, it is assumed that the aircraft/flight *(ACc)* enters the air route/track at FL *(i)* ≡ FL 310, and later on, after time: $\tau_{0/i} = 5$ min, needs to change these for FL *(j)* ≡ FL 390, respectively. The average fuel consumption rate of the aircraft B 737-800 on these levels is: $f_{c/310} = 5.56$ and $f_{c/390} = 53.4$ kg/min. This rate during changing FL 310 for FL 390 is equal to: $f_{c/310-390} = 79.6$ kg/min. The average climbing speed is: $w = 1470$ ft/min (EEC, 2000). If manages to change the given combination of FLs, the aircraft/flight *(ACc)* stays at the changed FL until leaving the sector. The aircraft/flight flows are assumed to continuously enter the air route/track during the average time of staying there, i.e., $\tau_k = 0.5$ h ($k = 1, 2,..., 9$). This gives the average density of aircraft/flights on particular FLs as: $D_k = (q_k \cdot \lambda \cdot \tau_k)/A_k = (q_k \cdot \lambda \cdot 0.5)/41000$ (ac/nm), where (λ) is the intensity of aircraft/flight flows entering the sector (ac/h) and *(qk)* is the probability of using FL *(k)* given in Figure 4.29. The results consisting of the probabilities of successful change of FL and related fuel consumption are shown in Figure 4.30(a, b).

Figure 4.30a shows that the probability of FL change is relatively high and decreases almost more than proportionally with an increase in the aircraft/flight flows entering the sector and consequently their density there. This has been intuitively expected given the proportion of use and consequent traffic density on a particular intermediate, including the requested FL. Figure 4.30b shows that the average fuel consumption in case of FL change would be higher than otherwise, i.e., if the aircraft/flight remained on the current one independently of the intensity of aircraft/flight flows. In addition, this

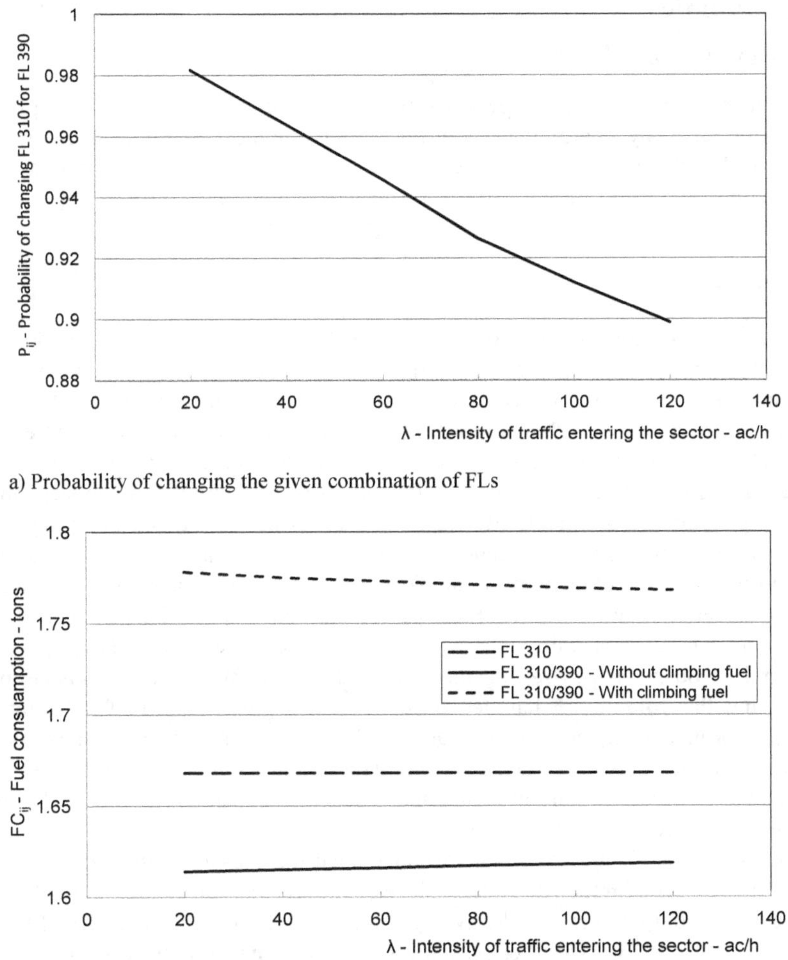

Figure 4.30 Relationships between the probability of changing FL (Flight Level), related fuel consumption and the intensity of aircraft/flight flows entering the sector (FRA – Free Route Airspace) in the given example.

fuel consumption would decrease at a decreasing rate with an increase in the intensity and density of aircraft/flight flows despite reducing the probability of FL change. This indicates that, under given conditions, the aircraft/flight *(AC_c)* would be "forced" in some sense to remain at the current FL all the time while in the sector. Some reasons are as follows: the difference in the average fuel consumption on the current (FL 310) and requested (FL 390) level for the given aircraft/flight (B 737-800) was relatively small. As such, the corresponding fuel savings would not be substantial due to the short flying time through the sector, (0.5 h) combined with a rather high fuel penalty for the FL change. In total, the results indicate that the flexible air route network would offer also flexible high probability of successful safe changing of FLs. However, the feasibility of such changes should be always checked, depending on the configuration and prevailing traffic conditions in the given sector.

4.5 Economics

4.5.1 Description

In general, most these systems operate at the zero-profitable entities, implying that the revenues obtained by charging users-airline flights should cover the operating costs. Therefore, dealing with the economics of these systems in the given context mainly relates to charges for the air navigational services provided to the aircraft/flights passing through a given (controlled) airspace under their jurisdiction.

4.5.2 System

As mentioned in Chapter 1, the economics of ATC systems or ANSP worldwide mainly relate to their revenues, costs, and their differences, i.e., profits. In the given context, economics of the ATC system relate to the costs of provision of ANS (Air Navigation Service) to the users, i.e., airline aircraft, in order to enable safe, efficient, and effective flights between their origin and destination airports under given conditions. In order to cover these costs, an ATC system or ANSP charges its users by the amounts enabling at least zero profitability of operations, i.e., at least covering the total costs during a given time period (usually one year). In managing the system's operations, similarly to airlines and airports, the prime objectives of an ATC system are to simultaneously maximise the profits (difference between the revenues and costs) and expectations of users (airline flights). This can be carried out either by maximising revenues, minimising costs, or by both simultaneously. In general, on the one hand, the adopted charging system/policy of ANS and the volumes of handled traffic during a given period of time influence conditions for increasing the revenues. On the other, the prices/costs of main inputs (such as generally labour, capital, and energy) and the internal system's efficiency in spending the available resources mainly influence the costs. Consequently, the profits depend on the differences between the two aforementioned factors.

The above-mentioned costs and revenues have three basic characteristics that are regarded as important for their analysis: (i) level, variability and structure; (ii) relationships with the system output; and (iii) the average and marginal cost/revenue (ISAVIA, 2019; Janić, 2000; NC, 2013). Specifically, the *total costs* primarily depend on the size and operational characteristics of the facilities and the volumes of other resources consumed (labour, energy) for provision of services. For example, larger and more sophisticated ATC units will have higher total investment and operating costs than their smaller and less sophisticated counterparts. Regarding the variability in dependence on the volume of output, the costs can be classified as fixed and variable. The former are not dependent on the volumes of output in the short-term. The latter change as the volumes of output change. The cost structure can be expressed by share of the costs of particular physical inputs (labour, capital and energy) in both total and variable costs. The average cost per unit of output can be estimated by dividing the total cost by the total volumes of output, both during a given period of time. The marginal cost per unit of output can be obtained by deriving an analytical form of the total cost function subject to the volume of output as the independent variable. In this case, the cost function has to be continuous, i.e., having at least the first derivative at each point within the range of possible outputs (Janić, 2000). At the ATC system, the number of flights handled during a given period of time under given conditions can represent its output. Again, similarly to the case of airports, the output from the ATC system reflects its capacity.

4.5.3 Modelling Economics

Modelling the economics of an ATC system in the given context deals with the principles and models of economies of scale, the models of charges for provision of ANS to users, i.e., aircraft/flights, and examples of their application.

4.5.3.1 The principles and model of economies of scale

The general relationships between the total costs and volume of output of an ATC or ANSP system may have different forms. Nevertheless, the three typical forms shown in Figure 4.31 are characteristic.

In each, it is assumed that the initial cost is required to serve any volume of output and that there is the output at the moderate level (O_1) where the average cost of producing an additional unit is relatively constant. Curve 1 assumes that the cost of producing an additional unit of output starts to rise beyond a certain volume of output, say (O_2). This means that production of each additional unit of output greater than (O_2) costs more. This operating regime indicates the system's inefficiency reflecting its diseconomies of scale. Curve *2* indicates that the cost of an additional unit of output remains constant for any volume of output (O), thus reflecting the system's constant return to scale. Curve *3* indicates that each additional unit of output will cost less beyond the volume of output (O_2), thus indicating increasing of the system's efficiency as economies of scale (Manheim, 1979). Based on Figure 4.32, the system's economies of scale can be defined as decreasing the average costs per unit of output while increasing the total volume of output in the long-term (five or more years). In addition, the system's *economies of density* is defined as decreasing the average cost per unit of output while increasing density (i.e., concentration) of output (Bannock et al., 1988). In the given ATC system, both economies of scale and economies of density can be modelled by specifying the analytical form of the cost function depending on the volume of output. The cost function allows the average and marginal cost per unit of output to be determined. Let (O), $(A(O))$, and $(M(O))$ be the volume of output, the average and marginal cost, respectively, of an ATC system. Detecting and measuring the economies of scale can be carried out as

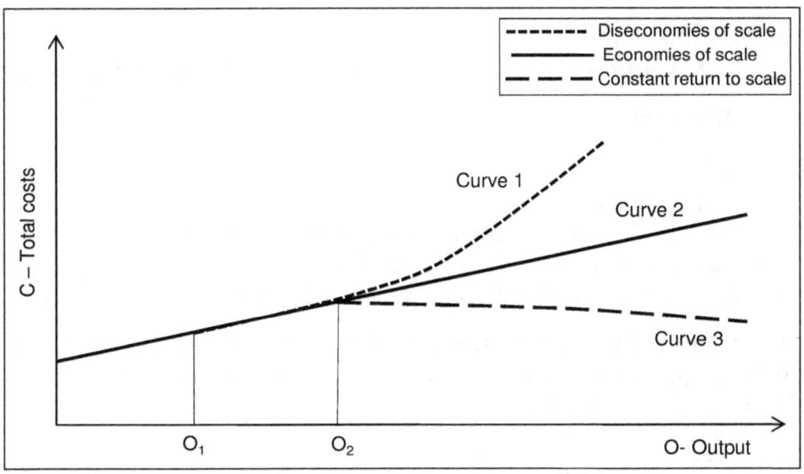

Figure 4.31 Relationship between the total cost and volume of the system output (Manheim, 1979).

follows: $l(O) = AC(O)/MC(O)$. Then, if $l(X) > 1$ there are economies of scale; if $l(X) = 1$ there is constant return to scale; and if $l(X) < 1$ there are diseconomies of scale at the given volume of output (Weisman, 1990).

4.5.3.2 The models of charges for providing ANSs (Air Navigation Service(s))

a) Background

The charges for provision of ANS are based on the cost characteristics of a given ATC system and the airspace under its jurisdiction, and the characteristics of users, i.e., aircraft/flights. For example, in Europe, the terminal and en-route charges can generally be distinguished (EEC, 2019a). In the U.S., these are terminal, en-route, and Oceanic charges (FAA, 2019a). The European terminal[6] charges are set up for departing aircraft/flights while in the U.S. these are set up for the arriving aircraft/flights. The en-route charges are set up for the aircraft/flights being en-route between their origin and destinations airports. In the U.S., the aircraft/flights flying over the country are charged by the en-route charges. The 'Oceanic' charges are also appropriately applied to the aircraft/flights flown over the Oceanic controlled airspace by the ATC/ATM units having jurisdiction there.

b) Assumptions

In dealing with the models of charging the ATC services, the following assumptions have been introduced:

- The models for charging the air navigational services are already known and, as such, widely applied by ATC systems or ANSP worldwide;
- These models have generally been based on the characteristics of aircraft carrying out particular flights, their routes, and the cost characteristics of ATC units providing the air navigation services; and
- The models are applied to the hypothetical cases of estimating terminal, en-route, and Oceanic charges carried out by different aircraft types.

c) Structure of the models

i) Model of TMA charges

The model of TMA charges has the following structure (EEC, 2019a; Sridhar et al., 2015):

$$C_{ta} = u_{tma} \cdot (MTOW/a)^b \tag{4.21a}$$

where

u_{tma} is the unit rate of TMA charge ($US or €/ton);
$MTOW$ is the aircraft certified Maximum-Take-Of-Weight (tons);
a is the factor commonly taking the values (1, 50, 100, ..);
b is the factor commonly taking the values (1.0, 0.5, 0.7.).

For example, EUROCONTROL sets up the TMA charges for departing aircraft/flights only by applying the factors: $a = 50$ and $b = 0.70$ (EEC, 2019a). Consequently, Eq. 4.21a is transformed as follows:

[6] TMA or 'Terminal' ANS charges are different from the airport charges. The latter mainly include landing, passenger, cargo, parking, hangar, and noise (EEC, 2019a).

$$C_{ta} = u_{tma} \cdot (MTOW/50)^{0.70} \tag{4.21b}$$

In the U.S. the factors *(a)* and *(b)* can differ between the airports, but some typical values applied to the arriving/landing aircraft/flights are: $a = 0.454$ and $b = 1.0$ (Sridhar et al., 2015). Both in Europe and the U.S., the unit rate of TMA charge *(u_{tma})* is usually set up by dividing the forecast number of chargeable TMA service units, i.e., aircraft/flights for the given year, by the corresponding estimated costs for provision of the air navigation services. In particular, in Europe this factor is ATC country specific (EEC, 2019a).

ii) Model of the en-route and oceanic charges

The model of en-route and Oceanic charges generally takes into account the unit rate charge, the aircraft weight, and flying distance as follows (EEC, 2019a; Sridhar et al., 2015):

$$C_{er} = u_{er} \cdot (MTOW/a)^{b} \cdot d_{er} \tag{4.22a}$$

where

u_{er} is the unit rate of en-route and/or Oceanic charge ($US/km or €/km);
$MTOW$ is the aircraft certified Maximum-Take-Of-Weight (tons);
a is the factor commonly taking the values (1, 50, 100);
b is the factor commonly taking the values (1.0, 0.5, 0.7);
d_{re} is the en-route and/or Oceanic great circle distance (km).

In Europe, EUROCONTROL uses the factors *(a)* and *(b)* in Eq. 4.22a as: $a = b = 50$. Consequently, Eq. 2a becomes (EEC, 2019a):

$$C_{er/o} = u_{er/o} \cdot (MTOW/50)^{0.50} \cdot (d_{er/o}/100) \tag{4.22b}$$

In the USA, FAA uses the factors *(a)* and *(b)* in Eq. 4.22a as: $a = 1.0$ and $b = 0$ for both en-route and Oceanic flights. Thus, Eq. 4.22b transforms as (FAA, 2019a):

$$C_{er_o} = u_{er/o} \cdot MTOW \cdot (d_{er/o}/100) \tag{4.22c}$$

Again, in both cases, the unit rate charge *(u_{er})* is set up to cover the costs of ATC units providing the services to aircraft/flights passing through the airspace of their jurisdiction. In addition, the great circle distance *(d_{er})* between the aircraft/flight(s) origin and destination airports is reduced by a certain amount in order to reflect the actual flying distance to be charged.

d) Application of the models

i) Economies of scale

As mentioned above, ANSP and their ATC system units have to provide safe, efficient, and effective aircraft/flight operations within the airspace of their jurisdiction. In order to be able to fulfil these rather complex tasks, the ATC system needs to be adequately equipped with the advanced facilities and equipment, both on the ground and in the air, operated by the highly trained staff—ATC controllers. In general, the corresponding investments and funds for covering the systems' daily operating costs have mostly been provided by the national governments, primary users of their services, i.e., airline aircraft/flights, and private companies, the latter to a lesser extent. Under such conditions, a question may arise as to whether there may be some economies of scale in the particular ANSP and/or ATC systems indicated, according to the above-mentioned definition, i.e., decreasing the average costs per ANS service while increasing the volume of services carried out.

To this end, the examples of the total and average cost vs the volumes of output are considered. In particular, the former reflects the generic cases on Figure 4.31. Therefore, Figure 4.32(a, b) shows the relationship between the total and average en-route costs of operating 37 European ANSP (Charging Zones under jurisdiction of EUROCONTROL) and the number of TSUs[7] (Traffic Service Unit(s)) handled during the given period of time (EEC, 2017; 2019a).

Figure 4.32a shows that the total costs have generally increased at a decreasing rate with increasing of the annual volumes of output-handled TSUs. This approximately

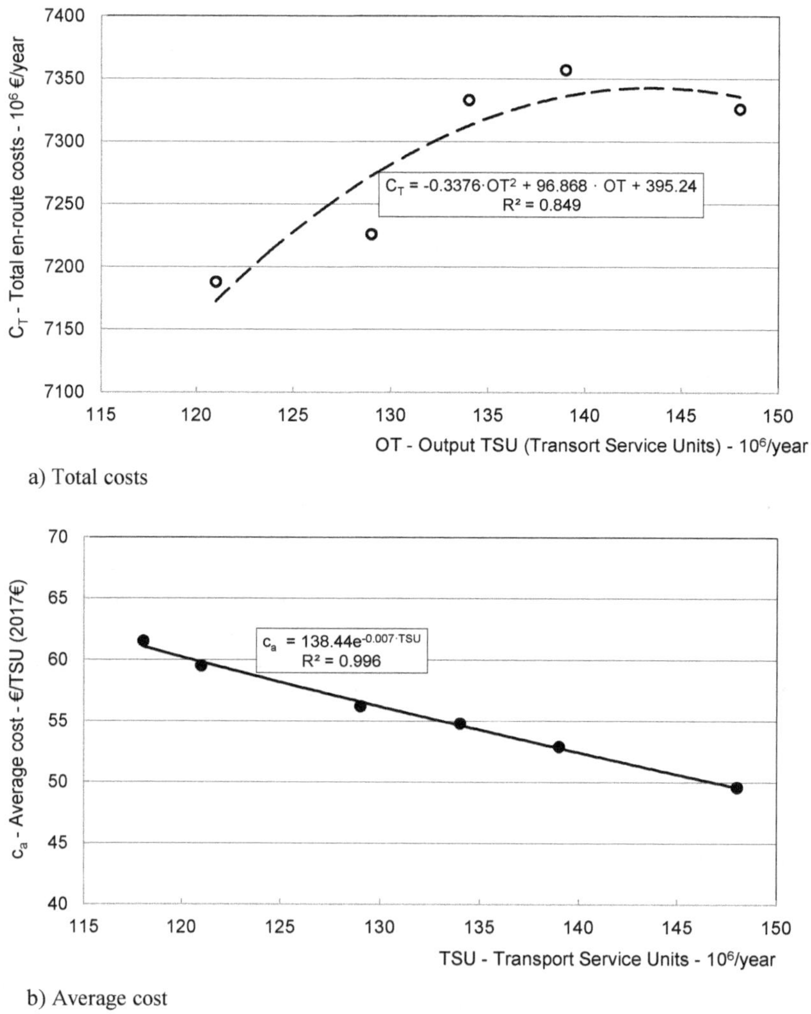

Figure 4.32 Relationship between the costs and the volume of output (the annual number of TSUs)—Case of European ANSP (EUROCONTROL) (Period: 2012–2017) (EEC, 2017; 2019a).

[7] En-route TSU (Traffic Service Unit) is calculated as the product between the distance factor and the weight factor of the aircraft/flight concerned (EEC, 2017; 2019; 2019a).

corresponds to Curve 3 on Figure 4.31, thus indicating the existence of economies of scale. Figure 4.32b shows that, during the observed period, the average cost per TSU has decreased more than proportionally while increasing the number of processed TSUs, thus again clearly confirming the existence of economies of scale in the given case. In addition, Figure 4.33 shows the relationship between the average cost per flight and the annual number of handled flights during the specified period of time for 37 European ANSPs (Charging Zones under jurisdiction of EUROCONTROL) and the U.S. FAA-ATO (Air Traffic Operations) ANSP is investigated.

These results indicate that, generally, the annual number of flight hours has been about 70% higher in the U.S. than in European ANSPs. Consequently, the average cost per flight hour in Europe for the average annual number of flight hours of: FH = 14 · 10^6/year was: $c_a \approx 551$ €/FH. The corresponding average cost per flight in the USA for the annual number of flight hours of: FH = 26 · 10^6/year was: $c_a \approx 314$ €/FH, which is about 40 fewer than in Europe. Furthermore, in both cases, these average costs have decreased more than proportionally with an increase in the number of flight hours, thus indicating the existence of economies of scale during the observed period of time. In the European ANSP, the rate of gaining economies of scale has been higher by about 20% than that in the U.S. FAA-ATO.

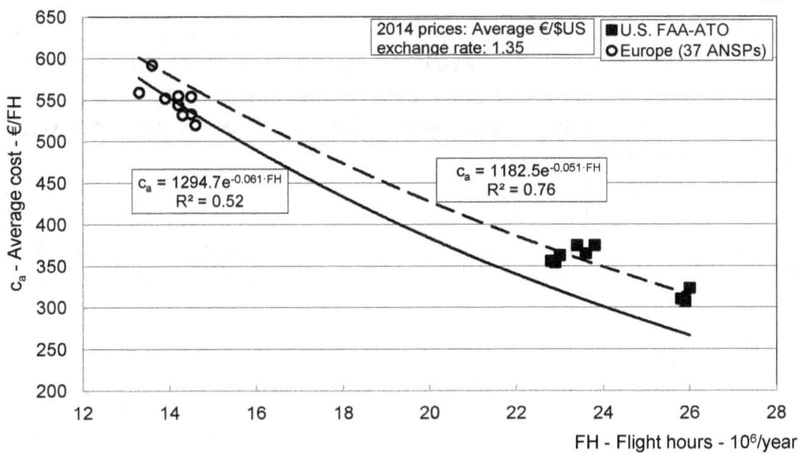

Figure 4.33 Relationship between the ANSP's average cost per flight hour and the annual number of FHs (Flight Hour(s))—Case of Europe and U.S. (Period: 2006–2014) (EEC, 2016).

ii) Charges for provision of ANSs (Air Navigation Services)

The above-mentioned charging models of ANSs (Air Navigation Services) are applied to two hypothetical cases. The first relates to the TMA charging model handled by the European ANSP – EUROCONTROL—on behalf of its 41 national/country's ATC system units. The other relates to the en-route and Oceanic charging model applied by the U.S. ANSP – FAA—to a hypothetical flying distance/route in the corresponding airspace under its jurisdiction. In both cases, it is assumed that the departures and flights, respectively, have been carried out by different aircraft types. In the former (European) case, the unit rate of terminal charge is adopted to be an average from the contracted countries of: u_{tma} = 56.78 €/departure (EEC, 2017). Figure 4.34 shows the relationship between the ATC

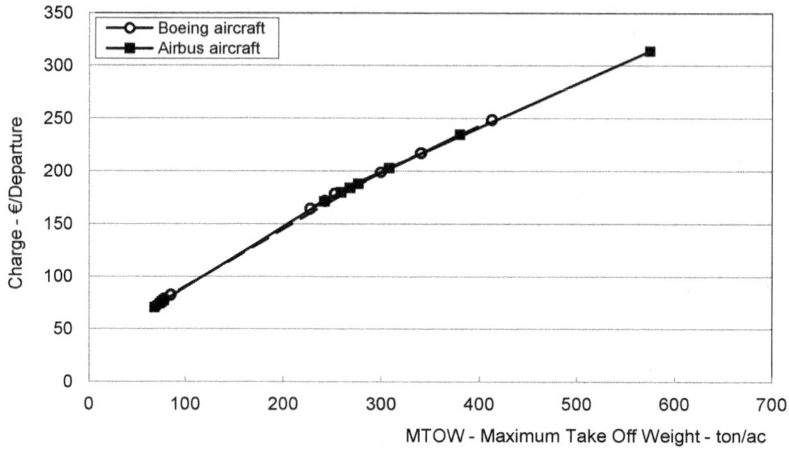

Figure 4.34 Relationship between the TMA charges and the aircraft MTOW (Maximum Take OFF Weight)—Case of Europe (EUROCONTROL) (EEC, 2017).

terminal charges and the aircraft MTOW if departures are realized by different (Boeing and Airbus) aircraft.

As can be seen, as intuitively expected, the charge per single departure has increased as the aircraft MTOW (Maximum Take-Off Weight) has increased at a decreasing rate, as expected regarding the structure of the above-mentioned models. At the same time, it has been similar for nine of Boeing's and twelve of Airbus' aircraft types within the corresponding range of their MTOW.

As far as charging the aircraft/flights operating in the en-route and/or Oceanic airspace is concerned, the U.S. FAA unit rates of: u_o = 61.75 \$US/100 nm or u_o = 0.333 \$US/km for the en-route, and u_0 = 26.51 \$US/100 nm or u_o = 0.143 \$US/km for the Oceanic flights are considered (FAA, 2019a). Figure 4.35 shows the 3application of this

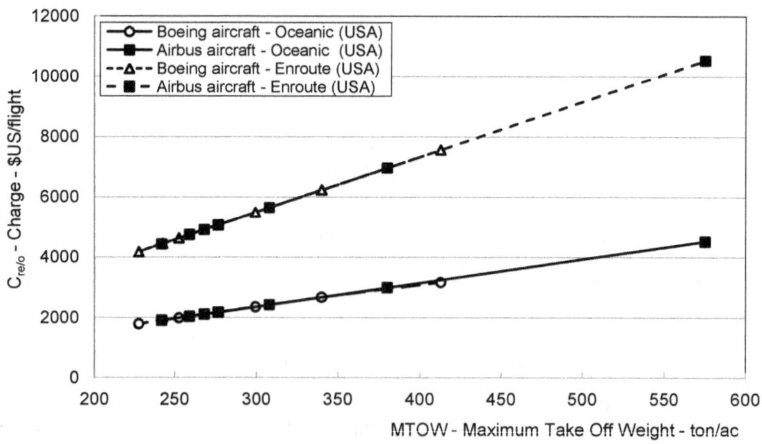

Figure 4.35 Example of the relationship between the en-route and Oceanic charges and the aircraft MTOW (Maximum Take-Off Weight)—Case of the U.S. FAA-ATO (Air Traffic Organization) (FAA, 2019a).

rate to a hypothetical flight carried out by different long-haul aircraft types along the en-route and Oceanic distance of: d_o = 5500 km (an equivalent to the distance between London and New York).

As can be seen, similarly to the case of terminal charges, the en-route and Oceanic charges have increased as the aircraft MTOW has increased at a decreasing rate, again as expected regarding the structure of the corresponding models. In addition, they have been similar for six of Boeing's and seven of Airbus' long haul aircraft types within the corresponding range of their MTOW.

References

Albalate, D., Bel, G. and Fageda, X. (2014). Competition and cooperation between high-speed rail and air transportation services in Europe. Journal of Transport Geography, http://dx.doi.org/10.1016/j.jtrangeo.2014.07.003.

Ang, H.-S. A. and Tang, H. W. (1974). Probability Concepts in Engineering Planning and Design: Basic Principles. Vol. 1, John Wiley and Sons, New York, USA.

Bannock, G. et al. (1988). Dictionary of Economics. 4th Ed., Penguin Books, London, England, UK.

Carlson, J. S. and Jensen, C. M. (1982). Reaction time, movement time, and intelligence: a replication and extension. Intelligence, 6: 265–274.

Dhief, I. (2018). Optimization of Aircraft Trajectories over the North Atlantic Airspace, PhD Thesis, Optimization and Control, L'Université de Toulouse, Toulouse, France.

EEC. (2000). Aircraft Performance Summary Tables for the Base of Aircraft Data (BADA), Revision 3.3, EEC Note No. 18/00, EUROCONTROL Experimental Centre Brétigny-sur-Orge, CEDEX, France.

EEC. (2003). Pessimistic Sector Capacity Estimation, EEC Note No. 21/03, Project COCA, EUROCONTROL Experimental Centre, Brussels, Belgium.

EEC. (2007). Capacity Assessment & Planning Guidance: An Overview of the European Network Capacity Planning Process, Edition September 2007, EUROCONTROL, Brussels, Belgium.

EEC. (2008). European Medium-Term ATM Network Capacity Plan Assessment 2009–2012, DMEAN - Dynamic Management of the European Airspace Network < Edition September 2008, EUROCONTROL, Brussels, Belgium.

EEC. (2009). Implications of End-to-End Communication for Air Traffic Control, EEC Technical/Scientific Report No. 2009–012, Project: LTI-EtEcom, EUROCONTROL Experimental Centre Brétigny-sur-Orge, CEDEX, France.

EEC. (2016). Market Segments in European Air Traffic 2015, European Organisation for the Safety of Air Navigation, Brussels, Belgium.

EEC. (2019). European Route Network Improvement Plan: PART 2 European ATS Route Network - Version 2019–2024, EUROCONTROL, European Organisation for the Safety of Air Navigation (EUROCONTROL), Brussels, Belgium.

EEC. (2013). Challenges of Growth 2013, Task 7: European Air Traffic in 2050, EUROCONTROL—European Organisation for the Safety of Air Navigation, Brussels, Belgium.

EEC. (2016). U.S. – Europe Comparison of ANS Cost-efficiency Trends 2006–2014—Update, Performance Review Unit, EUROCONTROL, European Organisation for the Safety of Air Navigation, Brussels, Belgium.

EEC. (2017). Report on the Operation of the Route Charges System in 2016, Central Route Charges Office (CRCO), EUROCONTROL, European Organisation for the Safety of Air Navigation, Brussels, Belgium.

EEC. (2018). European Aviation in 2040—Challenges of Growth, Annex 1: Flight Forecast to 2040, EUROCONTROL—European Organisation for the Safety of Air Navigation, Brussels, Belgium.

EEC. (2019a). Eurocontrol Central Route Charge Office: Customer Guide to Charges, Central Route Charges Office, EUROCONTROL, European Organisation for the Safety of Air Navigation, Brussels, Belgium.

EEC/FAA. (2017). Comparison of Air Traffic Management-related Operational Performance U.S./Europe Traffic Management-Related, US FAA (Federal Aviation Association, Air Traffic Organization

System Operations Services, Washington D.C. USA, EUROCONTROL, European Organisation for the Safety of Air Navigation, Brussels, Belgium.
FAA. (2005). FAA Aerospace Forecasts: Fiscal Years 2006–2017, U.S. Department of Transportation, Federal Aviation Administration, Office of Policy & Plans, Washington DC, USA.
FAA. (2018). Aerospace Forecast: Fiscal Years 2018–2038, Federal Aviation Administration, Washington D.C., USA.
FAA. (2019). Air Traffic by the Numbers, U.S., Federal Aviation Administration, Washington D.C., USA.
FAA. (2019a). Update of Overflight Fee Rates, 14 CFR Part 187, Docket No.: FAA–2015–3597; Amdt. No. 187–36, RIN 2120–AK53, Federal Aviation Administration, Department of Transportation, Federal Register, 81(229): 85843–85854, Washington, D.C., USA.
http://www.ainonline.com/aviation-news/air-transport/2011-05-09/longitudinal-airspace-separations-reduced-over-north-atlantic/
http://www.eurocontrol.int/articles/free-route-airspace/
http://www.faa.gov/nextgen/
http://www.sesarju.eu/
http://d-maps.com/carte.php?num_car=3211&lang=en/
https://www.anna.aero/2018/04/18/transatlantic-treks-europe-americas-us-majors-lead-way-s18-london-heathrow-tops-airport-table/
https://www.transtats.bts.gov/HomeDrillChart.asp/)
https://www.transtats.bts.gov/OT_Delay/ot delaycause1.asp?type=5&pn=1
https://www.skybrary.aero/index.php/Jet_Stream#Related_Articles
https://www.flightradar24.com/
ICAO. (2016). Procedures for Air Navigation Services—Air Traffic Management, Doc 4444, Sixteenth Edition, International Civil Aviation Organization, Montreal, Canada.
ICAO. (2017). North Atlantic Operations and Airspace Manual, v. 2017–2017-1, International Civil Aviation Organization, European and North Atlantic (Eur/Nat) Office, Neuilly-Sur-Seine, Cedex, France.
ICAO. (2018). Manual on Collaborative Decision-Making (CDM), Doc. 9971, International Civil Aviation Organization, Montreal, Canada.
ISAVIA. (2019). International Route Air Navigation Services—Charges, Isavia ohf, Reykjavik Airport, Reykjavík, Island.
Janić, M. (1993). A model of competition between High Speed Rail (HSR) and Air Transport (AT). Transportation Planning and Technology, 17: 1–23.
Janić, M. (1997). Model of air traffic control sector capacity based on air traffic controller workload. Transportaion Planning and Techology, 20(4): 331–335, DOI:10.1080/03081069708717596.
Janić. (2000). Air Transport System Analysis and Modelling: Capacity, Quality of Services and Economics. Gordon and Breach Science Publishers, Volume 16, Amsterdam, The Netherlands.
Janic, M. (2000a). An assessment of risk and safety in civil aviation. Journal of Air Transport Management, 6(1): 43–50.
Kleinock, L. (1975). Queuing Systems, Volume I: Theory, Wiley, New York, USA.
Kleinock, L. (1976). Queuing Systems, Volume II: Computer Applications, Wiley, New York, USA.
Larson, C. R. and Odoni, R. A. (2007). Urban Operations Research, 2nd edition, Charlestown, Massachusetts, USA.
Lehrl, S. (1988). The basic parameters of human information processing. Personality and Individual Differences, 5(9): 883–896.
Loft, S., Sanderson, P., Neal, A. and Mooij, M. (2007). Modeling and predicting mental workload in en-route air traffic control: critical review and broader implications. Human Factors, 49(3): 376–399.
Majumdar, A. and Polack, J. (2001). Estimating capacity of Europe's airspace using a simulation model of air traffic controller workload. Transportation Research Record, 1744: 30–43.
Manheim, M. (1979). Fundamentals of Transportation Systems Analysis, Volume 1 Basic Concepts, MIT Press, Massachusetts Institute of Technology, Cambridge, Massachusetts, USA.
Masiol, M. and Harrison, M. R. (2014). Aircraft engine exhaust emissions and other airport-related contributions to ambient air pollution: a review. Atmospheric Environment, 95(Oct.): 409–455.
McLeod, S. A. (2018). Selective Attention. Retrieved from https://www.simply psychology.org/attention-models.html/(Oct 24).

NATS. (2015). Trial Implementation of 25 Nautical Mile Lateral Separation Minimum in the ICAO North Atlantic Region. Aeronautical Information Circular Y 065/2015, NATS Services, UK Aeronautical Information Services, Cranford, UK.

NC. (2013). Customer Guide to Charges-Effective November 15, 2013, NAV CANADA, Otava, Ontario, Canada.

Sridhar, B., Ng, K. H., Linke, F. and Chen, Y. N. (2015). Impact of airspace charges on transatlantic aircraft trajectories. 15th AIAA Aviation, Technology, Integration, and Operations Conference, June 22–26, Dallas, TX; USA.

Siddiqee, W. (1973). Air route capacity models. Navigation, 20(4): 296–300.

Song, L., Wanke, C. and Greenbaum, P. D. (2006). Predicting Sector Capacity for TFM Decision Support, 6th AIAA Aviation Technology, Integration and Operations Conference (ATIO), 25–27 September 2006, Wichita, Kansas, USA.

Spiegel, M. L. R. and Bryant, D. (1978). Is speed of processing information related to intelligence and achievement. Journal of Educational Psychology, 70: 904–910.

Treisman, A. (1964). Selective attention in man. British Medical Bulletin, 20: 12–16.

USDD. (2008). Global Positioning System Standard Positioning Service Performance Standard, 4th Edition, United States Department of Defence, Washington, D.C., USA.

Welch, D. J. (2015). En-Route Sector Capacity Model, Final Report, ATC-426, MIT Lincoln Laboratory, Massachusetts Institute of Technology, Lexington, MA, USA.

Chapter 5

Sustainability of Air Transport System

5.1 Introduction

5.1.1 General

As mentioned in Chapter 1, assessment and operationalization of the concept of sustainable development of the air transport system is recognized as an increasingly important but complex research, operational, and policy task at both local and global scale in the short-, medium-, and long-term time period. In general, the air transport system can be considered to be sustainable if its net benefits increase with an increase in the system output, i.e., growth, either in the absolute (total) or relative terms (per unit of output). In this case, the system's benefits are expressed as the sum of differences between the system's effects as "benefits" and impacts as "costs" considered at different spatial scales—global (intercontinental), regional (national/continental), and local (community) (Janić, 2003; 2004; 2007).

5.1.2 Sustainability at Global Scale

At global scale, the growth of economy and air transport demand have strongly driven each other with beneficial effects like contribution to GDP (Gross Domestic Product), employment, and overall social-economic welfare on the one hand, and on the other, the evident impacts on society and the environment in terms of the increase in energy (fossil fuels) consumption and related emissions of GHGs (Green House Gases), congestion, and risk of air traffic incidents/accidents, and their internalized costs/externalities. Under such conditions, several options have been considered for balancing the system's effects/benefits and impacts/costs and consequently driving it towards more sustainable development. In this context, "balancing" implies setting up and maintaining a trade-off between effects/benefits and impacts/costs means by the following mutually exclusive scenarios (Janić, 2007):

a) Constraining the system's growth, which would include setting up an absolute limit/cap on the growth of air transport demand and the consequent associated impacts/costs on society and the environment;
b) Setting up a cap on the particular impacts/cost, which would, in addition to constraining them, also indirectly possibly constrain the system's growth itself;

c) Decoupling the system's growth and the economic growth, the latter in terms of GDP (Gross Domestic Product) through stimulating potential users to change their practice and habits of using the air transport services in the medium- to long-term; and

d) Trading-off between global effects and impacts, which generally implies introducing the policy mechanisms that enable the growth of the system's effects/benefits to be faster than the growth of impacts/costs;

e) Developing innovative operational procedures and new technologies, which would enable further growth of the system and the associated positive effects/benefits while simultaneously reducing the associated impacts/costs or maintaining them at the present level (Janić, 2014; http://www.faa.gov/nextgen/; http://www.sesarju.eu/).

Figure 5.1 shows a scheme of prospective development of the air transport demand, effects/benefits, and impacts/costs over time, supported by development and implementation of the innovative operational procedures and new technologies.

As can be seen, this scenario as a compromising one enables faster growing of the system's effects/benefits in line with growing of the air transport demand, while at the same time contributes to reducing the impacts/costs thanks to the above-mentioned technical/technological and operational innovations.

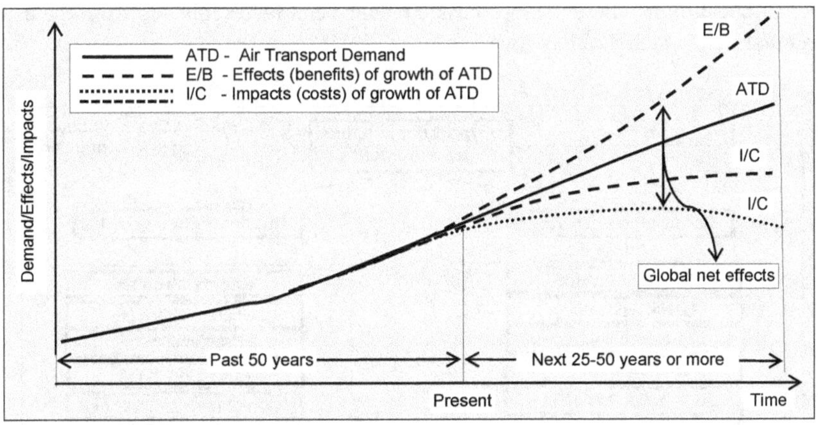

Figure 5.1 Scheme of the medium- to long-term sustainable development of the air transport system driven by innovative operational procedures and new technologies.

5.1.3 Sustainability at Regional/Local Scale

At the regional/local (national, continental) scale, particularly in the U.S. and Europe, the growth of air transport demand has been additionally driven by local forces, such as liberalisation of air transport market(s), increasing of the system productivity and diminishing airfares. Such growth has been faced with the constrained capacity of airports and the ATC/ATM (Air Traffic Control/Management) sub-system, which has resulted in increasing congestion and delays, and consequent compromising of the expected quality and costs of services. Under such circumstances, a balance between the system growth, including the effects/benefits and the associated impacts/costs, seems to be achieved by three scenarios as follows (Janic, 2007):

a) Affecting the regional air transport demand-driving forces in terms of compromising the market liberalisation and competition, productivity, and airfares;

b) Constraining the airport growth, either by setting up the cap on the annual number of atms (air transport movements) (1 atm = 1 landing or 1 take-off) in order to mitigate and keep the noise burden on local population under the prescribed limits, and/or due to the lack or additional land for physical expansion; both in turn could compromise the further growth of an airport; and

c) More efficient utilisation of the available infrastructure by using innovative operational procedures and new technologies, modification(s) of the airline operational practice, and stimulating intermodality, i.e., co-operation with other transport modes (particularly railways) (Janić, 2010; http://www.faa.gov/nextgen/; http://www.sesarju.eu/). Figure 5.2 shows as simplified scheme of balancing an airport's effects/benefits and impacts/costs under scenarios of constrained and unconstrained growth.

As can be seen, the scenario of constrained airport growth may compromise all airport effects/benefits and, thus, lead it to be considered as negative, while at the same time it may contribute to reducing the impacts/costs and, thus, make it positive. The scenario of unconstrained airport growth may influence quite the opposite on the effects/benefits and the impacts/costs compared to its counterpart, i.e., the constrained scenario. The actual question is where an appropriate balance can be established from the aspects of particular actors/stakeholders involved.

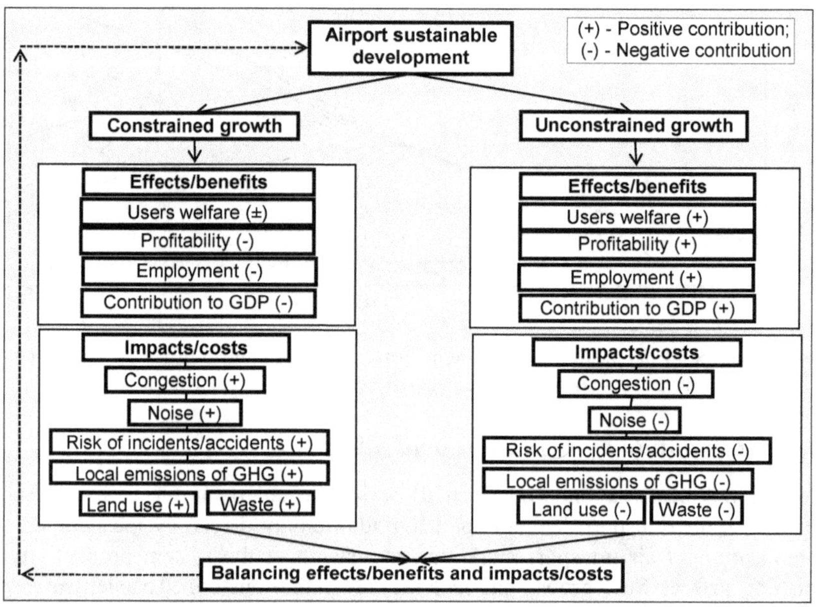

Figure 5.2 Scheme of sustainable development of an airport according to the scenarios of constrained and unconstrained growth.

5.1.4 Actors/Stakeholders Involved, Their Objectives and Preferences

Many actors/stakeholders have been directly or indirectly involved in dealing with the sustainability of the air transport system, as shown in Figure 5.3 (Janić, 2007). In particular:

a) Users of air transport services, such as air passengers and freight/cargo shippers, constituting the air transport demand usually prefer frequent, easily accessible, low cost, punctual, reliable, safe, and secure services.

b) Air transport operators providing the system's services by using infrastructure, facilities and equipment, such as airports, ATM/ATC (Air Traffic Management/Air Traffic Control, and airlines, prefer carrying services according to their business objectives in terms of profitability, safety and security on the one hand, and the users' preferences on the other.

c) Aerospace manufacturers producing the aircraft, ATM/ATC, and airport facilities and equipment prefer smooth selling of their reliable, safe, and profitable products to the system operators.

d) Local populations in the vicinity of airports usually tend to maximise their benefits and minimise the costs of the air transport system at a local scale. The employment opportunity and use of efficient air connections to other distant communities (regions) can be considered as the obvious benefits. The costs are regarded as exposure to the airport noise, air pollution, and risk of injury, loss of life and damage of property due to the aircraft incidents/accidents.

e) Local and central authorities creating the institutional regulation for the system operations at local (community) and central (national) level are mostly interested in the system's overall above-mentioned effects/benefits and impacts/costs or externalities.

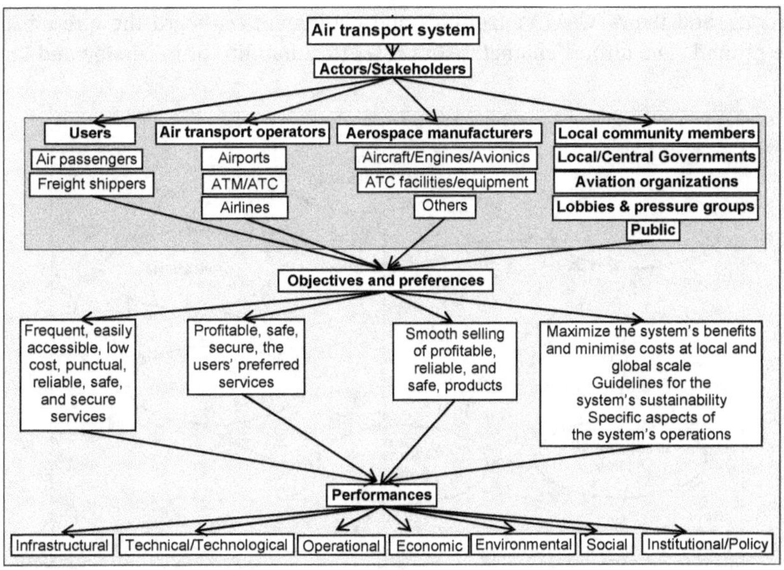

Figure 5.3 Structure of the air transport system used for dealing with its sustainability.

f) Aviation organizations, such as ICAO (International Civil Aviation Organization), IATA (International Air Transport Association), ECAC (European Civil Aviation Conference), AEA (Association of European Airlines) and ACI (Airport Council International), co-ordinate the system's development at a global (international) scale by providing the framework and guidelines for its sustainability at both regional (national) and global (international) scales.

g) Lobbies and pressure groups articulate the interests of people who may be for or against expansion of the air transport system infrastructure and operations by organizing campaigns against its global and local harmful impacts on people's health and environment; and

h) Public temporarily interesting in the specific aspects of the system operations use media such as radio, TV, Internet, and newspapers to get information about the system but also to deliver messages and opinions in the cases of launching innovations (aircraft, airports), severe disruptions of services and air incidents/accidents, and changes of airfares.

5.2 The System Performances

5.2.1 Categories

Sustainability of the air transport system has been frequently analysed using the concept of multidimensional examination of its performances specified for its three subs-systems—airports, airlines, and ATC/ATM. In general, these performances have been categorized as: (i) Infrastructural; (ii) Technical/technological; (iii) Operational; (iv) Economic; (v) Environmental; and (vi) Social (Janić, 2007). Figure 5.4 shows the scheme of mutual interrelationship between particular above-mentioned performances.

a) Infrastructural and technical/technological performances relate to the corresponding characteristics of the system's components—infrastructure (airports and airspace), aircraft, and the ATC/ATM facilities and equipment on-board the aircraft and on the ground. The airport characteristics refer to suitability of its airside and landside

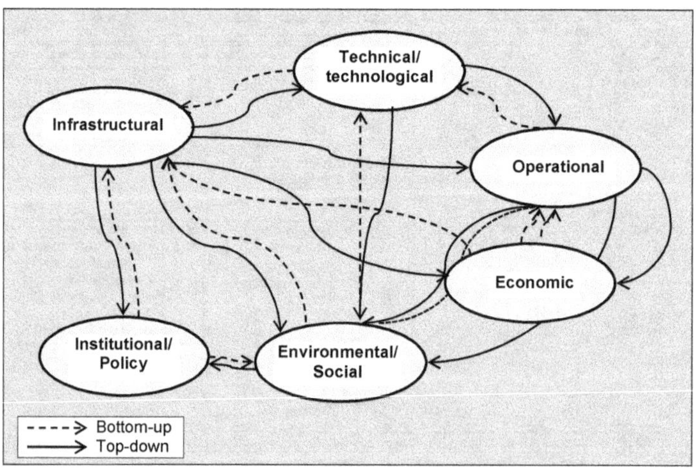

Figure 5.4 Scheme of the mutual interrelationships between the performances of the air transport system.

areas to flexibly (smoothly) handle different types/categories of aircraft and volumes and fluctuations of traffic up to the level of their saturation capacity. The aircraft characteristics include the engine technology, referring to the engine power (thrust), fuel consumption and related emissions of GHGs and noise. In addition, the number of aircraft of particular types/categories and engine technologies, their utilisation and share in the corresponding totals, and geographical distribution could also be taken into consideration at different geographical scales—local and/or global, depending on the purpose of analysis. The characteristics of ATC/ATM refer to the facilities, equipment and devices enabling (and supporting) the efficient, effective, and safe aircraft flights between origin and destination airports and consequently diminishing (or not additionally increasing) energy/fuel consumption and related emissions of GHGs, noise, and other impacts.

b) Operational performances relate to the processes and operations aiming at producing air transport services by using the above-mentioned system's components. Under such conditions, the common objective is to "optimize" utilisation of the disposable resources (aircraft fleet, labour/staff, and energy/fuel consumption) under given (technical/technological, operational/safety, economic, environmental, and social) constraints, which would result in a higher efficiency of the fuel/energy consumption and lower associated adverse effects, such as GHG emissions and noise. These performances mainly depend on and are driven by the technical/technological, economic, and environmental performances (Janić, 2007).

c) Economic performances include investments and costs for providing and operating the air transport system infrastructure, aircraft, supporting facilities and equipment, staff, and fuel/energy. In addition, they embrace revenues from operating the system, direct, indirect, and induced employment, contribution to GDP and equity. The economic performances mainly depend on the system's infrastructural, technical/technological, and operational performances.

d) Environmental performances reflect the physical impacts of the air transport system on people's health and on the environment in terms of local and global emissions of GHGs, noise, land use, and waste. Most of these impacts directly depend on the technical/technological and operational performances.

e) Social performances reflect the impacts of the air transport system on society in terms of congestion and delays, accessibility of services, traffic incidents/accidents, community cohesion, liveability, and social interactions. They mainly depend on the system operational performances.

f) Institutional/policy performances have become increasingly relevant and important because of the need for establishing the institutional forms for dealing more comprehensively with the system's sustainability. These forms were expected to bring the various external and the system's internal actors/stakeholders together, enabling the setting up and facilitating of rules of behaviour through the various institutions mediating their common and/or distinctive/diverse interests. In general, the other performances may influence these in terms of the number, structure, hierarchy and effectiveness of institutions and their objectives.

5.2.2 Indicator Systems

5.2.2.1 Definition

The multidimensional examination of performances of the air transport system is based on their indicators specified regarding their relevance for the particular actors/stakeholders involved. Their indicators constituting the indicator systems are based on the following assumptions:

a) Indicators of performances should measure the effects/benefits and impacts/costs of the air transport system operations in either absolute or relative monetary or non-monetary terms depending on the relevant system's output. If the "effect/benefit" (*BI*) indicator increases and the "impact/cost" (CI) indicator decreases (or remains constant) as the relevant output increases, the system will be considered as sustainable. Otherwise, the system will be considered as unsustainable. Figure 5.5 shows a general example.

b) As can be seen, setting up the limit-criterion on a particular indicator may have a two-fold effect. If the cost indicator *CI* is limited to CI_{max}, the output will be allowed to increase maximally to O_{max}. Such constrained output will affect the benefit indicator (BI), which will increase maximally to $BI(O_{max})$. Consequently, balancing (i.e., trading-off) between the effects and impacts should always be taken into account while setting up criteria on the particular indicators of performances.

c) Indicators of performances need to be sufficiently general, i.e., applicable to the system as a whole, its sub-systems, and their components;

d) Indicators of performances need to have a 'forecasting capability', i.e., they should be able to predict the system sustainability under different conditions, i.e., developing scenarios in the medium- to long-term period; and

e) Indicators of the particular performances should be estimated by using the available statistical databases.

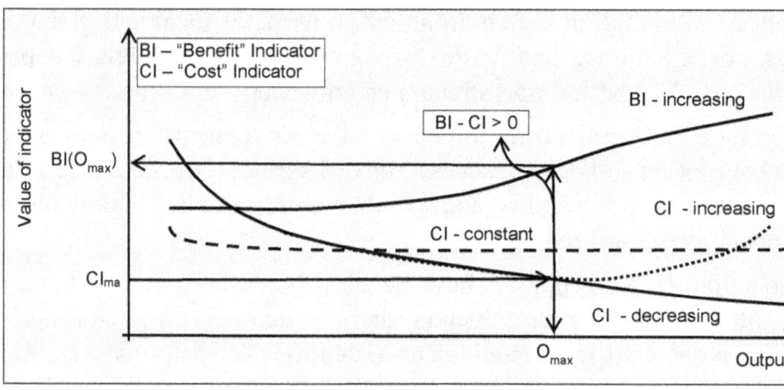

Figure 5.5 General relationships between the indicators of performances and output for assessing the system's sustainability (Janić, 2003; 2004).

5.2.2.2 Indicator systems for particular actors/stakeholders

a) Indicators for users - air passengers

In the given context, the indicator system for the air transport system users, i.e., air passengers, contains eight indicators of the operational, economic, environmental performances. They relate and can be quantified for an individual airline route and/or airport, as well as for the airline industry and the airport network of a given region.

i) Operational indicators

The indicators of operational performances are defined to be 'punctuality' and 'reliability' of service, 'mishandled baggage', 'safety' and 'security'.

- Punctuality of services refers to the users' perception of how the chosen airline carries out its flights-services according to the announced schedule, i.e., on time. This assessment can be carried out based on experience or by using the airline information. In the latter case, two measures of this indicator may be convenient. First is the probability that an airline flights are on time, estimated as the ratio between the number of flights on time and the total number of flights carried out during a given period of time (day, month, year). Another measure is the average flight delay.[1] Both measures may be relevant while choosing an airline, air route, and the air transport mode itself. Users usually prefer the former measure to be as high as possible and the latter one to be as low as possible under conditions of increasing of the number of flights or over time. Figure 5.6 shows an example of the punctuality of U.S. airlines over time.

  As can be seen, the proportion of on time arrivals has substantially fluctuated over the given period of time (2013–2019) ranging from the lowest (73%) in the year 2008 to the highest (82%) in the year 2012. As such, this does not enable judgement about sustainability of air transport systems regarding this indicator in the given case. Anyway, it is preferred to be as high as possible.

- Reliability of services reflects the users' perception of the chosen airline to carry out the scheduled flights. This indicator can be assessed by experience or from the airline information. In the latter case, the ratio between the number of cancelled (or diverted) and the total number of flights carried out by chosen airline during a given period of time (day, month, year) can be used as a measure. Independently of the causes of cancellation or diversion of flights, it is preferred to be as low as possible and to decrease as the number of flights increases.

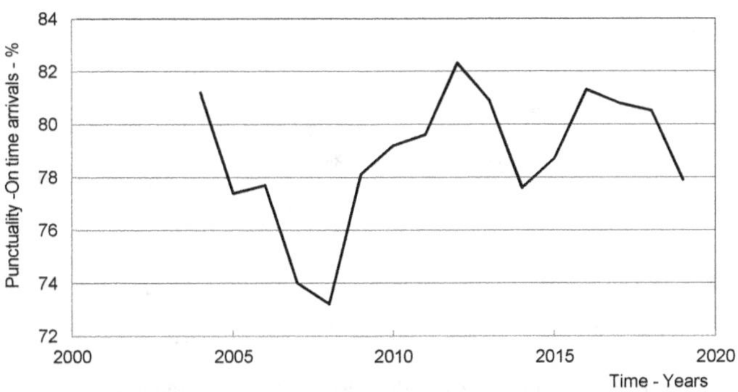

Figure 5.6 Punctuality of airlines over time relevant for air passengers—Case of 10 USA airlines and their branded code share partners (USDT, 2003/2019).

[1] Usually, delays are counted for the flight arrivals that are more than 15 minutes late according to the schedule (EEC, 2001; EEC/FAA, 2017; https://www.transtats.bts.gov/OT_Delay/OT_DelayCause1.asp/).

- Mishandled baggage includes delayed, damaged, pilfered, lost or stolen baggage. In addition to the users/air passengers experience, the information from the chosen airline can be used to assess this indicator. The ratio between the numbers of lost (or damaged) baggage and the total handled baggage during a given period of time (year) can be a measure preferred to be as low as possible and to decrease as the number of handled baggage items or enplaned users/air passengers increases. Figure 5.7 shows an example of the relationship between the rate of mishandled baggage and the number of enplaned air passengers per month at the selected U.S. airlines.

  As can be seen, the ratio of mishandled baggage generally increased as the number of enplaned users/air passengers increased, thus indicating unsustainability of the system regarding this indicator during the specified month (March) of the observed period (2003–2018).

- Safety emerges as a relevant indicator for users while selecting an airline among several airlines, as well as when selecting air transport over various alternative transport modes. This indicator measures the perceived risk of death or injury of an individual while on board. Again, in addition to the subjective judgements, the selected airline, CAA (Civil Aviation Authority) can also provide information on this indicator, which is usually expressed by the number of deaths (or injuries) per unit of output (RPK, RPM).[2] The users prefer this measure to be as low as possible and to decrease as the volumes of RPK (RPM) increase and over time (https://www.faa.gov/data_research/accident_incident/).

- Security relates to the perceived risk of an individual's exposure to threat from illegally carried weapons or other dangerous devices (bombs, firearms, guns, etc.) while at an airport or on board flights. The airport security service can provide information on this indicator. The ratio between the numbers of detected illegal dangerous devices and the total number of screened passengers can measure this

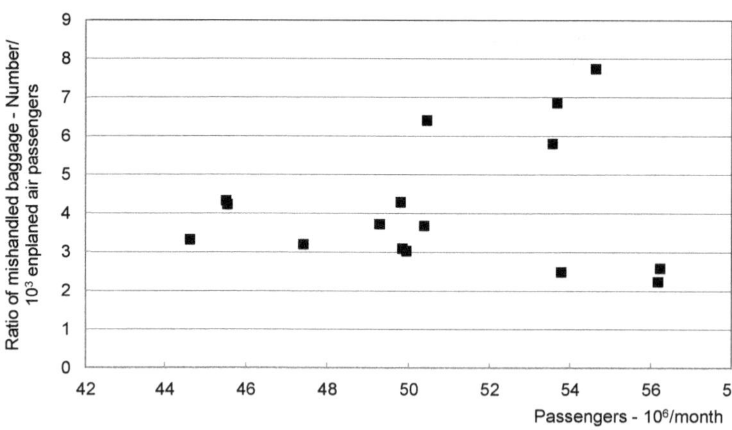

Figure 5.7 Relationship between the rate of mishandled baggage and the number of monthly enplaned users/air passengers—Case of 10 U.S. airlines and their branded code share partners in March each year; Period 2003–2018) (USDT, 2003/2019).

[2] RPK – Revenue Passenger Kilometer; RPM – Revenue Passenger Mile.

indicator. The users prefer this measure to be as low as possible and to, independently of the causes,[3] decrease as the number of screened passengers increases.

ii) Economic indicators

The indicator of economic performances is defined as the 'economic convenience' of carrying out an air trip.

- Economic convenience reflects the total generalised cost of the door-to-door trip where air transport has the largest share.[4] Under such circumstances, the average airfare per passenger, which dominates within the above-mentioned generalized costs, can be used as a suitable measure. The users always prefer the airfares to be as low as possible and to decrease over time (Janić, 2003). Figure 5.8(a, b) shows an example of development of air fares in the U.S. air domestic market over time.

Air fares are based on domestic itinerary fares, which consist of round-trip fares. They are based on the total ticket value, which includes the price charged by the airlines plus any additional taxes and fees levied by an outside entity at the time of purchase (https://www.bts.gov/explore-topics-and-geography/topics/air-fares/). Figure 5.8a shows that the airfares expressed in the constant $US have tended to decrease while, if expressed in the current $US, they have tended to be constant and even increasing during the observed period. This indicates both the system's sustainable and unsustainable development regarding this indicator. In addition, Figure 5.8b shows that the airfares expressed in the current $US have fluctuated higher than the average CPI (Consumer Price index), thus indicating the times (years) when it has been more and less convenient to spend on air travel compared to the other expenses (https://www.bls.gov/opub/ted/2019/cpi-increased-1-point-7-percent-for-year-ending-september-2019.htm/). This implies that there have been the years when the system has been ultimately sustainable and other years when it has been ultimately unsustainable regarding this indicator.

iii) Environmental indicators

The indicators of environmental performances are defined to be 'comfort and healthiness' of the airport and aircraft interior while assessing the quality of travelling environment. The indicator 'comfort and health' can take that into account.

- Comfort and healthiness reflect the users' feeling of comfort while at an airport terminal and on board the aircraft/flights. At an airport, in addition to the subjective judgement, the passenger density (the number of passengers per unit of space) and the experienced queuing time can be used to measure the comfort as a component of the airport quality of service. In addition to the individual experience, the airport operator can provide information on these measures (Janić, 2013). Configuration and size of the seats in economy class and the quantity of fresh air delivered to the aircraft passenger cabin per unit of time can be relevant measures of comfort and healthiness while being in the aircraft interior. The measures of airport comfort are preferred to be as low as possible and to decline as the number of passengers served

[3] This ratio is out of the full control of the airport security service since it cannot directly control the individuals bringing illegal and dangerous devices into airports. However, using the sophisticated facilities and equipment for screening passengers may provide a high rate of detection of such devices.

[4] Some airfares charged by the low-cost air carriers in Europe and the U.S. may represent the exceptions from this general rule.

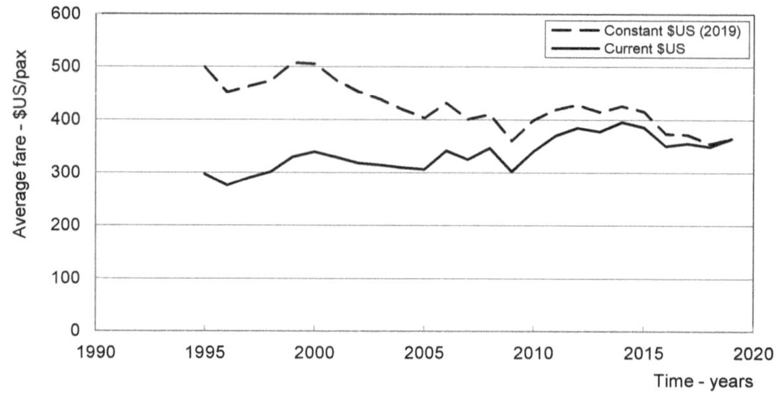

a) Air fares vs time

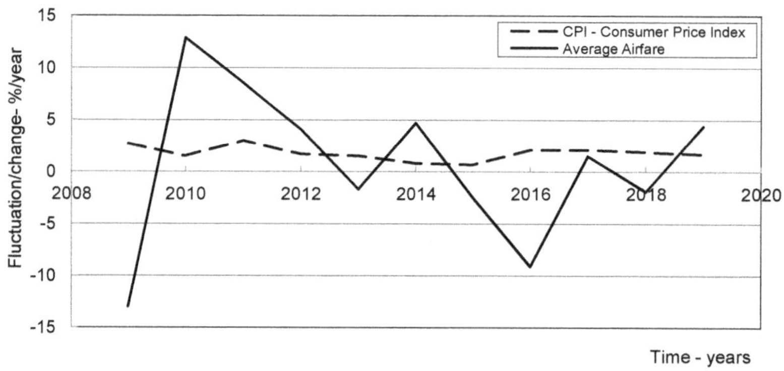

b) CPI (Consumer Price index) and average airfare

Figure 5.8 Development of airfares over time—case of U.S. domestic air transport market (https://www.bls.gov/opub/ted/2019/cpi-increased-1-point-7-percent-for-year-ending-september-2019.htm/; https://www.bts.gov/explore-topics-and-geography/topics/air-fares).

increases during a given period of time. Both measures of comfort while being in the aircraft interior are preferred to be as high as possible and to increase over time.

iv) Social indicators

The indicator of social performance relevant for users can be 'accessibility and connectivity'.

- Accessibility and connectivity reflects the opportunities to travel from a given airport by the selected airline(s) to the other medium- and long-distance areas (places). The number of destinations served from an airport (or region) by a given airline can be used as a measure. In addition, the connectivity by non-stop, one-stop or multi-stop flights with respect to the trip purpose (business, leisure) can be taken into account in order to refine this measure. Recently, this measure has become a global competitive tool of both airlines and airports, which both can provide relevant information. In general, the users, independently of the trip purpose, prefer this measure to be as high as possible and to increase over time.

b) Indicators for airports

The attitudes of airport operators with respect to sustainability are expressed by ten indicators of the operational, economic, environmental, and social performances. They can be quantified for an airport or an airport network.

i) Operational indicators

The indicators of operational performances, defined as 'demand', 'capacity', 'quality of service' and 'flexibility in relieving congestion', are considered as the main airport operational indicators.

- Demand indicates a scale of an airport operation. The number of passengers and atm (air transport movements),[5] and the volume of freight handled during a given period (hour, day, year) can be a measure. Sometimes, it is more convenient to use WLU (Workload Unit)[6] as a common physical measure for both passengers and freight volumes handled. The airport operator prefers these measures to increase over time.
- Capacity reflects the maximal physical capability of an airport to handle the given demand. Commonly, two measures are used: the airside capacity in terms of the maximum number of atm, and the landside capacity in terms of the maximum number of WLU handled during a given period (hour, day, year). Both measures are preferred to be as high as possible and to increase over time in order to cope with growing demand.
- Quality of service reflects the relationship between the airport demand and capacity. Generally, the average delay per atm or WLU, which occurs whenever the demand exceeds the capacity, can be used as a measure. This measure is preferred to be as low as possible and to decrease as the demand increases during a given period (hour, day, year).
- Flexibility in relieving congestion may be a relevant indicator for airports operating at the saturated capacity and connected to the surface regional—national and international—ground transport network(s). Generally, these airports aim to relieve congestion and the associated negative noise and air pollution impacts by any means (Janić, 2011). One of the options consists of substitution of some short-haul flights by adequate surface, usually high-speed rail, services.[7] In such cases, a measure of this indicator can be the ratio between the number of substituted short-haul and the total number of feasibly substitutable short-haul flights during given period (year). This ratio is preferred to be as high as possible and to increase as the number of feasibly substitutable short-haul flights increases.

ii) Economic indicators

The indicators of economic performance are defined to be 'profitability' and 'labour productivity'.

[5] An atm (air transport movement) is either an arrival or a departure.

[6] Workload Unit (WLU) is an equivalent for one passenger or 100 kg of freight (Doganis, 1992).

[7] For example, three European super hubs, Frankfurt Main, Paris CDG, and Amsterdam Schiphol are connected to the HSR (High Speed Rail) network. Partial substitution of the short-haul flights has taken place there (EC, 1998; HA, 1999; IFRAS, 2000).

- Profitability reflects the airport financial success measured by the operating profits (the difference between the operating revenues and the operating costs). They can be expressed in both absolute and relative terms, the latter per unit of output—WLU (Doganis, 1992). This measure is preferred to be as high as possible and to increase as the airport output increases. Figure 5.9 shows an example regarding Amsterdam Schiphol airport (The Netherlands).

As can be seen, the average profit per WLU has generally increased at a decreasing rate with an increase in the annual volumes of handled WLUs during the observed period. Therefore, it can be said that the airport has been developing in the sustainable way regarding this indicator.

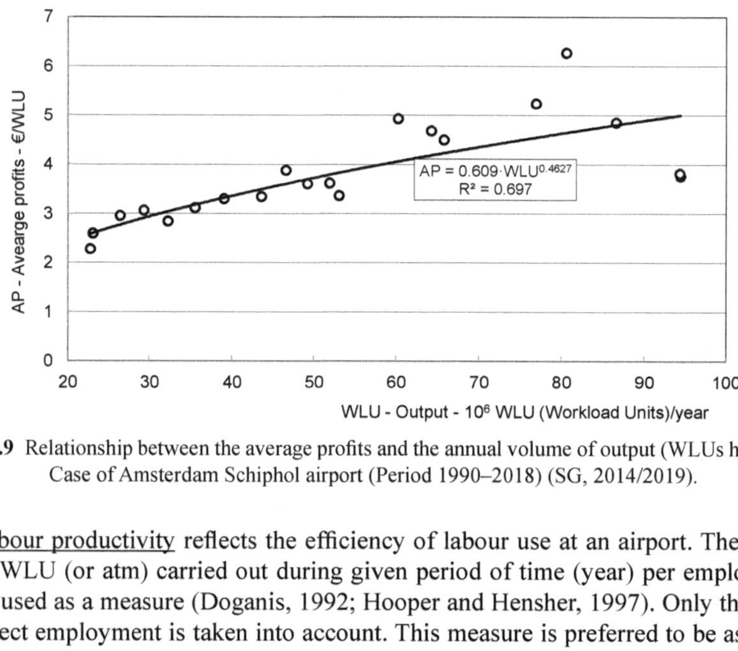

Figure 5.9 Relationship between the average profits and the annual volume of output (WLUs handled)—Case of Amsterdam Schiphol airport (Period 1990–2018) (SG, 2014/2019).

- Labour productivity reflects the efficiency of labour use at an airport. The number of WLU (or atm) carried out during given period of time (year) per employee can be used as a measure (Doganis, 1992; Hooper and Hensher, 1997). Only the airport direct employment is taken into account. This measure is preferred to be as high as possible and to increase as the number of employees increases.

iii) Environmental indicators

The indicators of environmental performances are defined to be 'noise', 'emissions of GHGs (Green House Gases)', 'waste' and 'land use'. These indicators relate to the physical impacts of an airport on the health of local people and the environment and become relevant while undertaking the mitigation measures.

- Noise relates to the noise energy generated by atm. A measure for this indicator can be the area determined by a certain equivalent long-term noise level L_{eq} expressed in decibels (dBA). The affected area is expressed in square kilometres (ICAO, 1993a). This indicator is preferred to be as low as possible and to decrease as the number of ATM increases during a given period of time (hour, day, year) (see Figure 1.14, Chapter 1).
- GHG (Green House Gas) emissions relate to the total quantity of all or specific GHGs linked to airport operations. In addition to atm, emissions of GHGs from

the other polluting sources, including the airport landside access systems, can be taken into account. In most cases, the quantity of GHG per an event, i.e., LTO[8] cycle, can be used as a measure of this indicator, which is preferred to be as low as possible and to decrease as the number of LTO cycles carried out during given period of time (year) increases.

- Land use relates to the area of land used for airport airside and landside area infrastructure and its utilisation. A convenient measure can be the number of atm carried out during given period of time (hour, day, year) per unit of area of taken land. This indicator is preferred to be as high as possible and to increase as the area of land increases. Figure 5.10 shows an example of the relationship between the intensity of land use and the area of land taken by the 33 selected world's airports.

As can be seen, the intensity of land use expressed by the annual number of passengers accommodated per unit of land taken generally decreases more than proportionally with an increase in the area of land. This indicates that more sizeable (spacious) airports are generally less efficient users of land and, consequently, less sustainable than their smaller counterparts regarding this indicator (see also Figure 1.13, Chapter 1).

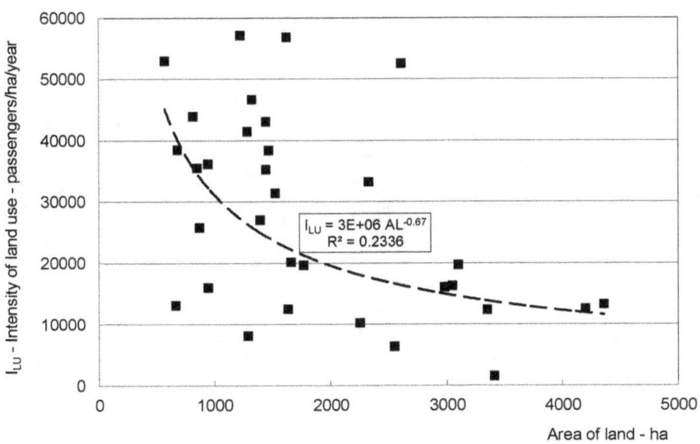

Figure 5.10 Intensity of land use vs. the area of land occupied by an airport—Case of 33 selected world's airports (Janić, 2016; http://en.wikipedia.org/wiki/World's_busiest_airports_by_passenger_traffic/).

- Waste relates to the quantities generated by an airport excluding the airline in-flight waste. The convenient measures can be the quantities generated per unit of the airport output (passenger or WLU) and their treatment, and the rate of recycling (TRB, 2019). The former measure is preferred to be as low as possible and the latter as high as possible as the airport output increases during the given period (year). Figure 5.11(a, b) shows an example of the waste management at Tokyo Narita International airport (Japan).

Figure 5.11a shows that the general waste incinerated per passenger, consequently generating emissions of GHGs, has decreased as the number of passengers has increased,

[8] LTO cycle—Landing/Take-Off cycle has been recommended as a standard framework for quantifying emissions of GHGs in the airport airside area (ICAO, 1993b; 2012).

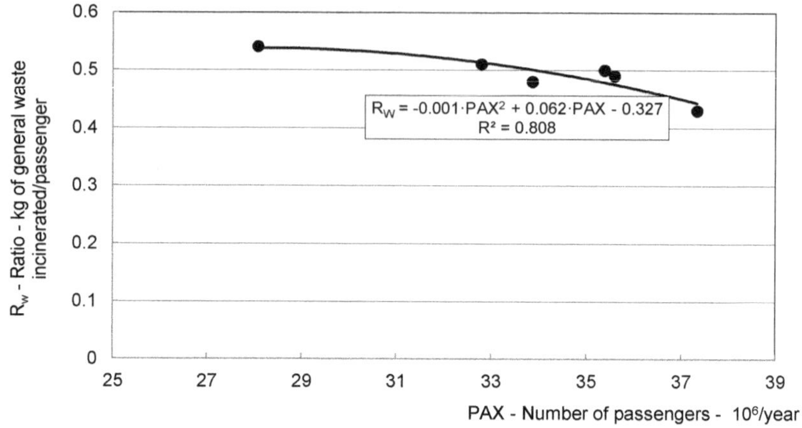

a) General waste incinerated vs number of passengers

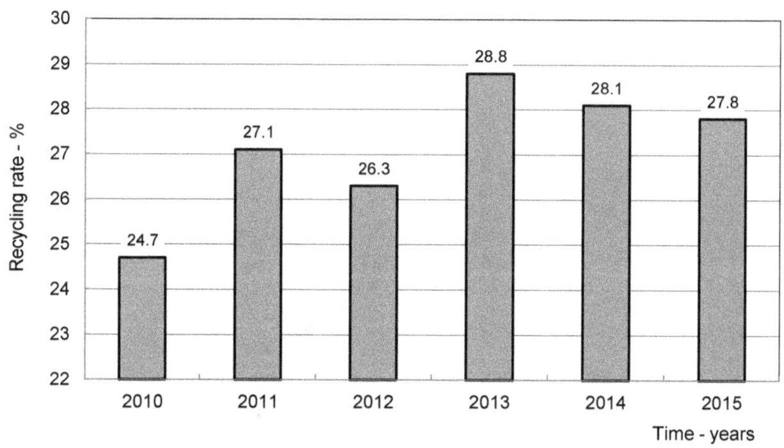

b) Rate of recycling waste vs time

Figure 5.11 Waste management—Case of Tokyo Narita International airport (Period: 2010–2015) (ACI Asia-Pacific, 2018).

thus indicating the airport's sustainable development regarding this indicator during the observed period. Figure 5.11b shows that the rate of recycling waste has fluctuated, but generally increases over time, thus again indicating the airport's sustainable development regarding this indicator.

iv) Social indicators

These indicators of performances are not defined in the present context.

c) Indicators for airlines

The airline performances regarding sustainability are expressed by eleven indicators, which can be estimated for an individual airline, airline alliance, or the entire airline industry of a given region (country or continent).

i) Operational indicators

The indicators of operational performances are defined to be 'airline size', 'load factor', 'punctuality', 'reliability' of services and 'safety'.

- Airline size reflects the volume of airline output in terms of the number of passengers and freight served in its air route network during a given period of time (month, season, year). The common measures are: the total RTK (RTK – Revenue Ton-Kilometre), total number of passengers, and total volume of freight carried out (Janic, 2000). In addition, RPK (Revenue Passenger Kilometres) and FTK (Freight Ton-Kilometres) can be used separately instead of the aggregate RTK. All the above measures are preferred to be as high as possible and to increase over time.
- Load factor measures the rate of utilisation of the airline available capacity during a given period of time (month, season, year). It is usually measured as the ratio between RTK and ATK (Available Ton-Kilometre). In addition, the load factor can be determined separately for passengers and freight. This measure is preferred to be as high as possible and to increase as the airline output RTK and/or RPK increases.
- Punctuality and reliability of services, and safety are the indicators analogous to those of users in terms of measurement and preference. The airlines use them as competitive tools on the one hand and as a measure of their operational efficiency on the other (Janić, 2003; 2004). Figure 5.12(a, b) shows the example of punctuality and reliability of services at two U.S. airlines – American and Southwest.

Figure 5.12a shows that, at American Airlines, the punctuality of arrival and departure flights was relatively constant as the number of flights (on average 75–80%) grew. At Southwest Airlines, the fluctuation of punctuality of both arrival and departure flights has been higher but decreased as the number of flights (from 65–85% to 75–80%) increased. At both airlines, regarding this indicator, the sustainability has been improving. Figure 5.12b shows the reliability of services, i.e., the rate of cancelled flights has a tendency to decrease for American airlines and increase for Southwest airlines as the number of flights grows. This indicates some improvements in sustainability at the former and worsens it at the latter airline regarding this indicator.

ii) Economic indicators

The indicators of economic performances are defined to be 'profitability' and 'labour productivity'.

- Profitability relates to the airline financial success or failure and is measured by the average profits (difference between the operating revenues and costs) per unit of output (RTK, RPK). This indicator is preferred to be positive, as high as possible, and to increase as the airline output increases over time. Figure 5.13(a, b) shows an example of the profitability of a U.S. LCC (Low Cost Carrier) airline.

Figure 5.13a shows that the average operating revenues and costs generally increased at a decreasing rate as the volumes of airline output increased. Two phases can be distinguished. The first being when the positive difference between the revenues and costs decreased as the volume of airline output increased, until becoming negative. The second coming just after that, when the positive difference between the revenues and costs has increased again as the volumes of airline output increase. Figure 5.13b shows that the first above-mentioned phase coincided with the period 2004–2009 and the second one with the period 2015–2019. Specifically, during the period 2015–2019, both

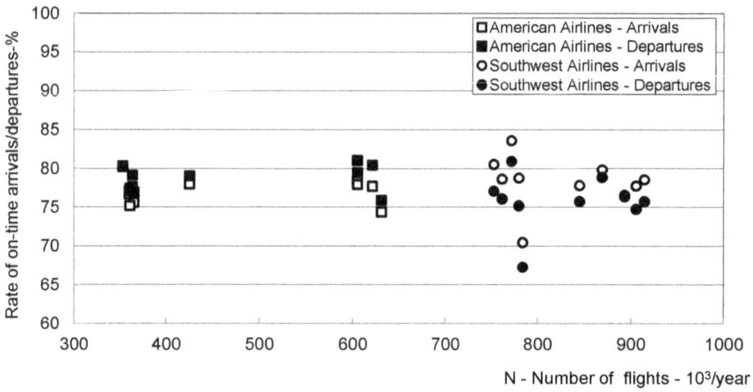

a) Punctuality of services - on-time arrivals and departures vs the number of flights

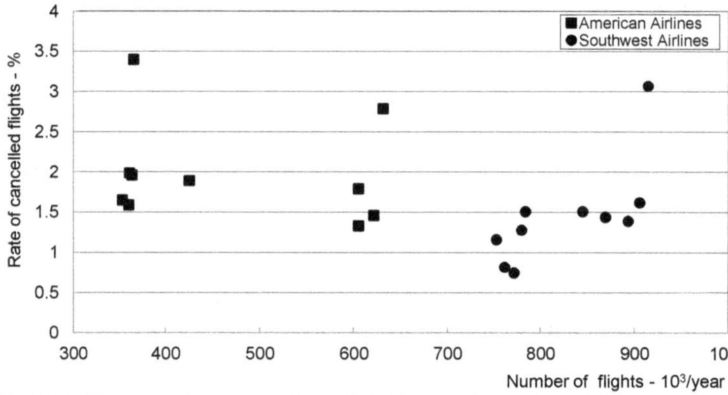

b) Reliability of services – rate of cancelled flights vs the number of flights

Figure 5.12 Relationship between the punctuality and reliability of airline services—Case of two U.S. airlines (Period: 2010–2019) (https://www.transportation.gov/briefing-room/dot 0718/).

revenues and costs have decreased while at the same time their positive difference has widened. The above-mentioned development indicates the general but fluctuating airline profitability regarded as an indicator of economic sustainability.

- Labour productivity reflects the airline's ability to use its available staff efficiently and is measured by the average output in terms of RTK (RPK) carried out during a given period of time per an employee. Total number of both direct and indirect temporary and permanent employees is usually taken into account. The preference for this measure is to be as high as possible and to increase as the number of employees increases (Janić, 2003; 2004).

iii) Environmental indicators

The indicators of environmental performances are defined as 'energy/fuel', 'GHG emissions', 'noise' and 'waste' efficiency.

- Energy/fuel and air pollution GHG emissions efficiency relates to the rate of modernisation and efficiency of utilisation of the airline fleet in terms of energy/fuel consumption and the associated GHG emissions. These indicators are measured by the average quantity of fuel and GHG emissions, respectively, per unit of output—

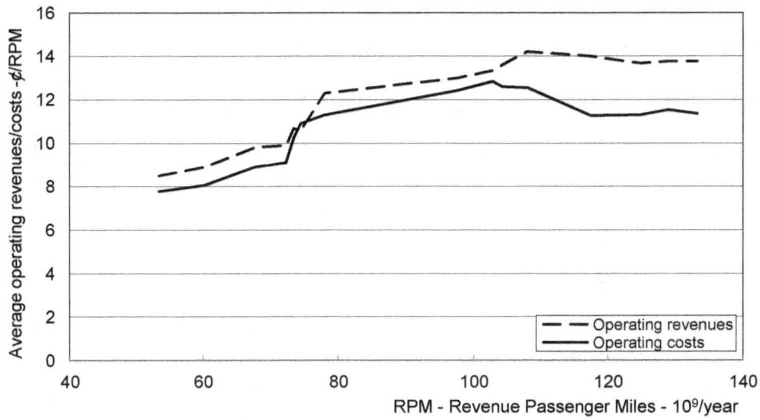

a) Operating revenues/costs vs annual output

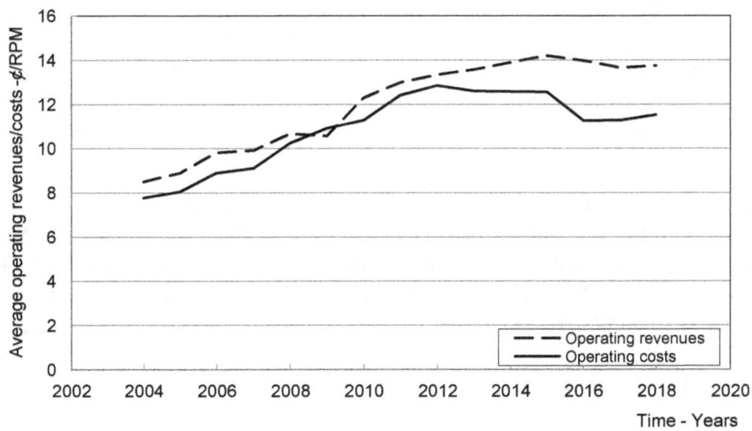

b) Operating revenues/costs vs time

Figure 5.13 Example of profitability—Case of Southwest Airlines (USA) (SA, 2004/2019).

RTK (RPK), D (Distance flown) and/or FH (Flying Hour). Both measures are preferred to be as low as possible and to decrease as the airline output carried out during given period of time (month, season, year) increases.

- Noise efficiency indicates the rate of modernisation of the airline fleet in terms of use of the aircraft of noise Stages 3 and 4 (ICAO, 2008). This indicator can be measured by the proportion of Stages 3 and 4 aircraft in the airline fleet, which is preferred to be as high as possible and to increase as the airline fleet[9] expands.
- Waste inefficiency indicates generation of the airline in-flight waste (BA, 2001). This indicator can be measured by the average quantity of in-flight waste per unit of the airline output—RTK (RPK). This measure is preferred to be as low as possible and to decrease as the airline output increases.

[9] After an airline fleet has been completely modernized by replacing all aircraft of Stage 2 with aircraft of Stage 3 and 4, this indicator becomes irrelevant.

iv) Social indicators

These indicators of performances are not defined in the present context.

d) Indicators for the ATC/ATM (Air Traffic Control/Management) system

The ATC/ATM (Air Traffic Control/Management) is assumed to consider eight operational, economic, environmental, and social indicators of performances. These indicators can be quantified for a part (ATC/ATM sector) or the entire system (the airspace of the country or a wider region – continent).

i) Operational indicators

The indicators of operational performances are defined as 'demand', 'capacity', and 'safety'.

- Demand is measured by the number of flights handled during a given period (hour, day, year) (Janić, 2003; 2004). This measure is preferred to be as high as possible and to increase over time.
- Capacity expresses the maximum capability of the ATC/ATM providers to serve the aircraft/flight demand under given conditions. It can be measured by the maximum number of flights served in the given airspace per unit of time (Janić, 2000). This indicator is preferred to be as high as possible and to increase in line with increasing of demand.
- Safety expresses the risk probability of occurrence of an air traffic incident/accident due to the ATM/ATC operational error(s). These incidents/accidents may happen at airports and in the airspace. A convenient measure of this indicator can be the number of individual aircraft accidents or the number of NMACs (Near Mid-Air Collisions) per unit of the ATC/ATM output, i.e., the controlled flights. These measures are preferred to be as low as possible and to decrease with increasing of the number of controlled flights and over time. Figure 5.14 shows the example for the U.S. ATC/ATM system.

As can be seen, the number of mid-air collisions and loss of separation has decreased as the number of handled/controlled flights during the observed period. On the one

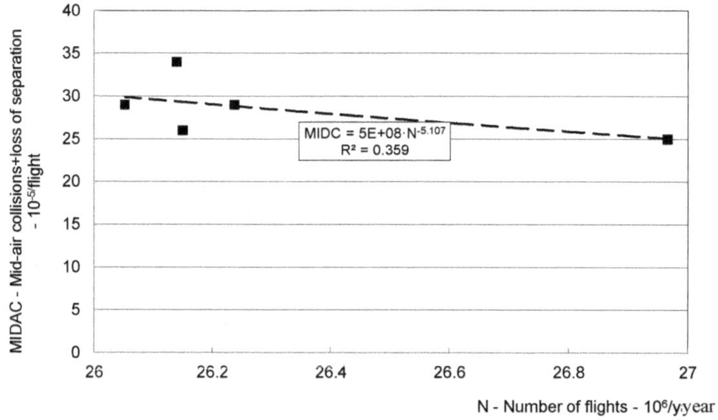

Figure 5.14 Example of the relationship between the number of mid-air collisions and loss of separation—Case of the U.S. ATC/ATM system (Period: 2013–2017) (FAA, 2018; 2018a).

hand, this indicates decreasing the risk of potential incidents/accidents and on the other improvement sustainability of the system regarding this indicator of performance.

ii) Economic indicators

The indicators of economic performances are defined to be 'cost efficiency' and 'labour productivity'.

- Cost efficiency[10] relates to the operating costs of ATC/ATM. It is measured by the average cost per unit of output, i.e., controlled flight. This measure is preferred to be as low as possible and to decrease as the number of flights during given period of time (year) increases (EEC/FAA, 2017; Janic, 2000).
- Labour productivity reflects the efficiency of the ATM/ATC providers in terms of labour (staff) use. A convenient measure can be the number of controlled flights per employee. This indicator is preferred to be as high as possible and to increase faster in relation to the increasing number of controlled flights than to the number of employees.

iii) Environmental indicators

The indicators of environmental performances are defined to be 'flight fuel efficiency' and 'flight GHG emission efficiency'.

- Flight fuel efficiency relates to the aircraft/flight additional fuel consumption due to deviations from the prescribed (fuel-optimal) trajectories dictated by the ATC/ATM safety requirements. This indicator can be measured by the rate of deviation of the aircraft/flight actual from the scheduled/planned trajectories between origin and destination airports due to the ATC/ATM reasons (EEC/FAA, 2017). As such, it is preferred to be as low as possible and decrease with growing of air traffic.
- Flight GHG emissions efficiency relates to the additional GHG emissions due to the above-mentioned fuel efficiency. The indicator is measured by the average quantity of GHGs emitted per flight. It is preferred to be as low as possible and to decrease as air traffic increases.

iv) Social indicators

These indicators of performances are not defined in the present context.

e) Indicators for aerospace manufacturers

The aerospace manufacturers produce aircraft (airframe, engines, avionics), and ATC/ATM and airport facilities and equipment. In most cases, these products are developed in collaboration with their main customers, i.e., airports, ATM/ATC providers, and airlines. Regarding sustainability, their performances are expressed by eight indicators that can be estimated for an individual manufacturer or for the specific sector as the whole.

i) Operational indicators

The indicators of operational performance are defined as 'innovations of aircraft', 'innovations of the ATC/ATM and airport facilities/equipment', and 'reliability of structures'.

[10] 'Cost' is considered as a more relevant indicator than 'profitability' because most ATC/ATM services providers charge their services on the cost-recovery principle. For example, EUROCONTROL member States and ATC/ATM providers in Canada, Australia, New Zealand, South Africa, etc. fully recover their costs by charges (EEC, 2017; 2019).

- Innovations of aircraft reflects the technological progress in terms of the aircraft speed, capacity, and cost efficiency. The progress in speed and capacity can be measured by the technical productivity determined as their product. The aircraft cost efficiency is usually measured by the average operating cost per unit of output (RTK, RPK). This cost has generally decreased with the increases in aircraft capacity, seat density, fuel efficiency, and range (Janic, 2000). Figure 5.15 shows the main steps in development of the aircraft technical productivity in terms of the number of TKM/h (Ton Kilometers per Hour).

As can be seen, this productivity has been increasing over time thanks to both airlines and their requirements, as well as capabilities of the aerospace manufacturers. After DC 3, the rise of technical productivity was primarily achieved by developing the larger aircraft and less by increasing their operating (cruising) speed. A culmination of development of this productivity has been reached by introducing A380 aircraft. This development has simultaneously included development and upgrading of the aircraft jet engines (after DC 3) in terms of their fuel and GHG emissions efficiency on the one hand and the sophisticated avionics on the other. This has resulted in the long-term sustainable development of the system regarding this indicator of performance.

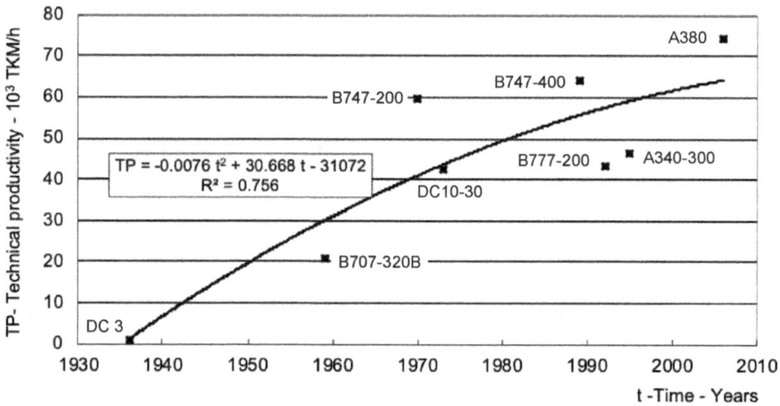

Figure 5.15 Development of technical productivity of the commercial aircraft over time (TKM-Tonne-Kilometer) (FI, 2000; 2001; http://www.boeing.com/airports/; http://www.airbus.com; http://www1.iata.org/Whip/Public/frmMain_Public.aspx?WgId=35/).

- Innovations of ATM/ATC and airport facilities and equipment express technical and technological progress in developing avionics, ATC/ATM and airport facilities and equipment. Progress in developing the former two can be measured by the cumulative navigational error in the aircraft position, which has significantly reduced over time (http://www.faa.gov/nextgen/; http://www.sesarju.eu/) (see also Chapter 4). This has brought gains in the airspace capacity and safety. Progress in development of the airport facilities and equipment can be measured by increased efficiency (capacity) of processing landings and take-offs in the airport airside area and users, i.e., passengers and freight shipments, in the airport landside area (Janić, 2013). This measure is preferred to be as high as possible and to increase over time.
- Reliability of structures reflects the ability of particular system components to operate without unexpected failures. This indicator can be separately measured for different

components, but in any case, the average number of failures per unit of operating time (hour, day, year) can be used as a measure. Due to safety and operational reasons, this measure, independently of the system component, is preferred to be as high as possible and to improve with the technical/technological progress that occurs over time.

ii) Economic indicators

The indicators of economic performance are defined to be 'profitability' and 'labour productivity'.

- Profitability, as in the case of airports and airlines, expresses financial success or failure of an airspace manufacturer. It is measured by the average operating profits (the difference between operating revenues and costs) per unit sold. With any type of airspace manufacturer, this measure is preferred to be as great as possible and to increase with an increase in the number of units sold.
- Labour productivity expresses the efficiency of airspace manufacturers in using their workforce. As in case of airlines, airports and ATM/ATC providers, the average number of units produced per employee can be used as a measure. This measure is preferred to increase with an increase in the total number of employees.

iii) Environmental indicators

The indicators of environmental performances are defined to be 'energy (fuel)', 'GHG emissions', and 'noise' efficiency.

- Energy (fuel), GHG emissions, and noise efficiency reflect absolute and relative reduction of fuel consumption, associated emissions of GHG and noise generated by the new aircraft and their engines. They can be measured by a relative decrease in these quantities per unit of the engine power or an aircraft operating weight. These measures are preferred to be as low as possible and to decrease with an increase in engine power and/or aircraft operating weight. Figure 5.16 shows the relationship between the average fuel consumption and seat capacity of commercial aircraft operating worldwide.

As can be seen, the average fuel consumption has generally decreased more than proportionally as the aircraft seat capacity has increased, thus indicating the larger aircraft

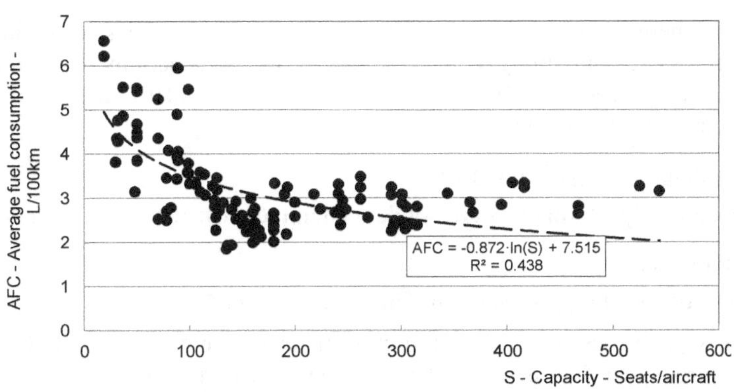

Figure 5.16 Fuel efficiency of commercial aircraft—relationship between average fuel consumption and seat capacity (L – Liter) (ICTT, 2015; https://en.wikipedia.org/wiki/Fuel_economy_in_aircraft/).

as being more fuel efficient in relative terms. If development of bigger aircraft was an objective in terms of sustainability, then this objective has been achieved in the long-term period, at least regarding this indicator. In addition, Figure 5.17 shows the aircraft noise efficiency in terms of the relationship between the average EPNdB (Equivalent Persistent Noise in Decibels) per unit of the aircraft maximum take-off weight and the maximum take-off weight.

As can be seen, the relative level of noise has decreased more than proportionally with the increase in the aircraft maximum take-off weight for both aircraft arrivals and departures. The arrival noise was slightly higher than the departure noise. Again, if development of bigger and relatively quieter aircraft was an objective, the progress has been sustainable with respect to this indicator.

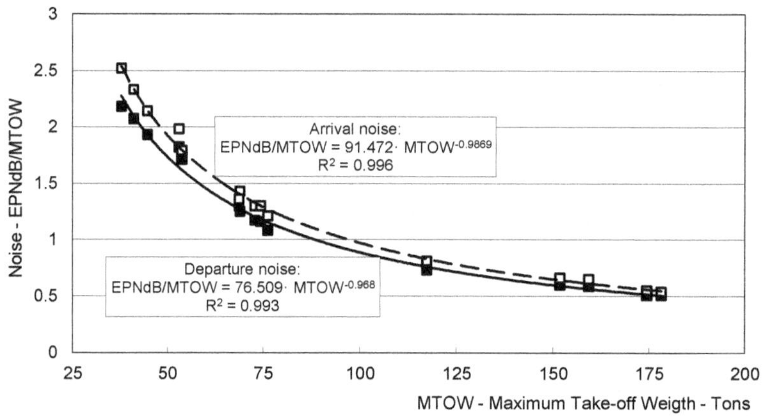

Figure 5.17 Relationship between the average level of noise and the aircraft take-off weight (FI, 2000; 2001; http://www.boeing.com/airports; http://www.airbus.com; http://www1.iata.org/Whip/Public/frmMain_Public.aspx?WgId=35).

iv) Social indicators

These indicators of performances are not defined in the present context.

f) Indicators for local community members

People living permanently or temporarily (tourist residential areas) near the airports represent the local community members. Regarding sustainability, they are mostly interested in the social and environmental and not particularly in operational and economic performances of an air transport system.

i) Operational and economic indicators

These indicators of performances are not defined in the present context.

ii) Social indicators

The indicator of social performances is defined to be 'social welfare'.

- Social welfare relates to the opportunity of local community people to get a job either directly or indirectly in the local air transport system (DETR, 2000). A convenient measure can be the ratio between the number of people employed by the air transport system and the total number of employed community people. This measure is preferred to be as high as possible and to increase as employment in the local community grows.

iii) Environmental indicators

The indicators of environmental performance are defined to be 'noise' and 'GHG emissions' disturbances and 'safety'.

- Noise disturbance reflects the level to which local people are annoyed by noise from atm at a nearby airport. This annoyance mostly depends on the subjective and objective factors. Subjective factors reflect the individual sensitivity and sensibility to noise. In such case, any noise being equal or exceeding a given individual threshold is considered as annoying. The most important objective factors are the amount of noise energy generated by aircraft flying over an affected area, the distance between a residential location and the aircraft flight path, and the quality of noise isolation of the houses. Bearing in mind both types of factors, two measures can be defined. First, the number of people exposed to the given level of airport, which is related to the volume of airport operations during the specified period of time. Second is the number of complaints per noise event (atm) carried out during a given period of time (day, night, month, year). Both measures are preferred to be as low as possible and to decrease as the number of atms increases. Figure 5.18 shows an example of the relationship between the number of people exposed to a certain level of noise and the volumes of demand over time.

As can be seen, the U.S. FAA (Federal Aviation Administration) has reported that, during the last 45 years, the number of people exposed to the current threshold for significant aircraft noise around U.S. airports has decreased from about 7 million in the year 1975 to less than 350000 in the year 2017. At the same time, the volume of air traffic, in terms of the number of enplaned passengers, has increased from about 200 (1975) to about 850 million (2017). Such progress has mainly been achieved through modernization of the aircraft fleets in terms of deploying primarily less noisy aircraft and complete exclusion of the particularly noisy (Stage 2) aircraft, and thanks to improvements in the arrival/departure operations. This indicates an obvious movement towards increasing the

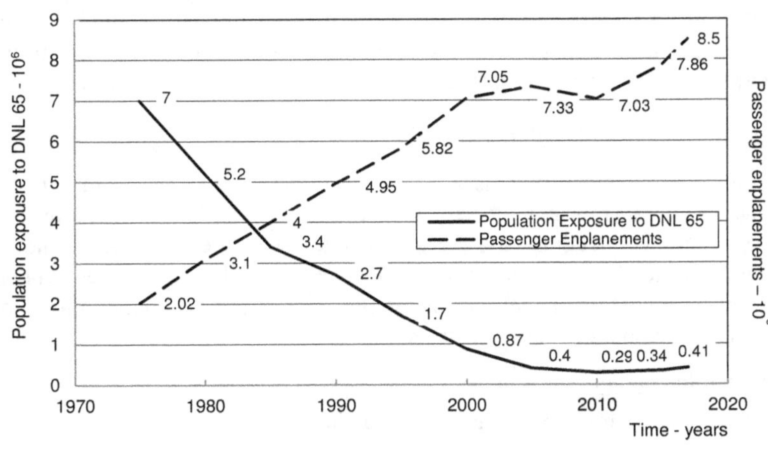

Figure 5.18 Relationships between the numbers of people exposed to the aircraft noise and the volume of air traffic over time—Case of the U.S. (HMMH, 2018).

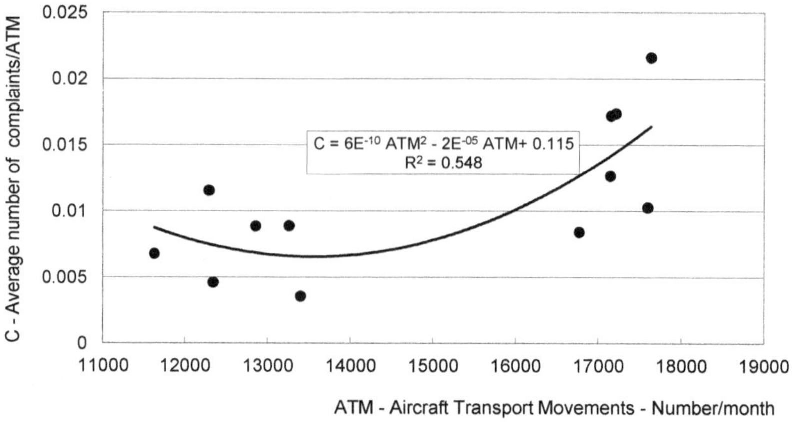

Figure 5.19 Relationship between the number of complaints to noise and the volumes of air traffic—Case of Manchester Airport (UK) (MA, 1999).

system's sustainability regarding this indicator (HMMH, 2018). In addition, Figure 5.19 shows an example of noise disturbance at Manchester Airport (UK) expressed by the average number of complaints per atm in relation to the total number of ATMs carried out during given period of time.

As can be seen, up to about 13 thousand movements carried out per month, the average number of complaints decreased, but beyond that it increased more than proportionally. This indicates that the airport has grown in an unsustainable way according to the attitudes of local community members.

- Air pollution disturbance relates to the exposure of local people to the harmful impacts of emitted GHGs from atms at a nearby airport. This indicator can be measured as the ratio between the quantity of GHGs emitted from atms and the total GHGs emitted from all local sources during a given period of time. This indicator is preferred to be as low as possible and to decrease as the total air pollution increases.
- Safety relates to the perceived risk of death or injury, or damage or loss of property due to an aircraft incident/accident (crash). It can be measured by the number of aircraft accidents per atm carried out during a given period of time (year). This measure is preferred to be as low as possible and to decrease as the number of atms grows.

g) Indicators for local and central authorities

The local and central authorities usually are not directly interested in the operational performances of an air transport system, except in the cases of significant interruptions. Otherwise, regarding sustainability, the primal interest of these actors/stakeholders is focused on the system's economic, social, and environmental performances, represented here by seven indicators: none for operational, three for economic, three for the environmental, and one for social performances.

i) Operational indicators

These indicators of performances are not defined in the present context.

ii) Economic and social indicators

The indicators of economic performances are defined to be 'economic welfare', 'internalization/globalization', 'externalities', and 'overall social welfare'.

- Economic welfare relates to the air transport system's contribution to the local (community) and regional/national welfare. A measure can be the share of air transport system's GDP in the total GDP. This measure is preferred to be as great as possible and to increase as the total GDP increases (see Figure 1.17, Chapter 1).
- Internalization/globalization relates to the air transport system's contribution to the internationalisation of the local and regional (national) business—investments, trade, and tourism. The measure can be the ratio between the number of long-haul business and/or tourist trips from/to the region carried out by air and the total number of such trips carried out by all transport modes during a given period of time (year). This measure is preferred to be as great as possible and to increase as the total number of business and tourist trips, respectively, increase.
- Externalities relate to the internalized costs of damages made by the air transport noise, GHG emissions, and air traffic incidents/accidents. The local (community) and central (regional, national) governments are both interested in these costs because of a general responsibility for creating policies that enable a healthy and environmentally friendly society (DETR, 2001; Janić, 2007). Once such policies are implemented, the causers, i.e., airlines and airports, together with their users, i.e., air passengers and freight shippers, as actual payers of the externalities will become more interested in these aspects of the air transport system operations. The externalities can be measured either individually or together by the average expenses used for preventing, mitigating, and/or remedying damages per unit of the system output (RTK, RPK). This measure is preferred to be as low as possible and to decrease as the system output increases.
- Overall social welfare represents benefits gained through direct and indirect employment by the air transport system at the local (community) and regional (country) levels. The measure of this indicator, preferred to be as high as possible and to increase over time, can be the annual total of people employed by the air transport system.

iii) Environmental indicators

The indicators of environmental performances are defined to be global 'noise disturbance', 'GHG emissions', and 'land use'.

- Global noise disturbance relates to exposure to noise at a global scale and can be measured by the total number of people exposed to excessive air transport noise during a given period of time (year). The measure is preferred to be as low as possible and to decrease over time (see Figure 1.17 and Figure 1.14, Chapter 1).
- Global GHG emissions are considered at the global scale and can be measured by the total quantities per unit of the air transport system's output—RTK (RPK). In this case, the total GHG emissions include those during LTO cycles, climb, cruise, and descent phases of flights (ICAO, 2012). This measure is preferred to be as low as possible and to decrease as the system's output increases (see Figure 1.12, Chapter 1).
- Global land use relates to total area of land used for the air transport infrastructure at global (national) and international scales. An appropriate measure can be the ratio

between the total area of land taken and the total volume of the system's output. In such case, the measure reflects the intensity of land use, preferred to be as low as possible and to decrease as the air transport system output increases (RTK, RPK) (see Figure 1.13, Chapter 1).

h) Indicators for others

The other actors/stakeholders involved, such as the international aviation organisations, different lobbies and pressure groups, and the public, can use the same indicators as those mentioned above for assessing the sustainability of the air transport system regarding particular performances. However, because of diversity of their objectives and preferences, the meaning and interpretation of particular indicators will generally be different.

5.3 Modelling Performances

5.3.1 General

Modelling the air transport system's performances is focused on an assessment of the potential of the conventional (oil-kerosene) Jet-A1/8, and alternative synthetic and biomass-derived SPK (Synthetic Paraffinic Kerosene) and LH_2 (Liquid Hydrogen) fuels for reducing fuel consumption and the related direct emissions of GHGs (Green House Gases) in the medium- and long-term. The convenient methodology, consisting of the analytical models for estimating this fuel consumption and corresponding direct emissions of GHGs, has been developed and applied according to a "what-if" scenario approach. These scenarios include the prospective development of the system and measures for mitigating fuel consumption and emissions of GHGs through development of the aircraft fleet and dynamism of introducing alternative fuels at the global scale in the medium- to long-term. The analytical models for estimating fuel consumption and related emissions of GHGs are scenario specific.

Despite making up only about 2% of the total man-made emissions of GHGs, the air transport system has come under increasing public pressure to become more sustainable, primarily by reducing its fuel consumption and related emissions of GHGs. As mentioned in Chapter 1, the system mainly consumes Jet A1/8 fuel derived from crude oil as a non-renewable resource, with emissions of GHGs (Green House Gases) contributing to global warming and climate change. The main GHGs are: CO (Carbon Monoxide) and CO_2 (Carbon Dioxide), H_2O (water vapour), NO (Nitric Oxide) and NO_2 (Nitrogen Dioxide), which together make NO_x (Nitrous Oxides), SO_x (Sulphur Oxides), NMHCs (Non-Methane Hydrocarbons), and PMs (Particulate Matter) (IPCC, 2013; 2015; Janić, 2007; PARTNER, 2009).

In increased concentrations and with the long lifetime of some in the Earth's atmosphere, these GHGs penetrate the air, water, and soil by constantly moving from one place to another, causing harm to plants, animals, and human health at the local scale and contributing to global warming and climate change at the global scale. In the case of an air transport system, the movement and related impacts of GHGs is both local and global. The local impacts of GHGs take place around airports. The global impacts take place in the en-route airspace during the cruising phase of flights at altitudes between 9 km and 12 km (below and above the tropopause) and latitudes from 40°N to 60°N. In addition, the lifetimes of various GHGs in the Earth's atmosphere are different. For

H_2O (water vapour) it is about nine days, for about 50% and 20% of CO_2 it is about 30–95 days and thousands of years, respectively, and for NO_x (Nitrous Oxides) it is about 114 years (IPCC, 2007; 2013; 2015; Janić, 2007; PARTNER, 2009; https://www.ipcc.ch/publications_and_data/ar4/wg3/en/tssts-ts-3-2-stabilization-scenarios-html/; http://www.climatehotmap.org/global-warming-effects/air-temperature.html). Since the lifetimes of particular GHGs in the Earth's atmosphere are different, they all affect global warming and climate change in different ways.

5.3.2 Characteristics of GHG (Green House Gases)

The main GHGs are generally CO_2 (Carbon Dioxide), NO_x (Nitrogen Oxides), H_2O (Water Vapor), SO_x (Sulphur Oxides), PMs (Particulate Matter(s)), and (NM)HCs (Non-Methane Hydrocarbon).

5.3.2.1 CO_2 (Carbon Dioxide)

CO_2 (Carbon Dioxide) is a gas with the above-mentioned very long residence time in the atmosphere, where it mixes well with other gases. For example, an instant doubling of the concentration of CO_2 relative to the present concentration would increase the average temperature on the earth's surface by about 1.4°K. This phenomenon can be explained as follows: increasing the concentration of CO_2 will reduce the earth's long wavelength radiation at the top of the atmosphere by a certain amount and consequently reduce the inward flux there by the same amount. The energy balance at the top of the atmosphere requires a constant flux. Therefore, the earth's surface temperature should rise in order to compensate such an imbalance. This effect is called "radiative forcing". The commercial air transportation has been responsible for about 0.02 W/m^2 (Watts per square meter). Any increase in the global temperature can cause additional impacts, i.e., increase or mitigation of the concentration of CO_2 as a reversible process. Some estimates suggest that the current concentration of CO_2 in the Earth's atmosphere is around 382 ppm and the tendency is that it will increase by an annual rate of about 1.2 ppm over the next forty years (until the year 2050) (ppm – parts per million). Some other estimates indicate that, when the total known reserves of crude oil of about 1650 billion (10^{12}) U.S. barrels are exhausted by the end of the 21st century, the concentration of CO_2 will contribute to an increase in the average global temperature of about 2.5°C (Boeker and Grondelle, 1999).

5.3.2.2 NO_x (Nitrogen Oxides)

An equally important gas in the earth's atmosphere is ozone O_3. Its presence protects the earth from the harmful solar UV radiation by absorbing all light with a wavelength less than 295nm* (nano-meter). The layer of O_3 in the earth's atmospheres is relatively thin, about 0.3–0.4 cm, under constant temperature and atmospheric pressure. The gas is present throughout the atmosphere but it is maximally concentrated at the altitudes of about 20–26 km from the earth's surface. It is permanently formed through the reaction of the molecular oxygen O_2 and the atomic oxygen O influenced by the solar UV radiation. Most of the ozone is formed above the equator where the amount of UV solar radiation is maximal. From there, it moves towards the poles where it is "accumulated" up to a thickness of about 0.4 cm during the winter period.

Ozone is sensitive to the free radicals, such as the atomic chlorine CI, nitric oxide NO, and hydroxyl radicals OH. They are formed from H_2O and CFCs (Chloro-Fluoro-Carbons), products of burning Jet A1/8 fuel, which escape from the troposphere (10–

12 km from the earth surface where most commercial flights take place) to the stratosphere where the ozone layer is formed. At these altitudes, free radicals, including NO_x, lead to the depletion of the ozone layer. Those that do not escape remain extremely stable in the troposphere where they, together with NO_x, contribute to a thickening of the ozone layer. The residence time of NO_x in these regions increases with altitude. Therefore, NO_x affects the ozone layer regionally if injected into the troposphere and globally if injected into the stratosphere. In any case, the increased concentration of NO_x generally depletes the ozone layer with inevitable impacts. For example, depletion of this layer by about 10% may cause an increase in the UV radiation by about 45%, which certainly could inflict damages to almost all biological cells and, in particular, cause skin cancer in exposed people (Boeker and Grondelle, 1999; IPCC, 2001; 2014; RCEP, 2003).

5.3.2.3 H_2O (Water Vapor)

H_2O (Water Vapor) in the form of clouds called contrails has recently been also identified as a contributor to the global warming. It has reduced the amount of the solar radiation returning to the atmosphere and increased the amount of the solar radiation reflected from the atmosphere. Consequently, the surface becomes warmer in order to keep the radiative forces in balance. According to some estimates, contrails contribute to radiative forcing by 0.007 to 0.06 W/m^2, with the expectation to increase to about 0.04–0.4 W/m^2 with the projected growth of the commercial air transportation industry by 2050 (IPCC, 2001; 2014).

Unlike CO_2, contrails are shown to have a substantial impact in regions with intensive air traffic, such as in Europe and North America. After being recognized, two proposals have been suggested to eventually mitigate their impact: (i) scheduling flights during the sunrise and sunset periods when the impacts of contrails in terms of blocking the earth-emitted radiation is less, and (ii) restricting the flight cruising altitudes to those where the contrails have a less significant impact. The latter option seems ambiguous for several reasons. First, restricting the aircraft fuel-optimal altitudes might increase fuel consumption and, consequently, the related emissions of CO_2, which in turn contribute to increasing of the total emitted quantities of GHGs. Second, restriction of the usable altitudes certainly diminishes the airspace capacity and consequently increases delays, which again might result in increased fuel consumption and related emissions of GHGs. Finally, the workload of ATC/ATM controllers might be increased, which would require more staff/employees to maintain the required level of safety, efficiency and effectiveness of air traffic.

5.3.2.4 SO_x (Sulphur Oxides)

Fuel consumed by commercial aircraft (JP1 or Jet A) may have considerable amount of Sulphur. In a chemical reaction with the water vapor in the atmosphere it creates acid rain which damages trees and other dependent natural habitats. In order to diminish its presence in the jet engine fuel and exhaustive gases after its burning, different catalysts are added.

5.3.2.5 PM_S (Particulate Matter(s))

The PMs (Particulate Matter(s)) from burning JP1 or Jet A fuel are emitted in the air as a mixture of solid particles and liquid droplets. Depending on the size, they can be PM10 (diameters that are generally 10 micrometers and smaller) and PM2.5 (diameters generally 2.5 micrometers and smaller) (1 micrometer = 0.001 mm). In general, increased concentration of PMs can compromise the health of exposed people (https://www.epa.gov/).

5.3.2.6 (NM)HCs (Non-Methane Hydrocarbon(s))

The (NM)HCs (Non-Methane Hydrocarbon(s)) emitted from burning Jet a or JP1 fuel contribute to the formation of smog and, consequently, global warming. The emitted quantities have not been recognized as particularly affective in comparison to their other polluting counterparts. Nevertheless, their contribution to global warming has proven to be through production of ozone O_3, extending the lifetime of methane CH_4, and conversion into CO_2 and H_2O, as the most important greenhouse gases (Janić, 2007).

5.3.3 Impacts of GHG (Green House Gases)

5.3.3.1 Physics

Generally, the life on earth is related to some physical properties of the sun-earth system. The surface temperature of the sun is about 5800°K, which results in an emission spectrum with the maximum wavelength of 500 nm* (nm* – nano meter; 1 nm* = 10^{-9} meter). This makes solar temperature of the exact magnitude to induce photochemical reactions. Depending on the radius of both sun and earth, their distance, and the above-mentioned surface solar temperature, one can estimate that the earth receives energy of about 1379 W/m^2 (Watts per square meter), although the solar constant is always taken a bit lower, i.e., S ≈ 1370 W/m^2. With the support of some gases in the atmosphere, this energy appears sufficient to maintain an average temperature on the earth's surface of T = 288°K (+15°C). A part of the received energy is reflected from the earth's surface back into the space. This is called albedo *a* from the Latin term "albus" meaning "white". Astronomers usually use this to express the brightness of the earth as seen from the space. Consequently, the energy equation can be set up as follows:

$$(1 - a)\pi R^2 S = 4\pi R^2 \sigma T^4 \qquad (5.1)$$

where

- a is albedo;
- R is the Earth's radius (6400 km);
- S is the Solar constant;
- σ is the Stefan-Boltzmann constant ($\sigma = 5.672 \times 10^{-8}\ Wm^{-2}K^{-4}$); and
- T is the Earth's surface temperature.

With an estimated albedo value of: a = 0.34, one can obtain the temperature of the Earth's atmosphere of: T = 250°K (Usually it is taken to be: T = 255°K). This is lower than the Earth's surface temperature (288°K), which is mainly due to the presence and concentration of gases such as CO_2, O_3 (Ozone), NO_2, CH_4 (Methane), and H_2O. Otherwise, this temperature would be lower by about 30°K. The above-mentioned gases absorb most of the heat radiation from the earth and re-emit it back towards the earth surface, which is the process known as the "greenhouse effect". Consequently, these gases are called GHGs "Green House Gases".

5.3.3.2 Scale of impacts

Contribution of the commercial air transportation to global warming through emissions of GHG from burning Jet A1/8 fuel could be roughly estimated by using the zero-dimensional GHG model based on the total energy flux at the top of earth's atmosphere (Boeker and Grondelle, 1999; IPCC, 2001). Under equilibrium conditions, the radiation flux vanishes, i.e., the inward and outward radiation is in balance. This balance can be disrupted by reduction of the earth's long-wave radiation ΔI at the top of the atmosphere, caused by an increase in the concentration of GHGs (for example CO_2). Consequently, the outward radiation will decrease by the same amount. Since the energy balance at the top of atmosphere requires a constant flux, the temperature at the earth's surface will increase by amount (ΔT) in order to compensate for the reduction of ΔI. This phenomenon, known as the radiative forcing, puts the variables (ΔI) and (ΔT) into the following relationship:

$$\Delta I = \frac{\partial I}{\partial T} \Delta T \qquad (5.2a)$$

where the term $(\partial I/\partial T)$ can be approximated as follows:

$$\partial I/\partial T = 4/T(1 - a) \cdot S/4 \qquad (5.2b)$$

where all symbols are the same as in Eq. 5.1. There, albedo: $a = 0.34$ as viewed from space and $a = 0.11$ at the Earth's surface; the Earth's surface temperature: $T = 288°K$; and the solar constant: $S = 1.370 \cdot 10^3$ J/s-m^2 (J/s/m^2) (J – Joule; s – second; m^2 – square meter).

Using the common values of the above-mentioned variables gives an estimate of Eq. 5.2b as follows: $\partial I/\partial T = 31\ W/m^2\ °K$ and its reciprocal value: $G = 0.334\ m^2\ °K/W$. From Eq. 5.2a it follows: $\Delta T = G \cdot \Delta I$. Typical value of the radiative forcing is: $\Delta I = 4 - 4.6\ W/m^2$ as contribution of the man-made GHG emissions. Some reports suggest that the air transport system might participate in this total up to about 3.5%, i.e., with $\Delta I = 0.035 \cdot \Delta I = (0.14 - 0.16)\ W/m^2$. Applying this to equation $\Delta T = G\Delta I$ gives: $\Delta T = G \cdot \Delta I = 0.334 \cdot (0.14 – 0.16) = (0.0468 - 0.0534)\ °K$. In addition, some other investigations indicate that the commercial air transportation could contribute to an increase in the earth's surface temperature of about: $\Delta T = (0.052 - 0.096)\ °K$ between the year 2010 and 2050 (IPCC, 2001; Boeker and Grondelle, 1999). Both above-mentioned estimations suggest that the air transport system will not significantly contribute to global warming and climate change as the current myths have suggested.

5.3.4 Characteristics of Air Transport System

5.3.4.1 Development

The global air passenger and freight/cargo transportation has been continuously growing over the past forty years, as mentioned in Chapter 1. The volumes of air passenger transportation were increasing from 500 in 1971, to 4157 in 2006, 5224 in 2011, 6664.5 in 2015, and 8258 billion RPKs (Revenue Passenger Kilometres(s)) in the year 2018. This latest figure required consumption of about 273 million tons of fuel and resulted in 1189 million tons of emissions of GHGs (CO_{2e}) (Carbon Dioxide Equivalents). These emissions of GHGs were about 4.7 times greater than that in 1990 (255.4 million tons of CO_{2e}) (IATA, 2016; ATAG, 2016; https://www.icao.int/annual-report-2018/Pages/the-world-of-air-transport-in-2018.aspx). Some long-term forecasts carried out by different international air transport organizations, ICAO (International Civil Aviation Organization), IATA (International Air Transport Association), and two large manufacturers of commercial aircraft (Boeing and Airbus), predict a continuation of the relatively stable growth of global air transportation over the next 20 years (Airbus, 2006; 2015; Boeing, 2007; 2016). For example, ICAO forecasts growth of global RPKs at an average annual rate of 3.7–5.2% (average 4.0%) over the period 2010–2030. Such growth is supposed to be mainly driven by global GDP (Gross Domestic Product), which is expected to grow at an average annual rate of 3.1% (Narjess, 2010). According to Boeing, the volumes of global RPKs are expected to grow at an annual rate of 4.82% over the period 2015–2035, i.e., to increase from 8258 in 2015 to 16734 billion RPKs in 2035 (Boeing, 2016). Air freight/cargo transportation is expected to grow at an average annual rate of 4.7% over the period 2013–2033, i.e., from about 220 in 2013 to about 510 billion RTKs in 2033 (Boeing, 2014). The other large aircraft manufacturer, Airbus, forecasts growth of global RPKs at an average annual rate of 4.6% over the period 2014–2034, which will result in about 16743 billion RPKs in 2034. Freight/cargo transportation is expected to grow at an average annual rate of 4.4% and reach about 460 billion RTKs in 2034 (Airbus, 2015). Again, such forecasted growth of both air passenger and air freight/cargo transportation will be primarily driven by the global GDP (Gross Domestic Product), which is expected to grow at an average annual rate of 2.9% over the same period—2015–2033/35. In order to support the above-mentioned current and prospective growth, the size of the global aircraft fleet will have to be increased. Boeing has forecasted demand for 38690 new passenger and 2440 new freighter aircraft during the period 2015–2035 (Boeing, 2016). Airbus has forecasted demand for 32600 new passenger and 2356 new freight aircraft during the same period (Airbus, 2015). On the one hand, the forecasts of air transportation demand can be considered 'optimistic' because of interests of both manufacturers to sell as many aircraft as possible. On the other, they can also be considered 'realistic' bearing in mind that these manufacturers have also taken into account some inputs from ICAO, IATA, and the various potential financial and business risks.

The above-mentioned forecasted development of global commercial air transportation will inevitably contribute to a further increase in the fuel consumption and related emissions of GHG (CO_{2e}). Figure 5.20 shows developments of the international air transportation according to the so-called 'baseline' scenario, implying that mitigating measures will not be introduced or effective.

As can be seen, the fuel consumption by international air transportation will increase from 140 in 2010 to about 850 million tonnes in 2050. In addition, the related emissions

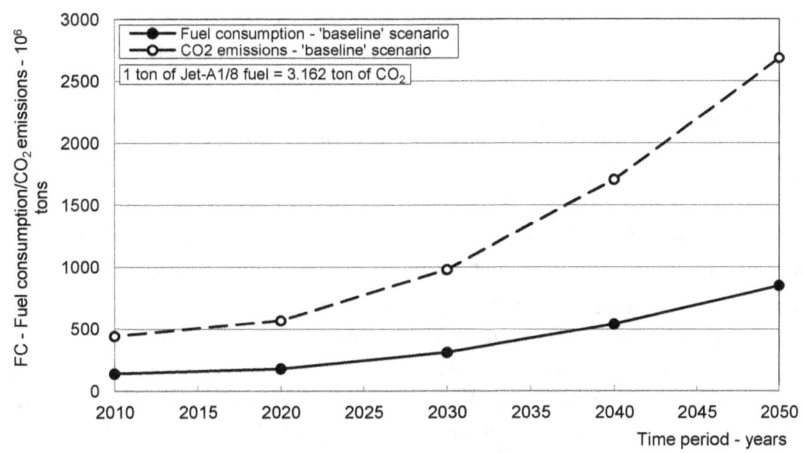

Figure 5.20 The fuel consumption and related emissions of GHG (CO_2) by the international air transportation according to the ICAO/IATA "baseline" scenario (Period: 2010–2050) (ICAO, 2013; 2016).

of GHG (CO_2) will increase from 442 in 2010 to 2686 million tonnes in 2050, i.e., six-fold (ATAG, 2015; 2015a; ICAO, 2013; 2016).

5.3.4.2 Mitigating impacts of GHG

In order to mitigate the growth of the above-mentioned forecasted fuel consumption and related emissions of GHG and the consequent impact on climate change in the medium- to long-term, the ICAO and IATA have undertaken various initiatives. These have been undertaken despite the above-mentioned relatively low expected impacts of the system on global warming and climate changes. These initiatives have been defined in the scope of the so-called "4-pillar" strategy of improving, developing, and implementing: (i) Technology, including increased deployment of alternative fuels; (ii) Efficiency of the system's operations; (iii) Air transportation infrastructure, including modernizing ATC/ATM (Air Traffic Control/Management) systems (examples are current European SESAR and U.S. NextGen program); and (iv) GMBM (Global Market-Based Measures) to fill in the remaining emission gap. The goals/targets to be achieved by full implementation of these initiatives are expected to be as follows (ICAO, 2013; 2016):

- Improvement of fuel efficiency and consequent reduction of emissions of GHG (CO_2) by 1.5% per year in the short-term by 2020;
- Reduction of the net emissions of GHG (CO_2) by 50% compared to those in 2005 in the long-term by 2050; and
- Setting up a cap on the net emissions of GHG (CO_2) in order to achieve "carbon neutral" growth in the medium-term after 2020; and
- Improvement of the fuel efficiency and consequent reduction of the net emissions of GHG (CO_2) means by technological measures have particularly related to increase BR (Bypass Ratio) of aircraft jet-engines and reducing SFC (Specific Fuel Consumption). This ratio is defined as the ratio of fuel burned per hour per ton of the net thrust (Janić, 2014). The SFC of the most contemporary aircraft jet engines amounts about 0.25–0.30 kg of fuel/kg of thrust/hour. After the year 2015, it has been expected to decrease to about or even less than 0.184 kg of fuel per hour per kg of thrust. In addition, the technological improvements in the aircraft airframe have related to the aerodynamic performances. Typical examples are Boeing B777-300ER (ER – Extended Range), B787-8/9, and Airbus A350 aircraft. On these aircraft, the weight has generally been reduced by using a greater proportion (30–60%) of much lighter composite materials. In addition, the more fuel-efficient "raked wing tip" design has replaced the winglets used previously on the other B777 family aircraft, as well as on B-747-400 and B737 NG (Next Generation) aircraft. Just these new winglets have been expected to improve the fuel efficiency by about 1–2% for the long-haul flights. Consequently, the average typical fuel consumption of these long-range commercial aircraft has generally decreased, as shown in Figure 5.21.

The above-mentioned aircraft technology improvements have been followed by improvements in optimizing the aircraft O-D (Origin-Destination) flights means by different air traffic and air transport operational-economic measures. The former measures include carrying out more direct flights and, consequently, reducing the total travel distance per passenger, optimizing the aircraft climb/descent profiles in terms of the fuel consumption, optimizing the cruising altitudes and reducing cruising speed(s).

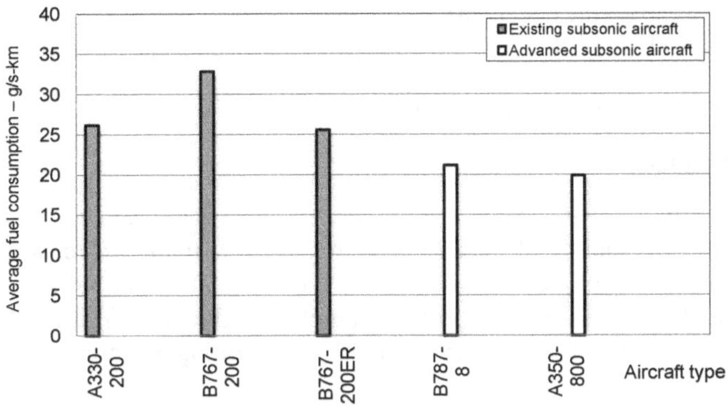

Figure 5.21 Average unit fuel consumption of the contemporary commercial aircraft (Janic, 2014).

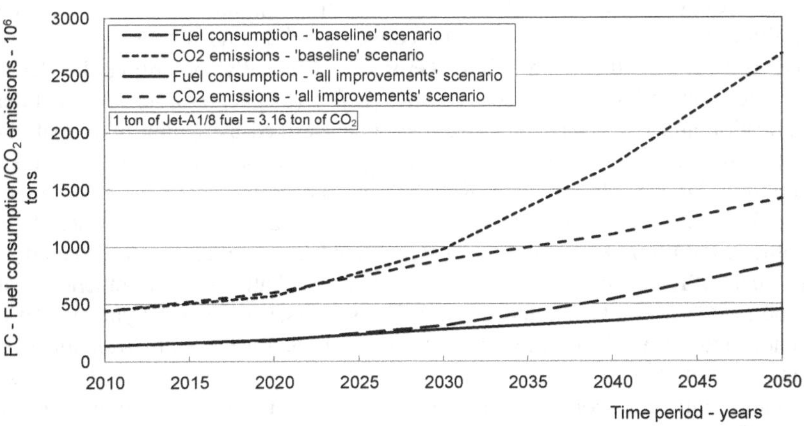

Figure 5.22 Effects of reducing net fuel consumption and related emissions of GHG (CO_2) by international commercial air transportation according to the 'all improvements' scenario, i.e., after achieving ICAO/IATA goals/targets (Period: 2010–2050) (ICAO, 2013; 2016).

The latter measures include improving load factor, reducing the long aircraft taxiing and towering at airports, and harmonizing the fuel prices as well.

Figure 5.22 shows the possible effects as scenarios of fulfilment of the above-mentioned initiatives on the international air transportation industry by the year 2050.

As can be seen, compared to the 'baseline' scenario characterized by the annual fuel consumption of about 850 and related direct emissions of CO_2 of 2686 million tonnes in 2050, the fulfilment of three goals/targets (i.e., 'all improvements scenario') would contribute to decreasing of the annual fuel consumption and related direct emissions of GHG (CO_2) by about 50% in the year 2050. Consequently, the impact of this segment of the air transport system on climate change will be reduced. However, regarding the future period of about 30 years and the above-mentioned long lifetime of CO_2 of about 95 years, it is realistic to expect that CO_2 will only accumulate in the Earth's atmosphere over time. If this is the case, the potential of gradual implementation of alternative fuels as a mitigating initiative and measure in the scope of the first above-mentioned ICAO/IATA "4-pillars" strategy needs further elaboration.

- Setting up a cap on the net emissions of GHG (CO_2) aiming at achieving the "carbon neutral" growth after 2020 has not been particularly considered.

5.3.4.3 Characteristics of air transport fuels

a) Conventional fuels

In general, kerosene or Jet-A1/8 obtained from a non-renewable feedstock, i.e., crude oil, has been used as the main fuel for air transportation over the past decades. Its burning produces the above-mentioned emissions of GHG. As mentioned above, they affect the environment at the local (around airports) and global (high altitude) scales. In particular, at the global scale, they directly contribute to global warming and climate change.

b) Alternative fuels

The alternative fuels for air transportation generally include synthetic fuels derived from other feedstocks than crude oil and LH_2 (Liquid Hydrogen). Since the former fuels consist of paraffinic compounds similarly to the conventional Jet A1/8 fuel, they are usually referred as SPK (Synthetic Paraffinic Kerosene). Currently, regarding the derivation method, they can be broadly categorized into: (i) F-T (Fischer-Tropsch) Jet Fuel derived by gasification and F-T (Fischer-Tropsch) Jet Fuel derived by synthesis of coal, natural gas, and/or biomass (solid waste, agricultural waste and forest waste, wood and energy crops), and (ii) HRJ (Hydroprocessed Renewable Jet) Fuel derived by the hydroprocessing of renewable oil. The U.S. FAA has approved several of these fuels: SPK-ATJ (Alcohol to Jet Synthetic Paraffinic Kerosene), SIP (Synthesized Iso-Parafins), which convert sugars into jet fuel; SPK-HEFA (Hydroprocessed Esters and Fatty Acids Synthetic Paraffinic Kerosene), which uses fats, oils and greases; SPK-FT (Fischer-Tropsch Synthetic Paraffinic Kerosene); and SKA-FT (Fischer-Tropsch Synthetic Kerosene with Aromatics) (http://www.atn.aero/article.pl/). In general, all these fuels are free from sulphur as well as from the aromatic compounds common to their conventional crude oil-based counterpart(s). The LH_2 actually being free of most of the above-mentioned GHGs is still under investigation as a long-term alternative (ASTM, 2014; Bossel and Eliasson, 2003; Dufour et al., 2009; Hileman et al., 2010; Karunanidhi, 2015; RAND, 2009).

Most of the considered alternative fuels are expected to contribute to the substantial reductions of both direct (during the flights) and total (during the life-cycle) emissions of GHGs, compared to the Jet A1/8 counterpart. However, in the present context, their total (life-cycle) emissions of GHGs are not considered due to the high uncertainty in the dynamism of their implementation. Tables 5.1 and 5.2 give the selected performance of these alternative fuels and Jet A1/8 fuel during their direct combustion, including the relative comparison, respectively (Hilleman et al., 2010; NASA, 2011; PARTNER, 2010; Puncher et al., 2011; Timko et al., 2001).

When direct combustion is exclusively considered, SPK fuels are characterized by a lower specific gravity and energy density, higher specific energy, slightly lower emissions of CO_2 (by about 2%), substantially lower emissions of NO_x (by about 11–22%), level of PMs (by about 97.5–99.9%) and NMHCs (by about 18–25%) compared to Jet-A1/8 fuel. In addition, they are free from sulphur. The specific energy of LH_2 is 120 MJ/kg, specific gravity 0.071 kg/L at 15°C, energy density 8.4 MJ/L, and freezing point 14°C (Chevron, 2006). In addition, LH_2 possess quite different performance compared to the two other types (Jet A1/8 and SPK) of fuels: lower specific gravity by about 98%, lower energy

Table 5.1 Performance of the conventional Jet-A1/8, alternative SPK and LH_2 fuels for air transportation (Hilleman et al., 2010; Karunanidhi, 2015; NASA, 2011; PARTNER, 2010; Pucher et al., 2011; Timko et al., 2001).

Performance	**Type of fuel**		
	Jet-A1/8 (1)	**SPK[1] (F-T Jet & HRJ) (2)**	**LH_2[2] (3)**
SG – Specific Gravity (kg/L)	0.802	0.757	0.071
ED – Energy Density (MJ/L)	34.9	33.4	8.4
SE – Specific Energy (MJ/kg)	43.2	44.1	120
FP – Freezing Point (°K)	226.15	223.15	14.4
SE – Specific Emissions (CO_2/MJ)[3]	73.2	70.4	0.0
UME 1 – 1 (kgCO_2/kg of fuel)[4]	3.162	3.105	0.0
UME 2 – 2 (kgH_2O/kg of fuel)	1.230	1.230	3.128
UME 3 – NO_x (g/kg of fuel)[5]	9.0–18.0	8.0–14.0	0.05–0.25
UME 4 – SO_x (g/kg of fuel)	0.084	0.00	0.00
UME 5 – PM (PM/kg of fuel)	$3.0–6.0 \cdot 10^{15}$	$1.5 \cdot 10^{12}–1.5 \cdot 10^{14}$	0.00
UME 6 – NM (HC) (g/kg of fuel)	18.0	13.5–14.7	0.00

[1] SPK – Synthetic Paraffinic Kerosene; F-T – Fischer-Tropsch Jet Fuel; Hydroprocessed Renewable Jet; [2] Liquid Hydrogen; [3] Direct combustion; [4] UME – Unit Mass Emissions; Obtained as the ratio: Specific energy MJ/kg)/Specific emissions (gCO_2/MJ); PM – Particulate Matters; [5] Engine power: 45–100%.

Table 5.2 Relative comparison of the performance of conventional, alternative SPK, and LH_2 fuel for air transportation (Derived from Table 5.1).

Performance	**Jet A1/8 (1)**	**SPK (F-T&HRJ) (2)**	**LH_2 (3)**	**GWP (5)**
Fuel				
Specific Gravity	1	0.944	0.0886	
Energy Density	1	0.957	0.24	
Specific Energy	1	1.02	2.78	
GHG				
CO_2	1	0.982	-	1
H_2O	1	1	2.543	0.1
NO_x	1	0.778–0.889[1]	0.005–0.0139	265
SO_x	1	0.00 [2]	-	
PMs	1	0.0005–0.025[3]	-	
NM(HC)	1	0.75–0.82[4]	-	

[1] Jet A1/8: 9.0–18.0 g/kg of fuel; (F-T&HRJ): 8.0–14.0 g/kg of fuel (Engine power: 45–100%); [2] Jet A1/8: 0.084 g/kg of fuel; (F-T&HRJ): Sulfur free; [3] Jet-A1/8: $3.0–6.0 \cdot 10^{15}$ PM/kg of fuel; (F-T&HRJ): $1.5 \cdot 10^{12}–1.5 \cdot 10^{14}$ PM/kg of fuel (Engine power 40–90/100%); [4] Jet A1/8: g/kg of fuel; (F-T&HRJ): 13.5–14.7 g/kg of fuel (PM – Particulate Matters); [5] GWP (Global Warming Potential) (Values for 100-year time horizon) (Derwent et al., 2006; IPCC, 2014).

density by about four times, and higher specific energy by about 2.8 times. In addition, LH_2 has about 2.5 times higher emissions of H_2O and about 75–95% lower emissions of NO_x. It is free from emissions of other GHGs, such as CO_2, SO_x, PMs, and NMHC. Emissions of NO_x are expected to be reduced by designing the appropriate combustion chambers in cryogenic jet engines. In general, LH_2 is considered a safe aviation fuel. Nevertheless, its main potential disadvantages are the explosive rate of 13–79% concentration in the air and its very low ignition energy (about only 0.02 mill joules). LH_2 also mixes faster with air than the conventional jet fuel vapour, and disperses rapidly through the air in contrast with Jet-A1/8 and SPK fuels, which pool on the ground. It burns with a nearly invisible, colourless and odourless flame, which is also an important safety concern (IEA, 2006; Janić, 2008; 2014; 2018).
The alternative SPK and LH_2 fuels have been developed and used as follows:

i) SPK (Synthetic Paraffinic Kerosene): Development and use of SPK (Synthetic Paraffinic Kerosene) fuels at the global scale was initially carried out through cooperation between airlines and fuel producers, and most recently through the involvement of other stakeholders close to the air transportation system. These initiatives have generally included coordination of the national stakeholders, and international cooperation, research and development. In many cases, both research and development initiatives have included cross-sectoral joint ventures between the fuel producing companies, aerospace manufacturers, and airlines as the end consumers. This has been shown to be feasible due to dealing with the logistics of these fuels during their life cycle, i.e., from choosing feedstocks, production, delivery and storage, i.e., supply, and final consumption/combustion (Hilleman et al., 2010; NASA, 2011; PARTNER, 2010; http://www.airbus.com/en/myairbus/headlinenews/index.jsp; http://www.omega.mmu.ac.uk/; http://partner.aero/). In addition, the numerous practical initiatives for providing the feedstock and deployment of alternative fuels have been carried out as shown in Figure 5.23 (IATA, 2015; ICAO, 2010; http://www.greenaironline.com/news/).

As can be seen, despite variations, the annual number of these initiatives has generally increased during the observed period from only one in 2006 to 22 in 2015. Of these 22 initiatives, six are related to the research and development, eight to deployment, and eight to stakeholders' action groups. The industry funded eight of them, the public

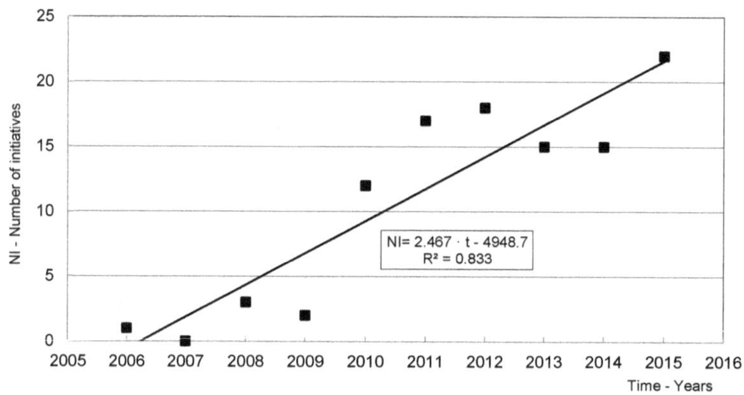

Figure 5.23 Development of initiatives for introducing alternative fuels over time (Period: 2006–2015) (IATA, 2015; ICAO, 2010).

five, and the public/private partnerships nine. The most numerous have taken place in Europe (11) and U.S. (four), but only five have related to using the alternative fuels on a commercial scale. In addition, these alternative fuels have also been used by the aircraft manufacturers and airlines in practice as follows:

On 1 February 2008, Airbus launched a research program into alternative fuels. As a result, the aircraft Airbus A380 carried out a three-hour flight with one of its engines powered by some kind of "alternative" fuel, a 40/60 GTL (Gas-To-Liquid)/kerosene blend. However, it should be mentioned that GTL is actually a fossil fuel, which had been approved for use in commercial air transportation well before the approval of biofuels.

On 24 February 2008, Virgin Atlantic used a 20/80% FAME (Fatty Acid Methyl Esters)/kerosene blend in one engine of its B747-400 flying between London and Amsterdam. These and many other experiments carried out subsequently, including the most recent ones, were intended to test the engine performance and prospective impact of GHGs on the environment from burning these fuels.

Finally, most recently, on 11 March 2016, United Airlines as the first U.S. airline started to use alternative biofuels at the commercial scale for regular scheduled flights from Los Angeles International Airport. The feedstock for these fuels includes agricultural waste and non-edible natural oils. These fuels have been used as a 30/70% blend with conventional Jet A1/8 fuel. This launch has been considered important since it has moved beyond demonstration flights and test programs to using alternative fuels for regular operations. The airline is planning to buy about 17 thousand tons/year of these fuels over the period of 3 years, which would enable about 12.5 thousand flights between San Francisco and Los Angeles to be carried out (https://www.united.com/web/en-US/content/company/globalcitizenship/environment/alternative-fuels.aspx/).

Cathay Pacific plans to use 100 thousand tons/year of alternative fuels over the period of 10 years starting from 2019. FedEx/Southwest has contracted for the delivery of 10 thousand tons/year of alternative fuels for the period of eight years starting from 2017. Finally, United Airlines plans to use 270 thousand tons/year of alternative fuels over the period of 10 years starting from 2019 (IATA, 2015).

Further introduction of alternative fuels at the larger commercial scale will be faced with many challenges. Some of the most important include the lack of policies supporting their introduction, the need for full development and completion of manufacturing technologies, price, which will have to be comparable to the price of the conventional Jet A1 fuel, and the provision of sufficient quantities to be applied at the larger scale. In particular, the indirect impact of alternative fuels through indirect changing of land use have and will have to continue to be subject to further investigation while considering their total (life-cycle) emissions of GHGs (IATA, 2015).

ii) LH_2 (Liquid Hydrogen): besides being considered by researchers and airspace manufacturers at the conceptual level for powering the future super and hypersonic commercial aircraft, LH_2 has not really been elaborated (and still not practiced) on as a fuel for air transportation to be used at the larger commercial scale (Koroneos et al., 2004; Okai, 2010).

c) Economics of fuels

The average share of fuel costs in the total operating costs of Jet A1/8 powered aircraft fuel operated by the airlines worldwide has varied over time. For example, it has been about 27% during the period 2004–2016 (IATA, 2017). For example, in the USA this share was mainly influenced by the average price of Jet A1/8 fuel of about 0.32 $US/kg during the period 2000–2007. Since that time, this price continued to increase and reached the maximum of 1.06 $US/kg in 2009. During the period 2011–2014, the average price varied from 0.60 to 0.71 $US/kg. In 2016 it was about 0.30 $US/kg (http://www.iata.org/publications/economics/fuel-monitor/Pages/price-development.aspx). Some forecasts indicate that the average price of Jet A1/8 fuel will continue to rise over the forthcoming medium- to long-term period and reach between about 0.43 $US/kg (low scenario) and 1.28US $US/kg (high scenario) in the year 2040. Therefore, in order to be able to compete with the conventional Jet A1/8 fuel, the alternative fuels will also have to be competitively priced. Some theoretical estimates show that these prices could be about 0.77–1.45 $US/kg for F-T and 0.78–0.87 $US/kg for HRJ fuel (IATA, 2015; Saynor et al., 2003). At the same time, the average prices of LH_2 have been estimated to be in range of about 1.00–1.73 $US/kg. Regarding the prospectively rising future prices of the conventional Jet A1/8 fuel and the lower SFC (Specific Fuel Consumption) of cryogenic engines by about 64% (as given in Table 5.2), some estimates indicate that the share of fuel costs in the total operating costs of the cryogenic aircraft could vary between 45% and 78%. If the prices of both Jet A1/8 and LH_2 fuel equalize at the level of about 1 $US/kg, the corresponding shares of fuel costs will amount about 60% and 35%, respectively, mainly due to the lower SFC of cryogenic aircraft. The prices of other inputs are assumed constant. This figure seems to be realistic under conditions of the prices of JetA1/8 fuel being expected to continue to rise (the dynamism of prices of SPK fuels is still uncertain) and that of LH_2 to fall, driven by improvements in its efficiency of manufacturing, storage, and distribution (Janić, 2014; Winchester et al., 2015). Therefore, in addition to being supplied in the required commercial quantities, the selling prices of all above-mentioned fuels should be quite similar in order to make them competitive and, thus, attractive for their users, i.e., airlines.

5.3.4.4 Characteristics of aircraft fleets

a) Conventional aircraft

The present aircraft powered by Jet-A1/8 fuel are considered as conventional aircraft. They are propelled by jet engines whose main performance in the given context is SFC (Specific Fuel Consumption), expressed in terms of g/s-km (grams of fuel per seat-kilometre) (Janić, 1999). Divided by the average load factor, such defined SFC gives the average fuel consumption, i.e., fuel efficiency of an individual aircraft and/or of an aircraft fleet, in terms of g/p-km (grams of fuel per passenger kilometre). The average fuel consumption expressed in g/p-km has decreased more than proportionally during the observed period. The total decrease amounted about 45% over the past 46 years (see Figure 1.12c in Chapter 1). The corresponding average emissions of GHG (CO_2) decreased in a similar way. As mentioned above, SPK fuels have just started to be used in commercial quantities in different blends with Jet-A1/8 fuel. Under such conditions, the conventional aircraft engines have not needed any modification, indicating that, if the share of alternative fuels increases, the conventional aircraft will accept such changes efficiently and effectively.

b) Cryogenic aircraft

Research on using LH_2 as fuel for air transportation has been carried out primarily in Europe, USA, and the Russian Federation. Consequently, different projects on the feasibility of aircraft design and scenarios for their eventual implementation have been completed. In particular, they have provided a concept of the prospective technical/technological, operational, environmental, and safety performances of cryogenic aircraft expected to be fully developed by around 2020 and, consequently, enter commercial service by around 2040 (EADS Airbus GmbH, 2002; Klug and Reinhard, 2001; Learmount, 2007; Janić, 2008; Sefain, 2000; Stadler, 2014).

i) Design and operational performances

Cryogenic aircraft will require about 4.3 times more fuel volume for an equivalent energy content than conventional aircraft, mainly due to the characteristics of LH_2 compared to Jet A1/8 fuel (2.8-times higher specific energy[11] and about 11-times less specific gravity or specific weight)[12] as given in Table 5.2. This will influence their design, requiring a relatively large volume of well-insulated fuel tanks that can be positioned differently within the aircraft configuration: above the payload (passengers and freight), above and aft of the payload, and fore and aft of the payload section. The wings, with no fuel storage space, could be smaller. This would increase the aerodynamic resistance and the aircraft empty weight compared to those of the conventional counterparts. However, the much lower specific gravity of LH_2 is expected to compensate for such an increase in the empty weight and consequently contribute to reducing the maximum take-off weight of cryogenic aircraft (EADS Airbus GmbH, 2002). Cryogenic jet engines will retain the basic structure of conventional jet engines with some necessary modifications, including the fuel pumps, fuel control unit, and combustion chambers. Experiments so far have shown that such LH_2 fuel-powered engines can have about 64% lower SFC (Specific Fuel Consumption) than their conventional Jet A1/8 powered counterparts (0.0976 vs. 0.2710 (kg/h)/kg for cruising and 0.0512 vs. 0.1420 (kg/h)/kg for the take-off phase of flight). In addition, these engines are expected to be 1–5% more efficient in generating thrust from the given energy content and operate with a slightly lower turbine entry temperature. This will, in turn, extend their lifetime and reduce maintenance costs (Corchero and Montanes, 2005; EC, 2003; Guynn and Olson, 2002; Janić, 2014; Winchester et al., 2015).

ii) Environmental performances

When introduced, the cryogenic aircraft will emit just two GHGs—H_2O and NO_x—as given in Table 5.2. The quantity of H_2O will be 2.543 times greater than that of conventional Jet A1/8 and/or the alternative SPK fuel-powered aircraft. This can act as a GHG in a certain sense if deposited at altitudes of 31000 ft and higher, but without severe consequences due to its above-mentioned very short lifetime in the Earth's atmosphere (Marquart et al., 2005). Cruising at the lower altitudes is one option to mitigate this impact, but this will then affect the other performances. In addition, the cryogenic aircraft will emit only about 0.005–0.0139% of NO_x of their conventional counterparts, which will be achieved through adequate design of the combustion chamber of the cryogenic engines. In addition, the cryogenic aircraft will have to be as safe as their conventional counterparts. In the case of an aircraft accident, LH_2 burns much faster (2.7–3.5 m/s compared to 0.85 m/s of Jet A1/8) and with low heat radiation, thus mitigating the impact

[11] Specific energy of fuel is the energy per unit of its mass or volume and is expressed in MJ/kg (MJ – Mega Joule; kg – kilogram).

[12] Specific gravity or specific weight of fuel is defined as weight per unit of its mass or volume and is expressed in kg/L (L – Litre).

if the fuselage collapses (Chong and Hochgreb, 2011). This contrasts to the corresponding impact of burning Jet A1/8 or SPK fuels. The burned LH_2 covers a much smaller surface area (EADS Airbus GmbH, 2002). The overall safety figure also includes an appropriate airport fuel supply system. LH_2 will predominantly be manufactured within the airport fuel area and its reserves will be stored in large tanks. Then, the fuel will be delivered to aircraft at the airport gates/stands through a dedicated underground pipeline system.

Table 5.3 summarizes the main relative differences between the conventional and cryogenic aircraft (Janić, 2008; 2014).

Table 5.3 The main relative differences in the selected performance of typical long-range conventional (Jet A1/8 and/or SPK) and cryogenic (LH_2) fuel-powered aircraft (EADS Airbus GmbH, 2002: Janić, 2014).

Attribute	Conventional aircraft (Jet A1/8)	Cryogenic aircraft (LH_2)
Fuel energy content	1	0.36
Volume of fuel	1	11
Volume of fuel tanks	1	4.3
MTOW	1	0.85–1.05
Aerodynamic resistance	1	1.1
Pollutants CO_2, SO_x, NM (HC)	1	0
H_2O	1	2.6
NO_x	1	0.005–0.0139

5.3.5 Methodology for Assessing GHG Potential of Air Transport Fuels

5.3.5.1 Assumptions

The methodology consisting of the analytical models for estimating the medium- to long-term direct emissions of GHGs (particularly CO_2) by air transportation using the selected types of fuels according to the specified "what-if" scenarios of their implementation and use is based on the following assumptions (Janić, 2018):

- The GHGs emitted from direct burning of all types of fuels are exclusively considered. This implies that those emitted during their life-cycle are not taken into account;
- Alternative SPK (F-T & HRJ) fuels are assumed to be available in sufficient quantities to completely replace the current Jet-A1/8 fuel in the considered Scenarios of their use;
- Use of particular fuel types in each Scenario implies their comparative/competitive prices; and
- Specifically, use of LH_2 is assumed to be possible under the following conditions:
  - Different categories of cryogenic aircraft are fully developed in terms of size-range (small-short, medium-medium, large-long);
  - Sufficient capacities for manufacturing the cryogenic aircraft and LH_2 fuel are available to satisfy the given rate of replacement of the conventional aircraft fleet;
  - The airport infrastructure and corresponding logistics processes for supplying and distributing LH_2 are fully developed and made operationally safe; and

– The direct emissions of GHG during manufacturing of LH_2 fuel from renewable primary sources are captured and stored.

5.3.5.2 Structure

a) Model for single fuel use

The model for estimating the annual quantities of global direct emissions of GHG by commercial air transportation using exclusively a single type of fuel is as follows (Janić, 2014; 2018):

$$E_k(i) = V_0 \cdot (1 + r_v)^k \cdot FC_0(i) \cdot [1 - r_f(i)]^k \cdot \sum_{i=1}^{L} e_l(i) \tag{5.3}$$

where

$E_k(i)$ is the total direct emissions of GHG from burning fuel type *(i)* in year *(k)* counted from the beginning of the observed period of *(N)* years, i.e., the base year "0" (tons);

V_0 is the volume of air transportation in the base year *(0)* of the observed period (RPKs – Equivalent Revenue Passenger Kilometer(s));[13]

$FC_0(i)$ is the average unit fuel consumption of the fuel type *(i)* in the base year *(0)* of the observed period (g/RPK);

r_v is the average annual growth rate of air transportation (in terms of equivalent RPKs) during the observed period (%);

$r_f(i)$ is the average annual rate of improvement of efficiency of using fuel type *(i)* during the observed period of time (%); and

$e_l(i)$ is the emission rate of *(l)*-th GHG from combustion of fuel type *(i)* ($l = 1,2,.., L$) (g/g of fuel).

According to Eq. 5.3, the total emissions of GHG, $E_k(i)$ could be affected by the influencing variables as follows:

- Increasing the average annual rate of improvement of the aircraft fuel efficiency compared to the growth rate of air transportation in the given (target) year, i.e., $r_f(i) \geq r_v/(1 + r_v) < 1.0$;
- Slowing growth of the air transportation according to the average annual rate of improvement in the aircraft fuel efficiency in the given (target) year, i.e., $r_v \leq r_f(i)/[1 - r_f(i)]$;
- Imposing a "cap" on the direct emissions of GHG in the given (target) year and consequently affecting the rate of growth of air transportation as follows: $r_v = \left[E_k^*(i) / [V_0 \cdot FC_0(i) \cdot [1 - r_f(i)]^k \cdot \sum_{l=1}^{L} e_l(i)] \right]^{1/k} - 1$, where $E_k^*(i)$ is the "cap" on the emissions of GHG in the target year *(k)*; and
- Affecting the growth rate of air transportation *(r_v)* by weakening its ties to/dependency on the main internal and external demand-driving forces in the given (target) year.

[13] Equivalent RPK is the sum of RPK and RTK (Revenue Ton-Kilometre) (1 RTK = 10 RPK).

b) Model for multi-fuel use

The model for multi-fuel use implies estimating the fuel consumption and related direct emissions of GHGs (again, particularly CO_2) under conditions of simultaneously using an aircraft fleet powered by three types of fuels during the observed medium- and long-term period of *(N)* years: conventional Jet A1/8, alternative (F-T&HRJ), and LH_2 fuel. It is assumed that the alternative fuels—SPK (F-T&HRJ) and LH_2—will gradually be introduced and replace their currently dominating conventional Jet A1/8 counterpart.

Based on Eq. 5.3, the direct emissions of GHG by such a "hybrid" fleet in the *(k)*-th year of the observed period of *(N)* years can be estimated as follows:

$$E_k(i) = V_0 \cdot (1 + r_v)^k \cdot \{\sum_{i=1}^{M} p_k(i) \cdot FC_0(i) \cdot [1 - r_f(i)]^k \cdot [1 - r_u(i)]^k \cdot \sum_{i=1}^{L} e_l(i)\} \quad (5.4a)$$

and

$$\sum_{i=1}^{M} p_k(i) = 1, \quad \text{for } k = 1,2,.., N \quad (5.4b)$$

where

M is the number of different types of fuels used (i.e., $M = 3$; $i = 1$ for Jet A1/8, $i = 2$ for F-T&HRJ fuel, and $i = 3$ for LH_2);

$p_k(i)$ is the average share of the total volume of air transportation (in terms of equivalent RPKs) carried out by the aircraft using fuel type *(i)* in the *(k)*-th year of the observed period ($0 \leq p_k(i) \leq 1$; $k = 1,2,.., N$); and

$r_u(i)$ is the rate of improvement in utilization of the aircraft fleet using the fuel type *(i)*.

The other symbols are analogous to those in Eq. 5.3.

The variable $(FC_0(i))$ in Eqs. 5.3–5.4(a, b) ($i = 1, 2, 3$) should be at the level achieved when the process of introducing the cryogenic aircraft starts, i.e., at the beginning of "transition" period, and would continue to improve later on (Janić, 2014; 2018). The cryogenic replacing the conventional aircraft will be introduced in a constant proportion each year, thus implying their constant increasing share in carrying-out the total volumes of RPKs.

5.3.6 Application of Methodology

5.3.6.1 Inputs

The proposed models are applied to the medium- to long-term development of global commercial air transportation and its direct emissions of GHG (exclusively CO_2). Three general hypothetical "what-if" Scenarios of using different types of fuels during the observed period 2006/2050 (Janić, 2018):

- Continuing exclusive use of Jet-A1/8 fuel during the entire observed period;
- Completely replacing Jet-A1/8 by SPK (F-T & HRJ) fuels during a significant part of the observed period; and
- Introducing and increasing the share of cryogenic aircraft and, consequently, the share of LH_2 fuel at a constant rate, starting from different times within the observed period until its end.

The first scenario is a rather hypothetical one since the alternative SPK (F-T&HRJ) fuels have already been used, albeit at a modest scale. The second scenario is again hypothetical since, at present and very likely in the future, there will not be sufficient quantities of these fuels to replace the conventional Jet-A1/8 fuel more substantially. The last scenario is also hypothetical since it is not certain if, when, and how the cryogenic aircraft and related logistics of using LH_2 fuel will be developed for the wider commercial use. In addition, the entire considered period is divided into three sub-periods: 2006–2020/25, 2020/25–2040, and 2040–50. Each sub-period is characterized by the attributes used as inputs for the above-mentioned Scenarios as given in Table 5.4.

As can be seen, the growth rates of RPKs are assumed to be constant during each sub-period, but decreasing over time. The average growth rate of air transportation over the entire time horizon is estimated to be about 3.2%, which is similar to the growth rate of 3.1% over the period 1990–2050 in one of the specified IPCC forecasting scenarios (IPCC, 2001). This rate, starting from the basic rates at the beginning of the particular sub-periods, will produce about 16.5 trillion RPKs. This is lower than the above-mentioned Boeing's forecast of 17.093 trillion RPKs in the year 2035, to be achieved at an average annual

Table 5.4 Inputs for Scenarios of using the conventional and alternative fuels during the observed period (2006–2050) (Hileman et al., 2010; Janić, 2018; NASA, 2011; Winchester et al., 2015).

Input variable	**Period**		
	2006–2020/25	**2020/25–2040**	**2040–2050**
• Basic annual air transport volume: V_0 (10 billion equivalent RPK)	6.26[1)]	13.78	22.61
• Average traffic growth rate: r_v (%)	5.4[1)]	3.5	2.0
• The number of aircraft at the beginning of the sub-period	18230[1)]	36420	48823
• Average utilization of an aircraft at the beginning of the sub-period (10^{12} RPK/yr)	0.3615	0.3784	0.4632
• Rate of improvement in the aircraft utilization: r_u (%/yr)	1.50	1.25	1.00
• Average unit fuel consumption of conventional aircraft fleet: $FC_0(1)$ (g/RPK)	27.7	19.66	16.28
• Rate of improvement in $FC_0(1)$: $r_f(1)$ (%/yr)	1.70	1.25	1.00
• Average unit fuel consumption of the aircraft fleet using SPK (F-T &HRJ) fuel: $FC_0(2)$ (g/RPK)	27.13[2)]	19.26[2)]	15.62[2)]
• Rate of improvement in $FC_0(2)$: $r_f(2)$ (%/yr)	1.70	1.25	1.00
• Average annual rate of introducing the aircraft fleet using F-T &HRJ fuel: $p(2)$ (%/yr)[3)]	100	100	100
• Average unit fuel consumption of the cryogenic aircraft fleet: $FC_0(3)$ (g/RPK)	-	-	5.86[4)]
• Rate of improvement in $FC_0(3)$: $r_f(3)$ (%/yr)	-	-	0.00
• Average annual rate of introducing the cryogenic aircraft fleet: $p(3)$ (%/yr)	2.00[5)]	2.00[6)]	2.00[7)]
• Dissipation rate of CO_2 from the Earth's atmosphere: DR_{CO2} - (%/yr)[8)]	1.0	1.0	1.0

[1)] Airbus, 2006; Boeing, 2007; [2)] Based on $FC_0(1)/FC_0(2) = (43.2/44.1)$; [3)] According to Scenario 2 starting from the year 2011; [4)] Based on: $SE(1)/SE(3) \cdot FC_0(1) = (44.3/120) \cdot 16.28 = 5.86$ (Table 5.1; Janić, 2008); [5)] Sub-scenario 3a – Starting from the year 2019/2020; [6)] Sub-scenario 3b – starting from the year 2025; [7)] Sub-scenario 3c – starting from the year 2040; [8)] Based on the life time of CO_2 in the Earth's atmosphere of 95–100 years.

rate of 4.82%, and Airbus's forecast of 15.2 trillion RPKs in the year 2034, to be achieved at an average annual growth rate of 4.6% (ATAG, 2015; 2015a; Airbus, 2015; Boeing, 2016; IPCC, 2001). The adopted decreasing growth rates of air transportation reflect increasing maturity of the air transport market (demand) in combination with weakening of its dependency on the main above-mentioned external driving forces. The rate of improvement of the average fuel consumption of conventional aircraft using Jet A1/8 and SPK (F-T&HRJ) fuel is assumed to be equal and constant during particular sub-periods, but also decreasing over time. Due to the lack of relevant data, it is still not clear if these rates of improvements will be different or remain the same for aircraft using both types of fuels. In addition, the utilization of an aircraft fleet using both types of fuels is assumed to increase over time at a decreasing rate. This implies that the number of aircraft and their average seating capacity will increase at a decreasing rate in order to adapt to the above-mentioned rates of growing demand during the observed period. As mentioned in Scenario 2, just for comparative purposes, the rate of using SPK (F-T&HRJ) fuels is assumed to be 100% during the entire period, implying the complete replacement of the conventional Jet A1/8 fuel. Introduction of the cryogenic aircraft and LH_2 fuel is assumed to be at constant rate between the time of starting introduction until the rest of the observed period (2050). Consequently, the composition of aircraft fleets and fuels LH_2/(Jet 1/8/SPK &F-T&HRJ)) will be 50/50% if introduction of the cryogenic aircraft/LH_2 fuel begins in the year 2040 at the constant rate of 2%/yr. (Sub-scenario 3a); it will be 80/20% if it starts in the year 2025 (Sub-scenario 3b); and it will be 90/10% if it starts in the year 2019/2020 (Sub-scenario 3c). In all the above-mentioned scenarios, the rate of improvement of the average fuel consumption of the cryogenic aircraft is assumed to be zero.

5.3.6.2 Results

The results from application of the models by using the inputs in Table 5.4 in terms of the development of air transportation and related direct emissions of CO_2 over time in relative terms (Index) are shown in Figures 5.24, 5.25, and 5.26.

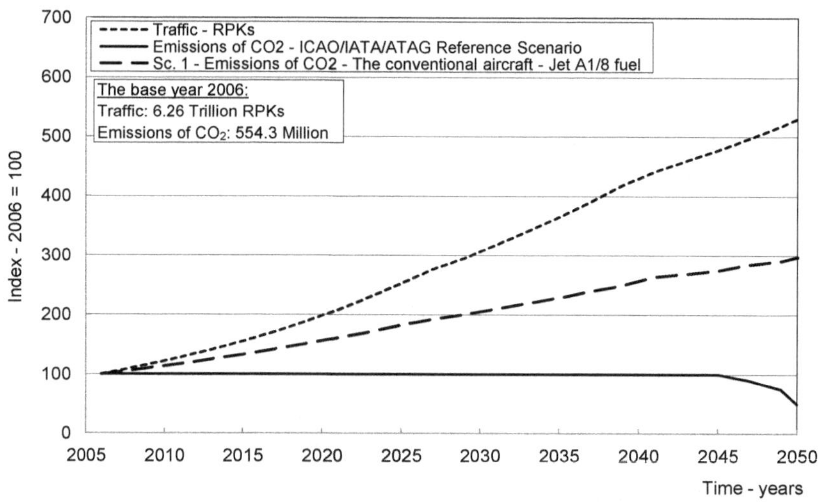

Figure 5.24 The medium- to long-term emissions of GHG (CO_2) by commercial air transportation: Scenario 1: Exclusive use of Jet-A1/8—Proportion over time 100% (Janić, 2014; 2018).

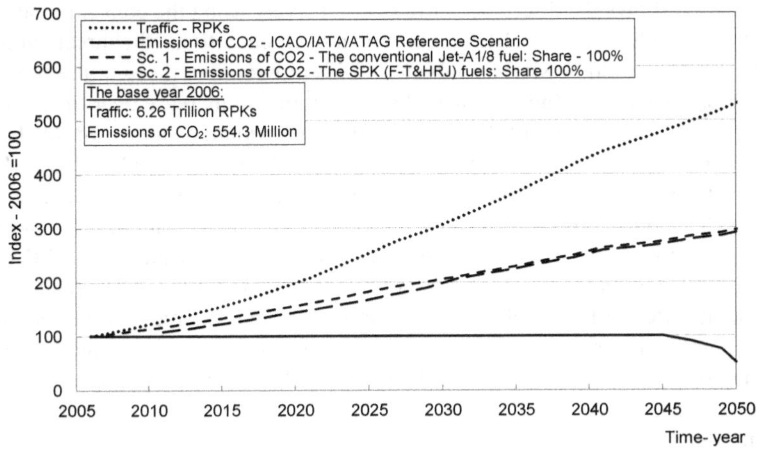

Figure 5.25 The medium- to long-term emissions of GHG (CO_2) by commercial air transportation: Scenario 2: Exclusive use of Jet-A1/8 or F-T&HRJ fuels: Proportion of each over time—100% (Janić, 2014; 2018).

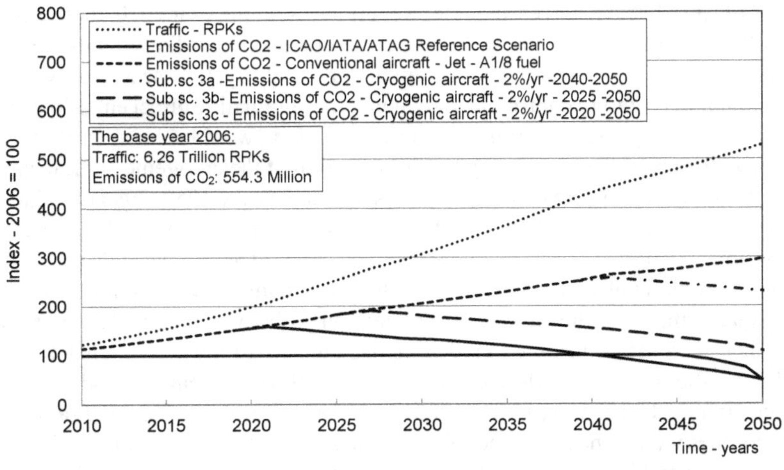

Figure 5.26 The medium- to long-term emissions of GHG (CO_2) by commercial air transportation: Scenario 3: Sub-scenarios (a), (b), (c) Gradual introduction of alternative LH_2 fuel (Janić, 2014; 2018).

Figure 5.24 shows the future Scenario of exclusively using the conventional aircraft and Jet-A1/8 fuel. Under such conditions, the direct cumulative emissions of CO_2 would continue to increase as the volumes of air transportation increase. This increase would be slower than that of the volumes of air transportation mainly due to permanent (although decreasing) improvements in the aircraft fuel efficiency and utilization of the aircraft fleets. For example, at the end of the period (the year 2050), the volumes of air transportation would increase about 6-fold and the related direct cumulative emissions of CO_2 about 3.5-fold, compared to the base year 2006. In particular, in the year 2050, these emissions will be 6-fold above the prescribed target of reduction compared to the year 2006, i.e., 300:50 (ATAG, 2015; 2015a; ICAO, 2013; 2016).

Figure 5.25 shows the future Scenario of exclusively using the conventional aircraft and the alternative SPK (F-H&HRJ) fuels. Under such conditions, the direct cumulative emissions of CO_2 would be slightly (about 2–2.5%) lower than those in the scenario of exclusively using Jet-A1/8 fuel, but would also continue to increase over time, similarly to their Jet A1/8 fuel counterpart. In the year 2050 they would be about 6-fold above the prescribed targets (–50% compared to that in the year 2006). This implies that, at least as far as CO_2 emissions by direct burning are concerned, the alternative SPK fuels would not bring substantial if any reduction of the cumulative emissions of CO_2 over time. However, this should be considered with caution since the contribution of SPK fuels to reduction of CO_2 during the life-cycle has not been taken into account. The main reason is that the use of different feedstocks as the primary renewable sources for producing these fuels is still highly uncertain. Therefore, in the present context, it can be said that the main contribution of SPK fuels could actually be reducing the dependency of the sector on the non-renewable feedstock—crude oil—of Jet A1/8 fuel and much less mitigating their impact on global warming and climate change.

Figure 5.26 shows scenarios of the gradual introduction of cryogenic aircraft powered by LH_2 at different times during the observed period (2006–2050). According to sub-scenario 3a (introduction starts in the year 2040 at an annual rate of 2%/yr), these aircraft would almost immediately start to contribute to decreasing the direct cumulative emissions of CO_2, but at a decreasing rate. Nevertheless, in the year 2050, these emissions would still be about 4.5-fold above the prescribed target. According to sub-scenario 3b (introduction starts in the year 2025 at an annual rate of 2%/yr), these aircraft would also almost immediately start to contribute to reducing the cumulative direct emissions of CO_2, again at a decreasing rate. Nevertheless, these emissions would remain a bit more than 2-fold above the prescribed 2050 target. If introduction of the cryogenic aircraft started in the year 2020 at an average rate of 2%/yr (sub-scenario 3c), the 2050 target of reduction of the cumulative emissions of CO_2 would be achieved just on time. In all above-mentioned scenarios, introduction of the cryogenic aircraft fleet would have a two-fold effect. On the one hand, it would contribute to decreasing the direct cumulative emissions of CO_2 and the consequent impact of the air transport system on global warming and climate change. On the other, it would contribute to reducing dependency of the system on Jet A1/8 fuel derived from the non-renewable feedstock—crude oil. In any case, introduction of the cryogenic aircraft fleet powered by LH_2 mainly obtained from the CO_2 neutral feedstock primary sources would enable a decoupling of the present dependency of the air transport system's growth and its direct cumulative emissions of CO_2, thus transforming it into the "carbon neutral" system with much lower if any consequential impact on global warming and climate change.

References

ACI Asia-Pacific. (2018). Waste Minimization, Green Airports Recognition 2018, Airports Council International Asia-Pacific, Airport World Trade Centre, Hong Kong International Airport, Hong Kong.

Airbus. (2006). Airbus Global Market Forecast, Airbus Industrie, Toulouse, France.

Airbus. (2015). Flying by Numbers: Global Market Forecast, Airbus Industrie, Toulouse, France.

Airbus. (2016). Certifiably Excellent. A350 XWB Shaping Efficiency Magazine, http://www.a350xwb.com/x-tra/magazines/.

ASTM. (2014). Standard Specification for Aviation Turbine Fuels Containing Synthesized Hydrocarbons, Designation D7556-14a, AST International, West Conshohocken, PA, USA.

ATAG. (2015). Aviation Climate Solutions, Air Transport Action Group, Geneva, Switzerland.

ATAG. (2015a). The Aviation Sector's Climate Action Framework, Air Transport Action Group, Geneva, Switzerland.

ATAG. (2016). Aviation Climate Solutions, The Air Transport Action Group, www.atag.org.

BA. (2001). From the Ground Up: Social and Environmental Report 2001, British Airways Plc., Waterside, (HBBG), Middlesex, UK.

Boeing. (2007). Current Market Outlook 2007: How Will You Travel Through Life? Boeing Commercial Airplanes: Market Analysis, Seattle, WA, USA.

Boeing. (2016). Current Market Outlook 2016–2035, Boeing Commercial Airplanes, Market Analysis, Seattle, WA, USA.

Boeker, E. and Grondelle, R. (1999). Environmental Physics, Sec. Ed., John Wiley and Sons, Ltd., New York, USA.

Bossel, U. and Eliasson, B. (2003). Energy and the Hydrogen Economy, Report ABB Switzerland Ltd. Corporate Research, Baden-Dattwil, Switzerland.

Chevron. (2006). Alternative Jet Fuels, Addendum 1 to Aviation Fuels Technical Reviews (FTR-3/A1), Chevron Corporation, USA.

Chong, T. C. and Hochgreb, S. (2011). Measurements of laminar flame speeds of liquid fuels: Jet-A1, diesel, palm methyl esters and blends using Particle Imaging Velocimetry (PIV). Proceedings of the Combustion Institute, 33: 979–986.

Corchero, G. and Montanes, J. l. (2005). An approach to use hydrogen for commercial aircraft engines. Journal of Aerospace Engineering, 219 G: 35–44.

DETR. (2000). The Future of Aviation, The Government's Consulting Document on Air Transport Policy, Department of the Environment, Transport and Regions, London, UK.

DETR. (2001). Valuing the External Cost of Aviation, Department of the Environment, Transport and Regions, London, UK.

Derwent, R., Simmonds, P., O'Doherty, S., Manning, A., Collins, W. and Stevenson, D. (2006). Global environmental impacts of the hydrogen economy. International Journal of Nuclear Hydrogen Production and Applications, 1(1): 57–67.

Doganis, R. (1992). The Airport Business, Routledge, London, UK.

Dufour, J., Serrano, D. P., Galvez, J. L, Moreno, J. and Garcia, C. (2009). Life cycle assessment of processes for hydrogen production: environmental feasibility and reduction of greenhouse gases emissions. International Journal of Hydrogen Energy, 34: 1370–1376.

EADS Airbus GmbH. (2002). CRYOPLANE – Hydrogen Fuelled Aircraft—System Analysis, European Commission 5th Framework Program, Hamburg, Germany.

EC. (1998). Interaction between High-Speed Rail and air Passenger Transport, Final report on the Action COST 318, EUR 18165, European Commission, Luxembourg.

EC. (2003). Liquid Hydrogen Fuelled Aircraft—System Analysis (CRYOPLANE), Final Report, European Commission, 5th R&D Framework Program (Growth 1998-2002), Brussels, Belgium.

EEC. (2001). ATFM Delays to Air Transport in Europe—Annual Report 2000, EUROCONTROL, European Organization for the Safety of Air Navigation, Brussels.

EEC. (2017). Report on the Operation of the Route Charges System in 2016, Central Route Charges Office (CRCO), EUROCONTROL, European Organization for the Safety of Air Navigation, Brussels, Belgium.

EEC. (2019). Eurocontrol Central Route Charge Office: Customer Guide to Charges, Central Route Charges Office, EUROCONTROL, European Organization for the Safety of Air Navigation, Brussels, Belgium.

EEC/FAA. (2017). Comparison of Air Traffic Management-related Operational Performance U.S./Europe Traffic Management-Related, US FAA (Federal Aviation Association, Air Traffic Organization System Operations Services, Washington D.C. USA, EUROCONTROL, European Organization for the Safety of Air Navigation, Brussels, Belgium.

FAA. (2018). Administrator's Fact Book, December 2018, FAA Office of Communications, Federal Aviation Administration, Washington D.C., USA.

FAA. (2018a). Aerospace Forecast: Fiscal Years 2018–2038, Federal Aviation Administration, Washington D.C., USA.

FI. (2000). Commercial Aircraft Directory Part I, II. Flight International, Oct/Nov, UK.

FI. (2001). Regional Aircraft: World Airlines: Part I. Flight International, Sept. UK.

Guynn, M. D. and Olson, E.D. (2002). Evaluation of an Aircraft Concept with Over-Wing Hydrogen-Fuelled Engines for Reduced Noise and Emissions, Technical Memorandum, NASA/TM-2002-211926, National Aeronautics and Space Administration, Langley Research Centre, Hampton, Virginia, USA.

HA. (1999). Environmental Impact of Air Transport in Comparison with Railways and Transport by Rail in the European Union, Intermediate Report, Hera - Amalthea, S.L., Madrid, Spain.

Hileman, I. J., Stratton, W. R. and Donohoo, E. P. (2010). Energy content and alternative jet fuel viability. Journal of Propulsion and Power, 26(6): 1184–1195.

HMMH. (2018). Airport Noise Management Benchmarking Study, HMMH Project # 309750, Burlington, Massachusetts, USA, www.hmmh.c.

Hooper, P. G. and Hensher, D.A. (1997). Measuring total factor productivity at airports. Transportation Research E, 33(4): 249–259.

http://www.airbus.com/en/myairbus/headlinenews/index.jsp

http://www.greenaironline.com/news/

http://www.iata.org/publications/economics/fuel-monitor/Pages/price-development.aspx

http://www.omega.mmu.ac.uk/

http://partner.aero/

http://www.sesarju.eu/

http://www.faa.gov/nextgen/

http://www.boeing.com/airports/

http://www.airbus.com/

http://www1.iata.org/Whip/Public/frmMain_Public.aspx?WgId=35/

http://www.atn.aero/article.pl/

http://en.wikipedia.org/wiki/World's_busiest_airports_by_passenger_traffic/)

http://www.climatehotmap.org/global-warming-effects/air-temperature.html/

https://www.transportation.gov/briefing-room/dot0718/

https://www.ipcc.ch/publications_and_data/ar4/wg3/en/tssts-ts-3-2-stabilization-scenarios. html/

https://www.united.com/web/en-US/content/company/globalcitizenship/environment/alternative-fuels.aspx/

https://www.icao.int/annual-report-2018/Pages/the-world-of-air-transport-in-2018.aspx

https://www.transtats.bts.gov/OT_Delay/OT_DelayCause1.asp/

https://www.faa.gov/data_research/accident_incident/

https://www.bls.gov/opub/ted/2019/cpi-increased-1-point-7-percent-for-year-ending-september-2019.htm/

https://www.bts.gov/explore-topics-and-geography/topics/air-fares/

https://www.transportation.gov/briefing-room/dot0718/

https://en.wikipedia.org/wiki/Fuel_economy_in_aircraft

https://www.epa.gov/

IATA. (2015). IATA Report on Alternative Fuels, 10th Edition, Effective December 2015, International Air Transport Association.

IATA. (2016). Annual Review 2016, International Air Transport Association, Geneva, Switzerland.

IATA. (2017). Fact Sheet—Fuel, Industry Economics Performance—Forecast Table (IATA Economics), International Air Transport Association, Montreal, Canada.

ICAO. (1993a). Aircraft Noise, Environmental Protection, Annex 16, Vol. 1, International Civil Aviation Organization, Montreal, Canada.

ICAO. (1993b). Aircraft Engine Emissions, Environmental Protection, Annex 16, Vol. 2, International Civil aviation Organisation, Montreal, Canada.

ICAO. (2008). Environmental Protection: Volume I – Aircraft Noise, Annex 16 to the Convention of International Civil Aviation, International Civil Aviation Organization, Montreal, Canada.

ICAO. (2010). Alternative Fuels, ICAO Environmental Report, International Civil Aviation Organization, Montreal, Canada.

ICAO. (2012). ICAO Aircraft Engine Emissions Databank, International Civil Aviation Organization, Montreal, Canada.

ICAO. (2013). ICAO Environmental Report 2013: Aviation and Climate Change, International Civil Aviation Organization, Montreal, Canada.

ICAO. (2016). On board a Sustainable Future: ICAO 2016 Environmental Report, International Civil Aviation Organization, Montreal, Canada.

ICTT. (2015). Fuel Efficiency Trends for New Commercial Jet Aircraft: 1960 to 2014, The International Council on Clean Transportation, Washington D.C., USA, www.theicct.org.

IEA. (2006). Hydrogen Production and Storage: Research & Development Priorities and Gaps, International Energy Agency, Paris, France.

IFRAS. (2000). Sustainable Aviation, Pre-Study, MM-PS, Zurich, Switzerland.

IPCC. (2001). Climate Change 2001: Synthesis Report. Contribution of Working Groups I, II, and II and III to the Third Assessment Report of IPCC, Intergovernmental Panel of Climate Change, Cambridge University Press, Cambridge, UK, New York, USA.

IPCC. (2007). Climate Change 2007: The Physical Science Basis, Contribution of Working Group I to the Fourth Assessment Report of the Intergovernmental Panel on Climate Change, Cambridge University Press, Cambridge, UK, New York, USA.

IPCC. (2013). Climate Change 2013: The Physical Science Basis Exit, Contribution of Working Group I to the Fifth Assessment Report of the Intergovernmental Panel on Climate Change, Cambridge University Press, Cambridge, United Kingdom and New York, NY, USA.

IPCC. (2014). Climate Change 2014, Synthesis Report—Contribution of Working Groups I, II and III to the Fifth Assessment Report of the Intergovernmental Panel on Climate Change [Core Writing Team, Pachauri, R. K. and Meyer, L. A. (eds.)], IPCC, Geneva, Switzerland, p. 151.

IPCC. (2015). Climate Change: Synthesis Report, Intergovernmental Panel on Climate Change, Geneva, Switzerland.

Janić, M. (1999). Aviation and environment: accomplishments and problems. Transportation Research D, 4: 159–180.

Janić, M. (2000). Air Transport Systems Analysis and Modeling, Gordon & Breach Science Publishers, Amsterdam, The Netherlands.

Janić, M. (2003). A methodology for assessing sustainability of air transport system. Journal of Air Transportation, USA, 8(1): 115–152.

Janić, M. (2004). An application of the methodology for assessment of the sustainability of air transport system. Journal of Air Transportation, 9(2): 40–82.

Janić, M. (2007). The Sustainability of Air Transportation: Quantitative Analysis and Assessment, Ashgate Publishing Company, UK.

Janić, M. (2008). The potential of liquid hydrogen for the future 'carbon neutral' air transport system. Transportation Research D, 13(7): 428–435.

Janić, M. (2010). True multimodalism for mitigating the airport congestion: substitution of air passenger transport by high-speed rail. Transportation Research Record (TRR), 2177: 78–87.

Janić, M. (2011). Assessing some social and environmental effects of transforming an airport into a multimodal transport node. Transportation Research D, 16(2): 137–149.

Janić, M. (2013). The Airport Analysis, Planning, and Design: Demand, Capacity, and Congestion, Nova Science Publishers, Inc. New York, USA.

Janić, M. (2014). Greening commercial air transportation by using liquid hydrogen (LH_2) as a fuel. International Journal of Hydrogen Energy, 39: 16426–16441.

Janić, M. (2016). Analyzing, modelling, and assessing the performances of land use by airports. International Journal of Sustainable Transportation, 10(8): 683–702.

Janić, M. (2018). An assessment of the potential of alternative fuels for "greening" commercial air transportation. Journal of Air Transport Management, 69: 235–247.

Karunanidhi, R. (2015). The Aviation Fuel and the Passenger Aircraft for the Future - Bio Fuel, Synthetic Fuel, Master Thesis, Department of Automotive and Aeronautical Engineering, Hamburg University of Applied Sciencies, Hamburg, Germany.

Klug, H. G. and Reinhard, F. (2001). CRYOPLANE: Hydrogen fuelled aircraft—status and challenges. Air & Space Europe, 3(3-4): 252–254.

Koroneos, C., Dompros, A., Roumbas, G. and Moussiopoulos, N. (2004). Life cycle assessment of hydrogen fuel production processes. International Journal of Hydrogen Energy, 29: 1443–1450.

Learmount, D. (2007). New-Technology Aircraft to Reduce Average Fuel Consumption, http://www.flightglobal.com.

MA. (1999). Manchester Airport Complaints and Community Disturbance: Report: 1998–1999, Manchester Airport PLC, Manchester, UK.

Marquart, S., Ponater, M., Strom, L. and Gierens, K. (2005). An upgraded estimate of the relative forcing of cyroplane contrails. Meteorologische Zeitschrift, Gebruder Bontraeger, 14: 573–582.

NASA. (2011). Alternative Aviation Fuel Experiment (AAFEX), NASA/TM–2011-217059, National Aeronautics and Space Administration, Washington, DC, USA.

Okai, K. (2010). Long term potential of hydrogen as aviation fuel. In Alternative Fuels, ICAO Environmental Report 2010, International Civil Aviation Organization, Montreal, Canada.

PARTNER. (2009). Aircraft Impacts on Local and Regional Air Quality in the United States, PARTNER Project 15 Final Report, The Partnership for Air Transportation Noise and Emissions Reduction Massachusetts Institute of Technology, Cambridge, MA, USA, http://www.partner.aero.

PARTNER. (2010). Life cycle Greenhouse Gas Emissions from Alternative Jet Fuels, Partnership for Air Transportation Noise and Emissions Reduction, e, Partner Project 28 Report, The Partnership for Air Transportation Noise and Emissions Reduction, Massachusetts Institute of Technology, Cambridge, MA, USA, http://www.partner.aero.

Pucher, G., Allan, W. and LaViolette, M. (2011). Emissions from a gas turbine sector rig operated with synthetic aviation and biodiesel fuel. Journal of Engineering for Gas Turbines and Power, 133: 111502-1–111502-8.

RAND. (2009). Near-Term Feasibility of Alternative Jet Fuels, Technical Report, RAND Corporation, Santa Monica, California, USA, http://www.rand.org/.

RCEP. (2003). The Environmental Effects of Civil Aircraft in Flight, Royal Commission on Environmental Pollution, Special Report, Department for Environment, Food and Rural Affairs, London, UK (www.rcep.org/uk).

SG. (2014/2018). Annual Report, Schiphol Group, Schiphol, The Netherlands, www.schiphol.nl/; www.annualreportschiphol.com/.

SA. (2004/2019). Annual Report to Shareholders, Southwest Airlines CO, Dallas, Texas, USA.

Saynor, B., Bauen, A. and Leach, M. (2003). The Potential for Renewable Energy Sources in Aviation, Imperial College Centre for Energy Policy and Technology, Imperial College, London, UK, www.iccept.ic.ac.uk.

Sefain, J. M. (2000). Hydrogen Aircraft Concepts and Ground Support, PhD Thesis, School of Engineering, Cranfield University, Cranfield, UK.

Stadler, P. (2014). Cost Evaluation of Large Scale Hydrogen Production for the Aviation Industry, Master semester project, Ecole Politechnique Federale de Lausanne, Lausanne, Switzerland.

Timko, T. M., Herndon, C. S., De La Rosa Blanco, E., Wood, C. E., Yu, Z., Miake-Lve, C. R., Knighton, B. W., Shafer, L., DeWitt, J. M. and Corporan, E. (2001). Combustion products of petroleum jet fuel, a fischer-tropsch synthetic fuel, and a biomass fatty acid methyl ester fuel for a gas turbine engine. Combustion Science and Technology, 183: 1039–1068.

TRB. (2019). Airport Waste Management and Recycling Practices (2018), Airport Cooperative Research Program, Transportation Research Board Washington, D.C. USA.

USDT. (2003/2019). Air Travel Consumer Report, A Product Of The Office of Aviation Enforcement and Proceedings, Aviation Consumer Protection Division U.S. Department of Transportation, Washington D.C., USA, http://www.bts.gov/; http://www. transportation .gov/airconsumer/.

Winchester, N., Malina, R., Staples, D. M. and Barrett, R. H. S. (2015). The Impact of Advanced Biofuels on Aviation Emissions and Operations in the U.S., Report No. 275, MIT Joint Program on the Science and Policy of Global Change, Massachusetts Institute of Technology, Cambridge MA, USA.

Summary

This is a unique book dealing with analysis and modelling of the demand, capacity, quality of services, economics, and sustainability of three main components of the air transport system—airports, ATC/ATM (Air Traffic Control/Management), and airlines.

The airport demand generally encompasses airline aircraft/flights as well as the passengers that are served and provided with a certain quality of services at certain costs and the consequent effects/impacts on the society and environment by the corresponding capacity of the airport airside and landside areas, respectively. The ATC/ATM demand is represented by aircraft/flights handled by the capacity of the given airspace; this factor is mainly influenced by the air traffic controllers' workload. The quality of services is provided to these aircraft/flights at given costs by assigning fuel-optimal routes and reducing impacts on the environment. The airline demand includes air passengers served by the corresponding capacity—flights carried out by given aircraft types—at a specified price, and the consequent effects/impacts on the environment and society.

The insight into the existing and details for the prospective innovative/new approaches to analysis and modelling in the given context is presented. The existing modelling approach embraces the illustrative analytical and simulation models of the airport, ATC, and airline demand, capacity, quality of services, economics, and sustainability supported by the corresponding applications and the real-world examples. The innovative/new modelling approach implies the prospective ways of improving knowledge about fast developing and changing air transport system and its components.

Graduates, researchers, consultants, engineers, experts, and other practitioners dealing with analysis, modelling, planning, design, and operations of the air transport system and its main components—airports, ATC/ATM, and airlines—will find this book of interest and useful.

Index

B

C

D

O

P

Q

R

T

U

V

W

Y

For Product Safety Concerns and Information please contact our EU
representative GPSR@taylorandfrancis.com
Taylor & Francis Verlag GmbH, Kaufingerstraße 24, 80331 München, Germany

www.ingramcontent.com/pod-product-compliance
Lightning Source LLC
LaVergne TN
LVHW011108300926
849160LV00008B/308
9780367632427